Ethical Governance of Emerging Technologies Development

Fernand Doridot
Centre for Ethics, Technics and Society (CETS), ICAM Lille, France

Penny Duquenoy
Middlesex University, UK

Philippe Goujon
*Laboratory for Ethical Governance of Information Technology,
   Facultés Universitaires Notre-Dame de la Paix, Belgium*

Aygen Kurt
Middlesex University, UK & London School of Economics (LSE), UK

Sylvain Lavelle
Centre for Ethics, Technics and Society (CETS), ICAM Paris–Sénart, France

Norberto Patrignani
Universita' Cattolica di Milano, Italy

Stephen Rainey
Facultés Universitaires Notre-Dame de la Paix, Belgium

Alessia Santuccio
Universita' Cattolica di Milano, Italy

Information Science
REFERENCE

T0338599

Managing Director:	Lindsay Johnston
Editorial Director:	Joel Gamon
Book Production Manager:	Jennifer Yoder
Publishing Systems Analyst:	Adrienne Freeland
Development Editor:	Christine Smith
Assistant Acquisitions Editor:	Kayla Wolfe
Typesetter:	Christina Henning
Cover Design:	Jason Mull

Published in the United States of America by
Information Science Reference (an imprint of IGI Global)
701 E. Chocolate Avenue
Hershey PA 17033
Tel: 717-533-8845
Fax: 717-533-8661
E-mail: cust@igi-global.com
Web site: http://www.igi-global.com

Library of Congress Cataloging-in-Publication Data

Ethical governance of emerging technologies development / Fernand Doridot, Penny Duquenoy, Philippe Goujon, Avgen Kurt, Sylvain Lavelle, Norberto Patrignani, Stephen Rainey, and Alessia Santuccio, editors.
    pages cm
 Includes bibliographical references and index.
 Summary: "This book combines multiple perspectives on ethical backgrounds, theories, and management approaches when implementing new technologies into an environment"--Provided by publisher.
   ISBN 978-1-4666-3670-5 (hardcover) -- ISBN 978-1-4666-3671-2 (ebook) -- ISBN 978-1-4666-3672-9 (print & perpetual access) 1. Technological innovations--Social aspects. 2. Technological innovations--Moral and ethical aspects. 3. Technological innovations--Management. I. Doridot, Fernand, 1972-
   HM585.E836 2013
   303.48'3--dc23
                           2012050349

British Cataloguing in Publication Data
A Cataloguing in Publication record for this book is available from the British Library.

All work contributed to this book is new, previously-unpublished material. The views expressed in this book are those of the authors, but not necessarily of the publisher.

List of Reviewers

Martin Meganck, *Katholieke Universiteit Leuven, Belgium*

Table of Contents

Section 7
Frame Development and Reflection

Section 8
Changing Methods and Procedures

Detailed Table of Contents

Section 1
Ethical Governance: Which Rationality?

Section 1 Introduction

Chapter 1

Laurence Masclet, University of Namur, Belgium
Philippe Goujon, University of Namur, Belgium

This paper describes the evolution of the discipline of governance study through a focus on the evolution of proceduralism. From a lecture and a critique of Maesschalck and Lenoble "contextual proceduralism" and their reconstruction of the evolution of the field, the authors aim at creating a more open theory ("comprehensive proceduralism") that would avoid some of the reduction of previous school of procedural governance, notably the reduction to argumentation, by opening the scope of governance to the different register of discourse. This paper aims at introducing the EGAIS book, and thus will introduce the theoretical (and critical) emergence of comprehensive proceduralism, leaving some of the consequences and application area to further investigation in the book.

Chapter 2

Pierre Livet, University of Provence, France

The paper addresses the relations between emotions, values and norms and argumentative debates, which makes possible to take into account the contextuality of the debates. The paper suggests that constraints and demands are mixed and that there are requirements on their articulation. Different demands and constraints make a contextual exogenous background. Before analysing attempts (mainly logical ones) to specify these dialectics between constraints and demands in social and public debates, this paper first gives a sketchy picture of the relation between values, norms and social emotions. The result of the analyses (of the debates) is that we need to take into account some dead-ends of the interaction, the ones related to unsatisfied demands, because in order to enforce norms, we need to take into account the frustration of other values. In order to take into account the social and emotional context of a debate, we have to look beyond the end of the debate, and to extend the reflection about unaccomplished proposals, in parallel with the execution of the decided project. This requires taking as constraints some attachments to other values (some demands).

Practical rationality, when collective choices are at stake, should certainly rely on principles. These principles are perhaps not without effect on our representation of the problems to be addressed in collective action. The authors investigate how this structuring role of pragmatic principles accounts for notable context-dependent features of governance procedures. In the field of social policies, for example, the enhancement of personal autonomy has come to the forefront of collective challenges. Capacity-based approaches indicate a way to put into question those conceptions of autonomy which lead to an excessively uniform treatment of individuals. Following these approaches, the beneficiaries of social policies should be treated as concrete beings with their personal history, living in specific social contexts and so on. The authors analyse the individualizing logic which is exemplified in interactive problem-structuring and institutional decision-making about the provision of apt, context-sensitive care and services for ageing handicapped persons. It is suggested that the sought-for adaptation to specific circumstances is made possible through a complex process of description of problems and challenges for collective action, in which procedural aspects are important. This process is by no means reducible to a passive process of adjustment to independent states of affairs. If the authors' analysis is correct, there is no such thing as the "real" nature of individual situations, as opposed to the fictions associated with ordinary social policies: the process under scrutiny really redefines the nature of institutional interactions, responsibilities and the underlying picture of the individual person.

Section 2
Emerging Technologies: Which Reflexivity?

Section 2 Introduction

This paper concentrate on a proactive engagement with emerging information and communication technologies (ICTs) with the goal of an early identification of the ethical issues these technologies are likely to raise. After an overview of the emerging ICTs for the next future (leveraging the results of the EU funded project ETICA), the paper identify the possible ethical consequences. Then the emerging ICTs are evaluated from different perspectives for prioritizing technical and policy intervention on them. The question of governance is then addressed with a final collection of recommendations for policy makers, industry, researchers and civil society.

This article compares the notions of security and privacy as they are treated in international cooperation relations. While security seems to benefit from a global consensus, it does not hold true for the concept of privacy. This article argues for not addressing the privacy in the context of the issue of human rights, marked by recurrent controversies about cultural relativism.

Section 3
Fields of Application (1): Genes, Atoms, Nanos

Section 3 Introduction

Chapter 6

Catherine Rhodes, University of Manchester, UK

This chapter covers the development and operation of norms within the international system and more specifically within the international governance of biotechnology. Through the use of case studies it highlights several key points which should be considered when analysing the role, application and implementation of norms in other governance areas.

Chapter 7

Sylvain Lavelle, Center for Ethics, Technics and Society (CETS), ICAM Paris-Sénart, France
Caroline Schieber, CEPN, France
Thierry Schneider, CEPN, France

Radioactive waste, remaining radioactive for very long periods up to hundreds of thousands years, introduces a new time dimension never experimented in the field of risk management. This situation led for more than 10 years, to reflections on the societal and organisational mechanisms for the development of protection systems, able to cope with those periods. Within the framework of the European research project COWAM 2, dedicated to the improvement of governance of radioactive waste management in Europe, a panel of stakeholders involving experts, authorities, waste managers, locally elected representatives and NGOs, opened a dialogue on the ethical considerations related to the long-term dimensions of this management. This article presents the main results of this panel, with specific emphasis on the meaning of the long term and what is at stake, the ethical dimension regarding long term issues, and the continuity and sustainability of the vigilance and surveillance of radioactive waste facilities.

Chapter 8

Fernand Doridot, Center for Ethics, Technics and Society (CETS), ICAM Lille, France

Based on an analysis of the current literature on nanoethics, this paper proposes to identify three different models for ethical governance of nanotechnology, respectively called 'conservative model', 'inquiry model' and 'interpretative model'. The propositions of the EGAIS1 Research Project in terms of ethical governance of nanotechnology are related to the latter model.

Section 4
Paradigms of Governance

Section 4 Introduction

Chapter 9

The elaboration of some paradigms of governance lies upon the opposition between the democratic and the non-democratic, namely, as will be shown and defined, the technocratic (skilled-based power), the ethocratic (virtue-based power), and the epistocratic (wisdom-based power). The point in this opposition is that, contrary to the democratic paradigm, the non-democratic ones assume that the condition for social rules or decisions to be valid is their reflecting, discussing and making by an elite of experts, virtuous or wise individuals or groups. There is no doubt in these paradigms a basic distrust as to the ability of the people to take in charge the public affairs and then to elaborate the appropriate standards and norms accounting for the regulation of actions and conducts. The re-construction of these four paradigms (the democratic and the non-democratic) can be illuminating as regards the interpretation of the actual expert and law-driven trends in the ethical governance of technology. It appears, indeed, that the paradigms of technocracy as well as that of ethocracy still operate in the design of governance settings aimed at regulating research and innovation projects.

Chapter 10

With an increasing focus on the inclusion of considering the ethical and social impact of technology developments resulting from research in the European Union, and elsewhere, comes a need for a more effective process in technology development. Current ethics governance processes do not go far enough in enabling these considerations to be embedded in European Union research projects in a way that engages participants in technology development projects. Such a lack of engagement not only creates a distance between the technology developers and ethics (and ethics experts) but also undermines the legitimacy of decisions on ethical issues and outcomes, which in turn has an impact on the resulting innovation and its role in benefitting individuals and society. This chapter discusses these issues in the context of empirical work, founded on a theoretical base, undertaken as part of the EGAIS (Ethical Governance of Emerging Technologies) EU co-funded FP7 project.

Chapter 11

Creative Commons oversees a system of common rights that provide creators and licensors with a simple method of indicating what freedoms they would like to pertain to their creative work. In this chapter, the notion of commons patrimony will be proposed in order to analyze whether Digital Commons necessitates a new type of governance: what we could call "patrimonial governance." The varying concepts of property, commons and patrimony will first be reviewed to give an insight into the fundamentals of the CC project. The various aspects of governance experienced through the CC Community will be analyzed and compared with research on patrimonial goods.

Section 5
Pragmatic Approaches to Governance

Section 5 Introduction

Chapter 12

Stephen Rainey, Facultés Notre Dame de la Paix, FUNDP, Belgium

European political life involves a productive tension between liberalist and communitarian tendencies. This 'Libero-Communitarianism' in the EU is the backdrop to various governance policies and potentials. This chapter develops a broad analysis of the governance setting in Europe and draws out some key areas of potential problems. This is based in the Ethical Governance of Emerging Technologies (EGAIS) project findings, and mirrors some of the issues flagged as ethically important in the field of emerging technologies. That such issues permeate European research and approaches to governance is testimony to their centrality and to the influence of Libero-Communitarian interactions.

Chapter 13

Matthias Kettner, Witten/Herdecke University, Germany

In the first and second part of the present article, the author provides a pragmatic reading of the very idea of governance. With the help of the late pragmatist Frederick Will's thoughts about the philosophic governance of norms, governance can be construed as a practice that is situated within other practices and whose aim is lending guidance to these practices. Since the point of establishing governance practices is to improve the targeted governed practices, governance is characterized by normativity, e.g. rationality assumptions, reflexivity and relativity to the general and particular significance of the governed practice. A schema is introduced for abductive inferences (as outlined by Charles Sanders Peirce) from observed defects in practices to expected improvements in governance practices. In response to the question, how governance itself is to be governed where it stands in further need of governance, I argue in the third section that there is an interesting problem of "polynormative" governance: Different forms of governance in different domains of practice may differ drastically in their advantages and disadvantages when compared from some particular evaluative point of view, and they will differ drastically across different evaluative points of view. The author argues that argumentative discourse, not in Michel Foucault's, but in Karl-Otto Apel's and Jürgen Habermas' sense of the term, is the governance practice of last resort for our giving and taking reasons. The relation of argumentative discourse to democracy (being the governance practice of last resort for political power) remains to be explored.

Chapter 14

Alain Loute, Université catholique de Louvain, Belgium

In this essay, the author focuses on what Jacques Lenoble and Marc Maesschalck call the "pragmatist turn" in the theory of governance. Speaking of pragmatist turn, they refer to recent work by a range of authors such as Charles Sabel, Joshua Cohen and Michael Dorf, who develop an experimental and pragmatist approach of democracy. The concept of "turn" may raise some perplexity. The author believes that we can speak of "turn" about these experimentalist theories because these theories introduce a key issue, what we may call the question of "self-capacitation of the actors." The author tries to show that this issue constitutes a novelty compared to the deliberative paradigm in the theory of governance. While the issue of collective learning is a black box in the deliberative paradigm, democratic experimentalism seeks to reflect on how the actors can organize themselves to acquire new capacities and to learn new roles. The author concludes in revealing the limits of this approach.

Section 6
Fields of Application (2): ICT's and Internet

Section 6 Introduction

Chapter 15

Jean-Michel Cornu, Next Generation Internet Foundation (Fing), France

The future Internet will be used by humans and objects. There are three main candidate solutions: the evolution with IPv6, the blank sheet with the Next Generation Networks and a rather different approach from the Pouzin Society. Last but not least, there is a way to avoid choosing. It names Virtual Routers.

Chapter 16

Paul Mathias, Collège International de Philosophie, France

Nowadays, life and culture are matters of technological interconnectedness. This is not to say that connecting computers equals to creating a cultural environment and we might just want to consider their day to day technical efficiency and productivity. But then, how can we assess a cultural, not just practical dimension of the Internet? The question of a cultural dimension of the Internet is not that of its usefulness. Though technical craftsmanship might be a requirement for an authentic cultural networked experience, instrumentalism, as a theoretical approach, connotes exclusively the necessity of controlling the flow of data and information. This paper aims to show that the Internet is not just a technical pattern and a normative framework aimed at producing and enacting otherwise defined cultural experiences. Information and communication technologies are complex programmed and interrelated functions eventually realised in the form of a cultural ecosystem – the Internet as such – and they do not simply form a web of industrially organised devices and services. It is then wrong to say that networks crush our minds and that nowadays acculturation prevails. The Internet is a cultural matrix generating new forms of meaningful interactions through the permanent and pervasive interconnectedness it allows for.

Chapter 17

Olli I. Heimo, University of Turku, Finland
Kai K. Kimppa, University of Turku, Finland

In this chapter, the authors present four cases of Finnish eGovernment application development in which the ethical approach has either been ignored, or mishandled grievously. The trend in Finnish eGovernment system acquirement has shown to lack reflexivity towards ethical issues. Between all the systems studied, the common denominators have been extended delivery times, increased costs and at best non-functional solutions. Testing, prototyping, and analyzing the systems have been lean and have only concentrated - interestingly enough - to the fields of functionality. The only discussion about the ethical issues has been between a minority of specialists and civil rights activists, and has been mostly ignored in the procurement of the systems.Some of these examples contain only clear mistakes and design failures which could have been – at least partially – avoided. Others, like the Case of Finnish biometric passports, show alarming features of lessening both privacy of the citizens and the security of the whole society. The reasons behind the lack of ethical thinking during these design processes are investigated in this chapter.

Section 7
Frame Development and Reflection

Section 7 Introduction

Chapter 18
    Craig Steven Titus, Institute for the Psychological Sciences, USA

This chapter argues that a developmental psychology based in a wider notion of reason and ultimate flourishing can employ both duty and virtue in the service of the common good. It identifies several important differences between cognitive structuralism and virtue-based approaches concerning the pre-empirical priority paid to either duty or virtue in moral development. It brings to light several challenges concerning the use of developmental psychology in ethics: (1) a weakness in schools of cognitive structuralism, such as that of Lawrence Kohlberg, inasmuch as they do not move beyond the theory of stages and structures that focus only on the cognitive judgment of justice and on duty; (2) a weakness in developmental virtue approaches, such as that of Martin Seligman, inasmuch as they do not employ moral content in the operative notions of virtues and values. This article concludes that a heartier notion of developmental psychology and normative ethics will need to recognize the interrelated nature of ethical acts (moral agency), ethical agents (moral character), and ethical norms (duties and law). Such an integrated approach must also attend to the input that diverse philosophical and religious presuppositions make toward understanding the place of developmental psychology in the practice of ethics.

Chapter 19
    Martine Legris Revel, Université de Lille Nord de France, France

The author intends to explore the work that can be done by somebody on his/her own history in order to modify its course. Those questions have been largely debated. The purpose of this paper is to analyze Pierre Bourdieu's thinking of those topics. He is concerned with the interiorization of social structures (habitus) and suggested self socioanalysis as a new way of highlighting them in 1991. Can somebody truly escape from social structure's interiorization effects or even have an action upon them?

Section 8
Changing Methods and Procedures

Section 8 Introduction

This chapter aims to construct a basis to move toward addressing lacunae in governance approaches in a way that is not merely ad hoc but rather is grounded in theory and that can affect practice in a positive way. Essential to this is the establishing of a problem clearly defined in order that it can be a problem understood. Here, the ground is cleared for proposals for ways to approach the specific problems we detect. In doing this, the ambition of treating them is rendered possible.

Chapter 21

Proceduralism has been a major philosophical stream that gathers some outstanding philosophers, such as John Rawls or Jürgen Habermas. The general idea of proceduralism, especially in the practical domains of morals, law and communication, comes from the need to provide some rational justifications to the rules, actions and decisions to be adopted or made by the society or the power in the context of highly 'plural-complex-developed' societies. It is particularly concerned with the governance of ethics, in other words, with the institutional and organisational conditions that the procedures of assessment must fulfil so that ethical questions can be addressed, especially in the domain of scientific and technical research and innovation. We show that classical proceduralism does not adequately address problems raised by an ethics of science and technology, and we take the context of Europe as a typical example of what a complex multicultural set of societies can be.

Preface

INTRODUCTION

This book is the outcome of a collective venture and effort within a European research project called EGAIS (Ethical Governance of Emerging Technologies). The project was funded by the Seventh Framework Programme (FP7) of the European Commission and gathered for about three years a set of researchers coming from several academic disciplines and several countries in Europe. The title of the book is quite similar to that of the project, nevertheless, the former is not a mere transfer from the latter, but rather a sort of encompassing translation of it, including chapters from outside contributors.

GENESIS OF A PROBLEM

In an increasingly pluralist Europe, there is widespread awareness of the importance of ethics in research and new forms of governance. Ethical governance is often carried out in a top-down, expert-driven manner, which has negative implications for its legitimacy and effectiveness and for the integrity of the research.

The aim of the EGAIS project was to elaborate a *new perspective on the ethical governance of research projects in Europe*, mainly in the technical field. It was not concerned only with the ethics of technology in the shape of a code of deontology or an ethics committee, for instance, which is rather the usual approach. The focus was on the question of the *effectiveness* of ethical norms, from the conditions of their emergence and production to their implementation within a social context. This effectiveness is rarely, if ever, questioned, and theoretical approaches to date implicitly presuppose the conditions that determine the effectiveness of the implementation of the norm. They all come from 'cognitive rules', are then a function of mental capacities, and are therefore independent of the external subjects' own context.

In fact, what is at stake in ethics is not the mere application of a procedure, nor the mere compliance of an action, a decision or a project with a set of rules. What is at stake is the question of the *normativity*, that is, the relationship of norms to the context and to the values of individuals or communities. In this respect, it is assumed that the process of *framing* of ethical issues is in a way more important and significant than the outcome of the procedure of ethical assessment as such. This focus on the framing calls for shifting in the procedural approach from an ethical analysis of the issues (privacy, dignity, discrimination, etc.) to a *meta-ethical analysis* of the governance process. In terms of procedure, this more reflexive method differs strongly from a 'check-list' of criteria to be fulfilled by projects officers or assessors, like in most operational procedural frameworks.

It was shown necessary to organise the *reflexive capacity of the actors* by constructing it in such a way as to not presuppose it as already existing. This presupposition is often due to the use of a formal method, such as argumentation, deliberation, debate or discussion that presupposes its own required conditions, often without any much reflexivity. It is important that every application of a norm presupposes a formal moment of choice of its acceptable normative constraints. But it is also important that it presupposes a selection of the possibilities according to the acceptable way of life within the concerned community. Without any negotiated reflexive momentum on the application of norms, there would be no control over the process of norm expression, and it will be left to the dominant common culture. Thus, even what is often presented as the only effective choice is always conditioned by an operation such as the above mentioned one, including in the elaboration of deontological codes.

Through the empirical analysis, data were collected from technical research projects, funded by the EU as part of the FP6 and FP7 programs, which employed Ambient Intelligence. The focus on the Ambient Intelligence projects allowed the EGAIS group to make a targeted yet representative study of ICT research projects. The different types of ethical governance and reflexivity that are used in EU projects were identified. The aim was not to create a quantitative or statistical study of reflexivity of governance within the projects, nor to compare the effectiveness of particular approaches or tools. Instead the interest was in the determination and the assessment of the limitations and the effectiveness of implementation of the different existing approaches.

EMERGING TECHNOLOGIES, ETHICAL GOVERNANCE

The notion of *emerging technology* suggests that technology is not a static business that is supposed to provide once and for all a range of useful products to millions of consumers. There is indeed a dynamic of technology, almost in the same sense as we have a dynamic of ethics or politics, with all kinds of uncertainties that are usually associated with the idea of an on-going process. The regulation of emerging technology is not simply a matter of governance, be it a democratic one, but, as we suggest, a question of *ethical governance*.

The word technology is not easy to define for it has different meanings according to the linguistic community and the cultural context in which it is commonly used. One can distinguish at least two basic meanings of technology:

- Technology designates the set of scientific techniques (e.g.: a car, a computer) as developed by engineers and experts and that differs from the more traditional techniques as produced in crafts and works and by some ordinary or lay people (e.g.: a bow, an arrow).
- Technology designates the complex process of socio-technical innovation encompassing several stages, such as those of research and development, or those of design and production of a technical artefact that more or less meets the need of a set of consumers or users.

It can be said that the notion of emerging technology emphasizes in the cycle of innovation the very stage of the development and the diffusion of a technical artefact than then appears for the user or the consumer as a product available on a market. The word 'appear' is certainly most appropriate to stress this impression for the user or the consumer of 'seeing' new objects for sale, but also for use, popping up in the course of his or her daily life. You wake up one morning, go for a walk, and you can see in a shop that, starting from now, you can phone your wife while walking in the street instead of calling her as discreetly as possible from your good old black telephone stuck onto your office's desk.

The dynamic of emerging technologies suggests that there is something a bit special in the innovation process that on a regular basis flows some new technical artefacts the use of which sometimes cannot be determined by a mere market rule or a mere purchase choice. It has been the case for some topical technologies, such as genetically modified organisms, nuclear radioactive waste, nano-products, and for an increasing set of new technologies (television, telephone, computer, audio/videos,…) that belong to the so-called information and communication technologies (ICTs). Those emerging technologies in their time raised a huge amount of critiques and controversies on the ground that they not only carried about some important potential risks for health, environment or society, but also that their legacy, their legitimacy and even their utility had not been fully evidenced.

In this respect, the kind of uncertainty that one can experience in the trajectory of an artefact or a product is not a technical one, but rather, it can be said an ethical one that requires a specific inquiry to overcome it. The problem is, alike in the technical side to some extent, the ethical investigation, exploration or advice is often achieved or produced by a panel of experts that are supposedly professionally skilled to solve this kind of difficulty. Then, the lay people or the 'man in the street' are deprived of their right to express their opinion or their position on those essential social and moral matters. The reason for that is they miss an ethical expertise, so to speak, in almost the same sense as they miss a technical expertise as far as the innovation process is concerned. This is typically a problem of governance, or, in other words, a problem of stakeholders' relations and interactions: namely, between the experts, the representatives, the researchers or the engineers on the one hand, and the users, the consumers, the citizens or the associations on the other hand.

The word governance also is not easy to define, all the more it is sometimes regarded - or rather, disregarded - as a fashionable misleading word that shadows the actions of the ruling class within firms or governments by giving a rosy picture of their actual strategies and tactics aimed at keeping the power and the wealth in their hands. One can distinguish at least two basic meanings of governance:

- Governance designates the set of rules and conducts (distribution of power and property, internal and external communication, etc.) that is required for a community of people to function in an efficient, legal, legitimate, or fair way.
- Governance designates the democratic process of opening up the power of institutions and organisations (government, parliament, councils, etc.) to the citizens or the civil society by making use of some interaction settings (information, consultation, deliberation, participation, etc.).

The notion of *ethical* governance suggest something more specific in that it does not equal democratic governance - or at least, it does not assume that democracy is in itself sufficient for ethics. The stake of ethical governance that basically questions the ethics of democracy is to elaborate some institutional arrangements for reflexivity processes to occur in order to overcome the limits of the existing ethical approaches that just ignore the issue of the norm application. The determination of those arrangements allows actors and institutions to go through a *learning process* while facing an ethical issue, in reflecting on the success of the learning process, and in reframing the context of the situation. The aim is to establish more effectively a norm within the context, and from a more official perspective, allow for assessing the effectiveness of that process' outcome.

To take into account the limitations and achieve second-order reflexivity, we need to escape the binds of formalism, which constrains ethics with its presuppositions and internal limitations. To more effectively incorporate ethical norms into contexts, we need to construct the framing of the context in relation to the norm (i.e. not presuppose it), then open up this context so that we can have a reflexivity on the opening of this framing (that is, a feedback mechanism). In order to do this, we need to reconstruct, from a normative perspective, how research projects should reconstruct the two-way relationship between the norm and the context to overcome the fundamental limitations in order to achieve a second-order reflexivity. Without the implementation of a reflexive capacity, and the construction of norms, normative injunctions risk remaining inefficient even if the objective is judged relevant and legitimate. The operation of judging the conditions of the choice of an improved way of life, facilitated by technological development is distinct and asymmetric. Asymmetry is the way in which the social meanings of a norm are conditioned by an operation that cannot be anticipated by formal variables of reasoning variables that condition the norm's relevance.

NORMS, CONTEXTS AND PROCEDURES

The link between norms, contexts and procedures can be summarized briefly in the following way: one the one hand, the formal method of proceduralism does not require that one pays attention to the *context* of application of norms, but only to the validity of arguments that offers a rational justification to them; on the other hand, if one do so, the lack of context reflexivity brings about the problem of the norm *relevance* as to the actors' context and then raises the correlated problem of the norm *effectiveness* in such a context. It can be suggested in this respect that the procedure in the philosophy of proceduralism can be questioned as to the level of relevance and effectiveness of norms that it allows in a governance process.

Concerning the *norm*, one must make the difference between a descriptive and a prescriptive interpretation of it, which refers to the issue of the *normativity* of a norm. It is one thing to assert in a descriptive way that smoking is prohibited (factual meaning of norm), it is another thing to commit or evaluate in a prescriptive way the binding force of the norm for one's conduct (evaluative/normative significance of a norm). We call the reductive stance towards normativity a *positivistic reduction*, and we suggest that, given the attachment of some scientists to what Weber calls 'axiological neutrality', the social sciences are necessarily the main reductive disciplines as regards normativity. This denial of normativity defined as the evaluative relationship to norms is almost a direct consequence of the requirements of the methodology of some perspectives on the social sciences.

Ethics, as a reflexive stance to norms, can be seen as the very approach concerned with the issue of the normativity of a norm, provided it is not subject to a positivistic reduction, as it is the case in descriptive ethics. We assume that normativity is the general and special property of a norm that is subject to a normative commitment or evaluation from an individual or a community of individuals. In that sense, normativity is not so much the content of a norm than the relationship that an individual or a community of individuals can experience as to the content of that norm. However, one must make a difference between the substantive 'norm' and the adjective 'normative', which means that normativity is more general than a norm. In this respect, an advice, a recommendation or a suggestion can be normative, whereas they are not a set of obligations. It derives from that definition of normativity that a moral norm is neither a social norm, nor a legal norm, even if the dividing line is not always very clear in the daily use of these notions:

- **A Moral Norm:** A norm that calls for a (ethical) problematic value judgement of an individual or a community as regards the prescriptive force of the norm.
- **A Social Norm:** A norm that calls for a fact judgement (so equalling an unproblematic value judgement) of an individual or a community of individuals given the descriptive relationship to the norm.
- **A Legal Norm:** A norm that does not call for a (ethical) problematic value judgement of an individual or a community of individuals despite the prescriptive relationship to the norm.

The differences between moral, social and legal norms are not absolute ones, but they suggest at least that they are not equivalent terms.

As to the *context*, in the broadest sense of the word, it can be defined as a temporary or permanent background, be it natural, artificial, social or cultural, shaping the modalities (necessity, obligation, possibility,…) of human thought, conduct and taste from the point of view of cognition, volition, action, judgement, experience and significance. The notion of context is often something perfectly simple to understand in objective terms, since there are instances when we can determine the characteristics of objective features of the environment. We instead suppose that it is a thickly connoting notion and that it is therefore very important to get an impression of what role context here plays.

Lenoble and Maesschalck describe three facets of 'context' to which the entire notion cannot be reduced: a context is not a set of factual constraints, neither a false representation or a framework offering resistance, nor a particular culture which cultural anthropologists could identify and which could be deposited in the individual minds of individual actors as continually adaptable conventions that would serve as capacitating structure for them. These three components exist but don't exhaust the function of context, so they are each reductions of the notion of context and then one misses the question of the relevance of the context. The very notion of context itself is very open regarding the purpose of circumscribing a domain even when it is objectively describable and the circumscription is steeped in subjective decision. It relies on prior decisions that characterise the context in a sense beyond the mere description of its elements, and these prior decisions are what constitute the framing of a perspective. The role of judgement is essential in all these operations: what the 'objective facts' mean for a given actor will be contingent upon their judgement of what the broader context is. In fact 'broader context,' used here to distinguish between the narrow description of elements of a scene and the relevance of that scene in the consciousness of the subjects to whom those elements are addressed, is something of an artifice – the broader context is the context in the sense we need to use it.

The context itself refers the norm to its effective possibilities, confronting it with all of the possibilities in which it can be accommodated within the world. This fundamental and radical reversal is contained by all normative elaborations, and, by itself, allows the norm to be translated in effective power in a given context. This is because the normative elaboration process integrates the reflexivity on the conditions that ensure the effective expression of the norm from the beginning. Deep-seated theoretical issues underlie the problems seen with ethics in research: *reason doesn't contain within itself its own determinations.* From pure principles of thought, logic, consistency we don't get substantive content: the reasons to accept an argument as valid aren't necessarily reasons to accept its conclusion as a maxim for action. We have to value the content of an argument, besides accepting its validity, to adopt its conclusions.

Dialogical *procedures*, while being a means to avoid the pitfalls of both deontological and teleological theories, face at least on some crucial points the same kind of limit. The major limit is that of the context, and more precisely, the relationship between the rational justification of norms and the context of application of norms. Dialogical procedures cannot avoid the objection of the limited relevance of norms to the context, all the more their supporters claim for the moral norms to be regarded as universal rules, or at least, universalisable ones. There must be something more than a mere procedural discussion to elaborate a rational justification of a norm and almost simultaneously adapt it to the specificity of a social and cultural context. In order to address the contextual blindness shown, one cannot but criticise the procedural reason behind it. An ethical norm needs to be conceived with a feedback mechanism on itself that is induced by the anticipation of the norm's insertion into the coherence of an application field. It must be questioned, from its point of view, of its capacity to participate in the emergence of a way of life. If we take the conditions seriously, all such Kantian schema approaches are fundamentally insufficient because they are still dominated by a logic of subsumption (*deductive relation*).

UNIVERSAL VS. CONTEXTUAL PROCEDURALISM

Proceduralism has been a major philosophical stream that gathers some outstanding philosophers, such as John Rawls or Jürgen Habermas. The principle of proceduralism, especially in the practical domains of morals, law and communication, has arisen from the need to provide some rational justifications to rules, decisions and actions in the context of 'plural-complex-developed' societies. One of the main characteristic of this kind of societies, indeed, for the institutions as well as for the citizens, is the lack of a homogeneous background that would offer a common framework for the argumentations or the justifications to be shown valid or not. In the traditional societies, be they Christian or Muslim ones, for instance, the existence of a common religious background make it possible for the community of people to refer to some shared obvious common principles, habits or experiences. This is no longer the case in contemporary societies and this requires a new method to be conceived of and implemented at the various stages and levels of the power structure or process.

Here, proceduralism plays the role of a minimalist method that is supposed to be neutral as to the substantive doctrines defining the good at individual or collective level within a society. Proceduralism as a method concerns most of the domains of philosophy, namely the epistemic, the technical as well as the political or the ethical. One can think of an epistemic or a technical proceduralism, but we would like here to concentrate upon the ethical proceduralism to be taken as both a reflexive and constructive stance about legal and moral norms. There exist a variety of procedural methods, but the main models are no doubt the rational discursive or deliberative ones as developed notably by Rawls and Habermas at a theoretical level. It should be mentioned that a set of practical applications has already come out of these models, such as the rules of 'positive discrimination' for Rawls, or the rules of 'fair communication' for Habermas.

All ethical governance cannot account for values, norms and contexts with respect to ethics, and this is an ultimate limit on the ethical governance of emerging technology. Procedural approaches to governance can lead to outcomes via overly formal means and contexts in which the perspectives of the governed can be lost. This can create problems for the legitimacy and effectiveness of governance processes for those who would be governed. The alternative, a substantialist approach, risks imposing preconceived ideals upon a citizenry with negative implications once more for legitimacy and effectiveness.

This is a problem between proceduralism and substantivism. We see the limits of proceduralism in substantial contexts (e.g. 'Privacy' is well established by argument as something worth protecting – but what does it mean here, now, in *this* context? Proceduralism (particularly that of Habermas) holds that a normative statement's semantic content is not important, but the approval process by concerned parties is the validating step. Thus free approval needs to be gained for a normative statement to be considered valid. Proceduralism seems to solve the problems outlined above, because it proposes an internalist, rather than externalist, point of view. The current Habermasian approach takes into account the reversibility of the operation of justification and the application of rules. This dimension of reversibility indicates a co-dependence between these two operations, that is, that the application of the rule retro-acts upon its justification. Therefore, the choice of a relevant norm necessitates a reflexive return on its context of application. This, however, leads to a supposition that the elements which condition the application of the rule are subsumable under rules that guide formal discursivity of the mechanism which is mobilised by the formal calculation of the relevant rule. However, procedural relations can resolve situations where two conventions of relations to the norm come into conflict, which is not a situation that can be resolved with the conventional relation.

Every judgement, and every norm, since a norm is the result of a judgement, can only be applied at the end of an operation. The operation is an exchange between the effects expected by the abstract norm and the possible effects raised by the coherences specific to the existing way of life. A norm can thus only be inscribed in reality, and can only make sense by being supported by particular perceptions of the way of life, and particularly those whom the rule affects. It fails to allow for any reflexive capacity for the actors to identify the various effective possibilities on which the operation of the selection of the relevant norm will be carried out. There is also an implicit limitation of this approach, which is that the discursive and rational construction of the definition of the relevant norm is capable by itself to take into account all possibilities of the social context to be regulated. Although there is a reflexive approach to rationality, there is no reflexivity relating to the context (social, economic, scientific contexts, etc.). Although this does not invalidate the discursive procedure for reasoned elaboration of relevant rules, it does require that the arrangements for normative production be more complex.

The choice of a relevant rule consists of a making a decision of a real-life solution that is supposed to optimise the ideal objective drawn out by the anticipation of an idealised way of life. The choice of this solution, however, rests precisely on an operation of the selection of possibilities that does not exhaust the possibilities of the context within which the idealised way of life would have been realised. Transforming the context with the view to incorporating an ideal norm within it will only result from non-formalisable operations of interpretation. Nothing within the formal mechanism can guarantee that the choice of possibilities taken into account in order to define the choice of a norm will ensure that the realisation of this ideal will correspond to any one of the diverse possibilities perceived by the actors concerned. It is therefore necessary for a rational procedure of the calculation of norms to be intersected by reflexive incentives. This would allow for the reconstruction of the problems that condition peoples' practical acceptance of the transformation of their way of life. Without this prior reflexive effort, the processes that would enable the effective expression of relevant norms would probably not be taken into account.

The operation of the selection of possibilities carried out by the production of each norm cannot therefore be restricted by the rules, because of the guarantee of the discursive operation of formal reason. The effective possibilities drawn out by the social application of a norm are a function of the conditionality that depends on the structure of the context. These can only be reconstructed within a concrete act of community reflexivity whose conditions are not anticipated by the formal rules of the discursive act. Using the Louvainist (Lenoble & Maesschalck, 2003; 2006) contextual pragmatic theory we can underline the limitations of every theory that presupposes the conditions that makes the exercise of reason possible. This is true for all the theoretical ethical approaches that are characterised, according to Lenoble and Maesschalck's terms, as "intentionalist, mentalist, and schematising" presuppositions. Even if individuals are able to revise or adapt these conditions, they do not take into account the reversible or reflexive character that allows for these revisions or adaptations.

The criticism we level here emphasises the necessity of understanding the reference to the background as a speculative and transcendental logical constraint of the operation. This allows us to better understand the consequence of our approach to the reflexivity of judgement on the level of the construction of governance arrangements. A *comprehensive proceduralism* is required in the sense that procedures must be deployed that are based in the perspectival comprehension of contexts by citizens.

TOWARDS A COMPREHENSIVE PROCEDURALISM

Contextual proceduralism fails to embrace the value-dimension of the agents' relationship to norms and of the reflexive stance of the judgement operation. The missing part in this method lies in the pragmatic operation of contextual judgement as well as in the genetic process of learning and identity-buiding of the agents. It appears that contextual proceduralism suffers from two weaknesses: first, an obliteration of the significance of norms; second, a disjunction between norms and values. The alternative methodological option that we propose can be termed *comprehensive proceduralism* insofar as it attempts precisely to take into consideration the value dimension in the governance processes. Comprenhensive proceduralism seeks to warrant a reflexive adjustment or rationally justified norms to the context of agents in testing, destabilizing and determining the value significance (or axiological commitments) of these norms.

Thus, proceduralism can be said 'comprehensive' in two senses:

- The word 'comprehensive' refers to the notion of *combination of options* in terms of procedures.
- The word 'comprehensive' refers to the notion of *value significance* of thought and conduct in the relationship to norms.

The first meaning suggests that comprehensive proceduralism is a method that pays attention to the variety of procedures to be selected according to their relevance as to the actor's context. The second meaning suggests that this method pays attention to the value dimension of the actors' judgement in the determination of the significance and the scope of norms. This option does not pretend solving all the harsh problems raised by the methods of proceduralism, but it attempts at least to overcome those raised by its contextual-reflexive version.

We can qualify this method a reflexive equilibrium through an adjustment of procedures to the contexts and the values of the actors, or, in other words, a *comprehensive procedure of reflexive equilibrium* that is context-adaptive and value-sensitive. The principle of judgement in the method of comprehensive proceduralism is not that of consensus, but that of a two-fold potential outcome: consensus or dissensus; consensus on some aspects, dissensus on some other aspects. In this line, a consensus can be reached about a norm, a rule or a principle (e.g: dignity, privacy, etc), while the significance of it, based upon each agent's context, is not the same at all. The main stake of a reflexive equilibrium is then to produce a *reflexive exploration* of the agent's context as well as a *susbstantive determination* of the value-significance of norms for each agent. This process of determination enables the actors to come out of the situation of substantive indeterminacy, in which they are misled by the appearance of formal agreement on a principle or a rule. We suggest that a governance process must be able to offer a variety of procedures best adapted to the actor's context (*context-adaptive proceduralism*) and to explore the diversity of value-significances given to a norm on the basis of the actors' contexts (*value-sensitive proceduralism*).

The governance of techno-ethics based on the idea of 'comprehensive proceduralism' pleas for a combination of approaches that are both procedural (rule-based), reflexive (context-based) and substantive (value-based). The ethical governance of research projects based in this idea of 'comprehensive proceduralism' thus seeks for a combination of approaches that are at the same time:

- **Procedural:** Rule-based
- **Reflexive:** Context-based
- **Substantive:** Value-based

In a more applied perspective, we are concerned with the *governance of ethics* and with the institutional and organisational conditions that the procedures of assessment must fulfil so that ethical questions can be addressed. The aim of what we call comprehensive proceduralism as far as applications are concerned is to elaborate a new perspective on the ethical governance as applied especially to technical development. We are not concerned only with a regulation of instrumental or strategic rationality in the shape of a code of deontology or an ethics committee, for instance. A series of concrete improvement can be achieved as a development of this view, like in the composition of the consortium (other experts, cross discipline committee, etc.), the education to ethics (ethical reasoning, ethical validating, etc.). Whatever the improvement, the important point is to warrant a connection between the criteria and the knowledge of the people who will use these criteria.

Section 1
Ethical Governance: Which Rationality?

Section 1 : Ethical Governance – Which Rationality?

It is pretty usual to make a difference between several forms of rationality: substantive rationality *versus* procedural rationality, universal rationality *versus* contextual rationality. The theory and practice of governance is often associated with the rational procedures of discussion as developed in the various streams of proceduralism, be it a "strong" one, or a "weak" one. In other words, the interactions between groups of stakeholders possibly leading to the elaboration of a common norm must be regulated by, and only by some principles or rules of discussion that function as a kind of "meta-norm" (freedom, equality, argumentation, inclusion, etc.).

The critique of classical proceduralism as embodied by Rawls and Habermas refers to the disjunction between the *rational justification* of norms on the one hand and their *contextual application* on the other. Thus, as suggested in Lenoble and Maesschalck's *contextual proceduralism*, a procedure of multi-stakeholder governance that neglects the problem of the actor's context can be reproached for not enabling any reflexivity process. The condition of reflexivity in ethics is required to be satisfied for the actor's context to be as far as possible examined and interpreted so that each actor becomes aware of its own judgement's assumptions, or presuppositions. For instance, if a Parliament enforces a law on the "positive effects of colonialism," it is important from an ethical perspective that each part is able to reflect on the cognitive and normative assumption or presuppositions that lie behind this law, whether you support it, or reject it. In this respect, it is essential to make a difference between a procedure of governance, a procedure of democratic governance, and a procedure of *ethical governance* on the ground that democracy does not necessarily fits the specific requirements of ethics. The ethics of governance calls for indeed that one can meet the conditions for ethical reflexivity of a governance procedure or process possibly leading to a common legitimate norm that does not equal a mere legal norm.

In proceduralism, the relations of the actors are bounded by the rational option of *procedural discussion* that requires from the actors to exchange and validate their *arguments*. However, the difference that is often made between a communicative rationality and an instrumental or a strategic rationality is not so easy to make when several types of rationality are mixed in a discussion. One can suppose that concrete situations mix the *different dimensions of rationality*, but their division does not make one able to say by which ways, neither which requirements ways of mixing dimensions of rationality are submitted to. Moreover, a set of other supposedly non-rational conditions are to be met for the arguments to be effective, among them the *emotions* and the *values* that play a significant role as conditions for the arguments to have some strength, or some significance. The result is that we have not only to consider the debate and its arguments, but also the *context* of the debate and the problems of social and emotional recognition of its participants as well as the absent or excluded people. We also have to consider the long-term effects of the conclusions of the debate on the emotional acceptance or resistance to the decisions that follow it, related in particular to the values that the normative conclusion of the debate has let aside.

In addition, the principles in *practical rationality* and *rational planning* have a structuring role for some context-dependent features of governance procedures. In the field of social policies, for example, the enhancement of personal autonomy has come to the forefront of collective challenges. Capacity-based approaches indicate a way to put into question those conceptions of autonomy, which lead to an excessively uniform treatment of individuals. Following these approaches, the beneficiaries of social policies should be treated as concrete beings with their personal history, living in specific social contexts and so on. It appears that in interactive problem structuring or in institutional decision-making, the adaptation to specific circumstances is made possible through a complex process of *description* of problems and challenges for collective action. Even if the procedure plays an important role in it, this process leads to re-defining *interactions*, *responsibilities*, and also *identities* of the actors involved in the governance.

Chapter 1
Governance in Perspective:
The Issue of Normativity

Laurence Masclet
University of Namur, Belgium

Philippe Goujon
University of Namur, Belgium

ABSTRACT

This paper describes the evolution of the discipline of governance study through a focus on the evolution of proceduralism. From a lecture and a critique of Maesschalck and Lenoble "contextual proceduralism" and their reconstruction of the evolution of the field, the authors aim at creating a more open theory ("comprehensive proceduralism") that would avoid some of the reduction of previous school of procedural governance, notably the reduction to argumentation, by opening the scope of governance to the different register of discourse. This paper aims at introducing the EGAIS book, and thus will introduce the theoretical (and critical) emergence of comprehensive proceduralism, leaving some of the consequences and application area to further investigation in the book.

INTRODUCTION

Ethics and governance are increasingly being recognized as a necessity in the technological and scientific fields. The issue is that those concepts are very broad and often vague, as they can cover a lot of different theories and practice. The field of governance studies is full of divergent theories, divergent practices that hold major contradictions between them. To give a perspective on governance, we will both have to choose between those theories and find a way to understand their own dynamic. The aim of this article is to give a philosophical perspective on the state of governance nowadays, the historical trend it comes from, the ways it is developing, the different challenges it

DOI: 10.4018/978-1-4666-3670-5.ch001

faces (gap between theories and practice, issues of decontextualisation, of instrumentalisation, and so on), and the possible ways to solve them.

To reach that aim, we have found a very useful theory in the theory of Marc Maesschalck and Jacques Lenoble, from Center of Philosophy of Law, Louvain University (forming what we call "the Louvain school," which has created the theory called "contextual proceduralism"). Their approach, despite some limits that we will point out (status of the theories, procedural presupposition, reduction to argumentation, etc.) has the interest of putting together fragmented currents in fields usually separated and finding a dynamic of the evolution of governance among them. The hypothesis of Louvain is that a lot of disciplines actually share, despite their separation, the same exigencies and intuitions. From that, they reconstruct in four steps (Maesschalck & Lenoble, 2007a) the dynamic of governance thoughts, from which their own theory is the final step. Indeed, the aim of their work, beside make light on some internal interdisciplinary dynamic, is to extend and deepen that dynamic into a united theory of governance.

The reason why we take this particular theory of governance as a way to approach governance in general is because they give us a meta-view on the development of governance, in which their own theory is integrated. Because they tackle every important governance theory, while constructing their own, following their work is a good way to understand the dynamic of governance, with the restriction of course, that their construction is indeed a construction, and, as such, has to be taken as an hypothesis, rather than as the only possible way of interpreting the different trends in the field of governance.

The Louvain approach gives a very complex reconstruction of the dynamic of proceduralism and neo-institutionalism while being very critical, but gives also an original theory to solve the insufficiencies of the field of governance. Their theory, along with the theory of Jean-Marc Ferry,

from the university of Bruxelles (ULB), has been the ground from which has been constructed the theory of governance called comprehensive proceduralism, which has been the theoretical result of the EGAIS project, and which has been since the ground of some other projects, continuing to test the hypotheses of that theory.

This paper is organised as follows. In section 2, we detail the theory of Maesschalck and Lenoble in its two steps (reconstruction of the procedural dynamics and the stakes it carries in various disciplinaries, and their own solutions to some of the challenges and limits of that dynamic). In section 3, we address the limitations of their solution and give hints for solutions to some of the challenges pointed out. The final section is the conclusion, in which we explore the possibility of further research, and the limits of our own approach.

THE LOUVAIN'S THEORY

The theory of Maesschalck and Lenoble aims at tracing a dynamic of the governance thoughts, splitting it into four steps, and examining the limits of each step, and, moreover, how each evolution takes the limits of the precedent and tries to give solutions by deepening the same intuitions with new tools and new hypotheses. Each step does not contradict the achievements of the precedent(s) but, on the contrary, tend to complexify its mechanisms to better comprehend the complexity of the governance situations, and by that, increase the number of tools to act.

Beyond complexification, the dynamic that Maesschalck and Lenoble point out is a process of *internalisation* of the resource for collective action. The first step still postulates a hierarchical institutional form as an external incentive, while, as they go further and explore the limits of that approach, the next steps go on to interiorise the mechanisms of regulation of collective social action, to put it in the hands of the agents (even if the hierarchical recourse is not completely destroyed or cancelled). The main hypothesis of Maesschalck and Lenoble is that "the existing approaches are affected by a limitation consisting of an inadequacy in how they deal with the theory of collective action" (Maesschalck & Lenoble, 2007-a, p. 6). Their first aim is not to judge or to classify the previous governance approaches, but to test their hypothesis on the trend, and, by doing that, integrate themselves in that evolution. This aim allows them to take in every field what fit with their hypothesis, which can be seen as a fundamental methodological flow (the old trap of only finding what you are looking for, in which most –if not every- research fall at some point). However, that method, if announced as such, can be very useful in order to construct new theories and new perspectives.

The Meaning of Governance

The Louvain theory proposes a very inclusive approach to governance, and linked it immediately to collective action.

As posed within the social sciences nowadays, the question consists of reflecting on the form of governance that should govern collective action in order to ensure that the best possible of all intentions is fulfilled. […] Use of the term 'governance' quite simply reflect the surmounting of the dual attitude mentioned previously: one branch of the attitude reflects on the question of the public interest in terms of spontaneous harmonization of various private interests while the other supposes that the requirements of the public interests are satisfied by virtue of the fact that action has been imposed or produced by an authority (the State or its agents) believed to represent or incarnate this interest. In the process of surmounting this dual attitude, the question of sole recourse to the market or forms of governance by hierarchical control becomes problematic. (Maesschalck & Lenoble, 2007-a, p. 6)

Governance is a combination of an inherent social mechanism, a social equilibrium that self regulates within itself, and of a top-down mechanism by which the state impose what is good for the public interest. This distinction is interesting, because, rather than talking about top-down approach versus bottom-up approach, it suppresses from the equation the possibility of participation, the freedom of the social actor to have an impact on policies, beyond a spontaneous self organisation. This is of course, as Maesschalck and Lenoble point out, a position that has to be surmounted. It still indicates a way of thinking that is still very much implanted in policies and in economy.

The notion of governance has been the object of conceptual mutations, which show how the way of thinking about society has evolved. From this balance between the spontaneous harmonisation of interest and top-down rules to guarantee the effectiveness of the harmonisation of interest, the term governance became increasingly used only to designate exactly what was left over in the previous use of the term, which is the cooperative mode of governing.

As R Mayntz rightly points out, '[T]oday the term governance is most often used to indicate a mode of governing that is distinct from the hierarchical control model characterizing the interventionist state.' 'Governance' she goes on, adopting a formulation characteristic of neo-institutionalist economists, 'is the type of regulation typical of the cooperative state, where state and non-state actors participate in mixed public/private policy networks.' Undoubtedly this formulation indicates very clearly that what is sought nowadays in the way of a form of governance for collective action cannot be reduced either to market self-regulation alone, or to traditional command-and-control forms of regulation. However, it should be handled with care, because it could imply the resolution of many questions that are in fact still unresolved, as is made clear by the current state of discussion on the theory of governance. (Maesschalck & Lenoble, 2007, p. 8)

This change in the meaning and use of the term governance certainly reflects the evolution of governance theories. However, say Maesschalck and Lenoble, we cannot assume that this change is completely integrated, because that would be considering as resolved what is actually at stake in the theories of governance. The passage is interesting in itself, and we can find in there some keys to resolve the new challenges for modern governance theories. This is why the Louvain school use a genealogical approach to the question of governance, and why we are following them on the matter.

An Evolution through Internalization

First Step: Critique of the Neo-Classical Economy by Neo-Institutionalist

The thought about governance reconstructed as the first step of Maesschalck and Lenoble's reconstruction begins with the critique of the neo-classical economy by the current of neo-institutionalism. Neo-institutionalists –and others- criticize the lack of realism of the neo-classical theory, which reconstruct equilibrium from the very restrictive hypothesis of the rational agent. In this neo-classical theory, the agent is perfectly rational and, as such, will always choose the optimal way to meet its interests. The equilibrium ensues from every optimal trajectory of the agents[1].

There is a need to complexify the neoclassical approach, notably at the institutional level, and try to take into account a more global view of the human behaviour, the uncertain context, and the failures of decentralised forms of governance. At this state, the need to get out of the model of perfect rationality, to enter a model of limited rationality is quite obvious (Arnsperger & Varoufakis, 2006)[2]. Nevertheless, the authors show us that this step out of the neoclassical approach is painful and that, finally, by proposing as solution the action of institutional mechanism that will encourage agents to take the best decisions according to

the norm, some theorists like Williamson (Williamson,1985), do not actually step out of the paradigm. If rationality is imperfect, Williamson add some mechanism external to the groups that will "correct" that rationality, which is to say, by some incentives, encourage the agents to act in a perfect way, and *optimal* way. The free play of market is not solely trusted anymore. There are "public" mechanisms to help, correct, and regulate the free play of the actor, in order for them to meet their normative expectations. However, those institutions are external, it is not supposed as given, they have to be added.

The novelty and the interest of the first approach is that it leads to the notion of learning, which will be the centre notion of the dynamic theory of governance by the school of Louvain.

This first approach also consider as given the ability of the agents to learn from the external institutional source in order to correct their action and, by learning, the ability to come back to those institutions and correct them. However, this learning process stays mainly natural in Williamson theory. It stays automatic, spontaneous, thus given in advance.

There are two related theories to the one of Williamson that try to respond to the epistemological contradiction of Williamson: the evolutionism(s) of Brousseau and David North, who try to see which conditions have to be respected for the actors to choose the most efficient solution, by combining learning and natural selection. Their theories postulate the existence of routines that are transmitted through generations. Unfortunately, we cannot enter to the discussion on those theories, as the goal of this article is not to summarize the construction of Maesschalck and Lenoble[3]. We can see that in those steps, there are elements of reflexivity and learning. The next steps, according to the school of Louvain will progressively abandon the idea of natural selection to focus on learning, in a dynamic of internalisation revealed by the authors.

Second Step: Aggregation of the Wills

The second step underlined by the authors is still in the neo-institutionalist framework, but can be defined by an accent put on relational and collaborative governance through dialogue. These kinds of theories are more frequently associated with juridical and political circles than economical theories.

The first version of this level of neo-institutionalism believes that the relational skills of the actor is sufficient to have an efficient collaborative strategy, on the only condition that it is put into some shape. The shaping of the collaborative relation is not external anymore, but is the result of cooperation and negotiation. Governance is seen here as the aggregation of relational skills and the goal of this approach is internal learning, as it is the "empowerment," or capabilitation. This level of neo-institutionalism could be attached to Habermas Discourse' theory (although, of course, Habermas has changed and evolved a lot).

The second version of this level of governance theory state that the aggregation of will and skill is not efficient, and that there is a need to pay attention to the conditions demanded by a real learning, the conditions of the empowerment.

From this perspective, this new approach to governance marks an effort to attend to the internalization of the conditions for adjustment required by the learning operation that conditions the successes of the governance operation. It remains the case, however, that this will to attend to the internalization of the conditions for satisfying the learning operation, although it explains why present-day research on governance theory has experienced the need to continue within the dynamic initiated by neo-institutionalist economists, also accounts for the internal dynamic of this second governance approach. What is more —and this thesis is at the heart of the present reflection- even in the 'pragmatist and experimentalist' version that is the most highly developed and most recent version of the second approach, this will to internalize the conditions for learning remains imperfect and assuredly insufficiently fulfilled. (Maesschalck & Lenoble, 2007, p. 53-54)

The process of internalization of the conditions of learning that the authors see as the main evolution in governance theory is running and will lead to the governance theories as we know them now. However, Maesschalck and Lenoble say it is still lacking. At this level, the capacity of learning is still presupposed, posed as given, and the relational approach is there to activate them (this is typically an example the mentalist presuppositions, which is the presupposition that there are rules in the mind of the participant that do not come from any external addition, but exit a priori in the mind of everybody).

Third Step: Pragmatism and Experimentalism

The slow evolution of governance approach in every field leads us to the third step, which is the pragmatist and experimentalist approach to governance, which goes further into the internalisation. Pragmatist and experimentalist approaches to governance aims at taking further into account the aim of the actors and their capacity of innovation. It explores the conditions of the transformation of behaviours, the manner by which individuals adapt, and the conditions and consequences of these adaptations. However, as we said, these approaches stay very much in the presupposition that an action on the institutional design can stimulate a rule of change already present into the mind of the actor. Once again, the governance approach falls in the mentalist presupposition and the other presuppositions that we will discuss later.

A first version of this level in the governance theories is exemplified by the experimentalism of Sabel (Sabel & Cohen, 1997, p. 313). Sabel starts by an analysis of non-standard firms, a firm that would not function by hierarchical power, but by collaboration and cooperation, which procedural support to manage collaboration. This is the first step for Sabel. The most important step is the enquiry. This enquiry is internal to the practice. A reflexive chock with a new situation or another group is the way to lead the agent to question their own normative frames. The stake of the question is to establish effective and lasting change of behaviour by devices internal to the groups, devices which integrate debates.

A second version of this level is the theories of Schön. For Schön (Schön, 1983), the shock alone is not enough. There is a need to pay attention to routines and stereotypes, to roads already traced, to defensive strategies, and so on. Schön acknowledges the range of obstacles towards behaviour change and negotiation. He understands the oppositions between two groups as opposition between two narratives, for example. The solution would be a reflexive back thinking on causal original stories for the actors to acknowledge the possible variation in the stories and understand the position of the other by reflexive shock. Schön states that a professional should be helpful to guide cooperation and overcome the obstacles.

The limits of Schön theory is much the same as the critique of Sabel theories: Schön stays in the model of the activation of adaptation possibilities within individuals and groups by confrontation to other logics and other contexts. For Maesschalck and Lenoble, it is still too external. Governance theories have to go further in the way of internalisation.

The Solution of Louvain: The Genetic Step

For Maesschalck and Lenoble, we cannot presuppose that there is a rule of identity that conditions the world view of the actors and that is already in their mind, ready to be activated by some procedures. The identity of the actor is something that has to be acquired, conquered. The first step, before recognition, is differentiation. The individual has to recognize himself in the image of himself in the other. The individual constructs its own identity before confronting itself to the other and adapting itself to it or not. The adaptative process is secondary. What is first is a process of *tierceisation.*

The genetic approach makes it possible to reveal critical operations by actors with respect to impasses of the past and the 'translation' operations required to go beyond them to a new positioning. This paradigm of self-capacitation is thus directly predicated on the identification of impasses that undermined similar strategies in the past, as in the case of possible alliances between public-service workers and those service' users that have been overlooked in the past. The value of taking a self capacitation perspective is that it makes one attentive to initiatives taken to recreate the ties between the actors involved. For example, in the case under examination, what would be needed is an operation of translation that links users's interests to the negotiation process and thus seeks to redefine the latter's operational frame. A concrete mechanism for terceisation of positioning is thus identified as a condition for learning. (Maesschalck & Lenoble, 2007-b)[4]

We can clearly see in this quotation how the genetic approach takes into account experiences of the people. The past experience of the persons has often a big influence on its decisions and actions[5], and has thus to be though of in governance. The way that Louvain wants to take that into account is by transforming it into argument, objectivise the past demarche for the actors to understand their own path of behaviour, their own strategies. Of course, this is an internal mechanism, (self-capacitation), done by the actor himself. However, and that is one of the critiques that we will develop later, the demarche of terceisation, of objectivation of past behaviour, element of its identity construction that is isolated by Maesschalck and Lenoble as "condition for learning" seems to redeploy the problem of the reduction to argumentation, and seems to close the mechanism of learning from anything else than the result of this terceisation. This is of course a problem because that was the critique of the experimentalist approach, and it cannot be solved easily, because is the mechanism of terceisation is suppressed from the theory, it is the entire mechanism of learning, and thus the entire difference with other theories that is cleared of.

Maesschalck and Lenoble introduce the theories of the construction of the self, some aspects of phenomenology (Ricoeur, Levinas, etc.) into the theories of governance, following their interdisciplinary approach, and more importantly, following the trend that they have reconstruct to increasing internalization, which has logically to go back to the construction of the self, within the individuals. The critiques of the previous step of the internalization is that they were still presupposing that mental dispositions, opinions, values, positions, ways of thinking and the capacity of adaptation, are something in the individuals, that can be change, (but with much difficulties) but that are there at the beginning. Thus, those theories did not investigate how the particular mind set of individuals are formed, but only how it could be changed. The theory of Maesschalck and Lenoble is supposed to give an answer to a latent limit of the trends they see in history. In that sense, their answer in conceptualised as the logical fulfilment of the dynamic.

This progression implies a simultaneous extension of these conditions [necessary for the success of the learning operation], because the requirement for internalization entails displacing the question of conditions. That is, the question of the conditions for learning becomes, as well, the question of the conditions required for the actors to be capable of learning; and it must ultimately arrive at the requirement for actors' self-capacitation, that is, the requirement that actors organize themselves to learn how to learn. [...] It constitutes the step further they require and thus, in a sense, their logical fulfilment. (Maesschalck & Lenoble, 2007-a, p. 100)

Maesschalck and Lenoble follow the dynamic of internalization that they have reconstructed and go to the very formation of the identity. And there, they discover that that identity is the result of a process of terceisation, which reintroduce exteriority in the equation. That exteriority is necessary to the birth of internality itself. The school of Louvain succeeds in this way to internalize the capacity of adaptation without considering it as given. Thus, it allies two exigencies that they found in the intuitions and limits of all the different approaches: explaining behaviours, including imperfects behaviour, and be able to modify them without recourse to purely external incentives, which would be replicating the approach of neo-classical economy, of which the first aim was to find an alternative; and stay away from to having presuppose, to avoid exteriority, a capacity of adaptation and relation already present within the mind and that it would be sufficient to discover and manipulate.

Nevertheless, the approach through the construction of the personal identity can also show some limits. Indeed, one can ask the consequence of making a governance theory which aim at managing collective action, rely on identity construction. Where is the place for action in the Louvain theory? How does the acknowledgement of the construction of personal identity can help manage collective action? What is, for example, the mechanism that leads to defensive attitude rather than cooperative attitude, and, if we can find where this mechanism stand, has there any action possible on that internal level?

The problem with going so far into internalisation is obviously that you are left with very few ways of action towards the mechanisms. It seems that the process of identity construction is very closed, because, even if they reintroduce externality as a primal source of the self, there are questions to be asked about the possibility to manipulate that primal external construction. It seems that, if it is possible to manipulate that, the theories then has to come back to the third step, with Schön and Sabel, or if it is not possible, than the risk of solipsism is important, and we fall outside of governance.

Let us explore more systematically some of the critique we can address to the school of Louvain.

LIMITS OF THE SCHOOL OF LOUVAIN AND POSSIBLE SOLUTIONS

Terceisation[6]

The approach of Louvain, the 'complete' learning[7] procedure in terceisation, involves what is essentially a decentring operation. By decentring the matter, Maesschalck and Lenoble bring the whole approach into the argumentative mode. The terceisation in their definition, seems to be very objectivising. Indeed, the construction of the individual's identity, that is the condition for any learning, is done by a projection of the self into an other, and a return to the self by the medium of this 'otherness', as a "thing", the identity, despite their own claim, is still a given, even if it is a given given by the self to the self. By making it something objective, a model of strategy that the individual can learn from, it reduce the all method to the medium of argumentation.

The metaphor of the mirror is, in that sense, remarkable as it shows us the mechanism of the terceisation, as a mechanism to transform identity into arguments.

Reference to a mirror clearly conveys the nature of the operation of representation. To represent oneself is to give oneself an image: it is to 'recognise' oneself in an image. But, as the metaphor of the mirror makes clear, it is necessary to 'recognise' oneself in the other that an image consists of. Thus it is necessary to first grant existence to this otherness, to this differentiation between oneself and this image that is reflected back from the mirror. Absent this operation of

differentiation, the operation of representation cannot achieve success. To put is in still more metaphorical terms, if I am to recognise myself in the mirror, the third factor represented by the mirror must be invoked. The frame of the mirror is the condition for possibility of the existence of the image that the subject recognises as self and identifies with. Thus the success of the operation of representation entails conditions for possibility closely associated with the process of differentiation of terceisation.

The faculty to recognised myself in the mirror is correlated with the faculty to recognise otherness. The exteriority is installed at the heart of the relationship that I have with myself, in Louvain's theory. That exteriority inscripts itself into the identity and is the key factor of the possible collective action. In that mechanism disappear any element of subjectivity that cannot be translated into an image reflected by the otherness, which is to say, any element that is not argumentative. What is given back to the individual after the meeting with otherness in the construction of his or her identity is objective element of identity that can be transformed into argument in a latter discussion. Identity is about to become a resource for the individual in which to go find resources for collective action[8].

Since governance is dependant of the learning operation and that operation turns out in Louvain to be argumentatively construed, the whole governance edifice built upon this foundation is threatened. Indeed, the argumentative mode cannot be the only mode that is valid, as showed by the very first critique of the neoclassical approach. Indeed, if we reduce discussion to rational argumentation, we are immediately stuck into the presuppositions that we have to overcome. To state it in a simple way, people are surrounded with dynamics that influence their views of the world and that cannot be reduced to reason. In other words, there are some areas that reason does not touch, but that are for a big part of the decision process. Somebody can agree on a norm, think it is the most rational decision, that it should be respected by everybody but still act without respecting the norm. There is a difference between accepting an argument as rational, and changing its own behaviour according to that argument. Actually, further than that, people can both agree and disagree in the same time with an argument, depending on the context, the way it is stated, the state of mind, the discursive situation, and so on. Individual have the capacity to hold contradictory idea in the same time, what psychologist called "cognitive dissonance"[9].

As a side-effect, moreover, the problematic parallelism between reason and morality is re-affirmed in Louvain, as argumentative reason takes over – valid argument is thought to be inherently motivating, thus the approach ultimately conflates the conditions for justification with conditions for application. This problem undoes the further critical work of Maesschalck and Lenoble as it destabilised the entire edifice. It does not mean of course that their critical work is not valid, but the solution and the focus of their questions are questionable. Their critical construction and the limits of their theory is the very basis on which has been constructed "comprehensive proceduralism"[10].

In the felt need to overcome the incompleteness of the experimentalist-pragmatist approach, what are the Louvain school thinking of? One problem they might see is the potential for caprice. Why would this be felt to be a problem? A plausible basis for this can be found in the longstanding division between public and private reason, and the distinction between value and norm upon which it is founded.

[…]Values are conceived of as the results of private experiences – based in the individual's experience of the world, including mediation of the same by culture, religion etc. In this sense, as mentioned before, they are possibly irrational. On the other side of the coin there are public norms, wherein subjective elements are suppressed. This is so as to make them capable of acceptance by as many as possible, owing to their relative value-freeness. The opposition of the two forms of reason introduces a gulf between the personal and the political. It is as if conviction and reason run on parallel tracks – but then how can the individual, convictions and all, be represented accurately? (Rainey & Goujon, 2009 p. 137)

One way of defending the theory of learning and the theory of terceisation that is supporting it may be to say that the reduction to argumentation is simply not a problem, and that nothing else is needed in the governance process. However, this argument would invalid the theory, because the main point of the critique by the Louvain School of their predecessors, notably discourse theory from Habermas, was to point out their reduction, in the process of the construction of the norm, of anything else than arguments (like the context of application of the norm, for example).

To help us overcome the self contradictions and reductions of the theory of Louvain, we found support in the theory of Jean-Marc Ferry (Ferry, 2002).

Integrating Other Registers of Discourse: Comprehensive Proceduralism

Jean-Marc Ferry states that the reasons that we accept or refuse a proposition in any given discussion are not necessarily the same as the reasons why we accept or refuse those reasons. For example, a frequenter of mediums accepts that astrology is predictive of his prospects, citing past successes, but refuses to let failures of prediction dent his conviction, even though the rational structure is symmetrical. Somewhere, reason runs out and narration steps in – the deep sense of self and all that my convictions connote. My being, in a thick sense that includes my upbringing, cultural religious convictions, feelings of indebtedness to a past, honouring legacies etc. The frequenter of mediums has a deep-seated need to feel the universe is not a blind, meaningless system, for instance, so favours confirmatory evidence over falsificatory. So, they employ deductive reason in matters of confirmatory 'facts', but narration trumps that same process in falsificatory eventualities.

While this is clearly important in comprehending who/what a person is, it is only comprehensible itself in argumentation if we step back from the primarily argumentative mode of discourse (with its positivist overtones) and regard narration not as an aggregative report of experiences had between t_1 and t_2, but rather as the authentic self-portrayal of a vulnerable human being – i.e. we need to use a *recognition* principle in order to cognise the information encoded in this. Via something like a Hegelian sense of recognition, the convictions held by another must be seen as constitutive of who they are. This calls for a re-synthesis of private and public reason, contrary to the privatising march of modernity in general. Modernity has religious and moral conviction boxed off in subjectivity, with civil and political society running at a remove (Rainey & Goujon, 2009, p. 137).

Narration cannot just be, as it seems to be in Maesschalck and Lenoble theory, element of the past that the individual transform into argument in the discussion about norm. Narration is not the only register of discourse that is forgotten by the procedural trend of governance (including of course Maesschalck and Lenoble). Narration has to be completed with interpretation, argumentation and reconstruction, according to Jean-Marc Ferry. Opening the scope of governance theories with a more flexible framing seems to be necessary. To do that, we do not actually have to go back to an idealised state of construction of the persons' identity. It is the particular situation on the field that will guide the theory, more than anything, and the theory is there to allow every kind of discourse to be heard. The strength of comprehensive proceduralism is to be an open system,

and not a system that aims to be closed, leading to some self contradiction, as we have seen with the contextual proceduralism of the Louvain school.

The openness of the theory has practical impacts, notably on the problem of guideline and expertise. The criticism of expertise and top-down approaches is a problem largely discussed in other places[11] and is not a specificity of this approach. The problem of guideline is more interesting, in the sense that guideline can be used in different ways, and the process of constructing the guideline is has to take into account the ways it will be, or wont be used, and other criteria that would replace it. The problem of guideline of course relate to the problem of norm construction, and the acknowledgment in the norm construction of the variety and the difference of the field of application. In that regard, guideline cannot be used without a feedback mechanism, and strong process of participation in their construction. Moreover, the procedure of participation cannot pre-emptively ban contradictory discourses based on other criteria than rationality, i.e. other register of discourse than argumentation.

The discussion about the meaning of the addition of the discourse register to the critical theory of the school of Louvain is the ground on which a new theory, called the comprehensive proceduralism, has been developed in the EGAIS project. The complete theory will be developed in other papers of this book.

CONCLUSION

This paper has attached itself to the task of retracing a dynamic in the theories of governance. This dynamic has been called by the school of Louvain "neo-institutionalism". The frame of the critique of Maesschalck and Lenoble of this trend, and the conclusion they take in their own theory is interesting, but unfortunately, carries on with the same inherent problem of the all trend, i.e. the reduction to argumentation. It seems that staying in the same dynamic reproduce the same presuppositions. The mentalist presupposition that there are rules within the mind that exist a priori, indifferently of the context of the actor, may not be the ultimate presupposition of the trend of governance (or it can be argued that the school of Louvain does not totally get rid of that presupposition), there is a rationalist presupposition in the contextual proceduralism of the school of Louvain that undoes their own critique of their predecessor by reproducing the reduction of the discussion to argumentative discourse, and by that, reducing the context, the values, the belief systems, the experience and the entire process of identity construction, everything that was not taken into account previously, to objective arguments, which is a way of immediately contradict the intuition behind the acknowledgment of those dimensions in the process of norm construction.

Our critique of the theory of the school of Louvain is a way to understand the limits of a framework maybe more vast than the one isolated by Louvain[12], the trend of proceduralism, and its focus on argumental rationality, at the exception of any other register of discourse, or indeed any other dimension of the human being behind the participant to the discussion. The question of the relationship between governance and procedure and rationalisation in general has to be asked. Of course it seems that the reduction to argumentation could be the inherent limit of the entire discipline of governance, but the acknowledgement of some of the limits of argumentation, and the testing of methods of inclusion of other element of the subjectivity of the actors, as well as element of the constitution of the group, the relationship between the group and the actor, the actor to the context, the norm create to its own context of application and so on, is a task that has to be undertaken by governance theories.

Louvain is of course very close from doing just that, which is why it is so interesting. By subordinating its very powerful conceptualisation of the learning mechanism to a process of terceisation,

it falls into a self contradiction that is difficult to overcome. However, lessons can be taken from that theory; and its conception of reflexivity, learning, the acknowledgment of the field of application of the norm and so on, is the foundation of the theory that we called "comprehensive proceduralism".

The originality of that theory is its immediate relationship to the context of application of the norms and the governance theories in general. Indeed, it has been developed by a direct confrontation to some fields of application, mostly in technology development, and is still testing its hypothesis in other fields, through diverse projects. This is a way for the theory to be taking into account from the beginning, the context of application of the norms created by procedures, and by doing that, taking into account the subjectivity of the people involved. Of course, the reduction of subjectivity to argument is still a concern, as it is a concern of all proceduralism, but the aim is to get a more complete picture, a more comprehensive picture of the context of application and conception of the norm, to be able to open the procedure enough to give a voice to every aspect of the human beings implies, in the limits of the possibility of the task at stake, without presupposing rules directly in the minds of participants, or a given metaphysic of the construction of their identity.

REFERENCES

Arnsperger, C., & Varoufakis, Y. (2006). What is neoclassical economics? The three axioms responsible for its theoretical oeuvre, practical irrelevance and thus, discursive power. *Post autistic economics review, 38.*

Callon, M., Lascoumes, P., & Barthe, Y. (2001). *Agir dans un monde incertain. Essai sur la démocratie technique.* Paris: Le Seuil.

Ferry, J.-M. (2002). *Valeurs et normes, la question de l'éthique.* Bruxelles: Edition de l'université de Bruxelles.

Joly, P. B. (2007). Scientific expertise in the Agora - Lessons from the French experience. *Journal of Risk Research*, *10*(7), 905–924. doi:10.1080/13669870701504533.

Maesschalck, M., & Lenoble, J. (2003). *Toward a theory of governance, the action of norms*. Amsterdam: Kluwer Law International.

Maesschalck, M., & Lenoble, J. (2007-a). Beyond neo-institutionalist and pragmatic approaches to governance. *Carnets du centre de philosophie du droit*, 130.

Maesschalck, M., & Lenoble, J. (2007-b). Synthesis report 2, Reflexive governance: Some clarifications and an extension and deepening of the fourth (generic) approach. *Reflexive Governance in the Public Interest (REFGOV)*, FP6 project.

Marengo, L., & Dosi, G. (2005). Division of labor, organizational coordination and market mechanism in collective problem-solving. *Journal of Economic Behavior & Organization*, *58*(2), 303–326. doi:10.1016/j.jebo.2004.03.020.

Masclet, L., & Goujon, P. (2011). *IDEGOV D.1.1. Grid of Analysis*. CIGREF Foundation.

Masclet, L., & Goujon, P. (2012). *IDEGOV D.3.2. Model of current and emerging governance strategies, Map of governance and ethics*. CIGREF Foundation.

Rabin, M. (1993). Incorporating fairness into economics and game theory. *The American Economic Review*, *83*, 1281–1302.

Rainey et al. (2012). EGAIS 4.3 New Guidelines Addressing the Problem of Integrating Ethics into Technical Development Projects.

Rainey, S., & Goujon, P. (2009). ETICA, 4.2. Governance recommendation.

Rainey, S., & Goujon, P. (2009). EGAIS 4.1 Existing Solutions to the Ethical Governance Problem and Characterisation of their Limitations.

Sabel, C., & Cohen, J. (1997). Directly-deliberative polyarchy. *European Law Journal*, *3*(4).

Sabel, C., & Zeitlin, J. (2008). *Learning from difference, the new architecture of experimentalist governance in EU*. European Governance Papers, Eurogov, C-07-02.

Williamson, O. E. (1985). *The economic institutions of capitalism*. New York: The Free Press.

Williamson, O. E. (1996). *The mechanism of governance*. Oxford: Oxford University Press.

ADDITIONAL READING

Argyris, C., & Schön, D. A. (1978). *Organisational learning: A theory of action perspective (Vol. 1)*. Reading, MA: Addison Wesley.

Brey, P. (1999). Method in computer ethics: Towards a multi-level interdisciplinary approach. *Ethics and Information Technology*, *2*(3), 1–5.

Duquenoy, P. (2005). *Ethics of computing. perspectives and policies on ICT in society*. Berlin: Springer & SBS Media.

Gregg, B. (2002). Proceduralism reconceived: Political conflict resolution under condition of moral pluralism. *Theory and Society*, *31*, 741–776. doi:10.1023/A:1021335112103.

Habermas, J. (1981). *The theory of communicative action*. Cambridge: Polity.

Habermas, J. (1991). *Erläuterungen zur diskursethik*. Frankfurt am Main: Suhrkamp.

Habermas, J. (1996). *Between facts and norms: Contribution to a discourse theory of law and democracy*. Cambridge: MIT Press.

Harris, I., Jennings, R. C., Pullinger, D., Rogerson, S., & Duquenoy, P. (2011). Ethical assessment of new technologies: A meta-methodology. *Journal of Information. Communication and Ethics in Society*, *9*(1), 49–64. doi:10.1108/14779961111123223.

Rawls, J. (1971). *A theory of justice*. Cambridge, MA: Harvard University Press.

Schön, D. (1983). *The reflective practitioner. How professionals think in action*. London: Temple Smith.

Sen, A. (2009). *The idea of justice*. Boston, MA: Harvard University Press.

Simon, H. A. (1972). Theories of bounded rationality. In McGuire, C. B., & Radned, R. (Eds.), *Decision and organization*. Amsterdam: North-Holland Publishing Company.

Stahl, B. C. (2008). Ethical issues of information and business. In Himma, K., & Tavani, H. (Eds.), *The handbook of information and computer ethics* (pp. 311–337). Hoboken, NJ: Wiley. doi:10.1002/9780470281819.ch13.

Van den Hoven, J. (2008). Moral methodology and information technology. In Himma, K., & Tavani, H. (Eds.), *The handbook of information and computer ethics* (pp. 49–68). Hoboken, NJ: Wiley. doi:10.1002/9780470281819.ch3.

von Schomberg, R. (2002). *Discourse and democracy: Essays on Habermas's between facts and norms*. New York: State University of New York Press.

Wiener, N. (1950). The human use of human beings: Cybernetics and society. Boston: The Riverside Press (Houghton Mifflin Co.).

ENDNOTES

[1] We must point out here that even within the neo-classical theory, there is an evolution to more complexity of the equilibriums, such as in the games theories. The prisoner dilemma, for example, show the existence of sub-optimal equilibrium, but can still be categorized within the neo-classical paradigm, as it still aims at finding equilibrium and optimal solutions, but probably one of the first effort towards complexity in the neo-classical theory. We must also underline that the neo-classical theory is not being held by the exact same hypotheses nowadays as it was in the 1960's. The model of perfect rationality has been criticised and overcome by tenant of the neoclassical approach, and a bunch of alternative theories has been developed, such as the psychological game theory for example (see Rabin, M. (1993). Incorporating fairness into economics and game theory. *American Economic Review*, 83, 1281-302.). However, as shown by Arnsperger and Varoukis, even those theories have common presuppositions that caracterised them as neoclassical, beyond any complexification in the theories. (see next footnote). The neoclassical theories, are also the mainstream.

[2] This criticism, which may appear as common sense, does not mean that we are leaving the neoclassical approach. As shown by Arnsperger and Varoukis (Arnsperger & Varoufakis (2006). What is Neoclassical Economics? The three axioms responsible for its theoretical oeuvre, practical irrelevance and thus, discursive power. In *Post autistic economics review,* issue n° 38), the neoclassical cannot anymore be defined by some characteristic like the rational agent, the use of Pareto equilibrium, the assumption that market will adjust automatically, the recourse of Say's Law, and so on, because there are a lot of neoclassical theories nowadays that distinguish themselves from those tools, while still be in the neoclassical framework. The real universally shared characteristic ("meta-axioms") of neoclassical economical theories are, for the authors, methodological individualism (observing agent as individuals, with individual tactics, and then, infer a global theory from the observation); methodological instrumentalism

(all behaviour is a manifestation of the ratio preference/satisfaction) and methodological equilibration (equilibrium are presupposed, *and then* tested: "neoclassicism cannot demonstrate that equilibrium would emerge as a natural consequence of agents' instrumentally rational choices. Thus, the second best methodological alternative for the neoclassical theorist is to *presume* that behaviour hovers around some analytically-discovered equilibrium and then ask questions on the likelihood that, once at that equilibrium, the 'system' has a propensity to stick around or drift away (what is known as 'stability analysis' " (p. 5).

because the neoclassical approach knows perfectly that it is playing with things that does not exist, such as perfect rationality, perfect knowledge, etc. The reduction is accepted as part of a methodological necessity in the neoclassical approach. To be valid, criticisms have to come up with models that would actually take into account more aspect of reality.

[3]   For more on the matter, go see the original text of Maesschalck and Lenoble. Lenoble, J., & Maesschalck, M. *Beyond neo-institutionalist and pragmatic approaches to governance* (pp. 18-47). Op. Cit. Or the discussion on the matter by Marengo and Dosi who are also a reference for the school of Louvain, C., Marengo, L., & Dosi, G. (2005). Division of labor, organizational coordination and market mechanism in collective problem-solving. *Journal of Economic Behaviour and Organization*, *58*(2), 303-326.

[4]   See reference Maesschalck, M., & Lenoble, J. (2007).

[5]   We saw the importance of experience plays in decision in many projects. For example, in the project IDEGOV (Identification and Governance of emerging ethical issues in Information Systems), by analyzing interview done by the De Monfort University, we have found that most of the professionals (in this case, it was Information Systems professionals) were relying first (and sometimes only) on their own experience, professional and personal, to deal with ethical issue.

[6]   See reference Rainey, S., & Goujon, P. (2009).

[7]   Just exactly what a complete learning procedure means is a problem in itself. Learning by its nature, one would imagine, is at least open in one direction – receptive of input – and therefore would require an incompleteness. Terceisation could be read as implying some manner of rational flight into omniscience-through-procedure.

[9]   This phenomenon is in itself the subject of an important literature, which very often tends to explain rationally the phenomena, whether by neurosciences, economy (it can be included in a neoclassical theory without having to discuss the axioms at the basis of the paradigm (cf. note 5 and 6 of the present paper, on Arnsperger and Varoukis' theories). This phenomenon can also be associated with a philosophical question treated since the beginning of philosophy: the question of the weakness of the will. The point here is to underline that we have to take into account on the formation of an individual decision or point of view or set of principles more than argumentation.

[10]   Although, of course, the construction of identity is a continuing process, but that does not resolve the problem.

[11]   See for example Joly, P. B. (2007). Scientific expertise in the Agora - Lessons from the French experience. *Journal of Risk Research, 10*(7), 905-924 and Callon, M., Lascoumes, P., & Barthe, Y. (2001). *Agir dans un monde incertain. Essai sur la démocratie technique.* Paris: Le Seuil.

[12]   We cannot develop here the possible difference between the trend of neo-institutionalism and the trend of proceduralism.

Chapter 2
Norms, Values, Argumentation, and the Limits of Rationality

Pierre Livet
University of Provence, France

ABSTRACT

The paper addresses the relations between emotions, values and norms and argumentative debates, which makes possible to take into account the contextuality of the debates. The paper suggests that constraints and demands are mixed and that there are requirements on their articulation. Different demands and constraints make a contextual exogenous background. Before analysing attempts (mainly logical ones) to specify these dialectics between constraints and demands in social and public debates, this paper first gives a sketchy picture of the relation between values, norms and social emotions. The result of the analyses (of the debates) is that we need to take into account some dead-ends of the interaction, the ones related to unsatisfied demands, because in order to enforce norms, we need to take into account the frustration of other values.

In order to take into account the social and emotional context of a debate, we have to look beyond the end of the debate, and to extend the reflection about unaccomplished proposals, in parallel with the execution of the decided project. This requires taking as constraints some attachments to other values (some demands).

INTRODUCTION

According to Habermas (1981), argumentative rationality goes beyond the limits of deductive and instrumental rationality. The closure of the deductive rationality is replaced by the openness of the principle of universalization. The requirement that a democratic debate has to be open to any people that it concerns is not simply a rule of procedure. Universalization is not a constraint but a demand. By contrast, deductive closure is a kind of constraint: if you want to select the propositions

DOI: 10.4018/978-1-4666-3670-5.ch002

that are related to logical operations and if your demand is that your logical system is complete, you need to satisfy the closure constraint. A constraint (which can be conditional to a demand) has to be entirely satisfied while we can only hope to satisfy the requirements of a demand inasmuch as possible.

Conversational rationality has to be confronted to the human reality of the debates, in which the constraints of the human psychology and the social demands are mixed. We have not only to take into account all the interests, motivations and demands of the concerned persons, as well as the arguments, counter-arguments, counter-counter-arguments and the ways for deciding which argument is the winner. We have also to take care of the emotions raised by the debate, the intensity of the will to defend some value to which a person identifies her quest, the need to show how attached she is to this value – a way of showing both that this value is deeply entrenched and that other people have to recognise her as a real supporter of this value, the desire of being socially recognised in the debate, the resentment against people that are better debaters that her and seem (unjustly in her opinion) to dominate the debate while their values are not so high than yours, and so on and so forth.

In Habermas' vocabulary, emotions belong to the domain of expressive rationality, values to the domain of conversational rationality, interests to the domain of instrumental rationality, and manoeuvre for being socially recognised to strategic rationality (related to instrumental one). This division makes difficult to analyse the impact of emotions as constraints on the argumentation, since "constraints" are supposed to pertain to the domain of instrumentality, as interests are. It is difficult to say whether the expression of values – in social recognition- is a strategic manoeuvre or a way of showing that one is sensible to the conversational rationality. One can suppose that concrete situations mix the different dimensions of rationality, but their division does not make one able to say by which ways, neither which

requirements ways of mixing dimensions of rationality are submitted to. Habermas assumes that conversational rationality integrates the others as it is the dominant one, but this kind of hierarchical integration is idealistic and simply stipulated, not analysed in the details of its operations.

In what follows we suggest that constraints and demands are themselves mixed and that there are requirements on their articulation. One could say that there are constraints and demands on the relations between constraints and demands. The advantage of such a perspective is that, as we consider directly the relations between constraints and demands, we have no difficulty to extend the results of our examination in one dimension of rationality in order to analyse the articulations between the different dimensions of rationality. Some of the constraints of one dimension are also constraints on the realization of the requirements of demands of another dimension and conversely the demands of this other dimension can impose some constraints on the demands of the first dimension or on another one. For example, strategic aims can impose constraints on the authenticity related to the expressive domain, and ethical values can impose constraints on the choice of means of the instrumental and strategic rationality. In our real life, not only constraints in a domain can impose demands onto another domain but also demands in one dimension become constraints for another one, constraints for the accomplishment of the combination of the demands of the two domains. Each domain has not only its own constraints and demands, but also its constraints on the demands of another domain and is submitted to the constraints related to the demands of other domains. These mutual constraints and demands make very difficult and even impossible to fully satisfy the demands of one particular dimension. Not only these mutual constraints are mutual limits for each dimension of rationality but also the mutuality of these constraints can be considered as a limit of rationality taken globally.

Before analysing attempts (mainly logical ones) to specify these dialectics between constraints and demands in social and public debates, let us first give a sketchy picture of the relation between values, norms and social emotions. We can define values as the qualities and properties of situations and actions that (if necessary) require that we give priority to them on other qualities and properties in our choices. Let us define norms as practical rules (most of the time collective ones, either implicit or explicit) that tell us, in case of conflict between two actions or choices, each of them satisfying a different value, to satisfy preferentially one of these values. Moreover, in a debate, it is easy to see that social emotions are interwoven with values. If the debate leads to a decision, this decision has the status of a norm, and deciding for a norm triggers negative emotions in the minds of the people attached to the disfavoured values. Norms introduce some kind of irreversibility in our choices, while we might hope to consider different values in turn and believe that this process is reversible.

The result is that we have not only to consider (1) the debate and its arguments, but also (2) the context of the debate and the problems of social and emotional recognition of its participants as well as the absent or excluded people, (3) the long-term effects of the conclusions of the debate on the emotional acceptance or resistance to the decisions that follow it, related in particular to the values that the normative conclusion of the debate has let aside. Point (2) belongs to background or exogenous context of the debate, and point (3) belongs to the context that the debate brings on by its own dynamics (endogenous context).

Some intertwined mutual limitations and relations between values, emotions and argumentation have just been evoked; now in what sense can we speak of limits in the case of an intertwined rationality? For example, is contextualisation (background and exogenous contextualisation as well as endogenous one) a limit of rationality? Surely it is a limitation for what could be called

"classical" rationality, related to the so-called Cartesian program, for example. But modern rationality (rationality after Gödel and the nuclear bomb, could one say) consists also in making us aware of the limitations of our theories, models, and of the powers of our technologies. In this sense, becoming aware of the context-dependence of an inference or an argumentation and its evaluation is both a limit of classical rationality and a result of modern rationality.

Nevertheless we cannot believe that simply claiming that rationality is limited by its context-dependence is all that has to be done. We have better to do: we have to identify what are the specific effects of this dependence – and by the way, what specific content is assigned in each case in order to make more precise the hand-waving notion of context. We have to work in order to make explicit what are precisely these limitations in such and such context. This requires that we have at our disposal some tests. The most obvious is the following one: trying to extend the inferences of some model or theory, or simply some assumption, and notice how different contexts block (in different ways) the extension of the inferential consequences of our assumption. Doing that is working on the limitations of rationality, not only taken as its limits but also considered as its own objects.

The problem of an argumentative debate is then the following: how to take into account in the debate these contexts, consequences and side-effects of the debate (mediated by the effects of emotions), while still taking seriously the argumentative requirements and constraints of the debate, including its endogenous context and first of all the values revealed in the debate as well as the values let aside or even rejected in its conclusion?

In what follows, I will limit my inquiry to the examination of the capacities of different logical approaches of argumentation to take into account the dialectics of constraints and demands. Choosing this logical approach can be justified by reasoning "a contrario." Logic is first supposed to impose constraints in order for a monologic discourse to be sound, and if possible complete. Apparently it is opposed to dialectics – related to dialogue or debate. So in opposition to the conversational values, demands and requirements, it could be taken as representing the constraint part of the argumentation (an illogical argumentation can be defeated. We could think that if we stick to these logical constraints, we could no longer be sensitive to the constraints proper to the emotional aspects of the debate and to the demands related to the desires for social recognition of the partners of the debate, and this would have bad consequences.

A contrario, if some logical approach is able to take into account the distinction between constraints and demands, and also the endogenously generated context of the debate in addition to the exogenous context, we could expect that at the same time it will give us rigorous ideas for giving specific content to our dialectical requirements on the relations between constraints and demands.

THREE LOGICAL APPROACHES

Dialogic

We have at our disposal four logical accounts for analysing argumentative debates: dialogic, non-monotonic logics (for example preferential systems), ludics (Jean Yves Girard) and the logical and philosophical approach of Brandom (1994; 2008).

Dialogic offers a nice idea: replacing the deductive apparatus by the rules of a game involving assertions, questions, and answers between the proponent of a thesis and its opponent. Unfortunately it is oriented only towards determining who is the winner of the game, and tries first to mimic classical logic. More recently R. Rahman (2005) has shown that dialogic can mimic apparently whatever logic you want (intuitionistic, basic

linear logic, preferential systems, para-consistent logics). The problem of dialogic was that mimicking another logic in dialogic implies sometimes to build rules the meaning of which is far from being intuitively understandable. The main task of Rhaman has been to simplify these mimicking rules. Even if his efforts and virtuosity have to be recognized, the dialogic game remains a game that defines a winner and a looser. At least classical logic does not necessarily trigger this possibly bad emotional aspect, the resentment of the looser, since its monological version does not raise the emotions of a game and a dialog!

Non-Monotonic Logics

Non-monotonic logics give us a tricky way of expressing the impact of emotions and values in the logical framework. Their notion of inference is the one of a "normal rule of inference": "A normally entails B." It is not a universally valid inference, but only a normal one, because exceptions can defeat the inference. Because of this duality with exceptions, a normal inference is related to a revision process: when A implies normally B, adding C to A could add an exception to the normal inference, implying the revision of the conclusion B. Moreover, if someone claims that "A Or B implies normally A & not B," (a formula that is not a logical inference), this means: "When considering A possibly in the context of B, or revised by B, and when I keep A and give up B, I reveal my preference for A over B, my reluctance to give up A, my tendency to resist the revision of A by B." For the observer, it is a way to reveal how attached the person is to A as a value (or to A as a fact and B as a counter-value). In this case we can presume that revising either "A" or "not B" would be for him a source of emotion. If A is a fact, and B a counter-value, having to revise "not B" is bad news. This interpretation relates defeasible inference, possible revision, attachment to value revealed by resistance to revision and emotion. Unfortunately, if non-monotonic logics have been used for formalizing argumentation, the other possibilities that we have just made explicit have not still been exploited.

We may note that in order to exploit these "emotional" and axiological aspects, we have to pay attention to non-logically valid inferences. This might lead to overpass the limits of rationality and to fall into irrationality. This can be avoided by considering the dynamics of the inferences. Formulas like "A Or B implies normally A & not B" are not logically valid inferences, but they can be taken as having a source exogenous to logic: the preferences and values of the person. Preferences have not to be supposed rational, but rationality can be found again in the inferential dynamics: if things occur that lead to the revision of some part of the formula, a rational person would have to make this revision (giving up some parts of the formula and saving other ones) in accordance not only with the purely logical rules but also with the preference for some value that she has shown by introducing in the logical game this formula the source of which is exogenous to logic.

Brandom

Brandom (1994; 2008) examines argumentation in an inferential and pragmatic perspective. He distinguishes classical deductive reasoning and inductive one. I am "committed" to the deductive consequences of my assertion, and "entitled" to draw some inductive consequences (deductive reasoning has to resist every counter-argument, while inductive one is justified as soon as I find some argument). For him, the meaning of a sentence is related to possible debates, as it is given by the set of all the assertions that exclude at least one reason for this assertion (the set of every counter-argument, of "incompatible" assertions). Necessity and possibility are also interpreted in this way.

The attractive property of this framework is that Brandom acknowledges the implicit normativity of meaning, its contextual and practical determination. Commitments could be considered as "constraints" imposed to the speaker by her assertions. Entitlements could be taken as possible answers to the demands of the addressee, at least some of these demands. The relation of incompatibility between some commitments and some demands could be interpreted as a relation between constraints and demands.

Unfortunately, Brandom does not examine seriously and extensively the network of such relations, neither the negative part of non-monotonic arguments: the possibility that their entitlements can be defeated. In a debate, in order to take care of the emotional and social consequences, we need to consider defeats as well as victories, and we cannot reduce the first to the complement of the second. Brandom reduces argumentation to these rules of the inferential games in conversation that enable us to give a meaning to our sentences, related to the dialectical relations between the inferential possibilities opened and the ones closed by the assertion of a sentence. When we are paying attention to the assertions of others only with the intention to grasp their meaning, emotions are not supposed to be raised – maybe only cognitive surprise. Defeats are then only negative tests, as useful as positive ones. But in a debate, more passions are involved, and behind meanings we suspect despite and anger. Brandom's inferential perspective orientates pragmatics towards semantics, not towards emotions and axiological demands.

LUDICS

Jean Yves Girard's ludic (2001; 2006; 2011) is at first sight very remote from actual debates, and from axiological and emotional aspects. In addition its formalism is not easy to understand. I will not stick to the logical details, not even to the strictly logical purpose. I want just to make the reader sensitive to the possibility offered by this formalism to take into account the duality between something like constraints and demands. I will extract from the logical form some principles and seminal ideas.

Remember that in order to show that a complex formula is a theorem, a proof consists in taking the complex formula at the bottom of the proof, and going up in the tree that builds the proof by decomposing the complex formula, using successively different rules of "elimination" of the connectors that compose this complex formula.

In ludics formulas are put together (packaged) in accordance with their connectors, divided into two kinds, positive or negative ones. Positive connectors are related with irreversibility, negative ones with reversibility. In the development of a logical proof, reversibility means that when you go down, from two premises to the formula that combines them, you have still enough information to go up, from the formula to the premises. Irreversibility means that when going up, you have to choose what contexts to put with each premise, and have no way to ensure that your choice is right: the combinations made on the way down could mask the needed information for the sharing out of the contexts.

For example, the additive Or is positive and implies irreversibility, because A in "A Or B" could be related to a different context than the one of B. Negative connectors ensure reversible choices like additive And, ("&"), that admits to develop A, then B, or the contrary, ad libitum, because A or B are both related to the same context C.

Irreversibility implies that even if the logical truths are assumed to be insensitive to the context, the dynamics of logical proofs is sensitive to the distribution of (formal) contexts.

In a classical research of proof, we successively focus on one subpart of the formula to prove and then on another one; we use successively the elimination rules linked with one connector and with another one, so that we go up and build a kind of tree of sub-formulas; the bottom formula is proved when we end only with axioms.

In ludics, (1) we have only two kinds of connectors, positive and negative, and only two "elimination" rules, one positive, one negative. (2) The proof needs interactions with tentative counter-proofs.

Let us explain the first point a bit further. Each rule tells us how to combine the part of the formula that we are focused on with some other parts in order to reconstitute the different premises of the formula. When the player named "I" make a positive move, she makes an irreversible choice of *some* contexts (a "ramification") in order to combine them with the sub-formula on which "I" is focused. But if "I" has to develop all of the combination of these selected contexts, she can do it in whatever order she want. When the player named "You" makes a negative move, he offers me the choice between different combinations of contexts (a set of ramifications, a repertory). Then "I" chooses whatever of these sub-sets of contexts she wants, and "You" is obliged to develop the combinations offered by this sub-set of contexts, and not another one.

```
Example:
011|- 012 |- 02 (positive, focus 01,
ramification {1,2}
             (« I » combines fo-
cus 01   with the elements of the
ramification {1,2}
       |- 01,02|-01,03  (neg-
ative, focus 0, Repertory:
{1,2},{1,3})
                     0|-
(« You » combines 0 with each element
of each ramification)
```

The reader can notice that even if the positive rule implies irreversibility, and the negative one reversibility – in the vertical sense, linked with the distribution of contexts when we go up from the formula to its premises- in another sense, the positive rule implies some reversibility (reversibility*), as I can develop each combination with the different contexts of the subset *in whatever order I want*. Conversely, the negative rule implies in another sense irreversibility (irreversibility*), as "You" is obliged – one step further- to develop the combinations chosen by "I," and cannot develop another ones.

Interpretation

We can reinterpret these relations between the positive and the negative rules (or moves) in terms of the dialectic between demands and constraints. By making a positive move, "I" satisfies one of my possible demands. A demand (related to the choice of one value) has to be freely chosen. This implies that "I" has a choice between different possible demands. By making his negative move, "You" offers the possibility to satisfy different demands, and give to "I" this freedom of choice. When "I" is focussed on her demand, she does not care of the order in which the constraints that the accomplishment of the demand requires are satisfied, but she has to satisfy each of these constraints. When "You" offers the choice between the different demands to satisfy, he is constrained, once "I" has chosen one demands, to satisfy all the constraints related to this demands.

We can see that some of these conditions are constraints implied by demands (demands imply freedom of choice) and others are demands implied by constraints (satisfying all the constraints). Other ones are the constraints that the realization of a demand triggers ("You" has to satisfy the specific constraints related to this demand), or the weakening of the demands on constraints that a demand triggers ("I" organizes the order of satisfaction of constraints as "I" likes). The logical framework gives us the possibility to determine constraints on relations between demands and constraints as well as demands on these relations.

These logical rules can be said to develop and articulate different aspects of freedom. Freedom implies choosing one action – and not another

one: irreversibility of the positive rule. It implies having a choice offered by another person (or by nature): reversibility of the negative rule. It can imply accomplishing the different elements of task ad libitum, without following an imposed order ("reversibility*" of the positive rule). It can imply being obliged to do what the other has chosen among the possibilities that you have offered (irreversibility* of the negative rule). Freedom implies constraints on its own, and ludics develop also these constraints.

We see that in ludics freedom is considered not as a mono-logic notion (the freedom of one people) but as a dialogic one (the intertwining of two freedoms). Notice that irreversibility is a manifestation of freedom, but introduces a constraint, while reversibility offers freedom, but implies the constraint of developing the subset chosen by the other. Notice also that demands are call for freedom and that freedom is a demand, but that the free choice of one person – in accordance with her particular demands- imposes constraints on the others.

In debates we can observe different regimes of argumentation between arguments invoking demands or requirements, and arguments invoking constraints. Presenting a project in a debate implies to have the freedom to make choices between the possible demands the ones that you want to satisfy, and the project is only supposed to satisfy some of the demands evoked in the debate. The presentation of a project follows the schema of a positive rule: the proponent of the project is free to choose the demands and values that he invokes, it is an irreversible choice, but he has also the liberty to organize the realization of the project and the satisfaction of its constraints. The opponents examine his project as only one of the possibilities that the situation offers. They consider all the demands evoked by the situation and can choose any one of them. The perspective of the opponents is the one of a negative rule. The proponent has then to show either that his project can satisfy this demand, or that the opponent – as well as the proponent- is unable to satisfy all the constraints implied by this new demand.

The innovation of ludics is that not only the proof is developed using an alternation of positive and negative rules. It is also that at the same time multiple *tentative counter-proofs* of the negation of the theorem are developed and interact with the development of the proof. Any branch of these developments that ends too soon, because in this interaction it has no longer sub-formulas to develop, is a defeated proof. But the spirit of ludics is that who is the winner does not matter. What matters are the symmetries between the different developments of the tentative proofs and the tentative counter-proofs. In debates also, the interactions between the agonists are more interesting than the designation of a winner. Another similarity between ludics and debates is the criterion for a defeat. In a debate also some argumentative paths are rhetorically and argumentatively dismissed because they have no new meaningful materials to bring to the debate (dead-ends).

This second main idea of ludic (developing all the tentative proofs, some of them failing when they cannot be longer developed in interaction with the others) makes us sensitive to what corresponds in a debate to dead-ends of the logical developments in ludics, the breakdowns of the debate, the sources of emotional social frustration.

In the interaction between two tentative proofs, you cannot keep on developing your proof if your combinations do not correspond to the one offered by the partner or adversary. Therefore a project that is fully realizable (satisfying every constraint) is not yet ensured of being the winner. It could not be able to interact with other demands that are also realizable. In this case, either it leaves the debate and is gave up, or it has to be revised in order for it to satisfy at least one more demand – if the demand is realizable. In this way, the project will go further in the development of the argumentative debate, because it offers new sub-formulas, new materials, allowing it to continue its interaction with other developments.

In ludics we need to develop the interaction between the tentative proofs as long as possible: if counter proofs give up too early, the remaining tentative proof would not be sufficiently tested. As a consequence, in a debate we have to encourage a project that aims at satisfying more demands and by the way cannot satisfy every constraint, and to help its authors to revise it in order to meet the unsatisfied constraints. It is a fairness demand in the argumentative debate to help axiologically rich ideas to meet realizability constraints, and this requires avoiding too short developments and trying not to be blocked in dead-ends.

Breakdowns in debates can be of two kinds. Let us suppose that two projects try to satisfy the same demands, but that one fails to satisfy all the required constraints. This dead end is just a factual failure, raising minimal emotions of frustration, because the demand can be satisfied by the other project. On the contrary, let us suppose that the two projects aim at satisfying different demands, and that one fails to satisfy the constraints attached to the satisfaction of its own demand. Then the demand is still active, even if not realizable for the moment. We have to compensate the negative emotions of the people who can rightly consider that they have been defeated in the debate, because of this defect of feasibility, while their demands are grounded on fundamental values. If this is not possible in the temporal limits of a public debate, it has to happen after the debate, giving the possibility to open again new debates. Breakdowns of this second kind could (and have to) be listed and counted as debts of social recognition for the people whose argumentations have been defeated, at least if they have not been defeated too soon, and can interact with the arguments of the supporters of others values. This is the only way to cool down the emotions of frustration that would otherwise give rise to resentment.

Among debates, we have to distinguish the ones of politically entitled assemblies. These debates lead to the formulation of norms. Norms also combine a demand part and a part related to the constraints of feasibility: the capacity of the norms so formulated of being enforced and becoming efficient. One of these constraints is often neglected. This particular constraint is related to the interaction with other demands. Some values and demands that the decision in favour of the norm has -implicitly or explicitly –gave up are still related to norms embedded in the social practices. Therefore, constraints of realization of the chosen norm imply to take into account the other demands, in order for the enforcement of the norm not to be weakened and countered by the reactions inspired by the unsatisfied values and demands. Counting as social debts of recognition the defeated values is just a demand when we consider only values, but it is a constraint when we consider the realization of norms, the conditions of their social efficiency.

CONCLUSION

This approach of the relations between emotions, values and norms and argumentative debates makes possible to take into account the contextuality of the debates. The contextual exogenous background is made of the different demands and constraints. The debate is supposed to evoke all the constraints, but only some demands, implicit in the contextual background. We can take also into account the endogenous context brought forth by the debate: the interactions between demands and constraints of the different positions that emerge from the debate. We can even pay attention to what can be called a forward context: the destiny of the different positions after the debate, when the decisions of the debate have become norms. We see that the duality of demands and constraints is decisive for each of these three contextualities, but in different ways. When building projects, people consider only some demands and constraints and examine only their interactions. During the debate, we have seen the positive and the negative moves are articulated and regulate

the interaction between demands and constraints. After the debate, we need to take into account some dead-ends of the interaction, the ones related to unsatisfied demands, because in order to enforce norms, we need to take into account the frustration of other values.

In order to take into account the social and emotional context of a debate, we have to look beyond the end of the debate, and to extend the reflection about unaccomplished proposals, in parallel with the execution of the decided project – and maybe afterwards changes of contexts lead us to revise it further. This requires taking as constraints some attachments to other values (some demands). Our duty to help supporters of other values to satisfy more constraints is an extra requirement – it belongs to the regime of demands, but its partial realization is a way to satisfy the social constraint.

Let us note that these considerations on the argumentative debates can themselves be seen for a part as demands or requirements (counting the dead ends of the debates as debts of social recognition) and for another part as constraints (for example, the resentment triggered by the absence or deficit of social recognition is considered as an obstacle – a constraint- to the success of the decided project). Therefore we can re-apply our approach to itself in the definition and revision of the flexible rules of the debates. This reflexive application of our conditions requires for example that all the argumentative and emotional constraints of one debate have to be taken into account, and at least some of its demands (e.g. helping the revision of the unaccomplished proposals) have to be satisfied.

The reflexivity of this perspective and the fact that contrary to a reflexive closure, it is compatible with openness and flexibility, seems a good feature of this approach for analysing argumentative debates in a more encompassing perspective, taking into account even the emotional effects of these debates, and integrating all the dimensions of an argumentative rationality.

REFERENCES

Brandom, R. (1994). *Making it explicit*. Cambridge, MA: Harvard University Press.

Brandom, R. (2008). *Between seeing and doing*. Oxford: Oxford University Press. doi:10.1093/acprof:oso/9780199542871.001.0001.

Girard, J.-Y. (2001). Locus solum. *Mathematical Structures in Computer Science*, *11*, 301–506. doi:10.1017/S096012950100336X.

Girard, J.-Y. (2011). *La syntaxe transcendantale, manifeste*. Retrieved from http://iml.univ-mrs.fr/~girard/Articles.html

Girard, J.-Y. (2006 - 2007). *Le point aveugle*. Paris, France: Hermann, (tome I & II).

Habermas, J. (1981). *Theorie des kommunikativen Handelns*. Frankfurt: Suhrkamp.

Rahman, S., & Keiff, L. (2005). On how to be a dialogician. In Vanderken, D. (Ed.), *Logic thought and action* (pp. 359–408). Berlin: Springer. doi:10.1007/1-4020-3167-X_17.

Schlechta, K., & Gabbay, D. (2009). *Logical tools for handling change in agent-based systems*. Berlin: Springer.

Chapter 3
Rational Planning.
Principles and Contexts[1]

Emmanuel Picavet
Franche-Comté University, Besançon, France

Caroline Guibet Lafaye
Maurice Halbwachs Center, CNRS, Paris, France

ABSTRACT

Practical rationality, when collective choices are at stake, should certainly rely on principles. These principles are perhaps not without effect on our representation of the problems to be addressed in collective action. The authors investigate how this structuring role of pragmatic principles accounts for notable context-dependent features of governance procedures. In the field of social policies, for example, the enhancement of personal autonomy has come to the forefront of collective challenges. Capacity-based approaches indicate a way to put into question those conceptions of autonomy which lead to an excessively uniform treatment of individuals. Following these approaches, the beneficiaries of social policies should be treated as concrete beings with their personal history, living in specific social contexts and so on. The authors analyse the individualizing logic which is exemplified in interactive problem-structuring and institutional decision-making about the provision of apt, context-sensitive care and services for ageing handicapped persons. It is suggested that the sought-for adaptation to specific circumstances is made possible through a complex process of description of problems and challenges for collective action, in which procedural aspects are important. This process is by no means reducible to a passive process of adjustment to independent states of affairs. If the authors' analysis is correct, there is no such thing as the "real" nature of individual situations, as opposed to the fictions associated with ordinary social policies: the process under scrutiny really redefines the nature of institutional interactions, responsibilities and the underlying picture of the individual person.

DOI: 10.4018/978-1-4666-3670-5.ch003

INTRODUCTION

Practical rationality is, among other things, a matter of pragmatic principles which have a guiding role for individuals and groups. The benchmark principles are usually quite general and the rationality of collective choices has to do, presumably, with the ability to turn the principles into reality with relevance, in an efficient way. How should we conceptualize this ability in the first place?

Pragmatic principles for collective choices could be identified, it seems, with the selection of desirable states of the world, chosen among the possible ones. Then we would be induced to look at the administrative or political implementation mechanisms as if they were more or less neutral instruments, by means of which we see to it that the world exhibits the desired patterns. But the following statements, if true, complicate the matter:

First of all, pragmatic principles undergo interpretative changes. This, of course, may impact representations of the collective implementation process, when it comes to spelling out the details of action problems with a view to the effectivity of principles[2]. Turning goals into reality depends on one's views about the meaning of those principles which help articulate the goals. It can be argued, in this respect, that pragmatic principles have distinctive properties when it comes to interpretation needs: for example, they have an unequal potential for being made precise in a useful way, or in an objective way[3].

In addition, pragmatic principles have a role to play in problem-structuring activities (and hence in decision-facilitation tasks at the prescriptive level), if only because they channel and format the information which is used in decision-making (this was emphasized in A. Sen's pioneering contribution to the information analysis of moral principles – see Sen [1979]). The chosen benchmark principles determine a selective awareness to specific features of the social context and personal situations; this enables them to play a crucial role in the development of joint work and inter-organizational (or inter-institutional) collaborative relationships[4].

Among these properties, it is perhaps fair to say that only the correlation of principles with their respective information needs has been the object of systematic inquiry up to now. In this joint research, we take a broader view. The noted characteristics are investigated with reference to the *autonomy*, *dependence* and *capacity* (or *capability*) concepts. The institutional use of these notions in social policies gives support, we believe, to our initial statements. Such notions, in their concrete use, are related to a constructive social process; this process, we'll argue, illustrates the characteristics we have just mentioned.

We'll highlight the notions of « dependence » and "autonomy" and their role in spelling out principles of collective action, with respect to the challenges of old age and the aging. We'll investigate the type of context dependence and some of the procedural features or governance which can be associated with dependence-based or autonomy-based principles for collective action. More particularly, we ask whether context dependence and the procedural features are impacted by those dominant interpretations of "dependence" and "autonomy" which rely on the "capability" notion (and the related principles for collective action)[5]. With this goal in mind, we'll examine the involvement of personal autonomy and individual capacities in governance processes. A case study will be provided by an institutional interactive process for answering the needs of the aging population of handicapped persons in France.

CAPABILITIES, CONTEXTS AND THE ENHANCEMENT OF AUTONOMY

General Intent of the "Capability" Approach to Personal Autonomy; the Interactive Side

The contemporary capacity or *capability* approach, as developed by A.K. Sen (and, along a different path, by Martha Nussbaum) has concentrated on the description, assessment or measurement of personal capacities for choice which contribute to the objective well-being of individuals. This approach is now widely recognized as a model which can be used to articulate collective goals which pertain to the enhancement of personal autonomy, or the mitigation of personal dependence. For this reason, it is advisable to look at a number of structural features of this approach, in order to elucidate how autonomy-based or dependence-based principles have a structuring potential in collective action tasks.

A. Sen has defined his "capability" set on the basis of "functionings," which are various features of doing and being for individuals. Capabilities (a special construal of the general notion of personal capacity) are envisioned, from the start, as opportunities for various types of achievement in life. Individual choices are the matter, but the perspective goes beyond the sheer availability of a number of alternatives and the selection operated among them. Sen's approach is remarkable on several accounts. It endorses a "freedom viewpoint" which brings together the choice faculties and the value an agent may locate in the different ways he can use these faculties.

In this perspective, it is allowed that various value judgments, including consequentialist evaluations, are constitutive of the worthiness, in the eyes of individuals, of those faculties which are involved in their freedom of choice, hence in their freedom generally speaking. In addition (and correlatively), this approach brings into close contact, in the joint assessment of freedom and well-being,

the "procedural" dimension of choice (the features of action - that is to say, of the process leading to results) and the "opportunity" dimension (the nature and value of choices in themselves). In such a perspective, it is hardly possible to overlook the social dimension of personal capacities: individual and social determinants of personal capacities are thus intimately associated. This accounts for the context-dependent features of the associated policy-making agendas.

It is recognized from the start (and how could we possibly deny it?) that the possible achievements of human agents are dependent upon the context, more specifically, the social environment in which their actions or initiatives take place. For instance, the ability or lack of ability of a handicapped person to engage in a university course may depend on the collective effort to see to it that handicapped persons who use a wheelchair are proper access to the amphitheatres and seminar rooms. This, in itself, gives a reason to look at a specific kind of information, namely, those features of the social world which explain the consequences of personal characteristics. It appears necessary, when it comes to assessing personal capacities, to bring some properties of the social environment into the picture. One may think of collective initiatives, public policies and their achievements. Indeed, the explicit amalgamation of pieces of information about personal characteristics and about the fitness of the environment can be considered a strong point of capacity-based approaches to social ethics.

There are limits to explicitness: being a descriptive framework or matrix, Sen's capability approach should not be expected to be ideally precise in its formulation[6]. It stands in need of interpretation but, compared with other principles of social ethics, the principles it puts forward strike one as relatively imprecise. This is not without consequence from a pragmatic point of view. Since the concrete use of capability-related principles is heavily context-dependent, it creates a need for a complex process of adjustment to the prevailing social context. Making these principles adequately precise in specific contexts is no easy task and this might impact the ways of collective action. Operationalizing the capability approach for the purposes of socio-economic field studies and the detailed analysis of collective options stands out as a major challenge for researchers. It should be noted, however, that generality and vagueness have merits of their own and create room for successive influential interpretations[7].

The capability approach can be used as a normative benchmark when it comes to assessing social policies as instruments for the enhancement of choice capacities or margins for action. To be sure, these values are not entirely consensual, as many people believe that restrictions on individual margins of action are valuable in themselves, especially as testimonies to the limits of individualism, or the individual endorsement of a social or traditional discipline. Nevertheless, the enhancement of personal capacities for choice is essential to all varieties of progressive thought. The approach is indisputably individualistic in character, as it concentrates on the situation of individuals with a view to evaluating (positively or negatively) the evolution of society, or those policies which have an impact on it. This brand of individualism, however, is attenuated by Sen's distinctive resolution to take full account of the complex interaction between the situation of persons, their social life and their environment.

This kind of approach can be used to assess the extent to which handicaps and capacity impairments are being compensated in an active way. For example, the handicapped persons may expect a better working life and everyday care both from improved medical services (which impact their personal situation) and from collective initiatives with an influence on their environment (for example, through the adaptation of workplaces or the setup of health-care and home-service procedures)[8]. Because of this association of personal, environmental and social parameters, the quest for the relevant information in problem-structuring

tasks benefits from being oriented toward the interaction of the person and the social, technical and material environment. In this respect, with practical purposes in mind, the capability approach can help.

The Interactive Side

Up to now, we have treated the social dimension as a set of parameters in the environment of personal initiatives. The interactive dimension has only been tacit. But should we not inquire into it in a more detailed way? One important interactive feature is related to the claims which are put forward in the aim of promoting the choice capacities (or the achievement opportunities) of persons in specific groups, or persons in specific situations. These "claiming" initiatives must be considered in association with the reactions they meet: the fact that they are heard (or not), understood (or not) and, possibly, successful. In the case of handicapped persons and those who must face "dependence" situations, this is often captured through the notion of "empowerment" for specific groups of persons with shared interests to uphold. There is a growing normative interest attracted by the empowerment of social and institutional actors in their mutual relationships.

This notable interactive dimension is involved in the claiming initiatives which aim at the development of choice capacities, or various lifetime achievements to be made eligible for specific groups in the population (or for people in specific situations). Such claiming initiatives should be viewed as correlated with the reception they meet, as theories of relational and collaborative governance have emphasized[9]: how they are heard, understood and, finally, followed by real effects in social life. The specificities of public choice (such as public-interest standards in deliberative activities) and public-image concerns may induce officials to develop their receptivity to the needs of handicapped or dependent persons. Insofar as reputational concerns are involved, the same applies to firms, in addition to immediate profit-making motives. Of course, this receptivity can be lowered by budget limits or by priority conflicts which involve other concerns.

One step beyond, we must encompass the fact that individuals adapt themselves to their environment. They develop useful skills and aptitudes in an environment which has been shaped by nature and by their fellow men and women. The resulting capacities have their role in shaping the *status quo* situation which provides a benchmark for evaluating collective initiatives: the usefulness of such initiatives is correlated, of course, with a judgment on the antecedent state of things. Thus, there is an interactive side to the determination of personal capacities. Personal capacities depend on individual adaptation to policies, and expectations about future policies. Policies, in turn, are launched in a way which may depend upon the situation of persons in the relevant population. But adaptation to impoverished conditions is no sufficient ground for a positive judgment on the state of society. Social criticism is useful in this respect, as a supplement to existing claims in society, especially when social demands reflect a more or less fatalistic acceptance of poor living conditions.

A paradoxical situation is possible with respect to public policy, as a consequence of individual expectations. Among other things, individual choices are influenced by expectations of future action (or the lack of it) at a collective level. For example, people with motricity problems may be tempted to limit their mobility through adjustments in their lifestyle and personal goals, because they don't expect favourable policy initiatives in the predictable future. After this pattern, individuals may improve their personal situation on their own initiative, in such a way that, all things considered, public authorities are distracted from acting in the required way in the interest of handicapped or "dependent" individuals. From a normative point of view, impoverished prospects in life are a predictable and problematic outcome. The other

way round, collective decisions may be reached on the basis of expectations about the selective adaptive efforts of handicapped or ageing individuals (and their circle of relatives, friends or employees). For example, limited care for the dependent elderly could be the consequence, in some cases, of optimistic estimates of the ability of these persons to rely on their own efforts for the enhancement of their living conditions.

A CASE STUDY

Dealing with the dependence situations associated with ageing is a collective and complex decision-making process. It involves successive administrative and political reports, plans and policies as well as detailed institutional steps. This process comprises the identification of emerging problems, the buildup of frameworks for analysis and description, the deliverance of prescriptive advice and collective action in the end. All this takes place against a background of deep uncertainty with respect to the future of the ageing population, in terms of size, qualitative needs and lifestyle. The selection of collective ways of dealing with imperfectly specified problems turns out to be connected, we hypothesize, with the limits of information and prediction, when novel situations keep emerging[10].

We now concentrate on the example of a CNSA 2010 report (the result of "practice exchange" workshops, Nov. 13th, 2009 to Feb. 4th, 2010, National Fund for Solidarity and Autonomy, France), entitled Aide à l'adaptation et à la planification de l'offre médico-sociale en faveur des personnes handicapées vieillissantes [Aid to adaptation and planning for medical/social supply in favour of ageing handicapped persons]. This example gives an opportunity to (1) examine (with a view to autonomy problems) the way "capacity" principles are mobilized, transformed into decision procedures and used in a context-dependent way, (2) examine how the capacity approach can be associated with (or favours?) individualizing strategies in the answer to collective problems, (3) examine the role of inter-institutional dialogue in interactive decision-making[11].

The report concentrates on three essential dimensions of interactive planning in the examined policy domain: (1) the elements of debate (what is at stake, really?); (2) the emerging consensual benchmarks; (3) examples and successful experiments. This last feature can be hold to be illustrative of a well-known stylized fact (established in comparative work by Robert Matland[12]): when policies have ambiguous goals or means, success or interesting results are typically dependent on successful experiments and initiatives. In the case at hand, the complexities of ageing as a process and the multi-dimensionality of "handicap" broadly conceived, certainly contribute to the ambiguity of goals and means. In addition, it is commonplace to observe that the notions of "autonomy" and "dependence" have complex meanings.

The CNSA report is the result of « practice exchange » workshops, aiming at the identification of the concerned population and its needs (in terms of accompanying actions or care). A major challenge was to characterize, on this basis, the necessary adaptations in collective answers, and the possible collective choices. Answering the needs is, by and large, identified with promoting autonomy: the whole point of the enterprise is to look for efficient ways to deal with the conditions of an autonomous personal lifestyle. The specific effects of ageing (at the individual level) are characterized in terms of autonomy losses. This has to do with the following factors: the negative evolution of functional capacities (which are already negatively affected by handicaps), the growing susceptibility to illnesses which are statistically associated with old age and, finally, the changing expectations associated with a new step in lifetime (which negatively impact the value of significant possible choices).

The technical report must provide guidance for decision-making. Indeed, it is conceived as some sort of problem-structuring and decision-facilitating device. There is a doctrinal side to the enterprise but the goals are practical ones and doctrine has a role in practical reasoning. The report promotes a specific step-by-step adaptation path. In addition, the report deals with the structuring of institutional dialogue. It delivers guidelines for appropriate queries and answers to assist the concerned persons in their ageing process. In the interactive dimension of institutional decision-making, the analysis of needs fulfils some of its most important functions: the enhancement of dialogue through the selection of appropriate common benchmarks (or focal points for attention) and the structuring of choices through priority-setting tasks.

The methodological concern for flexible adaptation turns out to be associated with the promotion of more substantial guidelines, concerning the appropriate collective choices for the country. Thus, it is suggested that collective organization should be compatible with a renewed attention to specific situations. It should be based, ultimately, on the revealed needs of individual persons – so the argument goes.

Providing for the needs of the elderly raises information problems. As Plato observed in The Statesman, policy-making for the city at large makes it impossible to adapt to the details of every individual situation. Collective goals and choices necessarily make use of the existing (and rough) categories which are otherwise used for descriptive purposes. For this reason, the project of placing individual situations at the heart of collective policy-making has difficulties of its own. None the less, if we follow the path of an individualistic approach, as the report recommends, the ways of collective action should rule out the more rigid sort of "answers" or "solutions" which are based on rough estimates of needs; as it turns out, they are always at risk of being too standardized and they are hardly able to adapt through time in order to accommodate the changing realities of individual lifestyles and environmental data.

Thus, we find methodological concerns at the root of the collective action process: the prevailing statistical categories are inappropriate for accurate descriptions and predictions of people's needs; if we start from such rough data, we are unable to give to pragmatic plans the desired flexibility in response to evolving contexts. This actually leads to prescriptive guidelines for the collective articulation of what a personal "life project" is about; the process exhibits a social buildup of the individual's "life project" for political (or institutional) purposes.

The chosen perspective on individual "life projects" highlights specific individual capacities, first and foremost, one's capacity to develop expectations about the future and to give shape to one's own future in accordance with personal wishes. This expresses personal autonomy of course, after the pattern of A. Sen's notion of "capability." Thus understood, autonomy concerns shed a new light on which capacities matter for the development of the whole "capability" of individuals. The subjective dimension of projects and needs comes to the forefront. As a result, this side of things is held to be the very foundation of needs-oriented dialogue with the relevant institutions.

The « life project » notion is thus somehow operationalized in a decision-facilitation perspective, with special emphasis on the buildup of institutional dialogue. It can be said to be embedded in a socially constructed dialogue situation between individual beneficiaries and institutional agents. Incidentally, this makes a difference with the philosophical notions of "life plan" (Joshua Royce) and "rational life plan" (John Rawls)[13]. Thus, the promotion of interactive decision-making and the concentration on a given picture of the individual (a more or less "liberal," autonomy-based picture) go hand in hand. The liberal perspective on persons and their choices gives weight to a number of directly relevant individual capacities, which are at the core of one's aptitude to express needs and to articulate expectations and claims. Let us note that the key notion of a personal and evolving "life project" offsets potentially rival notions, such as the continuity of an established lifestyle, or the good fit (or "harmony") between a person's lifestyle and the social environment. This can be hold to express definite, predominantly liberal values.

As a matter of operational planning at the collective level, "life project" appears to be essentially correlated with the notion of an "individualized compensation plan." This notion refers to an inclusive and coordinated intervention strategy which aims at dealing with all handicap situations, irrespective of the possible association with the ageing process or with old age as such. The whole enterprise is to let collective action and institutional cooperation revolve around personal needs; the ultimate aim is to help individuals fulfill their expectations with due assistance in order to compensate for capacity losses.

Although they are socially constructed, and designed as matrices for institutional exchange and decision-making, "life project" and "individualized compensation plan" are individualistic notions. Thus, the CNSA report exemplifies an individualizing approach to the assessment of situations and the elaboration of rational collective strategies. Probably, this is favoured by the insistent reference to personal capacities, such as the capacities to choose, to engage in activities, to elaborate and revise personal plans, etc. Claude Gamel (2007) has argued that capacity-based approaches of social needs tend to be associated with (or indeed, actively promote) an individualizing treatment of needs in social policies. The basic general idea is that institutional dialogue and interactive decision-making should aim at adapted answers to singular situations.

For all its problems of applicability, this trend in public policy is attractive to some degree. Indeed, it seems rational on the face of it, given the difficulty to predict the real needs of persons, as well as the appropriate ways of need-fulfillment, on the basis of general date about the ageing process, and age itself. This would seem to justify a tentative adaptation to fine-grained contextual conditions and personal needs. This is why the collective answers to capacity losses should be "modulated": we must pay due attention to the "complexity" and "diversity" of the ageing process in a case-based perspective.

The authors of the CNSA report stress that personal history (for example, whether one has lived or not in caring institutions) conditions personal needs for the ageing handicapped person. But it is not absolutely obvious that, starting from this departure point, we should conclude that taking real needs into account presupposes individualized answers and care, because the latter are associated with specific problems such as the difficulty and cost of truthful information gathering.

We have tried to characterize the "individualizing" logic which underpins the "life project"/ individualized compensation plan pair of notions. This logic, we believe, tends to favour institutional dialogue on the one hand (so that needs can be identified in a detailed way) and, on the other hand, interactive decision-making (in order to promote well-articulated, well-coordinated and efficient answers to existing needs, in the interest of personal capacities and autonomy). Such recommendations, if they are to be taken seriously in practice, involve a complex, continuous process of reallocation for institutional domains of action (or prerogatives). Thus, from the point of view of the involved institutional actors, attempts at rational planning are shown to lead, in this case, to a potential reshuffling of professional identifying characteristics[14].

It seems to us that the institutional dimension of interactive decision-making is well illustrated, in this case study, by the quest after a correct equilibrium between the social supply of general basic services and the provision of specialized, adapted care. It is expected, in this respect, that this kind of pragmatic breakdown of general and special issues could eventually coincide with a demarcation line between the needs which can be predicted from rough data and the needs which call for dialogue and the familiarization with concrete situations.

A further interesting feature of the recommendations is the invitation to look for reasons when it comes to relying on this or that decision-maker. Choices of this kind are best understood as answers to contextual elements, such as the comparative relevance of the acquired know-how of various institutional agents, when it comes to facing new situations. Here again, context-dependence is placed in favourable light. It is not portrayed as a source of instability or shifty expectations. Rather, it is positively associated with flexible decision-making.

Correlatively, it is suggested that the challenges of ageing call for new skills and new collaborative initiatives. This tends to confirm that the action domains of institutional actors should be flexible enough, so that they can adapt to evolving contexts of collective action. As it turns out, here again, methodological concerns impact substantial conclusions. The need for shared reasons in collective action is part of a methodology of collective-action planning. Among the relevant reasons, we find the reasons to invest some institutional agents (rather than others) with the responsibility to act in given classes of situations. Accordingly, consent to the flexibility of institutional responsibilities and prerogatives is advocated.

CONCLUSION

Our main conclusion is that context-dependence is, to some extent, shaped by the chosen principles themselves. Focal principles in collective action are not just means to adapt to changing circumstances or contexts (in a passive way). The chosen principles actively favour definite ways of adapting to circumstances, as exemplified by the development of "individualizing" social strategies to address social needs. These strategies involve a high degree of reliance on general principles, both methodological and substantial.

In our case study, we haven't examined the process of developing benchmark concepts and principles for collective action in its temporal development; rather, we have considered things as they are, even though we should remember at each step that the involved notions, political (and ethical) principles and evaluative judgments are, by and large, the result of constructive, often interactive social processes. As a matter of fact, the social use of general notions and the elaboration of the companion prescriptive judgments are dependent upon inter-institutional relationships, and it might be conjectured with some confidence that the latter are influenced by transitory circumstances or emerging contexts. This influence, however, does not necessarily rule out objectivity in evaluation or in the reasons which motivate choices.

REFERENCES

Ackroyd, S., Batt, R., Thomson, P., & Tolbert, P. S. (Eds.). (2005). *The Oxford handbook of work and organization*. New York, Oxford: Oxford University Press.

Alchian, A. A. (1990). Uncertainty, evolution, and economic theory. In O. E. Williamson (Ed.), Industrial organization (p. 23-33). Hants: Edward Elgar.

Calvert, R., & Johnson, J. (1999). Interpretation and coordination in constitutional politics. In Hauser, E., & Wasilewski, J. (Eds.), *Lessons in democracy. Jagiellonian: University Press & University of Rochester Press.*

Cropper, S., Ebers, M., Huxham, C., & Smith Ring, P. (Eds.). (2008). *The Oxford handbook of inter-organizational relations*. Oxford: Oxford University Press. doi:10.1093/oxfordhb/9780199282944.001.0001.

Da Fonseca, E. G. (1991). *Beliefs in action. economic philosophy and social change*. Cambridge: Cambridge University Press. doi:10.1017/CBO9780511628412.

De Munck, J., & Zimmermann, B. (Eds.). (2008). La liberté au prisme des capacités, vol. 18 in series Raisons pratiques. Paris: Editions de l'EHESS.

Gamel, C. (2007). Que faire de "l'approche par les capacités"? Pour une lecture "rawlsienne" de l'apport de Sen. *Formation-emploi*, *98*(April-June), 141–150.

Gilardone, M. (2007). *Contexte, sens et portée de l'approche par les capabilités de Amartya Kumar Sen*. (Unpublished doctoral dissertation). Lyon-2 University of Lyon, France.

Lenoble, J., & Maesschalck, M. (2010). *Democracy, law and governance*. Farnham, Burlington: Ashgate.

March, J. G., & Olsen, J. G. (1979). *Ambiguity and choice in organizations*. Oslo: Scandinavian University Press.

Matland, R. (1995). Synthesizing the implementation literature: The ambiguity-conflict model of policy implementation. *Journal of Public Administration: Research and Theory*, *5*(2), 145–175.

Miralles, C., González-Alcántara, O. J., Lozano-Aguilar, J. F., & Marin-Garcia, J. A. (2008). Integrating people with disabilities into work through OR/MS tools. An applied vision. In *Human Centered Processes Conference*. Delft.

Picavet, E. (2011). *La Revendication des droits. Une étude de l'équilibre des raisons dans le libéralisme*. Paris: Les Classiques Garnier.

Picavet, E., Dupont, G., Dilhac, M.-A., & Bolaños, B. (2009). *Identité et nouveauté des situations politiques*. Paper presented at Congrès des Associations des Sociétés Philosophiques de Langue Française. Budapest, ELTE University.

Rawls, J. (1971). *A theory of justice*. Cambridge, MA: Harvard University Press.

Reynaud, B. (2003). *Operating rules in organizations. Macroeconomic and microeconomic analyses*. London: Palgrave.

Royce, J. (1908). *The philosophy of loyalty*. New York: MacMillan.

Sen, A. K. (1979). Informational analysis of moral principles. In Harrison, R. (Ed.), *Rational action. studies in philosophy and social science*. Cambridge: Cambridge University Press.

ENDNOTES

[1] Our research on this topic has taken place within the CONREP Project (Franche-Comté Regional Council and Franche-Comté University, Besançon, France) and the CEEI project, Burgundy/ Franche-Comté higher education initiative (PRES Bourgogne/ Franche-Comté). The authors have benefited from joint work with Dawidson Razafimahatolotra at the *"Logiques de l'agir"* laboratory, Besançon. We have also benefited from discussions and presentations at the EGAIS-ETICA workshop in Brussels ("Investigating Contextual Proceduralism"), April 29th-30th 2011. Corresponding author: Emmanuel.Picavet@univ-fcomte.fr.

[2] This can be illustrated by some of the examples discussed by Calvert and Johnson. Calvert, R., & Johnson, J. (1999). Interpretation and coordination in constitutional politics. In E. Hauser & J. Wasilewski (Eds.), *Lessons in democracy*. Jagiellonian: University Press & University of Rochester Press.

[3] This has been the object of a full-length discussion in Picavet, E. (2011). *La Revendication des droits. Une étude de l'équilibre des raisons dans le libéralisme*. Paris: Les Classiques Garnier.

[4] See Reynaud, B. (2003). *Operating rules in organizations. macroeconomic and microeconomic analyses*. London: Palgrave.

[5] The notion of capacity has been variously used in ethics (as evidence by the work of Martha Nussbaum), in normative economics (Amartya Sen) and in the evaluation of social policies (Robert Salais).

[6] This has bee emphasized in Muriel Gilardone's PhD thesis. Gilardone, M. (2007). *Contexte, sens et portée de l'approche par les capabilités de Amartya Kumar Sen*. (Unpublished doctoral dissertation). Lyon-2 University of Lyon, France.

[7] Eduardo Giannetti da Fonseca thus writes: "[…] it must be observed that attempts to overcome the vagueness of ordinary language by setting up new and taylor-made technical terms, interposing rigid definitions and carrying as far as possible the formalism of the presentation are likely to run into fresh problems of their own. Even clarity and precision […] may be bought at too high cost" (chap. 10, p.143). Da Fonseca, E.G. (1991). *Beliefs in action. Economic philosophy and social change*. Cambridge: Cambridge University Press. See also: March, J.G., & Olsen, J.G. (1979). *Ambiguity and choice in organizations*. Oslo: Scandinavian University Press. And De Munck, J., & Zimmermann, B. (Eds.). (2008). *La liberté au prisme des capacités*, vol. 18 in series *Raisons pratiques*. Paris: Editions de l'EHESS.

[8] See Miralles et al. (2008).

[9] See Lenoble, J., & Maesschalck, M. (2010). *Democracy, law and governance*. Farnham and Burlington: Ashgate. (Specifically sec. II-4).

[10] See Alchian, A. A. (1990). Uncertainty, evolution, and economic theory. In O. E. Williamson (Ed.), *Industrial organization* (p. 23-33). Hants: Edward Elgar. And Picavet, E., Dupont, G., Dilhac, M.-A., & Bolaños, B.

(2009). *Identité et nouveauté des situations politiques*. Paper presented at Congrès des Associations des Sociétés Philosophiques de Langue Française. Budapest, ELTE University.

[11] For general background material on present-day research in the field, see Ackroyd, Batt, Thomson, and Tolbert (2005) and Cropper, Ebers, Huxham, and Smith Ring (2008).

[12] See Matland, R. (1995).

[13] See Royce, J. (1908).

[14] This side of planning tasks is addressed by March, J.G., Schulz, M., & Zhou, X. (2000). *The dynamics of rules. Change in written organizational codes*. Stanford: Stanford University Press.

Section 2
Emerging Technologies: Which Reflexivity?

Section 2: Emerging Technologies – Which Reflexivity?

We assume that *reflexivity* can be a diverse and multiple process of "becoming aware," ranging from consciousness to reflection, and whether it relates to the "external" world (the environment) or the "internal" world (the self). One can distinguish at least two basic meanings of reflexivity as regards the dynamic of reflexive modernisation in which technology has played a major role. On the one hand, *first-order* reflexivity refers to how modernity deals with its own implications and side effects, the mechanism by which modern societies grow in cycles of producing problems and solutions to these problems that produce new problems, in a dynamic of self-confrontation. On the other hand, *second-order* reflexivity refers to the cognitive reconstruction of this cycle in which problem solving through instrumental rationality generates new problems. The second-order reflexivity differs from the first-order reflexivity in that it entails application of rational analysis not only to the problems, but also to its own process, conditions, and effects.

The *reflexive* stance about technology is a question of both ethics and governance in that the concept of governance can open an alternative paradigm for doing ethics of technology and help shifting from technology assessment to a kind of *ethical accompaniment*. The ethical approach then is not meant to focus on determining whether specific technologies are right or wrong, but on accompanying their development in such ways that ethical considerations can play an important role in their design. The basic framework of an ethics of technology is often *externalist* in that humanity and technology are located in different spheres and the task of ethics is to protect humanity against technology. Current approaches to the roles of technology in society drives this model of ethics inadequate, for the separation it makes between a "human" domain and a "technological" domain is ultimately untenable. In contrast, we must develop an ethical framework that can be said *internalist* and that does not place ethics outside the realm of technology, but explicitly engages with its development. The crucial question in such an ethics of accompaniment is how we should give shape to the *interrelatedness* between humans and technology and anticipate from the design stage the *technological mediations*.

The anticipation of ethical issues as raised by the development of an emerging technology is perhaps best illustrated by the *information and communication technologies*. It must be noticed that there can be a proactive engagement with emerging ICTs with the goal of an early identification of the ethical issues these technologies are likely to raise. Such proactive engagements with novel technologies tend to concentrate on technologies that are either perceived to be problematic from the outset or that have already caused significant ethical issues. There are numerous examples of issues, questions or controversies that are directly related to or caused by new ICTs or new areas of application. Noteworthy examples include privacy issues in social network sites, intellectual property questions arising from the activity of search engines or the extent to which states can and should use the capabilities of novel technologies to store and analyse data on citizens.

Technological reflexivity is also to be taken in the context of *globalisation* and in considering the kind of selective agenda that the ideology of *cultural contextualism* brings about, especially concerning some issues like *security* and *privacy*. The first paradox is while we have adopted a stance that global problems require global solutions, we have regressed in some domains to discussing how we must limit global governance in cases where there are "cultural" aspects to consider. The second paradox is, unlike on the security agenda, there is little capacity building on human rights, and despite globalisation transforming the human rights agenda just as much as the security agenda, one is receiving more attention than the other. It seems that the spread and expansion of the security agenda has been enabled through international policy-making and conventions in ways that far outpace that of privacy. The most commonly preferred reason for this distinction is that privacy is a cultural issue, whereas we are left to draw the conclusion that security is universally valid. It is perhaps easier to argue for the universality of safeguards and protections rather than to engage in debates about *cultural relativism* and the definition of human rights and privacy. However, the debate about human rights cannot be avoided and universal protections for personal information, if not necessarily based on individual rights, needs at least for a *sense of the self*.

Chapter 4
Ethical Issues of Emerging ICT Applications

Bernd Carsten Stahl
De Monfort University, UK

ABSTRACT

This paper concentrate on a proactive engagement with emerging information and communication technologies (ICTs) with the goal of an early identification of the ethical issues these technologies are likely to raise. After an overview of the emerging ICTs for the next future (leveraging the results of the EU funded project ETICA), the paper identify the possible ethical consequences. Then the emerging ICTs are evaluated from different perspectives for prioritizing technical and policy intervention on them. The question of governance is then addressed with a final collection of recommendations for policy makers, industry, researchers and civil society.

INTRODUCTION

Proactive engagement with emerging information and communication technologies (ICTs) should allow an early identification of the ethical issues these technologies are likely to raise. This, one could continue the argument, would allow avoiding some of these issues and ensure that the beneficial consequences of technology research and development will outweigh the problematic consequences. This idea is by no means new and

in some incarnation or other has influenced the way in which scientific and technological progress is planned and governed. Such proactive engagements with novel technologies tend to concentrate on technologies that are either perceived to be problematic from the outset or that have already caused significant ethical issues. It is much less common in areas that are less headline-grabbing, including the area of ICTs.

Due to the increasing influence that ICTs have on most areas of life in western industrialized nations, there is a growing awareness of the ethical relevance of these technologies. There are

DOI: 10.4018/978-1-4666-3670-5.ch004

numerous high-profile examples of issues, questions or controversies that are directly related to or caused by new ICTs or new areas of application. Noteworthy examples include privacy issues in social network sites, intellectual property questions arising from the activity of search engines or the extent to which states can and should use the capabilities of novel technologies to store and analyze data on citizens. In addition to these larger issues which have caused significant public debates, there are numerous more localized issues, which a look in any newspaper will reveal. ICTs play a role in traditional issues of an at least partly moral nature from fraud to murder. They furthermore cause novel issues, as for example with regards to privacy or intellectual property issues.

This recognition of the moral relevance of ICTs was the starting motivation of the ETICA project (GA230318). The acronym ETICA stands for "Ethical Issues of Emerging ICT Applications." The project lasted from April 2009 to May 2011 and included 12 partners from 7 European Member States. ETICA had four main aims which will structure the present discussion. It set out to:

1. Identify emerging ICTs.
2. Identify ethical issues likely to be raised by those ICTs.
3. Evaluate and rank these issues.
4. Provide recommendations on appropriate governance structures to address these.

Before these individual aspects are discussed in more detail, it is important to briefly explain the claims that the project can raise and their limitations. ETICA aimed to discuss future developments. The future is, however, fundamentally unknown and unknowable. However, human beings always explicitly or implicitly make assumptions about the future and base their decisions on these assumptions. In many cases these assumptions and beliefs turn out to be correct. A key issue here is that we can in many cases reasonably assume that the future will be similar to the present and the past.

We can extrapolate from the past to the future. At the same time one needs to understand that this approach only works to some degree. A key factor that influences the reliability of extrapolations to the future is the temporal horizon in which such extrapolations take place. To put it differently, we can reasonably expect that tomorrow is going to be similar to today. Such an expectation is more likely to be disappointed in a month, a year, ten years and it is very unlikely to hold 100 years or even 1000 years from now. The ETICA project therefore had to define the temporal horizon in which it wanted to explore ICTs. It settled on the medium term future which was defined as approximately 10 to 15 years from the time of investigation, i.e. the time from 2020 to 2025. This was justified by the traditional view of technology innovation and development life cycle which suggest that research takes about this time to mature and become socially and economically relevant.

A temporal horizon of 10 to 15 years certainly exceeds the horizon in which statements about the future can be made with high certainty. Any projections of predictions of the ETICA project therefore need to be understood in this way. The best way of reading the outcomes and findings of ETICA is thus to see it as a foresight project. This means that ETICA does not claim to be able to predict the future and know what will happen with regards to ICT. Instead it is best understood as a foresight project (Georghiou, Harper, Keenan, Miles, & Popper, 2008; Martin, 2010). Foresight research recognizes the limits of the possibility of predicting the future but find the value of exploring the future in visualizing possible alternative futures and using such understandings to make present decisions about appropriate ways of dealing with and influencing such technologies (Cuhls, 2003). These considerations can also guide the reading of the present chapter. The reader should not so much ask herself whether the following accounts are true but rather whether they are enlightening and help her to form an opinion that will allow for current action.

Having now clarified the nature of the claims arising from ETICA, the following sections discuss the actual work. The first section describes how emerging ICTs were identified. This is followed by a description1 of the ethical issues associated with that technology. This leads to a discussion of the evaluation and ranking. The final section on the project outlines the governance analysis and recommendations arising from the project, leading to a general conclusion.

EMERGING ICTS

The methodology employed by ETICA to identify emerging ICTs was to do a dual discourse analysis of publications pertaining to these technologies. The first discourse that was looked at was the one emanating from the political level, from governments, funding councils and the like. This was aimed to ascertain what the political decision makers thought of ICTs that the deemed worthy of funding and likely to become relevant. Such political or politically inspired discourse is relevant because they have the political and financial wherewithal to make their visions into a reality. What they nevertheless tend to lack is the understanding of technical possibilities and current state of the art that ensures that their politically chosen aims are capable of being realized. ETICA therefore decided to complement this initial analysis by a second one that concentrated on the publications of individuals and organizations that are at the forefront of the actual technical development activities. To this effect it collected publications emanating from research centers, networks of excellence, universities and other sources of research in the area of ICTs. In many cases such organizations publish their plans and strategies including the technologies they intend to pursue.

By looking at these two different discourses, which are clearly not independent but present different takes on the question of future ICTs, the ETICA project managed to get an overview of expected ICTs that are based on a social and organizational reality that has the potential of realizing these visions.

During a number of iterations of the initial discourse analysis and readings of these discourses it became clear that a somewhat more fine-grained method of analysis was required. It was therefore decided that the discourses were to be analyzed with regards to technologies, application areas and artifacts. This distinction, which in practice was often difficult to uphold, allowed for a more differentiated understanding of the different entities discussed in the documents. It was furthermore decided that the analysis would look at what the documents said about social impact, critical issues (i.e. expected legal, ethical and other normative issues) as well as the more artifact-centered questions of capabilities and constraints of the technologies in question. Having engaged in analyzing the two discourses using these principles of analysis in an effort distributed across the consortium, it turned out that there was a significant number of all of these entities. The consortium identified more than 100 technologies, 70 application examples and 40 artifacts. The problem with these findings was that they contained considerable overlap. In some cases similar phenomena or technologies were described using different terms. In other cases the same terms were used for different phenomena. The consortium therefore spent significant efforts in reducing the initial list and overcoming redundancy with the aim of coming up with a list of technologies that captured the essence of current developments without duplicating descriptions and thereby ethical issues. This effort which was externally reviewed by subject experts in the different fields to ensure validity, eventually led to the following list of emerging ICTs that formed the basis of all subsequent steps of the project.

- Affective Computing
- Ambient Intelligence
- Artificial Intelligence
- Bioelectronics

- Cloud Computing
- Future Internet
- Human-Machine Symbiosis
- Neuroelectronics
- Quantum Computing
- Robotics
- Virtual/Augmented Reality

The reader who is familiar with the development histories and trajectories of ICTs may remark that this list is surprisingly unsurprising. Many of the technologies named here have been discussed for significant amount of times, in some cases several decades. Some of these technologies are already in wide-spread use and have an impact on the daily lives of numerous users. It is therefore important to recapitulate what this list represents. These are technologies that are currently being researched and developed and they display the potential to significantly change the way we live our lives, engage in social or economic activities or go about our business in ways that matter to us. They are technologies that are supported and funded because they appear to have the potential of having important consequences.

Having thus identified the technologies in question, it was clear that ETICA needed a more specific understanding of these technologies to provide a basis for the subsequent ethical analysis. The consortium therefore developed descriptions of each of the technologies that covered the following items:

- History of the technology
- About 5 examples of usage
- Defining features
- Related technologies
- Critical issues (ethical, legal, social, capacities, constraints).

These descriptions which are available on the project website were aimed at providing a basis of the subsequent ethical analysis. They also raised numerous methodological challenges because they required a detailed discussion of technologies which, by definition, do not yet exist. As it turned out, they provided a sufficient starting point for both the ethical analysis and the subsequent evaluation and ranking of the technologies despite numerous fundamental questions concerning the appropriate description of technologies (Stahl, 2011a).

ETHICAL CONSEQUENCES

Having thus achieved the first aim of the project by establishing a set of emerging ICTs, the next step was to explore the ethical consequences of these technologies. This step also raised numerous conceptual and methodological challenges. A fundamental question concerned the normative foundation of the ethical analysis. It was discussed whether the project should adopt a particular ethical stance, such as the traditional ethical positions of utilitarianism (Bentham, 2009; Mill, 2002), deontology (Kant, 1986; 1998) or virtue ethics (Aristotle, 2007). Alternatively the subject area would have lent itself to more recent ethical positions that were developed with information and technology in mind, such as information ethics (Floridi, 1999; 2008), disclosive ethics (Brey, 2000; Introna, 2005) or value sensitive design (Manders-Huits & van den Hoven, 2009). Due to the diversity of possibilities and the difficulty of justifying a well-grounded position, ETICA opted for a purely descriptive position. This allowed the project to avoid to engage in deep meta-ethical debates and to fulfil its aims and purpose of identifying ethical issues by drawing from a broad range of sources.

The starting point of the descriptive ethical analysis of the ethical analysis was a bibliometric approach that used a software tool to map concepts in a discourse (Heersmink, van den Hoven, van Eck, & van den Berg, 2011). This approach was used to gain an initial overview of the field and find interesting relationships between the technologies and the ethical issues they are likely to raise. On the basis of this heightened sensitivity with regards to ethical issues, a more detailed review of the literature with regards to ethics and each individual technology was undertaken. For each of the technologies a detailed description of the ethical issues found in the literature was developed. These findings are too extensive to be represented here but they are available on the project website.

For the purposes of this chapter, a brief overview of the characteristics of the technologies and the ethical issues they raise will suffice (Stahl, 2011b). Such an overview can by nature not be comprehensive but point to some of the interesting findings. Overall, the technologies, which in many respects still overlap and serve as enablers or amplifiers to one another, are expected to have some of the following effects. ICTs are expected to become easier to interact with by adopting principles of interfaces that are closer to natural interactions between humans. One aspect of this is that these technologies become invisible, either by shrinking or being integrated into backgrounds and infrastructure. The natural interaction with invisible technologies will allow a direct link between users and technologies, which may be implemented through invasive procedures (implants) as well as other approaches (e.g. seamless surveillance). The technologies, in order to fulfil their purpose, will need to be based on and incorporate a highly detailed model of the user. This is required to accurately predict the user's preferences and to allow the technologies to proactively engage with the user and the environment. Such technologies are furthermore expected to be pervasive and ubiquitous, being spread through all relevant aspects of users' environment.

An important aspect is that these invisible, ubiquitous ICTs will require less and less direct input from the user. Instead they will act autonomously. There are important philosophical questions to be asked what is meant by autonomy and whether it can be achieved by the type of ICT under investigation here. For the purposes of the ETICA project, such questions could safely be ignored because the meaning of autonomy in the literature surveyed is that technologies act on users' lives without these users' active involvement.

If these visions of technologies are realised, then it seems fair to suggest that they will structure human's spaces of action in novel ways. Technologies are always more than just neutral tools and the interaction between technology and human affects the way in which a human can act. The appropriate interpretation of this interplay is often subject to debate and public discussions. At least as important, however, and in many ways ethically more relevant is the way in which spaces of action are structured in a non-conscious manner. It is easy to envisage scenarios in which the technologies with capabilities as the ones just described could be used to structure a society that gives preferences to some types of action over others. At the extreme it might lead to a brave new world as described by Huxley (1994).

These shared characteristics of technologies are clearly ethically relevant in their own right. During the ETICA project the ethical analysis concentrated on the individual technologies, however, and not on the overall picture of emerging ICTs. It is nevertheless possible to develop a similarly general overview of the different ethical issues identified.

A look at the many ethical issues identified by the project shows that many of them are relevant to more than one technology. It is therefore legitimate to describe them as ethical issues likely to arise from the combination of the emerging ICTs outlined earlier.

When looking at these shared ethical issues, one can distinguish broadly between ethical issues that are currently well discussed and that one can plausibly assume to remain relevant in the future and such ethical issues that are currently less widely discussed where no solutions are as yet visible. The most prominent example of the former group of ethical issues that are currently being discussed in much detail is the issue of privacy. Privacy has been established as a technology-related problem in need of attention since the late 19th century (Warren & Brandeis, 1890). The use of ICT has raised numerous new issues and problems related to privacy, which in some jurisdictions have led to considerable legislative activity. The primary example is the European Directive 95/46/EC which required comprehensive privacy legislation in all European Member States. The technical developments described in the ETICA project render it more than likely that privacy will not only remain a problem but will intensify. Existing privacy concerns will be exacerbated by significantly increased amounts of data available, novel ways of linking and interpreting them as well as qualitatively new types of data, such as emotional data. Similar accounts can be given for several other ethical issues. The increasing interconnection of technologies and their immediate linkage to many areas of life will render security questions more pressing than ever. Related issues will arise from issues of trust and trustworthiness of these technologies. Ethical issues with strong legal connotations will arise from problems related to liability. If the new interconnected technologies malfunction or are intentionally misused, then how will liabilities be defined and enforced? On a social level it stands to reason that problems of digital divides will be exacerbates. Current divisions between the digital haves and have-nots can have ethical consequences that are already widely debated. In light of the new services and opportunities that emerging ICTs are likely to afford, such divides become even more pressing.

In addition to such problems that are already widely discussed an in many cases subject of legislative or regulatory intervention, there are important issues that are currently not explored in much depth. One core group of issues has to do with our collective view of humans, what constitutes and defines humans and how this is changed by technology. Pervasive and potentially invasive technologies blur the fuzzy distinction between therapy and enhancement further. Our expectation of what is considered normal both in terms of behaviour and capabilities may change. Just like a general expectation of availability via mobile phones is becoming a social norm, novel technologies may create new expectations of normality which are likely to affect humans' rights and expectations. Overall, there is a very open question what these technologies will do with our individual identities.

In addition to such general questions on the individual level, there are questions relating to the relationship between individuals. The pervasive, autonomous and directly linked technologies not only open or close new potential spaces for action but allow for the more or less subtle manipulation of such spaces. These ICTs therefore have a potentially deep impact on power structures and relationships within a society. This can refer to traditionally contested relationships such as those between employers and employees. In addition, there will be other types of relationships that will be transformed. A good present example of this is the relationship between personal acquaintances and friends, which has taken on new meanings in the age of social networks. These changes that are already reflected in general language, e.g. the novel term to "unfriend" somebody may find even wider applications in other social domains.

As these examples show, there is reason to believe that emerging ICTs will influence personal and social aspects of our lives that traditionally were not affected by technical change. The considerations so far render it plausible to assume that such changes may be even further-reaching than the individual consequences and effects on relationships outlined in the preceding paragraphs. While it is very difficult to predict or even just envision what exactly the ICT-induced changes on the macro-level will be or entail, it is probably reasonable to assume that such changes will materialise. They may be cumulative effects of individual changes or they may be emergent phenomena arising from the unforeseen interplay between different components of technical, social and economic systems. Such consequences may affect the way societies organise their functions and they may affect our collective self-image, our cultural affiliations and the way we interpret human societies. These are grand ideas and the actual change may only be minor. Due to the scale of potential consequences, however, these are important issues worthy of attention.

EVALUATION

Having thus outlined a very abstract and high-level description of the identified technologies and their ethical consequences, the chapter needs to point to the activities that the ETICA project undertook to evaluate and rank these technologies.

In parallel to the ethical evaluation outlined in the preceding section, ETICA aimed to evaluate these in order to provide a means for prioritising potential technical or policy intervention. To this end, the 11 technologies were evaluated from four different perspectives:

1. **Law:** The legal evaluation investigated whether and in which way the technologies were subject to legal regulation or were mentioned in legal texts. This evaluation concentrated on European law and aimed to establish whether there were important differences with regards to these technologies between European Member States.
2. **(Institutional) Ethics:** The second set of evaluation surveyed what was labelled European institutional ethics. This refers to the published outputs from European or national ethical bodies, such as the European Group on Ethics in Science and New Technologies or national ethics committees. This body of work was reviewed to find out whether and in which ways any of the technologies had been addressed.

3. **Gender:** One particular issue of interest was whether there was a discernible aspect of gender with regards to the different technologies. This was done by selecting themes from research on gender and technology and exploring the relevance of these themes for the emerging ICTs.
4. **Technology Assessment:** Finally and as an overarching approach, the technologies were evaluated from the perspective of technology assessment. This covered the earlier points as well as further research in the technology assessment area.

In addition to the mostly literature-based research outlined in the preceding paragraphs, ETICA held an expert workshop with experts from all four of the above areas to come to a general view on the relevance of the different technologies. The resulting evaluation was based on both the prior research in the four evaluation areas and the expertise of the participants in the workshop. As a result the following list of technology was compiled. This list shows the technology in order of relevance / need of attention. It has combined some of the technologies, either because they were evaluated identically or because they were perceived to be substantially similar in terms of consequences.

1. Ambient Intelligence
2. Augmented and Virtual Reality
3. Future Internet
4. Robotics, Artificial Intelligence and Affective Computing
5. Neuroelectronics, Bioelectronics and Human-Machine Symbiosis
6. Cloud Computing
7. Quantum Computing

This list represents the ETICA project's view on the order of relevance and it can build the basis for further policy advice.

GOVERNANCE

Having thus undertaken the empirical and descriptive part of the project, ETICA could now point to emerging ICTs, their ethical consequences and their evaluations. A final point of interest was the question what could and should be done to address these issues. This is the question of governance.

In an initial step the project reviewed governance arrangements. In the context of the present volume it will not be necessary to discuss current governance arrangements in any more detail. It is nevertheless important to point out that one important finding was that current ethics processes within the EU are not sufficient to identify and address the ethical issues outlined here. The EU currently deals with ethical questions in science and technology research and development via an ethics review system. This system is based on biomedical research ethics reviews and undertakes to identify and outline ways of addressing ethical issues at the project proposal stage. Through a series of screenings and reviews potentially problematic projects are identified and scrutinised in more depth. Ways of addressing ethical issues are then incorporated into the project contract and conduct (European Commission, 2011). This approach works well with regards to established ethical problems, in particular with regards to ethics arising from research conduct. It centres on questions of informed consent and ensures that research participants do not suffer disadvantage because of the project. The downside of this approach is that it is abstract and does not recognise the context of ethical issues nor does it develop sufficient sensitivity towards problems that arise from the development and use of the technology under investigation.

Given that many of the ethical issues outlined above fall within this latter category, the ethics review approach will need to be complemented. This supported by the approach taken by the European Commission itself which in its recommendation for research in nanotechnology clearly states that

"Researchers and research organisations should remain accountable for the social, environmental and human health impacts that their N&N research may impose on present and future generations" (European Commission, 2008, p.7).

The ETICA project therefore developed the recommendations detailed below. They are based on the recognition that an exact prediction of emerging ICTs or the ethical consequences they will entail is impossible. At the same time there are sufficiently robust ways of gaining an understanding of possible futures that allow present actions aimed at facilitating a desirable future. Governance structures that rise to the challenge of responsible research and innovation in ICT will need to combine sensitivity towards substantive ethical issues with procedures that promote sensitivity towards such substantive issues and structures that allow addressing them. The ETICA project therefore gives the following recommendations, which are aimed at policy makers on the one hand and industry, researchers and others involved in ICT research and development on the other hand[2].

RECOMMENDATIONS FOR POLICY MAKERS

Policy makers have an important role in creating the regulatory framework and the infrastructure to allow ethics to be considered in ICT. If emerging ICTs are to be developed in a responsible manner that allows identifying and addressing the social and ethical problems outlined above, then a framework and infrastructure for the development of responsibility needs to be provided. Such a framework should cover at least the following three main areas of policy activity:

- Provide regulatory framework which will support Ethical Impact Assessment for ICTs.
    - To raise awareness of the importance of ethics in new ICTs.

- To encourage ethical reflexivity within ICT research and development.
- To provide appropriate tools and methods to identify and address ethical issues.
- To address the wide range of current and new ethical issues arising from ICT, modelled along the lines of environmental, privacy or equality impact assessments.
- To allow ICT professionals to use their expertise in emerging ICTs to contribute to ethical solutions.
- To raise awareness of ethical issues regarding animals and environmental issues.
- To proactively consider legal solutions to foreseeable problems that will likely arise from the application of future and emerging technologies.

Overall, this set of recommendations addresses the institutional framework that will be required for further subjects to recognise responsibilities and develop mechanisms of discharging it. The idea of an "Ethical Impact Assessment for ICTs" was chosen because it can draw on precedent from areas of the environment, privacy, or equality. Such a framework is required to provide incentives to engage with issues of responsibility in innovation and emerging ICTs. It will thereby encourage discourses that will lead to the development of specific responsibilities.

- Establish an ICT Ethics Observatory.
    - To collect and communicate the conceptual, methodological, procedural and substantive aspects of ICT ethics.
    - To provide a community-owned publicly accessible repository and dissemination tool of research on ICT ethics.
    - To give examples of approaches and governance structures that allow addressing ethical issues.

- To disseminate past and current research in ethics and ICT including relevant work packages and deliverables and relevant National Ethics Committee opinions.
- To facilitate the Ethical Impact Assessment.
- To provide an early warning mechanism for issues that may require legislation.

While the first set of recommendations aimed at providing a procedural framework for identifying and addressing ethical issues in ICT, this set of recommendations aims to provide the content required for an Ethical Impact Assessment. The work undertaken by the ETICA project, for example, provides important pointers towards possible ethical issues to be considered. Individuals involved in technical development are often not experts in these matters. A shared repository of ethics-related theories, practices, methodologies etc. is a necessary condition of the development of widely shared good practice.

- Establish a forum for stakeholder involvement.
    - To allow and encourage civil society and its representations, industry, NGOs and other stakeholders to exchange ideas and express their views.
    - To exchange experience between these stakeholders to develop ethical reflexivity in the discussion.
    - To reach consensus concerning good practice in the area of ethics and ICT.
    - To build a bridge between civil society and policy makers.

This final recommendation for policy makers points to the necessity of institutionalising important discourses that allow civil society and other stakeholders to engage on a content level with the policy as well as the technical community. Such a forum is required to ensure that responsible innovation covers not only specific technical interests and perspectives but is allowed to reflect broader societal concerns.

RECOMMENDATIONS FOR INDUSTRY, RESEARCHERS AND CIVIL SOCIETY ORGANISATIONS

Industry, researchers and other individuals or organisations should adhere to the following recommendations in order to be proactive and allow innovation to be socially responsible. If the institutional framework, background, repository and societal discourses are there, then the conditions will be favourable for the incorporation of ethics and reflexivity into technical work and application usage.

- Incorporate ethics into ICT research and development.
  - To make it explicit that ethical sensitivity is in the interest of ICT users and providers.
  - To distinguish between law and ethics and see that following legal requirements is not always sufficient to address ethical issues.
  - To engage in discussion of what constitutes ethical issues and be open to incorporation of gender, environmental and other issues.

The points of this recommendation aim to ensure that ethical reflexivity is realised within technical work. It furthermore aims to sensitise stakeholders to the difficulties of discharging their responsibilities.

- Facilitate ethical reflexivity in ICT projects and practice.
  - To realise that ethical issues are context-dependent and need specific

attention of individuals with local knowledge and understanding.
  - To simultaneously consider the identification of ethical issues and their resolutions.
  - To be open about the description of the project and its ethical issues.
  - To encourage broader stakeholder engagement in the identification and resolution of ethical questions.

This final set of suggestions aims to ensure that the different stakeholders realise that ethics is not a pre-determined and fixed structure. Ethical issues are context-dependent and need to be interpreted in the particular situation. Interpretive flexibility of technology requires the participants in a technology development project to engage collectively in the initial definition of ethical issues to consider, but also to review this initial definition continuously and engage with stakeholders involved in other stages of the technology development process.

CONCLUSION

This chapter has provided a brief summary of the ETICA project which set out to identify emerging ICTs, their ethical issues and make recommendations on how to address them. It has shown that there are robust and rigorous ways of exploring possible futures and using them to draw conclusion concerning the present. The recommendations developed in the project demonstrate promising avenues of activity in order to ensure that our societies are alerted to the ethics of ICT and have ways of proactively engaging with it. This will allow these societies to enjoy the contributions and positive aspects of such novel technologies while retaining the necessary vigilance to pre-empt or address negative consequences.

ACKNOWLEDGMENT

The research leading to these results has received funding from the European Community's Seventh Framework Programme (FP7/2007-2013) under grant agreement n° 230318.

REFERENCES

Aristotle. (2007). *The nicomachean ethics*. Filiquarian Publishing, LLC.

Bentham, J. (2009). *An introduction to the principles of morals and legislation*. Oxford: Dover Publications Inc..

Brey, P. (2000). *Disclosive computer ethics: Exposure and evaluation of embedded normativity in computer technology*. Presented at the CEPE2000 Computer Ethics: Philosophical Enquiry. Dartmouth College. Retrieved October 22, 2010 from http://ethics.sandiego.edu/video/CEPE2000/Responsibility/Index.html

Cuhls, K. (2003). From forecasting to foresight processes - new participative foresight activities in Germany. *Journal of Forecasting*, *22*(2-3), 93–111. doi:10.1002/for.848.

European Commission. (2008). *Commission recommendation on 07/02/2008 on a code of conduct for responsible nanosciences and nanotechnology, No. C 424 final*. Brussels: European Commission.

European Commission. (2011). *Fourth FP7 Monitoring Report - Monitoring Report 2010*. Brussels. Retrieved October 12, 2011 from http://ec.europa.eu/research/evaluations/pdf/archive/fp7_monitoring_reports/fourth_fp7_monitoring_report.pdf

Floridi, L. (1999). Information ethics: On the philosophical foundation of computer ethics. *Ethics and Information Technology*, *1*(1), 33–52. doi:10.1023/A:1010018611096.

Floridi, L. (2008). Information ethics: A reappraisal. *Ethics and Information Technology*, *10*(2), 189–204. doi:10.1007/s10676-008-9176-4.

Georghiou, L., Harper, J. C., Keenan, M., Miles, I., & Popper, R. (2008). *The handbook of technology foresight: Concepts and practice*. Edward Elgar Publishing Ltd..

Heersmink, R., van den Hoven, J., van Eck, N., & van den Berg, J. (2011). Bibliometric mapping of computer and information ethics. *Ethics and Information Technology*, *13*, 241–249. doi:10.1007/s10676-011-9273-7.

Huxley, A. (1994). Brave new world (New ed.). London: Flamingo.

Introna, L. D. (2005). Disclosive ethics and information technology: Disclosing facial recognition systems. *Ethics and Information Technology*, *7*(2), 75–86. doi:10.1007/s10676-005-4583-2.

Kant, I. (1986). *Kritik der praktischen Vernunft*. Reclam, Ditzingen.

Kant, I. (1998). *Grundlegung zur Metaphysik der Sitten*. Reclam, Ditzingen.

Manders-Huits, N., & van den Hoven, J. (2009). The need for a value-sensitive design of communication infrastructures. In P. Sollie & M. Düwell (Eds.), Evaluating new technologies: Methodological problems for the ethical assessment of technology developments (pp. 51-62). The International Library of Ethics, Law and Technology. Springer.

Martin, B. R. (2010). The origins of the concept of "foresight" in science and technology: An insider's perspective. *Technological Forecasting and Social Change*, *77*(9), 1438–1447. doi:10.1016/j.techfore.2010.06.009.

Mill, J. S. (2002). *Utilitarianism* (2nd ed.). Hackett Publishing Co, Inc..

Stahl, B. C. (2011a). What future, which technology? On the problem of describing relevant futures. In M. Chiasson, O. Henfridsson, H. Karsten, & J. I. DeGross (Eds.), Researching the future in information systems: IFIP WG 8.2 Working Conference, Future IS 2011 (pp. 95-108). Turku, Finland. Heidelberg: Springer.

Stahl, B. C. (2011b). What does the future hold? A critical view of emerging information and communication technologies and their social consequences. In M. Chiasson, O. Henfridsson, H. Karsten, & J. I. DeGross (Eds.), Researching the future in information systems: IFIP WG 8.2 Working Conference, Future IS 2011 (pp. 59-76). Turku, Finland. Heidelberg: Springer.

Warren, S. D., & Brandeis, L. D. (1890). Right to privacy. *Harvard Law Review, 4*, 193. doi:10.2307/1321160.

ENDNOTES

[1] For a more detailed description of the methodology and findings of the identification stage of the project, see Deliverable D.1.2 at www.etica-project.eu.

[2] These recommendations have previously been published in several official ETICA publications.

Chapter 5
International Co–Operation and Intercultural Relations:
Reconciling the Security and Privacy Agendas

Gus Hosein
Privacy International, UK & London School of Economics, UK

Maria-Martina Yalamova
Privacy International, UK & London School of Economics, UK

ABSTRACT

This article compares the notions of security and privacy as they are treated in international cooperation relations. While security seems to benefit from a global consensus, it does not hold true for the concept of privacy. This article argues for not addressing the privacy in the context of the issue of human rights, marked by recurrent controversies about cultural relativism.

INTRODUCTION

There is a general consensus on the idea that in global times, global solutions are required. Globalisation is primarily the result of the unprecedented flows of people, goods, services, ideas, and data across borders. In turn, as the logic goes, national governance structures no longer suffice and global and international governance

DOI: 10.4018/978-1-4666-3670-5.ch005

structures are required. The challenge becomes one of resolving national values and laws with international standards.

Congresses and Parliaments are now caught in the throes of this globalisation of governance. Solutions to global challenges are now sought at international congresses, parliaments and conferences, and those solutions are then brought home for reconciliation with national laws. These forums have become part of the common language of policy-making. 'Kyoto' is synonymous with environmental policy, 'Doha' with trade policy, 'Vienna' with diplomacy. International treaties are popping up everywhere to regulate international activities. 'International obligations' were used to pressure the U.S. on trade and environmental policy just as 'international standards' have been used to place pressure on China on intellectual property. Globalisation is serving as a leveller of national laws.

Security is one of the global challenges for which global solutions are sought. We have written before on how globalisation pressures have been used to implement anti-terrorism, crime, and security policies (Hosein, 2004). Inter-governmental bodies like ASEAN, the Council of the European Union, and the Group of 8 meet regularly to discuss new policies and techniques to respond to new and emerging challenges. These international bodies also facilitate capacity building, so that standards can be established to enable governments to understand the framework of risks and policy options. International identity documents standards are established by the International Civil Aviation Organisation (ICAO); methods of dealing with online crime are considered at the G8 and the Council of Europe (COE); standards for communications surveillance are established by the European Telecommunications Standards Institute.[1]

In the domain of human rights, there is also a globalising force, though often weaker than the momentum granted to the security agenda. There are dozens of international and regional human rights statements and instruments, and even enforcement bodies. For instance, both the Bush Administration and the Obama Administration mention the Geneva Convention whenever they consider the treatment of detainees and enemy combatants. Similarly, the United Kingdom Home Office must now reconsider its DNA database after the European Court of Human Rights ruled against the UK in December 2008 on the retention of profiles of innocent individuals, despite the UK

Parliament and the national courts approving of the Government's practices.

None of this is new. This chapter discusses two issues that have not received sufficient attention.

1. While we have adopted a stance that global problems require global solutions, we have regressed in some domains to discussing how we must limit global governance in cases where there are 'cultural' aspects to consider.
2. Unlike on the security agenda, there is little capacity building on human rights. Despite globalisation transforming the human rights agenda just as much as the security agenda, one is receiving more attention than the other.

We will focus primarily on information privacy and security issues as they best illustrate this international dynamic.

We therefore focus on the following dynamic: there is an interesting lack of promulgation of safeguards and protections for individual citizens and consumers in the face of similar threats around the world. That is, if modern crime and anti-terrorism policies are required to deal with the threats of international criminals and terrorists without regard to borders, why is it that the safeguards against abuse are not also made universal?

GLOBAL SOLUTIONS FOR SECURITY

The global security agenda is certainly not new, but there are some recent changes worth noting. It is no longer the case that security agreements between states come at the expense of another state (Jervis, 1982). In fact, the increase of security is now seen as a good in itself, and it is being promulgated internationally. This agenda is stronger in its conviction and is spread further than ever before as countries around the world are seeking security solutions to shared problems.

A further change is that now governments are developing new ways to harness information and communications technologies for the advancement of their security agendas. For example, while passports and visas have long been required for international travel, governments are advancing *biometric* passports and identity documents in ways that were previously unimaginable. These techniques are also being used to monitor the movements of citizens and foreigners through border management and travel surveillance.

Just as governments are keen to respond to the global movement of people, they are also looking for solutions to problems introduced through the global movement of data. With Internet communications across borders, it is now possible to 'act at a distance' and as such, malicious activities by computer users in one country can certainly implicate the systems of another.

But the security agenda is not always just about responding to foreign threats. Because of new technologies and new developments, there are emerging threats within a country. Due to the pervasiveness of Internet communication technology, governments are all seeking solutions to criminal activities that were previously unthreatening or simply impossible. With computerisation across government services, critical national infrastructure may now be vulnerable to attack, both from within and beyond the borders. Consequently, all countries are seeking the capacity to protect their assets from abuse, wherever it may originate from.

The building of capacity is important to ensuring global security, as a single country can be a weak link in global enforcement. As we've seen with both the drug trade and anti-terrorism policies in Afghanistan, a single failed-state could lead to problems in other countries. Thus, the global community is keen to ensure that there are no safe havens for criminality. But policy-makers in all these countries cannot be expected to build policies from the ground up. Some countries don't possess the expertise to understand the full nature

of the threats. This is particularly true in the realm of anti-terrorism policy and cybercrime policies.

Rather, less capable governments can make use of the expertise and experiences developed in other countries. We see this often as some countries copy the laws of others; after all, if one country has devised a good set of laws on dealing with travel security, it would save time and effort if that country can help other countries develop similar laws.

International conventions have also served as useful models for countries to adopt. Rather than having to consult extensively and develop laws on piracy, terrorism, and international organised crime, governments can work together at the international level to create standards, and then adopt the language from these conventions into their own laws.

Even still, the integration of international conventions into national laws is a complex affair. The secretariats of some international organisations have played essential roles in aiding governments through gap-analyses and other assessments and evaluations to help them comprehend the nature of the proposed legal changes.[2]

In the security domain, the U.S. and the United Kingdom have been leaders in training other governments on how to implement control systems to monitor for terrorists, for instance. The Council of Europe has been particularly effective at promoting its Cybercrime Convention (ETS 185) internationally and training national policy-makers around the world on the nature of the likely threats.

These initiatives speak of the importance of a coordinated and cooperative approach to security as a single weak airport security protocol could threaten the lives of nationals of many countries. Similarly, the failure to adopt a cybercrime law could prevent the prosecution of a malicious hacker. Building the capacity of all countries is necessary for global security. Failing to do this would be negligent.

SAFEGUARDS ARE NOT UNIVERSAL

When the Council of Europe was negotiating its Cybercrime Convention, they were interested in advancing the state of procedural laws in the Council of Europe countries. That is, some countries had laws that criminalised 'hacking', and had empowered their authorities to investigate criminal activities enabled through computing, while other countries did not. Levelling the playing field in this domain compelled all governments to pass laws to define new forms of criminality and to grant new powers to law enforcement agencies.

The Cybercrime Convention contained no safeguards against abuse, however. While it promoted the increasing of law enforcement powers in the 'cyber' realm, it did not promote greater protections for individual and human rights. When civil society institutions pushed for change and for some minimum standards for these investigative powers, the Council of Europe responded that they were not necessary because all Council of Europe countries have signed the European Convention on Human Rights. That is, the Council of Europe argued that because each country that would sign the Cybercrime Convention had already signed the European Convention on Human Rights, governments would be bound by the ECHR to implement new Cybercrime Convention powers in ways that respect human rights. In an odd twist, however, the Council of Europe also argued that it would be too complicated to ensure that the exact same procedural safeguards existed in each country.

A similar reasoning was used in the European Union when it was discussing the retention of communications logs. In 2005, the European Parliament was debating whether to approve a Directive on Data Retention, requiring communications service providers to retain log data of the transactions of their customers for up to two years to aid police investigations. Some Members of the European Parliament joined civil society organisations arguing that the Directive increased

the powers of law enforcement agencies by granting them access to more information without implementing any safeguards. The proponents of the policy, including European Commission, responded that any safeguards were up to the Member States to decide for themselves; and again, all of them were signatories to the European Convention on Human Rights. It was felt that it would be too complicated to compel every country to adopt the same safeguards.

The dynamic can be summarised as this: increasing security is an imperative requiring universal common laws, but implementing safeguards is a national exercise that must be negotiated, although we must be mindful of international human rights obligations.

The contemporaneous complicating factor is that even if we accept the 'weak link' argument that we must spread these security laws around the world, some of these countries do not necessarily abide by international human rights obligations. The Council of Europe has been active in Asia in particular, travelling through the region to promote its international conventions dealing with terrorism and crime. But the Council of Europe has been far more reserved in promoting its conventions on human rights. It is impossible to reconcile this situation with the promise that was made during the drafting of the Cybercrime Convention: that safeguards are not necessary in the text of the convention because all signatories will have to abide by the ECHR. If the Council of Europe is promoting the adoption of the Cybercrime Convention in Asia but simultaneously promoting the European Convention on Human Rights with equal vigour, then it is failing to uphold its promise. Countries in Asia are keen to sign up to the Cybercrime Convention but no one is talking about implementing new human rights protection in accordance with the ECHR. In fact, to compel countries to act to promote human rights is now considered imperialistic, while to enable them to perform the former is considered a noble and necessary mission.

THE POLITICS OF CULTURAL ARGUMENTS ABOUT HUMAN RIGHTS AND PRIVACY

The debate about human rights and cultural relativism is long and rich. Indeed, it may be said that 'human rights' theory and practice originated in the 'West'. But there is a sense of universality just within the term itself: all human beings have the same rights (Donelly, 2003, p.65). It is not our purpose in this chapter to fight one corner of this fight, however. Rather, we would like to analyse the modern debate about rights and freedoms in the context of globalisation.

The Council of Europe has for some time now been promoting its Cybercrime Convention beyond Council of Europe members, alongside its other Conventions. For instance in 2007 it held a meeting with cybercrime experts from 55 countries. It celebrated that 'reforms ... based on Convention guidelines are already underway in Argentina, Brazil, Egypt, India, Nigeria, Pakistan and the Philippines' (COE 2007). It has also actively promoted the Convention at the United Nations, and participates at the yearly meetings of the United Nations Internet Governance Forum where it promotes its conventions on the global stage.

On human rights, the Council of Europe has been more reserved on the global stage. The Council of Europe does indeed promote the global protection of human rights. Its agenda on global Internet policy highlights the importance of 'safeguarding our rights and freedoms'. In fact, it highlights the importance of free expression repeatedly within its international programming, drawing on Article 10 of the European Convention on Human Rights.

In its promotional material, the Council of Europe then goes on to promote the Cybercrime Convention, the Convention on the Prevent of Terrorism, and Convention on the Protection of Children against Sexual Exploitation an Sexual Abuse. So the Council of Europe is not necessarily shy of human rights as it promotes it security-oriented conventions. The curiosity is that its documentation is relatively quiet about Article 8 of the European Convention on Human Rights, on the protection of privacy, and on its Convention on Data Protection.

It would be unfair to say that the COE is absolutely silent on privacy, but as people who have followed the COE's actions for nearly a decade we can say with some certainty that it is more active in spreading the Convention on Cybercrime than its own Data Protection Convention. And countries around the world are showing far greater eagerness to adopt the Cybercrime convention and implement it into national law.

The reasons for the relative quiet on privacy are probably numerous. One of them is the emergence of the notion that 'privacy is a Western value' that is not shared in other countries around the world. This line of argument follows much of the same logic that 'human rights are a Western value', except that you rarely hear this argument practiced as widely anymore. We certainly hear that 'security' must trump 'rights', and this appears to be a universal line of argument. But the globalisation of privacy has been much further behind the other human rights, which are all collectively far behind on the security agenda.

What is most interesting is that industry voices tend to agree with the culturally relativistic approach to privacy. One would be hard-pressed to find industry organisations arguing that child labour laws are a Western value and as such a sweatshop in Asia operates under a different cultural regime, so it is acceptable. On the issue of free expression, Internet companies, and particularly those based in the U.S., take a more careful line that promotes free expression in other countries by arguing that the U.S. constitutional approach to free speech dictates how it must manage this issue internationally. Yet on privacy issues, industry organisations have repeatedly argued that privacy is a cultural issue and that we must not thrust privacy protections on other countries.

The politics of privacy law contributes to this situation. Privacy regulations are onerous and require strong enforcement bodies to make them work. Government departments around the world tend to be nervous about privacy laws because they may restrain the government as they endeavour to perform public services such as policing, or impair the delivery of more efficient government services. Industry is worried about increased costs and restrictions that prevent them from collecting and moving data across networks and around organisations. Both sets of policy stakeholders are worried about the regulators who may interfere with their ways of doing business.

So while Cybercrime laws increase global business confidence in that companies know that perpetrators of hacking will be brought to justice anywhere in the world, there is much less enthusiasm about the spread of privacy rules. Business leaders speak of the different cultures of nations, and how many languages lack a word for 'privacy'. They speak of how some cultures are more collectivist rather than individualist, and as such privacy is not as relevant to these people.[3] So, their logic goes, why should countries adopt Western privacy laws and enforcement mechanisms when the citizenry and consumers are not that concerned about privacy?

This logic must not be easily dismissed. Just as the human rights movement has for years fought the 'cultural relativism' argument, perhaps so must the privacy movement. We imagine at some point in time, these debates were had about free expression, torture and slavery, and so it must for privacy. We may debate about religious and cultural understandings, and the importance for respect to those values; but we may also point to the spread of human rights since the 1940s in response to global atrocities.

Though likely to be interesting, such a debating exercise would be largely repetitive of previous debates, and would seriously distract from advancing the standards and conventions required to protect against global threats. Just as

the security agenda is based on current threats of a global nature, so must we move forward the privacy agenda. Therefore, for the purpose of this paper, we propose a slightly different approach to the issue: even if some do not believe that privacy is a shared value, the absence of safeguards may become a universal concern.

THE SHARED SENSE OF ABSENCE OF PROTECTIONS

Though we are putting aside human rights arguments for the purpose of this chapter, we believe it is still useful to break up the presentation of cases by identifying individuals as two different types of stakeholders: individuals as citizens and consumers. As a citizen, individuals certainly have some civil liberties and rights, but for the purpose of this chapter we are focussing only on the processing of personal information by the public and private sectors.

The Citizen and E-Government

In today's information-rich society governments rely increasingly on data gathering in order to perform their functions as modern nation-states (Giddens, 1990). Information about citizens, such as their birth dates, addresses, telephone numbers, fingerprints, iris scans, and even religious and political preferences is collected, stored, and processed on demand for the delivery public sector services. While a lot of this information has already traditionally been held by governments in paper form the availability and low cost of computer storage have enabled the collection of many additional pieces of information. Furthermore, the creation of vast databases allows governments to establish and exploit relational information about citizens, their social ties, as well as political and non-political interests, something that was not possible in the past. Naturally, it is easy to focus on the benefits of these developments, as they are indeed substantial. E-governance or the use of information and communication technologies (ICTs) to improve the activities of public sector organisations may significantly increase the efficiency of service delivery; facilitates access; and enhances democratic responsiveness (West, 2004). National policies on information collection also aid national security and law enforcement agencies in combating terrorism and crime prevention.

Information inclusion in government-managed databases used for e-service delivery can be seen as a key determinant in decision-making processes. Unlike information inclusion, information exclusion is usually done implicitly and can serve as proof of insensitivity towards the excluded information. Both processes affect evaluative judgements made on the basis of available information. Thus the categories of data held by governments, and blanket data collection in particular, could have a direct impact on the objectivity and fairness of citizen-state relations. Offering adequate informational privacy and data protection safeguards at the national and international levels is a reasonable counter-measure to prevent misuse and the problems that may arise from security problems.

The emergence of safeguards is part and parcel of sophisticated policy deliberations in many countries. Since the data breach in November 2007 in the United Kingdom, where the Revenue and Customs agency lost 25 million records on British families, the government has been far attentive to concerns about security abuses and negligence. In the United States, every new policy must include a 'privacy impact assessment' to audit the privacy safeguards embedded within. Proposals to create national registers are greeted generally with some concern as they create key vulnerabilities to attack and failures, aside from all the other concerns about redundancies and implementation costs.

Similarly to cybersecurity and anti-terrorism policies, some international organizations have been promoting e-government as the cure-all approach to solving both economic and social challenges in developing countries. Yet similarly,

they have failed to acknowledge and evaluate the risks of vast information collection and the possible creation of a power imbalance between citizens and the state. Under the auspices of the World Bank, developing countries have been competing in the deployment of e-government services and have so far ignored important questions related to privacy and human rights.

For instance, Sri Lanka, one of Asia's e-governance champions has created a vast e-Population Register database[4], which contains and interlinks a multitude of smaller service-specific databases holding data on immigration and emigration, pensions, foreign employment, civil registrations, national ID cards, and other. Once collected this information is available for further processing and thus highly susceptible to secondary uses.

Similarly, Pakistan has created a national smart ID card database, managed by the National Database and Registration Authority (NADRA)[5]. Currently this database holds 170 million fingerprints, 72 million facial images, and has already issued 70 million ID cards. The information stored within the system is used for highway toll collection, cash grant systems, national driver's license system, civil registrations, passport and visa issuing, passport insurance and control biometric refugee registration, and the ID cards and access control system for the Pakistani army. The mere scale of this database raises questions about access management, security, and interoperability, but these have yet to be answered. A centralised pool of information may be an invaluable tool to any government; especially one trying to raise its people out of poverty. But given how high the stakes are, we need to ask ourselves whether adopting a one-sided approach and focusing on the benefits alone will not harm those same people in the long run.

As mentioned earlier, 'the West' has historically offered human rights protections particularly in response to governments' use of personal information. However, the proliferation of legislative tools enabling governments to command the collection and retention of data for extended periods of time for purposes other than public service delivery has disturbed the informational power balance between governments and citizenry in recent years. Vast amounts of communications, health, travel, and financial data are made available to state agencies for different types of risk assessment and profiling not only of criminals but also of ordinary citizens, such as travellers on international flights. A good example is the UK's "identification, tracking and referral" system, which if implemented, would allow authorities to share information on vulnerable children and assess their potential for criminal activity. This is an illustrative case of secondary use of information originally collected for the protection of children vulnerable to physical and sexual abuse. Tasked with the impossible job of making evaluative judgements on what constitutes good or bad behaviour, social workers are likely to contribute to a whole new set of problems, such as stigmatization of children and an increased risk of social exclusion. Systems like these have been abused by internal staff members[6], third parties[7] police and intelligence agencies, and employers[8], amongst others. These abuses have been experienced in countries with e-government schemes like these, and so the response has been to ensure that there are clear and strong safeguards against abuse. If democratic governments, who are in theory less likely to abuse human rights, are implementing safeguards, why is it we do not demand the same of less stable governments overseeing even greater information resources?

Furthermore, when introducing data-rich e-services, politically fragile states are at even greater risk of creating conditions for ethnic, religious, and sexual discrimination. The existence of databases containing biometrics of refugees, child soldiers, rape victims, or HIV-AIDS positive people creates opportunities for discrimination and social exclusion. Access to such information could also be easily abused in conflict zones where people of a particular ethnic, religious or political

background become target of oppressive governments or extremist groups in the future. Yet the great irony is that the 'Western' institutions are promoting the use of these techniques and technologies in these contexts, in ways that not only introduce vulnerabilities, but using means that would not be acceptable in the 'West'. For instance, a database of HIV-AIDS patients would likely be deemed discriminatory in accordance with the European Convention on Human Rights, and the collection of fingerprints of an entire population of individuals would likely be dismissed in the North American as illiberal. But governments in both regions are promoting these practices in more precarious environments.

Consumers

Since the advent of Internet commerce, the protection of personal information has been considered integral to the advancement of e-commerce and e-trading. More recently, concern about the protection of personal information has spread to more general online conduct, where privacy policies and practices on social networking sites and online advertising have generated more and more user concern. Privacy laws, and information privacy laws in particular, protect the rights of consumers when they feel that their data is at risk or has been wrongfully collected and abused. There are two leading regulatory approaches to this problem.

- In the U.S. there is a loosely regulated and sectoral regulatory system. In commerce, companies post a privacy policy, and if they do not follow that policy they are in breach of laws protecting consumers against unfair or deceptive practices.
- In other countries, including European Union member states, Australia, Canada and New Zealand, there are 'comprehensive' privacy laws that place burdens upon companies to abide by the law, which is

enforced by a national (and some state/provincial) data privacy regulator.

If a consumer in any of these countries with either type of regulatory system believes there has been an abuse and harm to his or her privacy, then he or she may appeal to the authorities for assistance. The most striking differences between the two regimes are the ease with which a consumer may object to how his or her personal information is processed, and the powers and accessibility of the regulator.

In countries 'outside' of the 'West', consumers are increasingly engaging in electronic transactions. The growth of electronic commerce is likely to come to China and developing economies in Asia, Africa, and Latin America. Social-networking has already spread extensively in those regions as well, often based on services that are run by foreign companies (often based in the U.S.). What is odd, however, is that European and U.S. law offers protections (somewhat more limited in the latter case) to the information of European or American consumers but not to that of Chinese consumers, for example.

As all countries are interfacing with the same type of services and the same types of technologies that introduce similar power dynamics, is it still fair to say that privacy is merely a cultural value rather than a right? It is possible that we are over-simplifying the 'cultural value' argument? Polling is not necessarily a good assessment of the cultural norms of a society, but it can certainly assess the current mood: one poll in 2007 in China found that 90 percent of Chinese Internet users want the 'earliest possible enactment' of a law to protect personal information.[9] Our experiences in working with partners and speaking at workshops around the world, makes us believe that while there are varying conceptions of what privacy, there is a shared concern about the abuse of personal information by telemarketers, fraudsters, and other malicious agents.

As part of our research programme over the past year, we have spoken with legislators and policy-makers in the Philippines, Thailand, India, Pakistan, and Malaysia who have all been able to point to draft laws on information privacy. Some of these countries even explicitly mention privacy in their constitutions, though the follow-through protections are lacking.

But not all legislative initiatives are built equally. We reviewed a wide variety of draft legislation and initiatives in India, Pakistan, and the Philippines. The industry and government-backed proposals were primarily focussed on protecting the outsourcing community, i.e. the companies that provide tele-centres and data-warehousing facilities for Western companies. That is, because of European law and U.S. consumer pressures, there are concerns about personal information of 'Western' consumers being sent for processing to Asia where that data can be abused. To enhance consumer confidence and assuage regulators' fears, developing economies in Asia are keen to implement regulations to protect the personal information of these foreign consumers. However, these policy-makers and industry players are less keen to protect the personal information of domestic consumers, and often arguing that there isn't a culture of privacy in their country. While there is some momentum behind a comprehensive privacy law in many of these countries, the more powerful stakeholders have yet to lend their support. As a result, domestic consumers are left without protections from the same abuse that Western consumers are protected from.

China has been drafting a law since 2003, proposed in response to intrusive technology and marketing activities; the need for promoting the outsourcing economy; and the need for protection of financial data by credit companies. A 2005 draft law also applied these protections to personal information held by the public sector. Treacy and Abrams (2008) note that the momentum for a privacy law is certainly based on consumer and economic protections, rather than on concerns about fundamental human rights. They go on to dismiss any enthusiasm for a privacy law in China as they contend it is no longer imminent because of a change of policy-makers. But more recent news reports have pointed to legislative changes that criminalise the sale or unlawful use of personal data by government and industry, provincial laws to address internet privacy concerns, and the expansion of Tort law. This would however mean that despite the loss of momentum for the introduction of a privacy law within the political environment, there is certain eagerness displayed by the consumer movement. Furthermore, we need to urgently reconsider the logic of Chinese consumers' privacy not being granted protections equivalent to those of consumers from elsewhere in the world, especially given that they all face the same types of pressures and threats.

ANALYSIS

We can certainly sidestep the human rights argument by using a contextual approach: citizens and consumers around the world are increasingly facing the same technological environment, the same public sector interface, and the same economic environment. With the increased use of information technologies, and the emergence of a consumer-oriented middle class in many countries outside of the 'West', we can now start asking the critical question: why are some citizens and consumers protected whilst others are not?

And even if research were to show that there are indeed nuances in the perceptions of privacy amongst cultures, granting different protections to different human rights might prove a dangerous undertaking. This may give rise to a complete about-turn on other rights: informational privacy is directly linked to other fundamental rights, such as the right to freedom of expression, the right to liberty, the right to freedom of thought, conscience and religion, the right to freedom of assembly and freedom of association. If ignored,

this might impede citizen participation in political processes and equally hurt grass roots democracy or even prevent it from occurring.

Meanwhile, such a scenario would be relatively unthinkable in the field of security. How could we possibly secure less than half the population of the world from global threats? Since all countries are facing similar security threats; we must reach for the same security solutions, particularly those preferred by international conventions. This is a settled view amongst most key stakeholders. This settlement is particularly true in the realm of cybercrime, where now industry sees the Cybercrime Convention as a benchmark for countries around the world.[10] There is no talk of cultural values or cultural relativism in such affairs. In fact, generally, in international policy-making the expansion of the security agenda is considered a task for harmonisation and international agreements.

Yet on privacy issues, there is much contestation on the task of harmonisation and the role of international agreements. U.S. industry generally disagrees with the European approach to regulating informational privacy, particularly as it is based on a foundation of human rights and thus is seen as overly restrictive (even though in practice European business does not seem unduly disadvantaged). Governments, on the other hand, are also not so keen on discussing privacy as a human right because it limits their policy initiatives and creates burdens upon the governments to implement sufficient safeguards, protection and oversight mechanisms.

Therefore, although the policy environments are quite similar with very similar stakeholders in both the security and privacy agendas, the political dynamics of the two could hardly be more different. But if we look at this situation with the frame in which we opened this chapter, these policy environments are both creatures of globalisation and modernity. That is, we have trans-border data flows; information and people are placed at risk by often foreign actions; and there are economic incentives to providing a more amenable operating environment for citizens, consumers, and industry. To create policy tools to deal with only one of these policy agendas is therefore quite odd.

The reliance on the 'cultural relativism' argument only seems to affect criticism of privacy and human rights safeguards. This is an oddity as well because the security agenda must also be subject to local adaptation. Those who promulgate the 'values' argument presume that local adaptation is necessary and good. As we advance the security agenda we cannot be blind to the advancing of domestic interests. After all, these governments are not immune to their own political interests. It is therefore of little surprise that despite the Council of Europe's commitment to rights and freedoms, the Pakistani Ordinance on Cybercrime, promoted as being consistent with the Convention on Cybercrime, goes well beyond the Convention. The 2007 version of the ordinance contended that:

Whoever commits the offence of cyber terrorism and causes death of any person shall be punishable with death or imprisonment for life' and the criminalisation of communicating 'obscene, vulgar, profane, lewd, lascivious, or indecent language, picture or image'. Ironically, the Pakistani ordinance does include a privacy provision. Under the offence of 'cyber-stalking', it says that it is an offence to coerce or harass if you 'take or distribute pictures or photographs of any person without his consent or knowledge; or (e) display or distribute information in a manner that substantially increases the risk of harm or violence to any other person.[11]

When the ordinance was reintroduced in 2008[12] by the new government, the offending articles remained unchanged.[13] When we tell our colleagues in Pakistan that these components are not in the Convention of Cybercrime, they are shocked. When U.S. and European companies operating in these countries are compelled to hand over information on consumers that may result in life sentences or worse, it is unlikely they will continue to celebrate the security agenda in that country, or that country's 'right' to determine for itself how to interpret international conventions.

CONCLUSION

The spread of the security agenda has occurred in response to shared threats brought upon by modernisation and globalisation. The emergence of privacy is similar. Yet the spread and expansion of the security agenda has been enabled through international policy-making and conventions in ways that far outpace that of privacy. The most commonly preferred reason for this distinction is that privacy is a cultural issue. We are left to draw the conclusion that security is universally valid.

Failing to protect an economy and society from modern threats would be negligent, and so governments have eagerly approved and adopted international conventions on anti-terrorism, crime, and security. But there is no felt need to protect citizens and consumers from modern threats to their privacy. We explained this as a result of the different political environment between the security and privacy agendas. After all, if it was just because of the differences in cultural values about privacy and human rights, we would have to presume that the citizens of China, the Philippines, and Pakistan are not at all concerned with the use and abuse of personal information by the private sector or government departments. But knowingly or unknowingly, personal information is placed at risk and citizens and consumers outside of the 'West' have no rights of recourse.

Our conclusion is that if we accept that we all face the same threats, and that international policies and standards are useful benchmarks for responding to these threats, then it is negligent for governments and companies to lobby against the protection of privacy on cultural grounds alone.

This chapter was an attempt to avoid the debate about privacy as a human right and show that we don't need to conceptualise it as a human right in order to see its universal application. We reversed the situation and argued that the benefits gained from the protection of privacy should be applied equally around the world, and not just to a class of people (based on jurisdiction). Therefore, we conclude that it is perhaps easier to argue for the universality of safeguards and protections rather than to engage in debates about cultural relativism and the definition of human rights and privacy.

Despite this, however, the debate about human rights cannot be avoided. Universal protections for personal information need not necessarily be based on individual rights, but for these protections to be heeded and valued there is a need for a sense of the self.

An individual in Asia who is constantly interrupted by telemarketing calls, or whose information is lost by his or her bank, or has his or her personal information stolen and used by fraudsters must be given some form of protections against further abuse. Security laws will go after the malicious perpetrators, and privacy law would certainly enable subsequent protections against abuse. But these are just theoretical notions based on the premise that the individual feels a sense of violation and has a sense that something can and must be done. Without that, the mandate of regulatory bodies would be irrelevant, the roles of courts superfluous, and parliamentary debate would be merely philosophical. Unless the individual feels that personal information is in a sense his or hers, its subsequent use and abuse would be inconsequential to him or her.

This 'ownership' over personal information is certainly not universally recognised, even by Europeans who are endowed with the 'right' to privacy. Without a sense of ownership and concern about disclosure and violation, this discussion about security or privacy would be relatively irrelevant. In recent years, however, this domain has been improving. Previously the concern about credit card security gave rise to some concern, but in more recent times the rise of reputational privacy

online (through social networking) and concerns about identity fraud (where personal information is abused for financial and reputational value) individual users have grown more interested in their personal information and profiles. This awareness also came just as governments and companies began collecting more and more personal information, sometimes without consent, and particularly in some high profile cases, they began losing this information. The growing sense of indignation is not a result of an article within a convention on human rights, but is rather an emerging and modern conceptualisation of ownership and new types of concerns about trust. It is in this type of framework, rather than one that repeats old debates about rights and values, that we must start asking about the universality of our policies.

Yet, even though we may situate this entire debate in the 'modern era,' and in response to the 'security agenda,' we must also remember the importance of the older debates. For instance, in July 2009 the Indian High Court in Mumbai issued its opinion in the case of whether it should overrule the ban on homosexual intercourse. Section 377 of the Indian Penal Code was a construct of English law, and the Government argued that it was necessary based on Indian values:

In our country, homosexuality is abhorrent and can be criminalised by imposing proportional limits on the citizens' right to privacy and equality. ... the right to privacy is not absolute and can be restricted for compelling state interest. Article 19(2) expressly permits imposition of restrictions in the interest of decency and morality. Social and sexual mores in foreign countries cannot justify de-criminalisation of homosexuality in India.[14]

The Court reviewed Indian jurisprudence on the right to privacy. Although it is not in the constitution, jurisprudence from the 1960s onwards established privacy as a fundamental right under Article 21 on the right to liberty. The Court then reviewed the jurisprudence of foreign courts, including the U.S. and Europe to show that banning homosexual intercourse was an unjust interference with privacy. The Court then concludes that:

It is not within the constitutional competence of the State to invade the privacy of citizens lives or regulate conduct to which the citizen alone is concerned solely on the basis of public morals. The criminalisation of private sexual relations between consenting adults absent any evidence of serious harm deems the provision's objective both arbitrary and unreasonable. The state interest 'must be legitimate and relevant' for the legislation to be non-arbitrary and must be proportionate towards achieving the state interest. If the objective is irrational, unjust and unfair, necessarily classification will have to be held as unreasonable. The nature of the provision of Section 377 IPC and its purpose is to criminalise private conduct of consenting adults which causes no harm to anyone else. It has no other purpose than to criminalise conduct which fails to conform with the moral or religious views of a section of society. The discrimination severely affects the rights and interests of homosexuals and deeply impairs their dignity.

This case has nothing to do with 'modern' developments or globalisation; rather it was a classic questioning of the rights of individuals and the conflicts of values. This court decision came down in favour of the right to privacy without regard to any statements about 'cultural values' or even polling data. We thus we return inevitably to the question whether there is such a thing as a human right, and whether it is universal. The Indian High Court ruled that it has nothing to do with region, class, economy, or social attitudes. It is about the protection of individuals, and has less to do with modernity and globalisation than with individual dignity.

REFERENCES

Council of Europe. (2007). *Countries worldwide turn to Council of Europe Cybercrime Convention.* Press Release 413(2007), June 13, 2007.

Donelly, J. (2003). *Universal human rights in theory and practice* (2nd ed.). Cornell University Press.

Giddens, A. (1990). *The consequences of modernity.* Stanford: Stanford University Press.

Hosein, I. (2004). The sources of laws: Policy dynamics in a digital and terrorized world. *The Information Society, 20*(3), 187-199. ISSN: 0197-2243

Jervis, R. (1982). Article. *International Organization, 36*(2).

Krasner, S. D. (1983). *International regimes.* Cornell University Press.

Perri, G. (2004). *E-governance: Styles of political judgement in the information age polity.* New York: Palgrave MacMillan.

President promulgates ordinance to prevent electronic crimes. (2008, November 6). *Associated Press.* Retrieved from http://www.app.com.pk/en_/index.php?option=com_content&task=view&id=58277&Itemid=1

Sen, A. (2001). *Development as freedom.* Oxford: Oxford University Press.

Treacy, B., & Martin, A. (2008, May 29). A privacy law for China? *Complinet.*

West, D. M. (2004). E-government and the transformation of service delivery and citizen attitudes. *Public Administration Review, 64*(1), 15–27. doi:10.1111/j.1540-6210.2004.00343.x.

Xinhua. (2007, November 19). Nine in ten Chinese want law to protect personal information enacted soon. *Chinaview.cn*. Islamic Republic of Pakistan. (2007). *Cybercrime Ordinance of the Islamic Republic of Pakistan*. Islamic Republic of Pakistan. Khan, M. (2008, November 7). Pakistan unveils cybercrime laws. *BBC News*.

ENDNOTES

[1] See for instance http://www.quintessenz.at/etsi/etsi_intro.htm.

[2] For an example, see the Council of Europe's Project on Cybercrime report on the Republic of the Philippines, February 2008, available at
http://www.coe.int/t/dghl/cooperation/economiccrime/cybercrime/Documents/CountryProfiles/567-LEG-country%20 profile-Philippines_5feb2008_En.pdf

[3] For a recent articulation of these points, see Jay Kline's, *Opinion: You Say 'shameful Secret', I Say 'privacy'*, CIO.com, June 29, 2009 where he points to scholarship on China and Japan.

[4] For more information: http://web.worldbank.org/wbsite/external/topics/extinformationandcommunicationandtechnologies/0,,contentMDK:22187853~menuPK:2644022~pagePK:64020865~piPK:51164185~theSitePK:282823,00.html

[5] For more information: http://web.worldbank.org/wbsite/external/topics/extinformationandcommunicationandtechnologies/0,,cont entMDK:22142712~menuPK:2644022~pagePK:64020865~piPK:51164185~theSitePK:282823,00.html

[6] See the story on the US State Department passport database breach: http://news.bbc.co.uk/1/hi/world/americas/7309165.stm

[7] See the ICO *What price privacy now?* report on unlawful trading of personal information: http://www.ico.gov.uk/about_us/news_and_views/current_topics/what_price_privacy_now.aspx

[8] See the story on employee surveillance by Deutsche Bahn (http://news.bbc.co.uk/1/hi/business/7887017.stm); on staff spying by Airbus (http://news.bbc.co.uk/1/hi/business/7978713.stm), and by the German supermarket chain Lidl (http://www.spiegel.de/international/germany/0,1518,544372,00.html).

[9] See Xinhua (2007, November 19).

[10] e.g. Microsoft (2008) Asia pacific legislative analysis: Current and pending online safety and cybercrime laws. Retrieved from http://www.microsoft.com/asia/

[11] See Islamic Republic of Pakistan (2007).

[12] See President promulgates ordinance to prevent electronic crimes (2008, November 6).

[13] See Khan (2008, November 7).

[14] Statement of the Additional Solicitor General, quoted in the opinion *Nas Foundation vs. Government of NCT of Delhi*, WP(C) No.7455/2001, Decided July 2, 2009 in the High Court of New Delhi.

Section 3
Fields of Application (1): Genes, Atoms, Nanos

Section 3: Fields of Application – Genes, Atoms, Nanos

The idea of an ethical governance of emerging technologies can be examined on the basis of some examples taken from several key technological fields: genes, atoms, nanos. These examples of emerging technologies were chosen precisely because, together with the ICTs, they are among the most strategic domains in contemporary research and innovation. In addition, these technologies have significantly impacted the material and social basis of markets and societies in many developed (or less developed) countries.

First, the *gene technologies* are well known for being at the origin of a series of controversies - from the cloning of animals and humans, after that of bacteria and plants, to the genetic DNA recombination, then leading to the Genetically Modified Organisms (GMOs). It is pretty obvious that, despite a fierce opposition of the populations, especially in Europe, the governance of biotechnology has remained quite monopolised by some legal experts producing sometimes *ad hoc* regulations and by political institutions resorting sometimes to the precautionary principle. The regulators often refer to some advices and recommendations as elaborated by other experts (lawyers, physicians, sociologists, anthropologists, philosophers, …) belonging to ethics committees, but for some reason, they hardly listen to the "voice of the people."

Second, the *nuclear technologies* are also well known for long raising strong oppositions—from the development of military devices, such as atomic bombs, to that of civil devices, such as nuclear plants. Yet, perhaps more than in other technologies, and despite the tragedy of Fukushima, some years after Chernobyl, nuclear technologies are still monopolised by a community of experts, some of them still quite reluctant to open the discussion to the people. The case of the management of radioactive wastes coming mainly from the power production in nuclear plants is interesting insofar as, for reasons of radioactivity duration, it raises the issue of the long term in the governance of ethics, especially regarding problems of responsibility, justice, and democracy. It is also interesting by the use in this field of a method of cooperative research in a joint project gathering international experts and lay people in a process of co-production of ethical norms.

Third, the *nano-technologies* offer another example of an on-going process of research and innovation that is raising lots of contests, even if the controversy about them is not so advertised by the media as in some other fields. It might be that some inherent factors are likely to illuminate the socio-technical trajectory of the 'nanos', like the unusual size of the nano-elements, the lack of information and inquiries about their actual toxicity, and the strategic opportunity for some countries to take the lead in this field. Nevertheless, this field is of high interest in that it shows as far as normative regulation is concerned a variety of models of ethical governance and then does not restrict the ethical governance of technology to one single model.

Chapter 6
Lessons from the International Governance of Biotechnology

Catherine Rhodes
University of Manchester, UK

ABSTRACT

This chapter covers the development and operation of norms within the international system and more specifically within the international governance of biotechnology. Through the use of case studies it highlights several key points which should be considered when analysing the role, application and implementation of norms in other governance areas.

INTRODUCTION

This chapter examines the governance of biotechnology at the international level, looking at how norms have been incorporated in its development and how they are being used in emerging governance efforts. First, some particularities of governance at the international level are outlined to provide background and contextual information for understanding the governance of biotechnology. Case studies will be used to highlight particularly pertinent issues and processes in relation to ethical governance.

GOVERNANCE AT THE INTERNATIONAL LEVEL

There are differences between international, regional and national governance which may mean that some of the lessons from the international governance of biotechnology will not apply directly at these other levels. This chapter begins by outlining some key points about the international system and how governance operates within it, to set the context.

DOI: 10.4018/978-1-4666-3670-5.ch006

System Characteristics

The international system is characterised as anarchical because there is no supranational authority to govern it. The most important actors in international relations are states. Other actors such as international organisations, non-governmental organisations and multinational corporations are having increasing influence on international processes, but states remain dominant. They create and are the subjects of international law and are still the key decision-makers in international governance.

In the absence of supranational authority, the international system is shaped partly by norms and rules, but power relations play a significant role and the pursuit of power (particularly in military and economic terms) is still a core part of conceptions of national interest. Powerful states dominate the direction and content of international governance. For example, in law-making they have more influence on the creation of rules; can persuade or pressure other states to comply (or not) with rules; and can afford to ignore certain rules.

Powerful states can, based on their superior resources, dominate the processes of international law-making. They have the resources to be able to enforce the laws that they support and bear the costs of attempted enforcement action by others. (Rhodes, 2010, p. 55)

International treaties are unlikely to be agreed without the support of the main powers. Not surprisingly then, the rules and governance arrangements that exist tend to favour the interests of dominant states.

A Place for Ethics?

A major strand of international relations theory – realism – was dominant for most of the second half of the twentieth century and still retains some policy influence today. It holds that there is no place for moral considerations in international relations. Its arguments in this regard follow the lines of:

- Human nature as self-interested and power-hungry scales up to the international level.
- States, motivated solely by self-interest, are the only significant actors in the international system.
- States' interests are their own survival and the survival of the system.
- In the anarchic international system, the balance of power is the ordering factor.
- States' interests can therefore only be achieved through pursuit of power (for their own survival) and consideration of the balance of power (for survival of the system).
- Appropriate foreign policy will be determined solely by considerations of power and power relations.
- The inclusion of moral considerations in foreign policy making distracts from a focus on power and will lead to flawed decision-making, threatening state survival and the stability of the system.

Realism developed as a response to (what it labelled as) 'idealist' international theory, which held that a more peaceful and just international society could be created through the establishment of international institutions, norms and rules. This thinking prompted the formation of the League of Nations in 1919 and can be seen in the wording of its charter – the *Covenant of the League of Nations* – which mandated its member states to:

Achieve international peace and security by the acceptance of obligations not to resort to war, by the prescription of open, just and honourable relations between nations, by the firm establishment of the understandings of international law as the actual rule of conduct among Governments, and by the maintenance of justice and a scrupulous respect for all treaty obligations in the dealings of organised peoples with one another.

Realists blamed such thinking for the outbreak of the Second World War. The expectation that states, based on moral considerations such as justice and the prevention of human suffering, would cooperate to avoid war, had led to a failure to consider power dynamics and to correct a power imbalance. As argued by Hans Morgenthau, a leading realist theorist:

Neville Chamberlain's politics of appeasement were, as far as we can judge, inspired by good motives... he sought to preserve the peace and ensure the happiness of all concerned. Yet his policies helped to make the Second World War inevitable and bring untold misery to millions. (1978, p.6)

There are several arguments for dismissing the realist demand to exclude moral obligations from international relations. To start with, the choice to consider power alone was itself based on moral considerations, and though to be the 'right' course of action:

There can be no political morality without prudence; that is without consideration of the political consequences of seemingly moral action. Realism then, considers prudence – the weighing of the consequences of alternative political actions – to be the supreme virtue in politics. (Morgenthau, 1978, p.12)

State survival is not a sufficient goal for foreign policy. A state ought not to be considered successful in its international relations if it has military power and can maintain territorial integrity but has left its population to starve, has weak or oppressive governance systems, etc. Indeed such states are generally labelled as pariah or failed states. This suggests that national interest can be more broadly conceived to take into account

matters such as maintenance of public health, food security, environmental stability, and good governance. This broader conception of national interest has a clear place for moral considerations because, for example, consideration of justice and equity may produce better outcomes in relation to these goals.

Neither should national interest be conceived of as purely 'national'. The international system has been transformed by deep interdependencies created by global dynamics and by increased awareness of interconnections between groups and states. Neglect of common interests has global effects that can impact negatively on all states; therefore international cooperation will often be in the national interest and will require that interests of other states and actors be taken into account.

Finally, there are increased demands for moral action at the international level, assisted by increased capacities for groups across the world to see what is happening elsewhere, communicate about their situations, and to articulate combined demands. This includes demands for states to incorporate moral considerations in their foreign policies both in response to one-off events (e.g. natural disasters) and on a long-term basis (e.g. debt reduction). (These demands do not necessarily result in moral action, but they do place pressure on states to consider wider effects of political action and to give moral justifications for their policies.) Policies based solely on power politics are now viewed as less legitimate, even among other states – this can be seen, for example, in international reactions to the holding of detainees at Guantanamo Bay and policies of 'extraordinary rendition' (Council of Europe, 2006; UN Economic and Social Council, 2006).

EARLY DEVELOPMENT OF INTERNATIONAL NORMS

The development of norms within early international law making is closely connected to their use within current biotechnology regulation, not least because some of those rules have their origins in these early efforts. Here the term 'international norm' is used to describe accepted standards of conduct for state behaviour in their international relations. These usually take the form of prescribed rules or generally agreed principles.

There are two main views (stemming from international legal theory) on where international norms emerge from. The first, connected to natural law theory holds that norms have a moral basis and guide what *ought to be* (how states ought to act). They, and the rules based on them, can be rationally deduced from a pre-existing universal basis (Peters, 2001, p.27). The second view, connected to legal positivism holds that international norms develop from state practice and reflect what *is*. International governance appears to reflect a combination of the two: there are rules based on state actions and what they have agreed to (customary and treaty law); and there are well-established and widely accepted concepts of universal principles, rights, and obligations that inform state practice (e.g. human rights).

The idea of there being an international system or community of states did not develop until the mid-seventeenth century, when the concept of nation-states as sovereign independent units became established. Once this had happened, states developed various customs and rules to facilitate transactions between themselves.

In the early stages of their development, international norms and international rules were very closely connected. Most of this early development focused on the conduct of warfare, with the aims of reducing human suffering and establishing a more peaceful world order. For example, rules dealing with the treatment of wounded soldiers and the unacceptability of particular weapons began to be codified in the late nineteenth century, alongside the establishment of the International Committee of the Red Cross.

Among the earliest examples are the:

- 1864 Convention for the Amelioration of the Condition of the Wounded in Armies in the Field.
- 1868 Declaration Renouncing the Use in Time of War of Explosive Projectiles Under 400 Grammes Weight.
- 1880 Laws of War on Land.

The number of these declarations expanded rapidly in the early 1900s (the ICRC has an online database of international humanitarian law which includes the early agreements – www.icrc.org/ihl). The principles these declarations outlined form the basis of current international humanitarian law (also known as the laws of war) – the most well-known strand of which is the Geneva Conventions.

Establishment of these norms and rules did not prevent the outbreak of major global conflict and appears to have had limited effect on its conduct – there was, for example, widespread use of chemical weapons in the First World War, despite a prohibition on use of poison being outlined in declarations such as the 1880 Laws of War on Land and 1907 Hague Regulations Concerning the Laws and Customs of War on Land. This may partly be because most states considered the rules to apply only to first-use, i.e. an initial attack by an enemy justified retaliatory use. But it is also related to military expediency and the lack of practical recommendations to guide application of the norms and monitor implementation of rules.

There were further significant developments in international norm creation after the Second World War. This also occurred in response to a desire to avoid repetition of the horrors of warfare, and included the establishment of limited circumstances in which the use of force in international relations is considered to be legitimate and the statement of protection of human rights and fundamental freedoms as a core duty of states and of the United Nations. The UN Charter of 1945 begins with the following statement:

We the peoples of the United Nations

Determined

to save succeeding generations from the scourge of war, which twice in our lifetime has brought untold sorrow to mankind, and

to reaffirm faith in fundamental human rights, in the dignity and worth of the human person, the equal rights of men and women and of nations large and small, and

to establish conditions under which justice and respect for the obligations arising from treaties and other sources of international law can be maintained.

As the number and type of international transactions grew – because of processes such as expanding trade, travel and communications – norms and rules developed in other areas of concern, such as disease control, environmental protection and the reduction of barriers to trade.

State Sovereignty

State sovereignty is the overarching principle of the international system. Even though the interdependence of states has increased rapidly and the need for cooperation on issues of common concern is recognised, state sovereignty has remained central to international relations and is continually reinforced by state practice and within international law-making.

Weiss and Hubert (2001) provide the following definition of state sovereignty:

State sovereignty denotes the competence, independence, and legal equality of states. The concept is normally used to encompass all matters in which each state is permitted by international law to decide and act without intrusions from other states. These matters include the choice of

political, economic, social and cultural systems and the formulation of foreign policy.

State sovereignty is enshrined in the UN Charter and restated in many international treaties, including several of those relevant to the governance of biotechnology, for example:

States have in accordance with the Charter of the United Nations and the principles of international law the sovereign right to legislate and implement legislation. (Article 3.4 of the International Health Regulations)

States have in accordance with the Charter of the United Nations and the principles of international law, the sovereign right to exploit their own resources pursuant to their own environmental policies. (Article 3 of the Convention on Biodiversity)

The Contracting Parties recognize the sovereign rights of states over their own plant genetic resources. (Article 10 of the International Treaty on Plant Genetic Resources)

State sovereignty is connected to other important principles including:

- **Non-Interference:** No state can legitimately interfere in the affairs of another, except in very limited circumstances (i.e. the United Nations Security Council declaring a threat to international peace and security under Chapter VII of the UN Charter).
- **Sovereign Equality:** All states are nominally equal in international processes, such that they, for example, all have one vote in most inter-governmental forums.
- **Legitimate Expectation:** States are only bound by treaties to which they have consented, and
- **Pacta Sunt Servanda:** States should keep to agreements they have made.

These principles are, for example outlined in Article 2 of the UN Charter and in the Vienna Convention on the Law of Treaties.

BRIEF INTRODUCTION TO THE INTERNATIONAL GOVERNANCE OF BIOTECHNOLOGY

Definitions

The following definitions are used in this chapter.

- **International Governance:** The subset of global governance that refers to the international i.e. that occurring between states.
- **Global Governance:** Includes the rules, norms, institutions, procedures and mechanisms that order the behaviour of states and other international actors in the absence of a supranational government. A complementary definition, outlined by Finkelstein (1995, p.368), is: "any purposeful activity intended to 'control' or influence someone else that either occurs in the arena occupied by nations or, occurring at other levels, projects influence into that arena."
- **International Regulation:** Rules agreed between states to govern their relations and actions. The term covers both the 'hard' and 'soft' aspects of international law – that is both legally-binding treaties and voluntary standards, guidelines and declarations. As used in this chapter, it is limited to rules that are potentially universal, i.e. open to any state without restriction on geographic, economic, or other grounds.
- **International Organisations:** The term is used here to refer only to inter-governmental organisations and those which have potentially universal membership. They are often associated with particular rules.

Key Roles for Biotechnology Regulation

International regulation of biotechnology needs to fulfil the general role of international law – coordination of state behaviour – through functions such as: facilitating cooperation; imposing constraints; establishing and shaping expectations; dealing with common threats; and channelling conflict and providing mechanisms for its resolution (Rhodes, 2010, p. 57-58). More specifically, there are four roles that biotechnology regulation needs to serve (these will apply at local, national and regional levels as well):

1. Promotion of benefits.
2. Identification, assessment and management of risks.
3. Prevention or minimisation of negative impacts.
4. Promotion of capacity-building.

Some of the governance of biotechnology needs to take place at the international level because, among other reasons:

- Biotechnology as a scientific endeavour is global – knowledge, materials, equipment and people move rapidly across the globe and are not limited by national boundaries.
- It has applications and impacts in several areas of international concern – areas in which there is a high degree of interdependence between states and where separate actions will be inadequate to achieve common objectives.
- The global context is one of great inequalities and in the management of biotechnology a particular concern is that inequalities will be entrenched unless action is taken to build scientific, technological and regulatory capacities. This action is more likely to be effective if it is coordinated at the international level.

The World Health Organisation outlined such concerns in *Genomics and World Health* (2002, p.102):

If, as seems likely, genomics does produce major benefits for health, the lack of biotechnological expertise in the pharmaceutical industry in the developing world will lead to a major exacerbation of the inequalities of health care among different countries.

The Regulations

There are thirty-seven international regulations relevant to the governance of biotechnology and fifteen international organisations directly associated with them. Here, they are categorised within six issue areas: arms control; health (covering rules on disease control, laboratory biosafety and biosecurity, and food safety); environmental protection; trade (covering rules for free trade, protection of intellectual property rights, and access to genetic resources); drugs control; and social impacts (particularly related to human genetics).

Arms Control

- Rules concerned with preventing the misuse of scientific advances in causing deliberate harm to humans, animals or plants:
  - 1925 Geneva Protocol for the Prohibition of the Use of Asphyxiating, Poisonous or Other Gases, and of Bacteriological Methods of Warfare.
  - Biological Weapons Convention (Convention on the Prohibition of the Development, Production and Stockpiling of Bacteriological (Biological) and Toxin Weapons and on their Destruction).
  - Convention on the Prohibition of Military or any Hostile Use of Environmental Modification Techniques.
  - Chemical Weapons Convention (Convention on the Prohibition of the Development, Production, Stockpiling and Use of Chemical Weapons and on their Destruction).
- Organisation
  - Organisation for the Prohibition of Chemical Weapons.

Health

- Rules on human, animal and plant disease control:
  - International Health Regulations 2005.
  - Terrestrial Animal Health Code.
  - Manual of Diagnostic Tests and Vaccines for Terrestrial Animals.
  - Aquatic Animal Health Code.
  - Manual of Diagnostic Tests for Aquatic Animals.
  - International Plant Protection Convention.
- Rules on laboratory biosafety and biosecurity to protect against accidental or deliberate release of pathogenic organisms:
  - Provisions in the Terrestrial Animal Health Code and Manual.
  - Laboratory Biosafety Manual.
  - Biorisk Management: Laboratory Biosecurity Guidance.
  - Guidance on Regulations for the Safe Transport of Infectious Substances.
- Rules on food safety for the protection of consumers' life and health:
  - Principles for the Risk Analysis of Foods Derived from Modern Biotechnology.
  - Guideline for Food Safety Assessment for Foods Derived from Recombinant DNA Plants.

- ○ Guideline for Food Safety Assessment for Food Produced Using Recombinant DNA Microorganisms.
- ○ Guideline for Food Safety Assessment for Foods Derived from Recombinant DNA Animals.
- Organisations:
  - ○ World Health Organisation.
  - ○ World Animal Health Organisation.
  - ○ Food and Agriculture Organisation.
  - ○ Codex Alimentarius Commission.

Environmental Protection

- Rules for the conservation and sustainable use of biodiversity:
  - ○ Convention on Biodiversity.
  - ○ Cartagena Protocol on Biosafety to the Convention on Biodiversity.
- Organisation:
  - ○ Convention on Biodiversity Secretariat.

Trade

- Rules for the reduction of barriers to trade:
  - ○ Agreement on Technical Barriers to Trade (TBT Agreement).
  - ○ Agreement on the Application of Sanitary and Phytosanitary Measures (SPS Agreement).
- Rules for the protection of intellectual property rights:
  - ○ Agreement on Trade Related Aspects of Intellectual Property Rights (TRIPS Agreement).
  - ○ Patent Cooperation Treaty.
  - ○ Patent Law Treaty.
  - ○ Budapest Treaty on the Deposit of Microorganisms for the Purpose of Patent Procedure.

- ○ International Convention for the Protection of New Varieties of Plants.
- Rules facilitating access to genetic resources and benefit-sharing from their use:
  - ○ International Treaty on Plant Genetic Resources for Food and Agriculture.
  - ○ Nagoya Protocol on Access to Genetic Resources and the Fair and Equitable Sharing of Benefits Arising out of their Utilization.
- Organisations:
  - ○ World Trade Organisation.
  - ○ World Intellectual Property Organisation.
  - ○ Union for the Protection of New Varieties of Plants.
  - ○ Food and Agriculture Organisation.
  - ○ Convention on Biodiversity Secretariat.

Drugs Control

- Rules to combat the illicit drugs trade:
  - ○ Single Convention on Narcotic Drugs.
  - ○ Convention on Psychotropic Substances.
  - ○ Convention against the Illicit Traffic in Narcotic Drugs and Psychotropic Substances.
- Rules against doping in sport:
  - ○ World Anti-Doping Code.
  - ○ International Convention against Doping in Sport.
- Organisations:
  - ○ United Nations Office on Drugs and Crime.
  - ○ Commission on Narcotic Drugs.
  - ○ International Narcotics Control Board.
  - ○ World Anti-Doping Association.
  - ○ United Nations Educational, Scientific and Cultural Organisation.

Social Impacts

- Declarations:
  - Universal Declaration on the Human Genome and Human Rights.
  - International Declaration on Human Genetic Data.
  - Universal Declaration on Bioethics and Human Rights.
  - United Nations Declaration on Human Cloning.
- Organisations:
  - United Nations Educational, Scientific and Cultural Organisation.
  - United Nations General Assembly.

NORMS IN THE GOVERNANCE OF BIOTECHNOLOGY

Rather than attempting to produce an exhaustive list of the different norms involved in the international governance of biotechnology and how they have developed, the rest of this chapter uses some particular cases to illustrate interesting issues and problems.

Norm Development and Rule Development

The link between norm development and rule development was particularly strong in the early development of international law. This is likely to be based on several factors including: that it was explicitly norm-driven; it was made between a limited number of similarly powerful states; it developed in a context of limited issues of international concern, so that there was less probability of conflicting norms occurring; and it had limited substantive content.

Later rules often incorporate, either explicitly or implicitly, norms from earlier rules, rather than being based on newly developed norms. That said, there has also been more recent norm development,

for example in the area of sustainable development from 1972-1992.

Evolution of rules in a particular area appears to be less about responding to changing norms, than about continually trying to amend rules in order to achieve more effective implementation of the underlying norms. For example, such a pattern can be seen in the development of rules prohibiting biological and chemical weapons:

- The 1907 Hague Regulations Concerning the Law and Customs of War on Land and other prohibitions on the use of poison and causing superfluous injury did not stop the use of chemical weapons during the First World War.
- *This prompted negotiation of the 1925 Geneva Protocol for the Prohibition of the Use of Asphyxiating, Poisonous or other Gases, and of Bacteriological Methods of Warfare.*
- The 1925 Geneva Protocol did not deter the development, production, stockpiling and use of chemical and biological weapons (before and) during the Second World War.
- *This prompted negotiation of the Biological Weapons Convention (adopted in 1972) and later the Chemical Weapons Convention.*
- The Biological Weapons Convention did not stop the development and production of biological weapons, for example by the Soviet Union, South Africa and Iraq.
- *This has prompted ongoing attempts to strengthen the Biological Weapons Convention, including through development (though not adoption) of a verification protocol, agreement on confidence building measures and adoption of UN Security Council Resolution 1540 in 2004.*

While the later rules are still, at least partially, motivated by the moral problem of human suffering caused by war / use of certain weapons, and explicitly make this connection, they develop more substantive practical content about how the prohibition on such weapons should be applied and implemented. As a rough indication of practical content: the Geneva Protocol is twelve sentences long; the Biological Weapons Convention has a preamble longer than the Geneva Protocol and fifteen articles; and the Chemical Weapons Convention has a preamble, twenty-four articles, and over one hundred pages of annexes.

Statements of ethical norms often appear only in the preamble to such rules, which – while it can influence the treaty's interpretation – is not regarded as legally-binding. Several of these norms, however, are considered to be part of customary international law, binding on all states.

When seeking to identify norms relevant to the governance of biotechnology (or any other international matter) it will be necessary to look not only at the content of the most recent rules, but also of those that have preceded them in order to gain a complete picture.

Norms and the Interests of Powerful States

International norms, and the rules which incorporate them, are more likely to have effect when they are considered to favour the interests of powerful states or regions. This links to what was said earlier about powerful states' dominance of international law-making. While, nominally, there is sovereign equality in the negotiation of international agreements, powerful states are better resourced in terms of the number of negotiators and range of expertise they can employ, and have more leverage to influence the negotiating positions of other states.

This can be seen within the governance of biotechnology. One example is the strength of the rules of the World Trade Organisation (WTO) in comparison to rules in other areas. WTO rules such as the Sanitary and Phytosanitary Agreement,

Technical Barriers to Trade Agreement and Trade Related Aspects of Intellectual Property Rights Agreement are of particular benefit and relatively little burden to powerful developed states such as the United States (US) and members of the European Union. These countries played a leading role in developing the rules' content and influenced the issues that were included and excluded. The structure of the rules also favours these states – in particular the enforcement mechanisms connected with dispute settlement procedures.

Under the WTO's Dispute Settlement Understanding (WTO, 1995 - 1) if a party to the dispute is found to be contravening the rules it must either amend its national measures or face trade sanctions equivalent to what the other party feels they have lost. This presents a severe burden to states that rely on a few key exports (predominantly developing states) because more powerful states can inflict great damage on them by implementing such sanctions, but have little to worry about if such sanctions are imposed on them by weaker states.

Because other rules that have trade-related provisions, such as the Cartagena Protocol on Biosafety, contain no such measures for enforcing compliance, countries are likely to choose to apply WTO rules in areas where they overlap. These other rules tend to receive less support from powerful states – the US, for example, is not a party to the Cartagena Protocol or its parent agreement the Convention on Biodiversity.

A second example can be found in relation to provisions on capacity-building, technical and financial assistance. Biotechnology has significant implications for development and about half of the international regulations relevant to biotechnology contain development related provisions. These provisions relate to matters such as support for capacity building and provision of scientific, technical and financial assistance, and they are designed to promote implementation of the rules by developing countries that may otherwise lack the necessary resources, infrastructure and expertise to do so. These provisions – of little perceived benefit to powerful states – are poorly implemented. Examples of development-related provisions include:

Each Contracting Party shall promote technical and scientific cooperation with other Contracting Parties, in particular developing countries, in implementing this Convention. (Convention on Biodiversity, Article 18.2)

The contracting parties agree to promote the provision of technical assistance to contracting parties, especially those that are developing contracting parties, either bilaterally or through appropriate international organisations, with the objective of facilitating implementation of this Convention. (International Plant Protection Convention, Article XX)

In order to facilitate the implementation of this Agreement, developed country Members shall provide, on request... technical and financial cooperation in favour of developing and least developed country Members. (TRIPS Agreement, Article 67)

Powerful states have also been observed to undermine such provisions. For example, the US and European Union (EU) have pressurised weaker states to give up concessions based on their economic status, such as differential treatment clauses and flexibilities in the TRIPS Agreement, during negotiation of bilateral free trade agreements (see, for example: Collins-Chase, 2008; Ovett, 2006; United Nations, 2004, July 5; Valdivieso, 2009).

The important thing to note here is that not all norms will be equal in international governance. Simply identifying the presence of a norm in a rule does not indicate that it has any practical effect. Power relations will need to be taken into account in any analysis of the operation of particular norms in international governance.

Adaptation to Scientific and Technological Developments

The biosciences and related fields that contribute to developments in biotechnology advance much more rapidly than regulatory responses do. International negotiations are time-consuming, with the agreement of legally-binding treaties generally taking at least several years. Combined with the fact that newer rules often incorporate norms from previous rules, it is not surprising that problems arise in regard to normative relevance being maintained.

A particularly clear example has arisen in relation to the rules on biological and chemical weapons. Some future development of such weapons is likely to contravene the rules but not the norms on which they are based. There is concern that this will weaken the international prohibitions.

Existing reasoning for banning biological weapons as a particular type of weapon is based on the following factors:

- They fail to discriminate between civilians and combatants
- They cannot be limited to military objectives
- They will cause unnecessary suffering and superfluous injury
- They exceed the permitted objective of disabling enemy combatants, and
- They are not proportionate to permitted military objectives

This reasoning draws on the following norms found in international humanitarian law:

- "The parties to the conflict must at all times distinguish between civilians and combatants. Attacks may only be directed at combatants."
- "Indiscriminate acts are prohibited."
- "Launching an attack that may be expected to cause incidental loss of civilian life, in-

jury to civilians, damage to civilian objects or a combination thereof, which would be excessive in relation to the concrete and direct military advantage anticipated is prohibited."

- "The use of means and methods of warfare which are of a nature to cause superfluous injury or unnecessary suffering is prohibited."
- "The use of weapons that are by nature indiscriminate is prohibited." (ICRC, no date given, Rules 1, 11, 14, 70 and 71)

And have resulted, *inter alia*, in the following prohibition: states parties to the Biological Weapons Convention are prohibited from developing, producing, stockpiling, acquiring, retaining or using:

- Microbial or other biological agents, or toxins, whatever their origin or method of production, of types or in quantities that have no justification for prophylactic, protective or other peaceful purposes;
- Weapons, equipment or means of delivery designed to use such agents or toxins for hostile purposes or in armed conflict. (BWC, Article 1)

Scientific advances point to future biological, biochemical and pharmaceutical agents being developed that will not breach the normative criteria, and may in fact be viewed as better than conventional weapons in this regard. For example, there is substantial military interest in agents that target the nervous system, e.g. through effects on neurotransmitters, and may induce such responses as unconsciousness, calm, fear, panic and depression (Dando & Wheelis, 2005; Lakoski et al., 2000).

Such agents could be used to temporarily incapacitate enemy combatants. They may not in themselves discriminate between civilians and combatants, but they could facilitate discrimination in follow-on military action. As their effects would be intended to be reversible, the goal of disabling the enemy could potentially be achieved with less injury and suffering than use of conventional weapons, and the proportionality criteria would be met.

There are several downsides to permitting the development of such weapons. They may well form a prelude to use of lethal force, they are likely to require development of delivery mechanisms that would also be suitable for lethal agents, they could easily be used for oppressive purposes, and they may well encourage arms racing. If some states decide to withdraw from current treaties in order to legitimately develop such agents, then the treaties may lose relevance and the prohibition as a whole may be undermined.

Analysis of the role of norms in biotechnology governance should include assessment of whether the norms retain relevance to the scientific and technological context or whether there may be significant gaps where normative development and/or adaptation is urgently required.

Incorporating Norms in Research Culture

Recent years have seen some interesting moves in international organisations and negotiating processes to develop alternative governance initiatives, particularly aimed at closing the gap between scientific and regulatory development. One strand of these focuses on promoting bottom-up ethical governance, through the incorporation of particular norms in scientific research culture, rather than relying solely on top-down governmental implementation of international rules. These efforts are most developed in the areas of laboratory biosecurity and arms control. This is not surprising because these rules have direct relevance to scientific conduct.

Laboratory Biosecurity

The World Health Organisation has developed guidance on laboratory biosecurity as an extension of its work on laboratory biosafety. (Laboratory biosafety concerns the prevention of accidental release of biological agents and laboratory biosecurity concerns the prevention of deliberate release, loss or theft of biological agents.) The WHO document *Biorisk Management: Laboratory Biosecurity Guidance* gives detailed reflection on the need for and context of bottom-up implementation of biosafety and biosecurity measures. It refers throughout to the need for development of a 'biorisk management culture' in laboratories, combining biosafety, biosecurity and bioethics.

The Guidance resulted "from careful thinking, comprehensive study of prevailing practices and recommendations, review of international norms and standards, and relevant ethical considerations" (p.1). It notes expectations of the general public in regard to safe conduct and responsible practice by laboratory personnel, who will also "follow an ethical code of conduct" (p.2).

Development of laboratory biosecurity programmes is expected to take place at the local level in order to increase feelings of ownership and ensure relevance to particular conditions and activities. It should be "representative of the institution's various needs" and be based on input from all staff in coordination with law-enforcement agencies (p.7). This focus on local development and avoidance of setting prescriptive rules is expected to: foster creativity and innovation; enable quick and easy response to unexpected events; allow easy incorporation of new findings and considerations; and promote broad acceptance of the outlined responsibilities.

Consideration of ethical issues is expected to form a core part of programme development and ongoing conduct of laboratory work: "Researchers, laboratory workers and biosafety and biosecurity managers should communicate and collaborate,

and strive to find the correct ethical balance for the activities performed" (p.21).

While providing some general advice on principles and best practice, the nature of this Guidance suggests that it is expected that the most effective way in which norms can be incorporated in governance is by supporting local development and implementation of specific standards of conduct.

The Biological Weapons Convention, Codes, and Education

States parties to the Biological Weapons Convention (BWC) are developing a similar approach to improving national implementation of treaty provisions. Following failure to agree a verification protocol to the Convention in 2001, the states parties have worked through a series of intersessional meetings to develop other means and methods of strengthening it. These have included consideration of development of codes of conduct and education for life scientists "covering the moral and ethical obligations incumbent on those using the biological sciences" with recognition that such "'bottom up' oversight by scientific establishments and scientists themselves" is a necessary complement and balance to 'top down' governmental controls (*Report of the Meeting of States Parties*, 2008, p.7).

Recommendations have included involving "relevant stakeholders in all stages of the design and implementation of oversight frameworks" and ensuring that outcomes are usable, relevant and appropriate (*Report of the Meeting of States Parties*, 2008, p.10). Codes of conduct are expected to "cover ethical and moral obligations through the scientific life-cycle"; be "built on existing arrangements and practice and/or derived from general overarching principles" where possible; "tailored to precise national or institutional requirements taking into account relevant cultural and social backgrounds"; and "be discussed and reviewed at international, regional and national scientific conferences and workshops, as well as in relevant publications." (*Report of the Meeting of States Parties*, 2008, p.16).

Again, this development indicates that combining international recommendations to create a degree of harmonization in approaches, with local implementation activities is expected to have greater effectiveness for incorporating ethical considerations in scientific practice.

Conflicts between Norms and Values

The international regulations relevant to biotechnology developed largely in separation from one another (although there are some connections within issue areas), at different times (from 1925-2010), and for a range of different purposes. Only a few have been specifically designed to apply to biotechnology; most do so only as a side issue to wider purposes. As such, it is unsurprising to find that there are some conflicts between the norms and values expressed in the rules.

One example occurs between the rules on intellectual property rights and the rules on access to and benefit-sharing from the use of genetic resources. The latter agreements include the right of farmers, local and indigenous groups to benefit from the commercial exploitation of resources that they have helped to develop, or knowledge they have about those resources. However, these rights are not incorporated in the rules on intellectual property protection. In those rules it is only the person / institution that took the 'innovative step' that is assigned the right to receive reward from an invention's commercial exploitation. Scholars also point to conflicts between protection of intellectual property rights and the value of open dissemination of scientific results (see, for example, iSEI, 2009) and, through their effects on the cost of essential medicines, with the right to health (WHO, 2002, p.135).

A second example concerns use of the precautionary principle. The Convention on Biodiversity and its Cartagena Protocol on Biosafety incorporate the precautionary principle as stated in the 1992 Rio Declaration: "Where there are threats of serious or irreversible damage, lack of full scientific certainty shall not be used as a reason for postponing cost-effective measures to prevent environmental degradation." This means that, under the provisions of the Cartagena Protocol – which addresses transboundary movements of genetically modified organisms – a decision not to allow import may be imposed in the absence of scientific evidence of harm. The trade rules apply a different standard – trade measures are only justified if based on scientific evidence:

Members shall ensure that any sanitary or phytosanitary measure is applied only to the extent necessary to protect human, animal or plant life or health, is based on scientific principles and is not maintained without sufficient scientific evidence. (SPS Agreement, Article 2.1)

Unifying Norms

There are also some occurrences of norms that span different regulatory areas. For example, in the areas of health and trade the key agreements on disease control incorporate the principle of not allowing health measures to unjustifiably restrict international trade. The Sanitary and Phytosanitary and Technical Barriers to Trade Agreements recognise that some trade measures for health protection purposes will be justified, but these should only be used where necessary and should be designed to have minimal impacts on trade. This can be seen in the following passages from the International Health Regulations and the Sanitary and Phytosanitary Agreement:

The purpose and scope of these Regulations are to prevent, protect against, control and provide a public health response to the international spread of disease in ways that are commensurate with and restricted to public health risks, and which avoid unnecessary interference with international travel and trade. (IHR, Article 2)

No Member should be prevented from adopting or enforcing measures necessary to protect human, animal or plant life or health, subject to the requirement that these measures are not applied in a manner which would constitute a means of arbitrary or unjustifiable discrimination between Members where the same conditions prevail or a disguised restriction on international trade. (SPS, preamble)

The existence of both common and conflicting norms within a set of rules points to a need for awareness of normative complexity in governance efforts, and for careful analysis of particular areas to draw out core norms and understand how they might interact with one another. The subjects of governance are likely to appreciate clarity on how they ought to prioritise compliance in situations where there are competing norms.

CONCLUSION

The international governance of biotechnology can provide some lessons for understanding the role of norms in technological governance. These include: that for understanding the use of norms it can be helpful to look back to earlier agreements to trace the basis on which current rules developed; it is worth being aware of core norms that influence all international governance efforts (e.g. state sovereignty); the fact that governance responses can be slow to adapt to scientific advances may mean that the norms used are no longer appropriate and points to the need for new approaches to governance that can be more flexible and responsive to scientific practice; and finally, that in areas with multiple rules and multiple actors there is likely to be a complex normative landscape that will require careful analysis to be fully understood.

The majority of international governance efforts maintain a top-down approach and reflexivity is unlikely to be applied in these areas. Powerful states have little interest in transformation of an international system from which they benefit or in reflecting on the moral implications of the context in which they operate. This is clearly observable in state behaviour, which despite recognition of interdependencies, continues to reflect narrow conceptions of national interest with the prioritisation of power politics and short-term economic gain. The WHO's *Laboratory Biosecurity Guidance*, which, as noted above, promotes a bottom-up approach to ethical governance, refers in both its preface and introduction to crisis events (such as the anthrax attacks of 2001 and SARS outbreak of 2003-04) which prompted increased awareness, review and renewed political commitment to laboratory biosafety and biosecurity (2006, pp. 1-4). It is unclear whether or not it will require crises to prompt similar developments in other regulatory areas.

REFERENCES

Biological Weapons Convention. (1972). Retrieved March 24, 2011, from http://www.opbw.org/convention/conv.html

Charter of the United Nations. (1945). Retrieved September 08, 2011, from http://www.un.org/en/documents/charter/

Collins-Chase, C. (2008). Comment: The case against trips-plus protection in developing countries facing AIDS epidemics. *University of Pennsylvania Journal of International Law*, *29*(3), 763–802.

Convention on Biodiversity. 1992. Retrieved August 17, 2009, from http://www.cbd.int/doc/legal/cbd-en.pdf

Council of Europe. (2006). *Secretary-General's report under Article 52 ECHR on the question of secret detention and transport of detainees suspected of terrorist acts, notably by or at the instigation of foreign agencies.* Retrieved March 02, 2007, from http://www.coe.int/T/E/Com/Files/Events/2006-cia/SG-Inf-(2006).pdf

Covenant of the League of Nations. (1919). Retrieved September 12, 2011, from http://www.iilj.org/courses/documents/CovenantoftheLeagueofNations_000.pdf

Dando, M., & Wheelis, M. (2005). Neurobiology: A case study of the imminent militarization of biology. *International Review of the Red Cross, 87*(859), 553–571. doi:10.1017/S1816383100184383.

Finkelstein, L. (1995). What is global governance? *Global Governance, 1*(3), 363–372.

Food and Agriculture Organisation. (1997). *International Plant Protection Convention.* Retrieved August 27, 2009, from https://www.ippc.int/file_uploaded//publications/13742.New_Revised_Text_of_the_International_Plant_Protectio.pdf

Food and Agriculture Organisation. (2001). *International treaty on plant genetic resources for food and agriculture.* Retrieved April 15, 2010, from ftp://ftp.fao.org/docrep/fao/011/i0510e/i0510e.pdf

Institute for Science. Ethics and Innovation. (2009). Who owns science? *The Manchester Manifesto.* Retrieved September 12, 2011, from http://www.isei.manchester.ac.uk/TheManchesterManifesto.pdf

International Committee of the Red Cross. (1868). *(St. Petersburg) Declaration renouncing the use, in time of war, of explosive projectiles under 400 grammes.* Retrieved September 08, 2011, from http://www.icrc.org/ihl.nsf/FULL/130?OpenDocument

International Committee of the Red Cross. (n.d.). *Customary IHL – Rules.* Retrieved September 08, 2011, from http://www.icrc.org/customary-ihl/eng/docs/v1

Lakoski, J. M., Murray, W. B., & Kenny, J. M. (2000). *The advantages and limitations of calmatives for use as a non-lethal weapon.* Retrieved March 24, 2011, from http://www.sunshine-project.org/incapacitants/jnlwdpdl/psucalm.pdf

Morgenthau, H. J. (1978). Six principles of political realism. In *Politics among nations: The struggle for power and peace* (5th ed.). New York: Knopf.

Ovett, D. (2006). Free trade agreements (FTAs) and human rights: A serious challenge for Latin America and the Caribbean. *PUENTES, 7*(1). Retrieved September 12, 2011, from www.3dthree.org/pdf_3D/Dovett_PUENTESarticle_Feb06_Eng.pdf

Peters, A. (2001). There is nothing more practical than a good theory: An overview of contemporary approaches to international law. *Jahrbuch fur Internationales Recht. German Yearbook of International Law, 44*, 25–37.

Report of the Meeting of States Parties [to the Biological Weapons Convention]. (2008). Retrieved September 08, 2011, from http://daccess-dds-ny.un.org/doc/UNDOC/GEN/G09/600/07/PDF/G0960007.pdf?OpenElement

Rhodes, C. (2010). *International governance of biotechnology: Needs, problems and potential.* London: Bloomsbury Academic. doi:10.5040/9781849661812.

United Nations. (2004, July 5). *Press release – US-Peru trade negotiations: Special rapporteur on right to health reminds parties of human rights obligations.* Retrieved September 12, 2011, from http://www.unhchr.ch/huricane/huricane.nsf/0/35C240E546171AC1C1256EC800308A37?opendocument

United Nations Economic and Social Council – Commission on Human Rights. (2006). *Situation of Detainees at Guantanamo Bay*. Retrieved September 12, 2011, from http://www.unhcr.org/refworld/docid/45377b0b0.html

United Nations General Assembly. (1992). *Annex 1 – Rio Declaration on Environment and Development* to *Report of the United Nations Conference on Environment and Development*. Retrieved March 03, 2004, from http://www.un.org/documents/ga/conf151/aconf15126-1annex1.htm

Valdivieso, L. V. (2009). Need to guard against TRIPS-plus enforcement agenda. *South Bulletin, 41*. The South Centre. Retrieved September 12, 2011, from http://www.southcentre.org/index.php?option=com_content&task=view&id=1077&Itemid=279

Weiss, T. G., & Hubert, D. (2001). State sovereignty. In *The responsibility to protect: Research, bibliography, background*. International Development Research Centre.

World Health Organisation. (2002). *Genomics and world health: Report of the advisory committee on health research*. Geneva: World Health Organisation. Retrieved May 29, 2003, from http://www3.who.int/health_topics/genetic_techniques/en/

World Health Organisation. (2005). *International health regulations*. Retrieved August 17, 2009, from http://whqlibdoc.who.int/publications/2008/9789241580410_eng.pdf

World Health Organisation. (2006). *Biorisk management: Laboratory biosecurity guidance*. Retrieved September 08, 2011, from http://www.who.int/csr/resources/publications/biosafety/WHO_CDS_EPR_2006_6.pdf

World Trade Organisation. (1995-1). *Dispute settlement understanding*. Retrieved September 12, 2011, from http://www.wto.org/english/docs_e/legal_e/28-dsu.pdf

World Trade Organisation. (1995-2). *Agreement on the application of sanitary and phytosanitary measures*. Retrieved August 18, 2009, from http://www.wto.org/english/docs_e/legal_e/15-sps.pdf

World Trade Organisation. (1995-3). *Agreement on trade related aspects of intellectual property rights (TRIPS)*. Retrieved April 15, 2010, from http://www.wto.org/english/docs_e/legal_e/27-trips.pdf

Chapter 7
Ethics and Governance of Nuclear Technology:
The Case of the Long Term Management of Radioactive Wastes

Sylvain Lavelle
Center for Ethics, Technology and Society (CETS), ICAM Paris-Sénart, France

Caroline Schieber
CEPN, France

Thierry Schneider
CEPN, France

ABSTRACT

Radioactive waste, remaining radioactive for very long periods up to hundreds of thousands years, introduces a new time dimension never experimented in the field of risk management. This situation led for more than 10 years, to reflections on the societal and organisational mechanisms for the development of protection systems, able to cope with those periods. Within the framework of the European research project COWAM 2, dedicated to the improvement of governance of radioactive waste management in Europe, a panel of stakeholders involving experts, authorities, waste managers, locally elected representatives and NGOs, opened a dialogue on the ethical considerations related to the long-term dimensions of this management. This article presents the main results of this panel, with specific emphasis on the meaning of the long term and what is at stake, the ethical dimension regarding long term issues, and the continuity and sustainability of the vigilance and surveillance of radioactive waste facilities.

DOI: 10.4018/978-1-4666-3670-5.ch007

INTRODUCTION

Due to the half-life of some radionuclides, their chemical nature and concentration, radioactive waste introduces a temporal dimension that has never been experienced so far in existing risk management. According to the waste, time scales of the order of thousands or even millions of years have to be considered. In order to identify management devices adapted to this situation, discussions were initiated over ten years on societal issues associated with radioactive waste management (NEA/OECD, 2010[1]; Hériard Dubreuil et al., 2007[2]; Lavelle, 2006[3]; ICRP, 2012[4]). These reflections have particularly highlighted the issue of the transfer of radioactive waste management system to the next and future generations. These rules concern not only the transfer of protective devices but also the means to ensure the continuity of vigilance and surveillance over the long term.

In the framework of the European Commission research project "COWAM2" (COmmunity WAste Management[5]) related to the governance of radioactive waste management, the ethical considerations associated with its long-term dimension have been addressed, in a dedicated panel of stakeholders, involving experts, authorities, waste managers, locally elected representatives and NGOs. This panel opened a dialogue to identify, discuss and analyse the institutional, ethical, economic and legal considerations raised by the long-term management of radioactive waste (Schneider et al., 2006[6]). Its aim was to identify a set of practical recommendations in order to better address long-term issues in decision-making processes.

This article presents the results of the work performed during this 3-year project, with specific emphasis on:

- **What is at Stake with Long-Term Dimensions?:** Focussing on the meaning of "long-term" from technical and societal points of view.

- **Ethical Considerations Regarding Long-Term Issues for Radioactive Waste Management:** Addressing the rights and duties of current and future generations, long term responsibility, democracy and justice.

- **Continuity and Sustainability of Vigilance and Surveillance:** Providing recommendations related to the memory and knowledge conservation and transfer, the local and regional economic development, the distribution of responsibilities between territories and generations, as well as the efficiency of financial schemes.

WHAT IS AT STAKE WITH LONG-TERM DIMENSIONS?

There is not "a unique" definition of the long-term and it is essential to first delineate what is at stake in terms of time dimension when dealing with long term governance. Regarding radioactive waste management, two long-term perspectives have to be considered: a technical perspective and a societal perspective. This section points out the key issues associated with these two perspectives and identifies the main features for the long-term governance.

Long-Term Dimensions from a Technical Point of View

The use of radioactive materials in any activity generates radioactive waste. In the case of the nuclear fuel cycle, part of the radioactive waste remains radioactive for very long periods up to hundreds of thousands years. Because the wastes concentrate the radioactivity, they are dangerous and need to be dealt with care to avoid or reduce as much as possible the risks for human and the environment. People would be affected by direct external exposure to the radioactive waste if they are standing in the vicinity of the wastes, and by internal contamination if they inhale or ingest radioactive substances transferred from the waste to the air or the foods.

Different technical options are envisaged in order to provide an adequate protection for public, workers and the environment. Without considering the recycling options or potential specific treatments of the radioactive waste, two main technical options can be distinguished: interim storage, and geological disposal.

Interim Storage

Their design aims at providing technical conditions for cooling the packages and ensuring the radiological protection of workers, the public and the environment. Generally, they are planned to last several decades, before envisaging new destinations for the radioactive waste. Recently, new designs have been developed to cope with longer durations: about one century (or more) without significant operation and maintenance requirements.

Geological Disposal

The key feature of the geological disposal is clearly to provide a protection system able to cope with very long-time periods. Different phases, with specific features for the protection, have to be considered regarding a geological disposal:

- The operational phase is the period covering waste emplacement and repository closure, and it can last several decades.
- The thermal phase is the phase during which the heat generating waste significantly increases the temperature in and around the repository; its duration lasts several centuries to 2,000 years.
- The isolation phase refers to the phase where the radionuclide releases from the disposal system are negligible; this phase

is situated between 1,000 and 10,000 years after the repository closure according the type of waste.

- The geological phase considers the period for which the repository enters the geological timescales (10,000 till million years after closure).

The first component of the timescale refers to the radionuclides half-life. For vitrified radioactive waste, the rough reduction factors are about 10 after a century, 500 after a thousand years and 10,000 after a hundred of thousand years (Lagrange, 2005[7]). Due to the reduction of their activity, the external dose rate associated with the radioactive waste is also reduced. It is expected that the external dose rate at 1 meter of the radioactive waste package will be reduced by a factor 100 after a century, and a factor 30,000 after a thousand years (Lagrange, 2005).

Because of the long-timescale related to geological disposal, the assessment of the level of protection provided by the disposal is subject to a series of uncertainties as described in Figure 1 and discussed by international experts:

It illustrates that, at least for a well-chosen site, the evolution of the broad characteristics of the engineered barrier systems (EBS) and the host rock are reasonably predictable over a prolonged period (105 or 106 years, say, in the case of the host rock). There are uncertainties affecting the engineered barrier systems and the host rock over shorter timescales, but these can, in general, at least be bounded with some confidence. The patterns of groundwater flow (the hydrogeological system), in particular near the surface, can be affected by climate change and are thus somewhat less predictable. Surface environmental processes and radiological exposure modes are not generally considered to be parts of a deep geological repository system, but are relevant for evaluating dose and risk. These are less predict-

Figure 1. The limits of predictability of various aspects of a geological disposal system

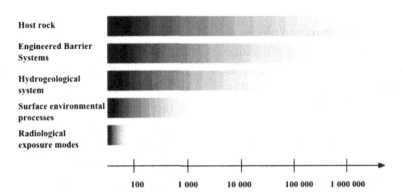

able still, being affected by ecological change, human activities and individual habits, which are highly uncertain, even on a timescale of a few years" (OECD/NEA, 2004[8]).

Long-Term from a Societal Perspective

The timescale considered from the technical point of view are outside the current field usually considered for the prediction of the evolution of the society. It is essential to recognize that nevertheless, given the potential risks associated with radioactive waste and the protection strategies, their long-term governance cannot be reduced to a technical issue. Their management induces a new complexity regarding the participation to the decision process and calls for a new organisation of the institutional and societal vigilance and surveillance over timescales never experimented before. The waste producers, the waste operators and the authorities are, of course, responsible for the implementation of the waste management options but it is the whole society which is embarked now into a long-term management process, introducing a responsibility toward future generations and a need for access to information and involvement into decision making processes.

From the societal perspective, considering timescales of the order of several thousands of

years is meaningless. It is not possible to envisage how the society will be organised in the far future. Even in several decades, a lot of evolutions may occur, reducing the ability to predict the economic, social and environmental situations. The current generation is however concerned by the possible future, even in several thousands of years. This is notably the core of the ethical reflections regarding the precautionary principle and the sustainable development in order to preserve the resources and the environment for the future generations. On this basis, the obligation for the current generation to avoid "undue burdens" on future generations regarding radioactive waste management was notably introduced. Although the duty to protect future generations is of prime importance, the capability to really achieve this obligation is largely impacted by technical and scientific uncertainties, and depends also on the evolution of the society. Furthermore, the decision to impose a specific behaviour to the future generations is questionable. In that perspective, it has been acknowledged within the participants of the COWAM2 project that a reasonable approach to cope with the long term duration of waste radioactivity is, for the current generation, to delineate and create management and governance processes favouring a continuous transmission to the next generation(s) of a "safety heritage" (know-how, protection options, procedures, resources,...) in order to ensure the continuation of waste management.

These management or governance processes may evolve with time, but the current generation needs to consider how they can be set up in order to achieve a number of "missions" which will be transmitted to the next generation. It will be the choice and the responsibility of the next generations to continue and/or reconsider these processes and to adapt them with the aim of ensuring the realisation of these different missions. The future generations being unable to participate to the current decision-making process, the organisation of the transfer of information and vigilance and surveillance processes is essential.

Having introduced the need for a transfer between generations, it appears obvious that the timescale to be considered from a societal point of view differs from the one adopted from the technical point of view. The key features of the time dimension rely on the legacy including the transfer of a safety patrimony for ensuring the protection of future generations and their environment. Therefore, the consideration of the long term from a societal point of view implies to cope with the past, the present and the future organisation of the radioactive waste management.

Towards a Common Approach for Long-Term Issues

In its future recommendations to be published on geological disposal, the International Commission on Radiological Protection considers that "one of the important factors that influence the application of the protection system over the different phases in the lifetime of a disposal facility is the level of oversight or 'watchful care' that is present. The level of oversight directly affects the capability to reduce or avoid some exposures" (ICRP, 2012). It is interesting to mention that for this purpose, there is a key concern to involve all the stakeholders at the different phases of the disposal facility

and to call for maintaining forms of oversight and the memory of the disposal as long as possible.

It is also acknowledged that due to scientific and societal uncertainties in the long-term, it is not possible for the present generation to ensure that societal action will be taken in the future but at the same time it is requested to provide the means for the next generations to cope with the long-term surveillance. The spirit of these recommendations is to design the protection system as a combination of technical options and vigilance and surveillance processes relying on stakeholders from present and future generations.

ETHICAL STAKES OF LONG TERM RADIOACTIVE WASTES MANAGEMENT

The importance of ethics in the long term management of radioactive waste was pretty well expressed fifteen years ago by the Board on Radioactive Waste Management (BRWM), a taskforce created by the US National Research Council. This taskforce gathered various independent experts from the National Academy of Science, from the National Academy of Engineering and from the Institute of Medicine. The BRMW 1990 report stressed an interesting point (BRWM, 1990[9]): «In the area of radioactive waste, ethical issues are as important as management and technical decisions. Interested parties approach the issues with different views about the right way to proceed, often due to differences in moral and value perspectives. As a result, an exploration of ethical issues can illuminate the fundamental policy debates in this field by showing the technical issues in their political and social context. Such an exploration also provides scientists with an opportunity to explore their own ethical responsibilities as they provide society with technical advice on controversial subjects.» The consideration of intergenerational issues has also been explored within the radioactive waste management community by: the International

Atomic Energy Agency (IAEA) (IAEA, 1992[10]), the Nuclear Energy Agency (NEA) of OECD (OECD/NEA, 1995[11]), the Swedish Consultative Committee on Radioactive Waste Management (KASAM) (KASAM-SKN, 1988[12]), and the Seaborn Commission in Canada. One of the main conclusions of these investigations was the statement that the driving principle for the elaboration of waste management options is to avoid «undue burden» for the future generations. The NEA 1995 report set up several principles (protection of health, of environment, beyond national borders, of future generations) but insisted on the principle of undue burden ('Principle 5: Burdens on future generations: Radioactive waste shall be managed in such a way that will not impose undue burdens on future generations'). For the NEA, the first concern was the achievement of «intergenerational equity» by choosing technologies and strategies which minimize the resource and risk burdens passed to future generations by the current generations which produce the radioactive waste. The second concern was the achievement of «intragenerational equity» and in particular an ethical approach to the handling, within current generations, of questions of resource allocation and of public involvement in the decision-making process. The conclusions were the following:

1. The liabilities of waste management should be considered when undertaking new projects.

2. Those who generate the waste should take responsibility, and provide the resources, for the management of these materials in a way which will not impose undue burdens on future generations (see Final report of COWAM2 - Work Package 4).

3. Waste should be managed in a way that secures an acceptable level of protection for human health and the environment, and affords to future generations at least the level of safety which is acceptable today; there seems to be no ethical basis for discounting

future health and environmental damage risks.

4. A waste management strategy should not be based on a presumption of a stable societal structure for the indefinite future, nor of technological advance; rather should it aim at bequeathing a passively safe situation which places no reliance on active institutional controls.

In COWAM2, these NEA principles, especially the principle of 'undue burden', are given much attention, but they are regarded as insufficient ones. Therefore, the COWAM2 developments were oriented towards the creation of the best conditions to favour the transfer to the next and following generations of the whole waste management system. One of the responsibilities of our generation is to organize a vigilance of the repository sites, but it is also to give the future generations a responsibility for the determination of their own future.

The issue of long term responsibility, especially in the case of radioactive waste, suggests that there is no reciprocity between generations. Indeed, the generation n will no longer be alive at the time of generation n+50, even if its actions can have long term consequences. But the lack of reciprocity is not a reason for putting aside any responsibility, and on the contrary, it is an opportunity for creating a new kind of responsibility between generations. The ethical criteria formulated within the work package according to the responsibility principle are presented hereafter.

Methodology of COWAM2

The originality of the COWAM2 project is that it addressed the various issues within working groups made up of stakeholders from different origins and European countries and a research team. About 20 participants, having an interest on long term issues, attended twice a year the meetings of the work package on long term governance.

They were members of local liaison committees, NGOs, operators, regulators and experts from research and public institutes. They originated from the following countries: Belgium, France, Germany, The Netherlands, Romania, Spain, Sweden, Switzerland, and United Kingdom. Although different countries and categories of stakeholders were involved in the work package, the driving force for their participation was their willingness to address the issues of long term governance.

The research team involved four different institutes from Belgium, France, and Switzerland and included expertise on ethics, radiation protection, economics, environmental assessment, and social sciences.

In order to perform the work, the approach adopted relied on the following steps:

- Establishment of the topics to be developed: for this purpose, the first meeting (in April 2004) was dedicated to the identification of participants' expectations which allowed to define the work programme.
- Preparation of topical documents and reviews of case studies by the research team for discussion during the work package meetings, and notably the discussion of ethical issues.
- Contribution of stakeholders: presentations of reflections on ethics, presentations of national and local contexts regarding long term issues, presentation of financing mechanisms for long term governance.
- Preparation of a draft final report by the research team.
- Comments and validation of the final report by the participants (including a dedicated meeting).

It has to be mentioned that two meetings of the work package were held in Gartow (Germany, near Gorleben) at the invitation of Pastor Eckhard Kruse and included a visit of the underground exploratory mine of the Gorleben salt dome (thanks to Jürgen Wollrath from BfS - Federal Office for Radiation Protection - Germany) and a debate with local stakeholders on the issue of geological disposal of high-level radioactive waste in Germany. Another meeting was held in Barcelona (Spain) with specific emphasis on the reflection of COWAM Spain regarding long term issues.

It is also important to mention that at the beginning of the project, several participants notified that, for them, a pre-requisite to their involvement in the governance of radioactive waste management would be to clearly address the articulation between energy policy scenarios and long term waste management scenarios. Although this was considered to be an important issue, all the WP4 participants acknowledged that COWAM2 - WP4 was not the place to open this debate. It was considered that an adequate forum has to involve different local and national stakeholders and energy policy-makers. Therefore, it was agreed to clearly quote this need in the final report and, knowing this pre-requisite, to engage with all the participants of the work package the reflections on the long term governance.

The objective of such a final report is not to present a consensus on all the topics. A lot of discussions during the meetings as well as the stakeholders contributions allowed to identify the topics on which there is an agreement between the different participants and those where there is disagreement or a need for further reflections. Therefore, the report tries to reflect as much as possible this situation and mentions when necessary the different points of view.

Elaboration of Ethical Criteria

The analysis performed within the COWAM2-WP4 on long term governance processes for radioactive waste management led to the identification of three major ethical principles as key issues for the long term governance of radioactive waste: responsibility, justice and democracy. These principles have been analysed by the stakeholders participating in the project in order to formulate ethical criteria specific to radioactive waste management to be used as aiding tools to evaluate on an ethical ground the various technical processes which can be proposed.

Some additional reflections were proposed within the work package, relating to the difficulty of a inter-generation equity, arguing that any fair system can only achieve an intra-generation one. Furthermore, the issue of financial compensation was studied pointing out the conflict between financial short term strategies and ethical issues regarding long term responsibilities (see Annex Report, contributions from G. Bombaerts and M. Bovy).

Long Term Responsibility

The issue of long term responsibility, especially in the case of radioactive waste, suggests that there is no reciprocity between generations. Indeed, the generation n will no longer be alive at the time of generation $n+50$, even if its actions can have long term consequences. But the lack of reciprocity is not a reason for putting aside any responsibility, and on the contrary, it is an opportunity for creating a new kind of responsibility between generations. The ethical criteria formulated within the work package according to the responsibility principle are presented hereafter.

In order to make future generations able to make relevant decisions about the future of radioactive waste, from their own point of view:

1. The future generations should be provided with some appropriate sustainable means (processes, money, institutions, knowledge, know-how,...) for the implementation and the assessment of radioactive waste management systems.
2. These sustainable means should also be designed to guarantee the long term protection of health and environment.

3. Regarding long term public health and environmental protection, the public sector should regulate the distribution of responsibilities between public and private sectors and its evolution over time.
4. Appropriate policy, organization or network should be designed to keep information, knowledge and skills about the radioactive waste.
5. These elements should be sustainable and available for the actors and for the education of the future generations.
6. A responsible long term radioactive waste management policy should articulate in a flexible way the current decisions with the future capacity of actions.

The implementation of such criteria implies to address various issues such as:

- **Ownership:** Who is the current owner and who will be the future owner of radioactive waste and storage/disposal sites? What are the conditions at present and for the future for the legal and financial responsibility? Who will be held responsible in case of further damages?
- **Surveillance:** Who is in charge of and who participate to the maintenance and the surveillance of the radioactive waste management facilities? Is the need for technical maintenance and surveillance coherent with the duration of institutions?
- **Education:** How are knowledge and skills on radioactive waste management transmitted through generations?

Long Term Justice

Justice is an evaluation of actions on the basis of a principle of equality or proportion as far as the relationship of individuals to the community is concerned. The issue of long term justice suggests that a generation n responsible for the increase of radioactive waste must give a proportionate contribution for the people affected by this waste. The people can belong to the same generation n (local people), or to future generations (for instance, $n+50$). This leads the participants of the work package to formulate the following ethical criteria.

1. The fairness of the situations should be evaluated in terms of advantages and disadvantages on the basis of intra/inter/trans-generation relationships.
2. This evaluation should integrate quantitative and qualitative aspects of the living conditions, and of probable economic, social or technical trends or backgrounds.
3. Our generation should provide a contribution that takes into account our current advantages compared to the disadvantages of the future generations.
4. This contribution should be proportionate to the efforts (research and development, etc) needed to manage the radioactive waste and to optimise the cost of the radioactive waste management systems.
5. A municipality accepting to manage the radioactive waste of a country should benefit from a long term solidarity in all respect from the rest of the nation.
6. In case that there is an agreement on the construction of a radioactive waste management facility, the local populations and municipalities should be entitled to socio-economic development funding.
7. The funding is aimed at supporting a sustainable development of the territories in order to ensure the continuity of the vigilance of local population and the surveillance of the site of the radioactive waste management facility.

The implementation of such criteria implies to address various issues such as:

*   **Fairness:** How to maintain a form of equity between generations as far as consideration and recognition of the role of local people is concerned? Does the relation between generations correspond to an intra-generation, an inter-generation, or a trans-generation relation?
*   **Compensation:** How to avoid that the compensation is just a financial bargain, a way of buying local consent? Will the money be dedicated to an appropriate local development?
*   **Recognition:** How to give sense to and to recognize the effort of local people accepting radioactive waste on their territory? Does the valuation of the effort of people imply for them to have a special political representation or participation concerning the decision process in radioactive waste management?

It is important to realize that the issue of the recognition of the effort of a local territory accepting the waste and the question of the solidarity of the nation will not only be solved with the financial contribution to a sustainable development of the region. There is also a need for a sort of "moral" recognition to be built among generations and through generations, using various means like the empowerment of citizens, the consideration for their demands or the recognition of their role in the long term process.

Long Term Democracy

Democracy (*demos*, people, *kratos*, power) is a political regime whose legitimacy lies in the representation or the participation of the people into the collective deliberation and the decision-making process. Regarding the long term governance of radioactive waste, the following ethical criteria were proposed by the work package participants:

1. A system of long term democratic governance requires a balanced flexible political procedure or organization combining representation, participation and deliberation of the people.
2. The long term governance of the radioactive waste implies that the technical options and participatory democracy are linked.
3. A democratic organization or procedure of governance should gather a variety of people belonging to several generations and to various backgrounds(local/national/international,authorities/experts/citizens/NGO/operators.).
4. An organization which is at least financially independent from political or technical authorities is more likely to guarantee the continuity of participation as well as the plurality of expertise, of information and of conceptions and values.
5. The institutions in charge of the radioactive waste management should be subjected to a democratic control and be counter-balanced by the political empowerment of the citizens through generations.
6. In order that the issue of radioactive waste remains a permanent topic of the democratic debates, it must be scheduled regularly on the political agenda of governmental and non-governmental organizations at local, national and international level.
7. All citizens must be provided with the means and information needed to fully participate in the process and to exercise their rights.

The implementation of such criteria implies to address various issues such as:

- **Participation:** How to organize the participation of citizens in the long term consultations and decisions concerning the management of radioactive waste? Should the participation be limited to the local people,

or enlarged to some representatives from the national community?
- **Control:** What could be the long term democratic control by the citizens over the decisions made by the institutional representatives? Will there be some confrontations between institutional and non institutional expertise? What kind of institutions may guarantee the continuity of the waste' surveillance?
- **Consultation:** Is the issue of the management of radioactive waste scheduled regularly on the agenda of the Parliament? Will there be regular consultation of local/national people for key issues to be addressed or key decisions to be made?

The need to elaborate a democratic process ensuring the participation of stakeholders in the decision making process is thus essential to open the question of radioactive waste management to the non-technical issues and build sustainable decisions including ethical and social aspects. These stakeholders come as well from the local level (elected people, economic actors, members of local commission of information, NGOs, ...) as from the national level (State representatives, nuclear industry, waste management organizations, ...) or European level.

SUSTAINABILITY OF THE VIGILANCE AND THE SURVEILLANCE AROUND RADIOACTIVE WASTE DISPOSAL FACILITIES

One of the key issues for long-term governance is to elaborate and set up a protection system lasting as long as possible including mechanisms for allowing its evolution in the future according the prevailing circumstances and the expectations of the future generations. Even if it is not possible (nor relevant) to predict how the society will be organised in the future, it is however possible to identify some key issues which should be addressed in order to favour the sustainability of the protection system and to create the conditions for its transfer to the next generations (See Table 1).

Hereafter is the proposed list of topics with a short explanation of their meaning.

Technical Processes

- Categories of radioactive waste. (i.e. Low Level, Intermediate Level, High Level, Short Lived or Long Lived)

- Storage/Disposal/Transmutation. (what is the device/strategy planned for future management of radioactive waste?)

- Combination of options for radioactive waste management in time.

- Development/reduction of nuclear energy. (what is the long term trend for the nuclear energy?)

- Sustainable energy policy. (what are the alternative energy devices or policies?)

Table 1. Topics to be considered for the elaboration of long term governance devices

Technical Processes	Institutional Conditions	Financial Conditions	Societal Conditions	Ethical Stakes
• Category of radioactive waste	• International/ national agencies and programmes	• Specific fund for the long term management of the waste	• Intra-inter-trans generation relations	• Long term protection of health/ environment
• Storage/Disposal/ Transmutation	• Public/private ownership over time and its evolution over time	• Provisions made by the operators or the state and their evolution over time	• Networks of territories/ municipalities/ citizens involved in radioactive waste management	• Freedom of choice for the local population over time
• Combination of options over time	• Co-operative management of the waste	• Financial support for the local development of municipalities and districts	• Involvement and empowerment of local population	• Conservation of memory and transfer of information, knowledge and skills
• Development/ reduction of nuclear energy production	• Robustness of institutions in charge of information transfer	• External control of the fund evolution and its sustainability	• Availability and accessibility of International/national/ local expertise on radioactive waste management	• Socio-economic benefit and development of local communities
• Sustainable energy programme and link to the nuclear energy policy	• Procedures of transparency and access of official information		• Co-operative inquiry and management of radioactive waste	• Control of energy consumption and waste production

Institutional Conditions

- International/national agencies and programmes. (what are the reference agencies or programmes?)
- Public/private ownership and its evolution over time. (what is the structure of the ownership of the waste over long term periods?)
- Co-operative management of the waste. (are the institution involved in radioactive waste management open to co-operation?)
- Robustness of institutions in charge of information transfer. (are the institutions in charge of the transfer of information to the future generations reliable?)
- Procedures of transparency and access to official information. (are the information procedures of the institutions transparent and is the necessary information accessible, to what extent?)

Financial Conditions

- Specific fund for the long term management of the waste. (are such funds available and is the institution managing the fund reliable?)
- Provisions made by the operators or the State and their evolution over time. (has the operators made sufficient provisions to finance the management of the waste in the long term?)
- Financial support for the local development of municipalities and districts where radioactive waste management facilities are installed. (what is the financial device for the municipalities holding the waste?)
- External control of the fund evolution and its sustainability. (how to control the use of the specific fund?)

Societal Conditions

- Intra-inter-trans generation relations. (what kind of relationship between generations?)
- Networks of territories/municipalities/citizens involved in radioactive waste management. (of which type are the networks implemented?)
- Involvement and empowerment of local population. (what is the strength of the local people?)
- Availability and accessibility of international/national/local expertise on radioactive waste management. (can the local people resort to an external expertise?)
- Co-operative inquiry and management of radioactive waste. (which co-operation with the authorities?)

Ethical Stakes

- Long term protection of health/environment. (how long is the protection of health and environment?)
- Freedom of choice for the local population over time. (are the local people free to reject a project?)
- Conservation of memory and transfer of information, knowledge and skills. (which long term devices exist or are planned?)
- Socio-economic benefit and development of local communities.
- Control of energy consumption and waste production. (what efforts applied to lower the quantities of waste produced?)

In conclusion, one has to keep in mind that the objective of the proposed guidelines is to favour a dialogue between the various categories of stakeholders in order to set up the key principles for developing long term governance devices relevant for their own context. Furthermore, it should be mentioned that the elaboration of these devices should be envisaged as a continuous process, largely influenced by the past and present situations. In that perspective, the devices should be regularly revisited and updated in order to cope with the evolution of the context.

The sustainability of the protection system relies notably on the organisation of the "vigilance and surveillance". This term can include several aspects of the protection system, such as:

- The monitoring of the facility and its environmental surrounding;
- The technical maintenance of the site, the management of any actions on site, including possible retrieval of waste;
- The preservation and transmission of know-how concerning waste management, and the training of the generations who will take over;
- The organisation of the stakeholders' vigilance at different levels.

To favour the sustainability of the protection system over long-term periods, the following key actions have been identified with the stakeholder group in the COWAM2 project covering the following fields:

- The organisation of surveillance and vigilance;
- The development of a centre of competence;
- The integration of the radioactive waste management facility in a local/regional economic development;
- The distribution of responsibilities between territories and generations;
- The efficiency of financial mechanisms.

The following paragraphs summarise the proposals identified with the COWAM2 stakeholder group.

- **Organisation of Surveillance and Vigilance**
  - The transfer between generations of the surveillance and vigilance system has to be studied to favour an active conservation of the memory. For this purpose, it is notably necessary to allow an evolution of the waste management and the surveillance and vigilance systems with time.
  - Local stakeholders should be involved in the surveillance and vigilance system of the site, as they are key actors for the transfer between generations.
  - The long-term surveillance and vigilance programme has to be clearly organised (distribution of responsibilities, reporting procedures, ...). Regular meeting points between the administration/state, the organisation in charge of the surveillance, and the local stakeholders should be planned in advance to ensure its efficiency and identify the needs for evolution.
  - Sustainable financing systems should be elaborated for the structure in charge of the surveillance and vigilance. The capability to mobilize international resources should also be studied.
- **Development of a Centre of Competence**
  - A centre of competence should be created for the operation, maintenance, vigilance and surveillance of the radioactive waste management facility in the long-term.
  - A system should be elaborated in order to maintain, develop and create knowledge and know-how to ensure an efficient vigilance and surveillance of the radioactive waste management facility with time.
  - The capacity to mobilize expertise (from local, national and interna-

tional level) should be studied and integrated in the functioning of the centre of competence. The capabilities to use the expertise of the centre of competence in various places or in other fields than radioactive waste management should be favoured.

○ The involvement of concerned stakeholders to the definition and follow-up of the activities of the centre of competence should be facilitated and the conditions to ensure a transfer of expertise between generations should be created.

- **Integration of the Radioactive Waste Management Facility and its Vigilance and Surveillance in a Local/Regional Socio-Economic Development**

  ○ The vigilance and surveillance function should be integrated within a global project for a sustainable territorial socio-economic development. This project should be elaborated, mainly by the local stakeholders, notably with a view to maintain the "life" around the radioactive waste management facility. The stability of the local and regional demography is recognised as one of the key issues for the sustainability of the vigilance and surveillance.

  ○ The development of economic activities linked with the environmental surveillance and monitoring, and in interaction with the scientific and technological competence at the regional level should be studied.

  ○ Dedicated systems should be set in place in order to guarantee that the storage/disposal is compatible with the territorial development.

- **Need for an Equitable Distribution of Responsibilities between Territories and Generations**

○ An efficient protection system needs a clear distribution of responsibilities between local, national and international actors and a regular up-date.

○ The notion of "safety heritage" should be developed in order to create a "safety link" between local, national and international actors, and between generations. To favour the participation of the stakeholders, the development of the radiation protection culture should be favoured.

○ Reflections on the interest of an international convention on the "protection of radioactive waste management facilities" should be developed to provide meaningful solidarity and sharing of competences among the different levels.

- **Efficiency of Financing Schemes for the Long-Term Management of Radioactive Waste**

  ○ The location of the responsibilities/liabilities regarding the management of waste should be clearly defined, and their transfer over time should also be planned in advance.

  ○ The decision-making process for defining the level of the funds or provisions and its use should be explained, as well as the waste management scenario used to determine the level of the financial needs in the future.

  ○ The ability of the funds to evolve with time should be clarified and take into account the possible evolution of the waste management options (reversibility, adaptation to new norms, ...).

  ○ External audit should be done on a regular basis in collaboration with national and local stakeholders.

  ○ The financing schemes should integrate guarantees to be used if the cost of waste management is higher than expected or if there is a bankrupt of a waste producer. The financing scheme should comprise specific systems to ensure (as much as possible) that the provisioned money will be available when necessary.

CONCLUSION

Long-term issues are inherent to radioactive waste management. The reflections of the panel of stakeholders within the COWAM2 project pointed out the key responsibility of the current generation to create the conditions for transferring a safety heritage to future generations through the elaboration of a long-term governance system. These investigations lead to propose a set of guidelines for a common technical and ethical elaboration of long-term radioactive waste governance devices.

The objective is not, of course, to define how future societies will be organized to manage waste, but to put in place provisions to promote transmission of the entire protection system to the generations to come. Thus, the performance of a facility for radioactive waste management (storage or disposal) should be evaluated by considering a complete system of protection including of course the technical dimensions but also on the transfer of knowledge and know-how, the organization of vigilance and surveillance and its evolution over time and the integration of the system in a socioeconomic development at the territorial level.

To this end, it is essential that the design options for radioactive waste management is not just a matter of scientific and technical experts, but it also involves other stakeholders that will be affected directly or indirectly by the existence of these waste management facilities. Those stakeholders can play a significant role in maintaining the efficiency and memory of these facilities over time. It is therefore important to develop procedures for coordination in the long run (meeting points, place for dialogue, identification of responsibilities, ...) between the different actors (authorities, experts, citizens, associations, operators, ...), based on the involvement at the local, national and international levels.

The aim of the reflection presented in this article is to favour the elaboration of long-term radioactive waste governance devices by a set of stakeholders (local, national and/or European), taking into account the technical, institutional, financial, societal and ethical considerations. The purpose is not to be prescriptive but to promote a common reflection and elaboration on this issue in a specific context, based on a structured approach.

Finally, a key dimension regarding long-term governance relies on the existence of networks at local, national and European levels involving different categories of stakeholders. In fact, the dissemination and sharing of feedback experience regarding long term governance could play a key role for improving the current governance systems as well as for ensuring a continuity of the surveillance and a solidarity between the different stakeholders and territories involved in the long term management of radioactive waste. In that respect, the existence of European networks is crucial for addressing the issues of long-term governance and favouring the emergence of innovative approaches.

REFERENCES

BRWM. (1990). *Rethinking high-level radioactive waste disposal: A position statement of the Board on Radioactive Waste Management*. Washington, D.C.: National Academy Press.

Cowam.com. (n.d.). Website. Retrieved from http://www.cowam.com

Heriard-Dubreuil, G., Gadbois, S., Mays, C., Espejo, R., Flüeler, T., Schneider, T., & Paixa, A. (2007, June). COWAM 2 - Cooperative research on the governance of radioactive waste management, final synthesis report. Mutadis, Paris, EC contract FI6W-CT-2003-508856.

IAEA. (1992). *Radioactive waste management, an IAEA source book*. Vienna.

ICRP. (2012). *Radiological protection in geological disposal of long-lived solid radioactive waste*. International Commission on Radiological Protection, Consulting Document.

KASAM-SKN. (1988). *Ethical aspects on nuclear waste – Some salient points discussed at a seminar on ethical action in the face of uncertainty in Stockholm, Sweden, September 8-9, 1987*. SKN Report 29.

Lagrange, M. H. (2005, September). *Modèle d'inventaire de dimensionnement (MID). Données descriptives du colis type C1*. Projet HAVL, Note technique ANDRA C.NT.AHVL.02.109, Indice B.

Lavelle, S. (2006). *Science, technologie et éthique - Conflits de rationalité et discussion démocratique*. Ellipses Edition, Collection Technosup.

OECD-NEA. (2010). Main findings in the international workshop "towards transparent, proportionate and deliverable regulation for geological disposal." Tokyo, Japan, 20-22 January 2009. OECD-NEA.

OECD/NEA. (1995). *La gestion des déchets radioactifs à vie longue: Fondements environnementaux et éthiques de l'évacuation géologique. Opinion collective du Comité de la Gestion des Déchets Radioactifs de l'AEN*. OECD.

OECD/NEA. (2004). The handling of timescales in assessing post-closure safety lessons learnt from the April 2002 workshop in Paris, France. Organisation for Economic Co-operation and Development/Nuclear Energy Agency, NEA n°4435.

Schneider, T., Schieber, C., & Lavelle, S. (2006). *Long term governance for radioactive waste management*. Final Report of COWAM2 - Work Package 4, Report COWAM2-D4-12/CEPN-R-301.

ENDNOTES

[1] See OECD-NEA (2010)

[2] See Heriard-Dubreuil et al (2007)

[3] See Lavelle (2006)

[4] See ICRP (2012)

[5] See Cowam.com (n.d.)

[6] See Schneider et al (2006)

[7] See Lagrange (2005)

[8] See OECD/NEA (2004)

[9] See BRWM (1990)

[10] See IAEA (1992)

[11] See OECD/NEA (1995)

[12] See KASAM-SKN (1988)

APPENDIX

Members of the Working Group on Long Term Issues Associated With the Management of Radioactive Waste

Research Team:

- ○ **France:** Thierry SCHNEIDER (Coordinator) and Caroline SCHIEBER (CEPN); Sylvain LAVELLE (ICAM).
- ○ **Belgium:** Michel BOVY, Gunter BOMBAERTS and Gaston MESKENS (SCK-CEN Mol).
- ○ **Switzerland:** Thomas FLÜELER (ETH Zurich).

Stakeholders:

- ○ **Belgium:** Hugo CEULEMANS, Jacques HELSEN and Joss PROST (MONA-Mol).
- ○ **Germany:** Eckhard KRUSE (Gartow Church representative); Jürgen WOLLRATH (BfS).
- ○ **Europe:** Laurent FUREDI and Mark O'DONOVAN (FORATOM, Europe).
- ○ **France:** Geneviève BAUMONT (IRSN); Eric CHAGNEAU (GIP Objectif Meuse); Joël CHUPEAU (EDF); Robert GRANIER (CLI du Gard); Benoit JAQUET and Jérôme STERPENICH (CLIS de Bure); Olivier LAFFITTE (CSPI La Hague); Alain MARVY (CEA); Chantal RIGAL (ANCLI); Wolf SEIDLER (ANDRA - projet ESDRED).
- ○ **The Netherlands:** Herman DAMVELD (chercheur indépendant).
- ○ **Romania:** Stela DIACONU (ANDRAD).
- ○ **Spain:** Felisa GARCÍA (ENRESA); Miquel FERRÚS SERAR (GMF); Fernando GARCÍA (Maire de Jarafuel); Jose Luis GÓMEZ (Maire de Frias); M. HERNÁNDEZ (Maire de Almaraz); Meritxell MARTEL (ENVIROS); Alfredo NAVARO (Maire de Valencia); Alfredo ROMERO (Maire de Mesas de Ibor).
- ○ **Sweden:** Olov HOLMSTRAND (Avfallskedjan - The Waste Network).
- ○ **Switzerland:** Pius KRÜTLI (ETH Zurich).
- ○ **United-Kingdom:** Shelly MOBBS (HPA).

Chapter 8
Three Models for Ethical Governance of Nanotechnology and Position of EGAIS' Ideas within the Field

Fernand Doridot
Center for Ethics, Technology and Society (CETS), ICAM Lille, France

ABSTRACT

*Based on an analysis of the current literature on nanoethics, this paper proposes to identify three differ-
ent models for ethical governance of nanotechnology, respectively called 'conservative model', 'inquiry
model' and 'interpretative model'. The propositions of the EGAIS[1] Research Project in terms of ethical
governance of nanotechnology are related to the latter model.*

INTRODUCTION

Abundant literature is nowadays devoted to the
ethical governance of nanotechnology. It is an
interesting attempt to position EGAIS' ideas and
recommendations within the field of the most
current theories on nano-ethics.

DOI: 10.4018/978-1-4666-3670-5.ch008

A particular western tradition turned the *issue*
into the preferred place for ethical questioning. In
this tradition, ethics begin by an issue or a dilemma
which impedes action, and which expresses itself
as a "what to do?" or a "how to act?" within a given
situation. Unless we want to remain at the level
of a casuistry very prone to turning complex, this
kind of essential discomfort leads to a reflection
on the principles, values and ethical theories to
implement or to follow in the given case, which
is this way related to other cases of the same kind.

Ethics' last stage lies in the implementation of the rules called forward to the particular given case, and, therefore, lies in the ethical choice which follows from these latter, and lies well in a return to action. Although it seems trivial maybe, this tri-partition seems to be found in most of the attempts at defining an ethical governance for nanotechnology. Nevertheless and since it is a matter of governance and, besides, for technologies most of which are *yet to come*, the first moment of a governance theory will most of the time be that of the *identification* of the issues to come, which will imply problematic cases requiring an action break and a subsequent adapted reflection.

Concerning ethical governance of nanotechnology, we will thus suggest distinguishing the following questions:

Question Q1: How to identify ethical issues related to the development of nanotechnology?

Question Q2: On which principles, norms, values, ethical theories, etc. shall we found and establish an answer to ethical issues related to the development of nanotechnology?

Question Q3: How to implement the retained solutions to resolving the issues related to the development of nanotechnology?

As we will see, most of the theories dealing with the ethics of nanotechnology place themselves one way or the other within these three questions. In addition, it seems possible to classify the nanotechnology's ethical governance theories in three wide categories or models, which distinguish themselves by the way they suggest answering these three questions. We will therefore distinguish a 'conservative model' (First Chapter), an 'inquiry model' (Second Chapter), and an 'interpretative model' (Third Chapter). And we will show that EGAIS' theoretical references along with its practical recommendations tend to place themselves in the latter model.

A FIRST MODEL OF ETHICAL GOVERNANCE OF NANOTECHNOLOGY: THE CONSERVATIVE MODEL

A first model we will suggest calling 'conservative' seems to sketch itself within all the reflections on the ethical governance of nanotechnology. Although this model does not benefit from a unified presentation or from a well-established consensus, it lies on certain hypotheses common to numerous authors, and which already find shared incarnations in numerous concrete governance processes. The chief of these hypotheses lies in the premise that ethical issues raised by nanotechnology are not particularly novel and that even if they were novel, the battery of principles, norms, and ethical values already available is largely enough to tackle them. This model is therefore conservative in that it answers question Q2 above in a conservative manner: the ethical issues related to the development of nanotechnology can and must be resolved by the implementation of principles, norms and values already known and implemented.

Some authors support a very extreme form of this model by disputing any novelty brought to the ethical questions raised by nanotechnology. Allhoff for instance (Allhoff, 2007; Allhoff & Lin, 2006), or even Holm (2005), in spite of a few nuances, both maintain that nanotechnology do is raise and restate recurrent and almost *a priori* identifiable ethical issues in a new form – such as the ones revolving around questions of safety, sustainability, privacy, dignity, equity, right to know or not to know, etc. These issues are viewed as no different from those identified from other technologies, and are also seen as already benefiting from reflections within well-established disciplines such as bio-medical ethics, business ethics, environmental ethics, neuro-ethics, etc. Other authors like Grunwald (2005), even if in overall agreement with the previous ones, admit nonetheless the existence of certain relatively novel issues, such as those related to the enhancement of human beings through technological means, or those deriving from a potential 'convergence' (similar to the convergence among scientific disciplines) within ethical issues handled by traditional disciplines of ethics. For instance, as noticed in Swierstra and Rip (2007, p.17), we may think that the prospect of the medical use of nano-devices introduced into the body and capable of autonomously deciding on the attack of detected cancerous cells is radically new, and is very different from ethical cases related to autonomous macro-devices (such as those that will probably be raised by the emergence of fully automated vehicles).

For other authors adhering to this model but ready to admit a larger emergence of novel ethical issues, their identification must nevertheless resort to traditional methods. Laurent (2010) describes this trend very well under the name 'truth-ethics'. Within this trend, the identification of ethical issues falls within the domain of the ethicist, helped by the scientist. It is done on the basis of a scientific clarification of situations which allows clearing the fantasies that often go with emerging technologies. It can benefit from a more or less systematic exchange with scientific practices, for instance in an 'interactionist' version of truth-ethics as Laurent describes it (Laurent, 2010, p.124). In any case, they all join in the judgement formulated by example by Chris MacDonald according to which "ethical reflection on nanotech requires that we apply ethical principles to new domains, but it does not demand new principles" (MacDonald, 2004, quoted in van de Poel, 2008, p.31). Thus, in this model, once question Q1 above is made clear, the second phase of the ethical reflection often brings itself back to the implementation of certain well-established and already known ethical principles. We can appreciate here this model's connection with the bio-ethics one, which is claimed by some to be the most fit for nanotechnology (see Ebbesen, Andersen, & Besenbacher, 2006 for example). The principles, norms and values set forward and sup-

posedly sufficient are often the ones inherited from bio-ethics. MacDonald thus mentions informed consent, risk minimization, and the protection of vulnerable populations (MacDonald, 2004). In Ebbesen, Andersen, and Besenbacher (2006), the principles of autonomy, charity, and equity, are called forth among others (see p.456).

This perspective does not go without problems. In particular, this latter remains largely dependent on a deductive conception of applied ethics, in which a set of preexisting principles, methods, and normative concepts is simply applied to a new domain. This implies that this perspective meets all the critics traditionally opposed to deductive ethics. Van de Poel (2008) summarizes them very well. On the one hand, it can be illusory to seek a basic reference framework allowing an absolute consensus. Thus in the case of nanotechnology, one can deduce from different ethical traditions some moral judgments that are contradictory to each other. (For example it is well-known that the right to the 'enhancement' of a person by technological means can be justified by certain systems of ethics and denied by others.) On the other hand, the search for a normative and theoretical framework beyond any concrete consideration could also be illusory. For example, a distinction such as that between the treatment of a disease and the 'enhancement' of a person can be *a priori* theoretically relevant, but can be blurred in the concrete situations caused by the use of nanotechnology.

To counter these type of complications, the supporters of this model seem to insist nevertheless on a refined appreciation of the context of a situation by the ethicist in charge of applying the predetermined principles. The most delicate question here is obviously the one pertaining to situations that demonstrate authentic value conflicts and where all the different principles and norms to apply seem to contradict each other. For instance, the case of nano-robots introduced in the body and deciding in an autonomous manner to attack cancerous cells by liberating particular substances seem to set in conflict the principle of nonmaleficence (administering medication is never trivial and there is always a risk of losing control involved) and the principle of beneficence, as well as the principle relating to the respect of the patient's autonomy. In this case, Ebbesen, Andersen, and Besenbacher (2006) suggests that it goes back to the ethicist's duty to wisely ponder and balance the priority and the importance of the principles in order to go beyond and settle the contradictions (p.457). The ethicist can also be faced with deciding between two differing interpretations of the same principle. For instance, when it comes to issues related to human enhancement, the same principle of respect for human dignity can benefit from a conservative interpretation forbidding any intervention on human beings, or from a liberal interpretation promoting rather the informed consent of individuals (Laurent, 2010, p.123). Even if they are usually built as more or less definitive systems (Ebbesen, Andersen, & Besenbacher, 2006, p.456), principles are nonetheless prone to evolution under the effect of a constant adjustment with the real issues they must apply to, this way achieving some sort of 'reflected equilibrium' à la Rawls, of which the derived procedure can be more or less institutionalized (*Idem*, p.457). There is nevertheless a wide inertia of principles systems. For Grunwald for instance, any amendment to the principles remains useless as long as there is a normative framework sufficient to answer to identified ethical issues and which is also unambiguous, locally consistent, in accordance with the laws and regulations in place, and largely accepted by everyone (Grunwald, 2000; 2001). Van de Poel (2008, p.33) notes that the last condition is rarely fulfilled in practice.

Despite its limitations, the conservative model is in fact much represented in the practices of governance. Numerous 'ethics committees' or other 'expert panels' are inspired from it, in Europe as well as in the United States. Its success can be explained by its simplicity, and by its capacity to extend the traditional conceptions of moral labor's division which, despite some evolutions, remains common to numerous actors (see for instance Rip & Shelley-Egan (2010) for an illustration of that point).

A SECOND MODEL OF ETHICAL GOVERNANCE OF NANOTECHNOLOGY: THE INQUIRY MODEL

In a second model, a need is sensed for a more specific approach of ethical issues related to the development of nanotechnology as well as of the consideration of a larger spectrum for the principles, norms and values' explanations and choices to tackle these latter. Two questions are therefore raised: how to identify on a larger scale the emerging ethical issues? On which principles, norms and values shall we build and establish the ethical choices relating to nanotechnology? Given the agreement that the traditional repertoire is not enough anymore, and that ethicists are not considered the sole depositary agents of the answers to these questions anymore, this second model will often take the form of an *inquiry*, diversely guided, but always aimed at answering those two questions.

This model can then be qualified an as *inquiry model* in the sense that it aspires to answer questions Q1 and Q2 above through inquiry: the ethical issues related to the development of nanotechnology can and must be identified through diverse forms of inquiry, and their resolution can and must be done through the implementation of principles, norms and values which can and must be identified through diverse forms of inquiry.

Such a model has found for a long time echoes in terms of governance in the sense that certain traditional practices naturally borrow from it. The different forms of proceduralism (whether it belongs to Habermas, Rawls, or Maesschalck and Lenoble which represent criticism of the previous ones) thus join within an inquiry model: it is a matter of identifying through the completion of a procedure allowing the expression of a plurality of actors, the ethical norms able to answer the challenges of real situations. The case of nanotechnology is nonetheless interesting insofar as the literature which is devoted to it offers a clear partition between answers aiming at the facts (answering this way question Q1) and answers aiming at principles (answering this way question Q2).

The Identification of Ethical Issues through Inquiry

First of all, the simple identification of ethical issues related to the development of nanotechnology can be itself the object of different types of inquiry. It is deemed possible to identify numerous ones in the literature.

A first solution (a) consists in taking seriously the anticipation discourses (whether they are produced by scientists themselves, or politicians, or science-fiction authors, etc.), and use them as a basis to identify aspects of the future which appear to be problematic from an ethical point of view. Drexler (Drexler, 1986) was probably the first representative of this trend by leaning directly his visions of the nano-technological future on a warning against the issues which could impede them – and by popularizing in particular the 'grey goo' risk, which was put into perspective subsequently. This trend was then implemented in the United States within the practice conducted by ethicists enrolled by NNI to make explicit the 'ethical obstacles' to the development of nano-technology. This trend is wide-spread nowadays and constitutes a natural *habitus* for numerous actors reflecting on the ethical aspects of nanotechnology. Nevertheless, with the arrival of more professional ethicists and social scientists, this approach suffered growing criticism. This trend was particularly discredited by Alfred Nordmann under the label 'speculative nanoethics' (see Nordmann (2007a), and also Nordmann & Rip (2009)). For Nordmann, the ethicist identifying ethical challenges from technological anticipations

– often deceiving ones – accepted without any critical evaluation prohibits himself from taking into account more concrete and real issues, and involuntarily supports illusory processes serving above all strategic interests. In Lucivero, Swiestra, and Boenink (2011), this criticism is extended in a very interesting manner by attempting to refine the 'reality checks' advocated in Nordmann & Rip (2009). It is obvious that one of the major drawbacks of this approach lies in that its purely hypothetical basis only allows for unwarranted ethical diagnosis, or justified on *a priori* criteria and independent from any consideration of real contexts at the very least – which, in a sense sends us back to the conservative model and its limitations.

Another solution (b) benefits factually from a large audience in terms of nanotechnology's ethical governance. It consists in the implementation of large, very diversified 'participatory panels' called forth to express themselves on issues which appear to be problematic to them from an ethical point of view within the development of nano-technology. In the end, it thus involves enrolling as many 'stakeholders' as possible in the most precocious identification possible of 'ethical issues' related to the evolution of nanotechnology. This governance method seems common at last, in spite of some nuances, to multiple theoretical thought frameworks of governance. The promoters of 'responsible development of nanotechnology' have made it their natural way. It is also the most frequent form of expression for 'anticipatory governance'. This latter aims at a collective construction of the technological future, in particular through the development of different anticipation scenarios, and recently found expressions to the case of nanotechnology (see Karinen & Guston (2010)). We could without doubt show that the concepts of foresight, of preparedness also often express themselves through this method.

But despite its political successes, this method suffers from numerous critics nowadays. Nordmann and Macnaghten, especially, stigmatize it under the label 'Conversational Mode' (Nordmann & Macnaghten, 2010, p.135). They see in it a 'game of language' tied to the indefinite production of different 'concerns' by a wide directory of actors carrying contradictory interests, called forth to participate to an endless and mainly formal conversation. With its temporality perpetually renewed, this latter prevents itself from any radical questioning of the nanotechnology's program, and, being a prisoner of a 'lure of the yes' (Nordmann & Schwarz, 2010), tacitly supports the idea of the inevitable development of nanotechnology. Its only production lies in the quantitative listing of large categories of 'concerns', such as privacy, safety, distributional justice, ownership, etc. (Nordmann & Macnaghten, 2010, p.137). These are not any more questioned and exist mostly as concerns to avoid neglecting or even as infringements to resolve when the time comes. The whole set carries out a parody of proceduralism which Nordman qualifies as 'symbolic proceduralism' (Nordmann, 2011).

The Identification of the Ethical Principles through Inquiry

The attempts mentioned above remain most of the time only at the level of the identification and the census of aspects or effects problematic from an ethical point of view arising out of the development of nanotechnology. They provide few leads for an analysis and resolution of the identified issues. In particular, they do not tackle the question of the principles, norms or values on which to build and establish an ethical approach of nanotechnology, and of the conflicts likely to arise among them. Yet, this question is critical especially if, contrary to the hypotheses of the 'conservative model' above, we assume that those principles and values are not *a priori* known and require a specific reflection. This reflection can, from then on, take also the form of an inquiry. Which principles, norms and values should be promoted in an ethical examination of nanotech-

nology, and where to find them? This question is even more difficult as even a slightly well-informed observation seems to offer the entertainment of a wide diversity of potential conceptions. This way, Shummer for instance, identifies a multiplicity of socio-cultural factors (in particular language, cultural heritage, economics and politics) likely to influence on the priorities and the meanings granted to the ethical challenges of nanotechnology (see Schummer (2006)).

A Diachronic Inquiry

In fact, literature offers various contributions to the ethics of nanotechnology which seem to us that they can be interpreted as moments of such an inquiry. The inquiry can first be diachronic and focus on past controversies. This is how, for instance, Swierstra & Rip (2007) examine the major types of arguments rallied to the occasion of the debates which were carried in the past on emerging technologies. Swierstra and Rip identify consequentialist, deontological, justice arguments, or even arguments related to a 'Good Life' ethics (*Idem*, pp. 11-16). The confrontation of these arguments draws different 'patterns of moral argumentation'. The whole constitutes a certain form of recurrent and common 'grammar' to all these debates which, we may think, will be carried over by the debates on nanotechnology. If these authors dispute an ethical specificity to nanotechnology while referring rather to the larger category of emerging technologies, they nevertheless advise us on the principles and values likely to clash regarding the ethical decisions on nanotechnology.

"Vision Assessment" and "Analysis of the Metaphysical Program" as Examples of Ethical Inquiry

Another solution to identify new principles, norms, and values on which to build and establish the ethical reflection around nanotechnology is the one suggested by the so-called 'Vision Assessment' (see for instance Grin & Grunwald (2000)). The Vision Assessment aims, among other things, at unveiling the principles and values encoded in the 'visions' or the 'expectations' expressed by different actors, analyzing and bringing to light the potential conflicts and controversies among them. The visions and expectations are not handled as objective productions but rather as points of views implicitly nurtured by certain world views and values. Even if in principle the Vision Assessment devotes itself to a plurality of protagonists, it seems to produce mostly analyses of discourses from scientists, politicians, or promoters of nanotechnology. It distinguishes visions according to their temporality (fictions, visions, guiding visions, goals, etc.) (Grunwald, 2004, p.56), or according to their thematic (Problem Related Visions, Assembler Bases Visions, Visions of Product Improvement, Material Based Visions, etc.) (Fiedeler, Grunwald, & Coenen, 2005, p.3). It can for instance seek to clarify which latent world visions express themselves in the discourses on NBIC convergence, molecular machines, etc. The unveiling thus realized, and the confrontation of visions, must in the end feed the collective deliberation on nanotechnology (see Coenen (2010)).

We can in a certain way relate to this tendency all the analysis devoted to what J.P. Dupuy labeled the 'metaphysical research program' of nanotechnology (see Dupuy & Grinbaum (2006)). Recapturing this notion from Popper, Dupuy sets to himself the ambition to unveil all the metaphysical assumptions – which means by definition not testable or falsifiable as scientific theories are – which implicitly lead the research on nanotechnology. Dupuy identifies in particular, among these assumptions, the promotion of 'bottom-up' methods which, breaking from the classic analytical methods of the 'top-down' kind, *de facto* transform the scientist and the engineer into sorcerer's apprentices, or even the ambition of

a reconstruction of nature, of life and mind within the framework of a materialistic and mechanistic paradigm. As well summarized in Macnaghten (2010), despite the differences according to the cultural contexts of expression or the scientific disciplines considered, it seems possible indeed to identify within the nanotechnology's programs a recurring core of common hypotheses blending visions of control and precision, of abundance and escape from scarcity, of emulation and/or improvement of nature, as well as a "'dream of reason' that is to overcome once and for all every given that is a part of the human condition" (*Idem*, p.25, which sends back to Dupuy (2007) for the last point).

It seems clear that all of these 'metaphysical hypotheses' carry a normative meaning and, as such, constitute the hidden expression of some particular ethical norms and values. The underlying values to the metaphysical program can be the object of a 'normative assessment' which will make their criticism in the name of other values (such as those of traditional humanism for instance), and which, one can guess, will essentially see in them the manifestations of a kind of contemporary *hybris*. Dupuy's thesis seems to be that these values are carried by the nanotechnologies themselves (it is a lesson which some virulent nanotechnology opponents have popularized while drawing the thesis that nanotechnologies are intrinsically bad). We will not rule here on the question of knowing how a technical object can in itself be carrying norms or values or even « do moral work » (for instance in the sense that according to Latour seat belts or speed ramps in roads do moral work). If there is an indisputable point to conclude here from these analyses of the metaphysical program, it is that in all cases these norms and values are endorsed by the scientists and promoters of nanotechnology. These norms and values are those which attract them in nanotechnology, which they point up and set forward, and which they seek to develop, to assert and exacerbate. From this point of view, the analysis of the metaphysical program of nanotechnology reveals itself to be a fruitful moment of the inquiry model regarding the answer to question Q2 above.

Deepen Project, "Narratives'", and Ethical Inquiry to Laypeople

Even if the inquiry on the principles, norms, and values has often turned towards the scientists and promoters of nanotechnology in the first place, it has nevertheless also targeted the lay people. We must here mention, within the lineage of these inquiries, the ones conducted in the framework of the European project Deepen (see Nordmann & Macnaghten (2010), Davies, Macnaghten, & Kearnes (2009), Ferrari & Nordmann (2009), Macnaghten, Davis, & Kearnes (2010)), which have acquired a certain fame. Deepen focuses on the discourses and positions expressed by 'lay people' without any particular experience in terms of nanotechnology. Its primary objective is to open a new way to "discover the ethical issues associated with nanotechnology" (Nordmann & Macnaghten, 2010), which implies, in our terms, to answer first and foremost question Q1 above. Nevertheless, along the way, Deepen will go beyond just the simple identification of new 'ethical issues' to reach the clarifying of principles and values incorporated more or less explicitly in the views of the 'laypeople', which it will set in comparison with the principles and values conveyed by the grand visions produced by the scientists and promoters of nanotechnology. Deepen resorts to various empirical strategies (theater, focus groups, etc.) to confront 'lay people' to scientific realities related to the development of nanotechnology as much as to the scientific discourses and the grand visions which present and justify them. It thus inspires discussions in which 'lay people' question certain key-concepts (such as those of human life, freedom, etc.) and position themselves according to them. These discussions are also the place for the expression of a certain

number of preoccupations related to the development of nanotechnology or even of visions quite pessimistic of this development which sharply contrast with the optimism of its promoters. In the end, these voicing of opinions and reflections seem to organize themselves according to certain motives or recurring patterns which they reproduce identically, and which Deepen suggests qualifying as 'narratives'. These narratives are thought as of cultural nature, as some kinds of archetypes deeply anchored in the European culture of which the dimension is affective and intuitive before being intellectual, and act as thought and imagination resources: "These are 'master-narratives' in Agnes Heller's sense: "guides of imagination" and "references to a shared tradition" which are "not just cognitively understood but also emotionally felt, without footnotes, without explanation or interpretation" (Macnaghten, Davies, & Kearnes, 2010, p.20, which quotes Heller (2006), p.257).

Deepen identifies five narratives to which he gives eloquent labels: 'Be careful what you wish for'; 'Opening Pandora's box'; 'Messing with nature'; 'Kept in the dark'; 'The rich get richer and the poor get poorer'. Deepen sums up itself the contents of the five identified narratives by the following succinct statements:

The 'be careful what you wish for' narrative builds on the age-old notion that getting exactly what you want may not ultimately be good for you, and may, inadvertently, lead to unforeseen disaster and catastrophe. This narrative was especially potent in structuring public resistance to the seductive and apparently boundless promises provided by nanotechnology. 'Opening Pandora's box' draws on the Pandora's box myth to provide a repertoire for articulating public unease about the hubris of meddling with things that should be left alone, of the danger of proceeding without limits, and of the likely subsequent release of a whole range of human evils. 'Messing with nature' is a further narrative deployed to express public concerns over nanotechnology's potential to disrupt long standing distinctions and boundaries – such as 'sacred' distinctions between the living and the non-living – in the face of the possibility of re-designing nature to our own needs. The 'kept in the dark' narrative expresses participants' sense of powerlessness in the face of nanotechnology's troubling but inevitable development, while 'the rich get richer and the poor get poorer' draws on how real-world drivers of commerce and consumption were seen as likely to further exacerbate injustice and inequality. Ultimately, the story goes, promises of green or socially relevant technology are likely to result only in the rich – big business and the already-powerful – benefiting, while the poor or excluded remain so. (Davies, Macnaghten, & Kearnes, 2009, pps.18-19)

It is clearly recognized by Deepen that the narratives thus identified are not specific to nanotechnology. They are seen more as establishing a recurrent framework of the relationship of the western man to technology on a long term ("from Icarus to Chernobyl"). A certain number of projects and reflections parallel to Deepen and inspired from a similar process have also identified other narratives which remain *grosso modo* of the same type and tally with them at certain points. One can mention among others the "'slippery slope' narrative, that technological advances that seem beneficial now will inevitably evoke further technological steps and applications that are morally doubtful; the 'colonisation' narrative, that technology will spread out and ultimately colonise life denying autonomy and agency; the 'Dr Strangelove' narrative, that advanced science designed for 'good use' will become corrupted and manipulated by evil people for evil purposes; the 'Trojan Horse' narrative, that innovations developed for progressive purposes will in the long term have unforeseen and potentially irreversible effects; the 'it's out' narrative, that involves the accidental release of harmful substances often due to technological and/or human failure" (Macnaghten, 2010, p.33). And also the "'bodily invasion' narrative, that

involves the introduction of invisible substances that subsequently violate natural processes; the 'Promethean' narrative, involving nature taking retribution on nanoscience's hubristic sense of its ability to transform both nature and humans to its own will and in violation and disregard for evolutionary process; the 'artificialist' narrative, that inadvertently instrumentalises life and human relationships trough conceiving of biological and mental life purely as machines; the 'false modesty' narrative, involving the pretense that nothing special is being undertaken" (Macnaghten, 2010, pps.33-34). Still in the same vein, Pavlopoulos, Grinbaum and Bontems (2010) refer as for themselves to multiple myths and fictions supposed to shape the relationship of man (and in particular of the 'layman') to technology, such as those of Prometheus, of the Pandora's box, of Daedalus, of the Golem of Jeremiah, of Frankenstein and of Matrix.

A Critical Insight on the Ethical Content of the So-Called "Narratives"

Numerous things could be said regarding the 'narratives' thus presented. First of all, their whole is very heterogeneous, going from grand classical myths which already benefited from mythological and literary forms (such as those of Prometheus and Pandora's box), to other constructs obviously more *ad hoc* (such as the 'false modesty' narrative, which evokes at its most a verbal scheme fit for certain contemporary aspects of the scientific communication, and which we could moreover set in comparison with the adverse scheme of 'false pretension' which seems as much represented). On the other hand, their nature would require them to be more precise. We can without a doubt and without effort recognize the status of cultural resource which has become an *habitus* of narratives such as the 'Be careful what you wish for' or the 'Messing with nature'. Western culture seems moreover to have associated to them a network of psychological complexes deeply anchored which makes them intuitive and molds our natural attitudes. Other narratives on the other hand (such as the 'The rich get richer and the poor get poorer' for instance) seem to be more political and to arise from a more rational effort and attitude. Others at last (such as the one of 'Bodily invasion') belong to yet another nature: their connotation is more imaginary than moral. They shall rather be related to studies dealing with the images (mental, dream-related, historical, mythological, etc.) and the stories through which nanotechnologies present themselves, or through which we apprehend them more or less spontaneously. Even if very interesting themselves, these studies seem to have less relevance for issues at the ethical level and with which we are concerned. One of the factors contributing to the confusion is that they often qualify themselves as 'narratives', with maybe more legitimacy than the previous ones (see Mordini (2007) for instance).

In the end, we must now ask the crucial question: what do the 'narratives' tell us in terms of ethics? How do they shed light on questions Q1 and Q2 above? Can we for instance find in the identified narratives some information on the principles or values promoted or carried by the laypeople, or on the way in which the principles or values' conflicts raised by the development of nanotechnology must be examined and resolved? It is with no doubt possible to see in the narratives the hidden expression of certain principles and norms, and of certain values. The 'Messing with nature' could without much effort express itself in the form of an incantation to respect nature for what it is, to not go beyond certain barriers, and to guarantee the continued existence of the 'sacred' distinctions retained by our long history, etc. The 'The rich get richer and the poor get poorer' could interpret itself as an affirmation of the values of justice and fairness, and insistence on their respect in an era which often ridicules them. From this point of view, the 'narratives' seem to present

certain analogies with the ethical principles and norms on the application of which is grounded the first conservative model. The temptation can even be there to try to organize the narratives in 'systems' comparable to 'ethical systems of reference' as they are promoted in the first model. It seems to us that this is to this type of ambition that, for instance, Dupuy's attempt to organize narratives in quasi-axiomatic systems owning certain consistency and completeness characteristics, and themselves decomposable in sub-systems of more natural coherence, points to (see for instance Dupuy (2010)). Nevertheless, a more advanced identification of narratives with ethical norms (and of narrative systems with norm systems) seems difficult. We must recognize that we do not find anything in the narratives' system which can shed light on the analysis and ethical choices while equaling the precision of the 'conservative model's' norms, as rigid and uncontextualized as they are. Nothing for instance, in the system of narratives, allows us to settle an ethical point of view in favor or not in favor of the possibility of human enhancement thanks to nanotechnology. In terms of ethics, the narratives' harvest is rather thin, or at least rests below the norms. A perspective on the true ethical nature of narratives can maybe be provided by Jean-Marc Ferry's theory of reconstructive ethics (see Ferry (1996)). Ferry suggests a model for the processes of ethical inter-comprehension made up of four moments which he respectively designates as the narration, the interpretation, the argumentation and the reconstruction. The narration is for Ferry the moment of the simple description of facts and events. It is followed by interpretation; this first reflective moment is a return on the experience, and aims at drawing the lesson or the moral from it, as in a tale. As such, it constitutes a sort of pre-norm. It is in the case of conflict of interpretation that speakers enter in argumentation, which is the very first true moment of rationale, and which can lead to agreement. Any completed inter-comprehension process must then benefit from a reconstruction phase in which the true rationales of the evolution and agreement of the speakers dawn and express themselves. Yet it seems to us that narratives belong to the *interpretation* phase. They express a sort of pre-ethics, a moral through dictums and through grand formulas, outlined in broad from the events' sight and the lessons of the past, and which can therefore collide with other interpretations. In terms of ethics traditions, it is moreover in the end the theory of virtues to which the narrative-encrypted ethics seem to come closer. Their general message is a message of balance, of equilibrium, of caution, reminding that everything carries an excess and a lack and that one must remain on a narrow crest line.

The Use of the Inquiry Model

We ought to finish addressing the question of practical implementation of this second model. In concrete terms, what should we do with the discoveries and results of the inquiry conducted as such, and how can we turn them useful to the benefit of ethical governance for nanotechnology? If it seems difficult to draw directly good deed rules from them, the general tone seems to be the will to make them serve the promotion of a more authentic form of *deliberation*. It is for instance the major recommendation of the Deepen project (see Davies, Macnaghten, & Kearnes (2009)). The general idea is that a more refined knowledge of norms and values defended by all the parties involved in the discussion around the development of nanotechnology should allow everyone to put his or her positions in perspective. The highlighting of values considered important by communities or by speakers not much listened to so far must ensure their consideration as well as their representation in the definitive decisions. The deliberation process must moreover open itself to less selective and disqualifying forms of expression compared to the traditional ones, such

as for instance the 'narrative' form. Besides, the unveiling of the narratives' system seems to be thought as likely to rehabilitate the 'lay-people's thought'. This latter seems to be seen as a kind of novel 'wild thought' (to wink at Levi-Strauss) unjustly stigmatized and discriminated as 'irrational', but which in fact carries its own logic, which deserves to be recognized and taken into account (Davies, Macnaghten, & Kearnes, 2009, p.23). In all cases, the ethicist in this model is the one who reconstructs the positions' system, reveals to each one the architecture of his personal normative conceptions, and highlights the latent oppositions (of principles, of values, etc.) in between the clashing world conceptions. Some authors seem nonetheless to foresee the possibility of more direct uses of the 'inquiry's' results. For Dupuy and Grinbaum for instance (Dupuy & Grinbaum, 2006), the clarification of the normative base of nanotechnologies and of their intrinsic values appears as a direct pledge for collective and desirable actions, as it allows the evocation of futures either 1) sufficiently desirable so they can be efficiently researched, or 2) sufficiently worrying so they can be efficiently rejected. For others the inquiry seems to be rather instrumentalized, and thought essentially as being capable to facilitate the communication and strategy of the scientists, who are made more familiar, thanks to the knowledge of the 'narratives', with the numerous biases which influence the troubled perception of the lay people (see Pavlopoulos, Grinbaum, & Bontems (2010) for instance).

In the end, it seems to us nevertheless that the inquiry's major interest must be researched somewhere else. While unveiling the narratives, it especially unveils their contingency. It also shows how the rigid 'principles' of the first model are in fact carried by representations and embodied by dynamic systems of world visions which rejoin, influence and oppose each other. It is, among other ones, this point which Nordmann and Ferrari (Ferrari & Nordmann, 2009) highlight when they draw the "lessons for nano-ethics" of the Deepen project. As they underline it, the laypeople's narratives allow unveiling all the vanity of the dominant narratives (those of the scientists, politics, nanotechnology's promoters, etc.), such as for instance their illusory quest of always win-win situations, or their delirious ambition of a recreation of the world. But this also invites to recognize that the laypeople narrative is not as it is more objective than the expert narrative. Moreover, the former only explains itself through a reaction against the latter: it is a point recognized in Ferrari & Nordmann (2009), and which also appears clearly if we really want to critically analyze the experimental protocols presiding over the identification of the laypeople narratives (as for instance they are mentioned in Macnaghten (2010)). From then on, every fair decision-making in terms of nanotechnology can without a doubt only reside in the transcending and the resolution of the antinomies implied by the confrontation of the two types of narratives. From this point of view, it seems, with regard to our second 'inquiry model', that the possible contradictions among the narratives, or even among the different parts of a same narrative, come to undertake a little the same role that the possible contradictions among the different principles or norms of a one and the same ethical reference system could undertake in our first 'conservative model'. Or, to say even better, what the contradictions among norms or principles reveal sometimes are precisely contradictions among underlying narratives, for which we must find a way to transcend. The inquiry model seems moreover to find here all its limits. How, in the end, settle between conflicts of narratives without having recourse to expert interpretations, as it is the case in the conservative model described above? This seems to direct us towards the need for considering a 'third model' capable to fulfill this challenge.

TOWARD A THIRD MODEL OF ETHICAL GOVERNANCE OF NANOTECHNOLOGY: THE INTERPRETATIVE MODEL

Conflicts between Ethical Norms Seen as Conflicts between Narratives, and Ultimately as Conflicts of Values

An interesting article from Wickson, Grieger and Baun (Wickson, Grieger, & Baun, 2010) seems to provide some answering elements to that question. In the same line of research as Deepen's, the authors take from a previous work of Wickson the identification of nine different narratives regarding the potential relationships between nanotechnology and nature, which they respectively designate as follows: 1) Nanotechnology *As* Nature; 2) Nanotechnology *Inspired By* Nature; 3) Nanotechnology *Improving On* Nature; 4) Nanotechnology *Using* Nature; 5) Nanotechnology *Transgressing* Nature; 6) Nanotechnology *Restricted By* Nature; 7) Nanotechnology *Controlling* Nature; 8) Nanotechnology *Threatening* Nature; 9) Nanotechnology *Treating* Nature (*Idem*, p.9). While analyzing the oppositions between narratives 8) and 9), which can seem antinomic, the authors prove that these latter can only be understood in light of the particular and tacit "ideologies" on which these two narratives are established. Each of these narratives, in particular, depends on a very particular interpretation of what nature is, of what a good or bad impact on nature is, of what is and what should be man's place in relation to nature, etc. In the end, every narrative supports a set of beliefs, convictions and above all a set of particular *value judgments* which its usual expression tends to hide. And it is only through *the interpretation* of the concepts rallied by the narrative (here in particular those of nature, care, etc.), and through the expression of value judgments which express themselves in their respect, that the oppositions between such narratives can be transcended.

This analysis is important for us. First of all, it draws lessons from the 'inquiry model' while assuming there is no ethical position or system (nor there is therefore a political position or system) which does not merge within a 'story' or a 'narrative', which forms its base and anchors it in a more general view of the world. In the example discussed above, the two narratives set in opposition can give rise to different systems of ethical norms. The first one, expressing the need to protect nature, would promulgate for instance ethical norms such as "every human being must be respected as such by the applications of nanotechnology." The second, expressing the need to treat or even to improve nature, would promulgate for instance ethical norms such as "every human being must be able to benefit fully from the opportunities of nanotechnology." Actually, the two norms above could even find themselves within one and the same ethical system. They would undoubtedly get into conflict on the occasion of their application on real cases, the same way the norms of rigid systems of the first 'conservative model' would get into conflict. However, this confrontation would only reveal, in fact, the opposition of the two narratives considered above. As perfectly shown moreover in Wickson, Grieger & Baun (2010), this opposition cannot reduce itself to a scientific issue, that new data or new technical or scientific experiences would be able to settle. It is indeed an opposition of world views, which therefore can only explain and transcend itself fully through the expression of the deep value judgments in which these views are anchored. And it is interesting to notice that, *mutadis mutandis*, everything that precedes could also apply to other types of narratives, norms and values. This way, for example, the opposition between two narratives of the "nano-technological chips threaten individual privacy" kind and of the "the nano-technological chips strengthen individual privacy" kind could come under the same analysis. Each of these narratives constitutes in fact a base of different norms and ethical systems,

using concepts such as 'privacy' with different meanings. Only the in depth interpretation of concepts like the one of 'privacy', especially *via* the expression of values through which they are invested and the value judgments which are in confrontation regarding them, can then allow to transcend the oppositions. It is moreover the need of such an analysis which is also pointed towards by, as a last resort, Nordmann and Ferrari in their conclusion of the Deepen project (see Ferrari & Nordmann (2009)). They raise in it the need of a deeper comprehension of the narratives identified by Deepen and call for a more in-depth examination of the concepts of risk, power, justice, agency, responsibility, accountability, etc., which could "render them more definite and meaningful" (*Idem*, p.59). If this examination must benefit from historical and philosophical perspectives for them, it seems clear that the challenge lies also in exploring the way in which these concepts along with the normative declarations produced about them, are apprehended by the actors within the fabric of subjective values and judgments which constitute their 'meaning', and therefore render them improperly suited to any literal and formal interpretation.

Prequisites for an Interpretative Model of Ethical Governance

The whole set of these propositions rests on a base of ideas and orientations around which the possibility of a 'third model' of governance seems to draw itself. This latter can be qualified as interpretative in the sense that it calls for the *interpretation* of the actors to answer questions Q1 and Q2 above. The idea is not just anymore identifying and revealing to the actors the underlying narratives which determine their preference for such or such ethical norm. The ambition is from now on to make try and test this narrative by the actors, as well as try and test these preferences, and to encourage the actors to interpret this narrative and these preferences *via* the expression of the values and the value judgments which condition their choice. This model carries as a first merit the fact that it makes tangible the plurality of the subjective contents able to be associated to a same norm. It is a phenomenon to which Nordmann points in Nordmann (2007b), by showing for instance how the preference in the nano-medicine field for the same 'efficiency' norm can lie on very different interpretations and values (such as the "the medical notion of individualized treatment, the scientific notion of root cause and targeted intervention at the cellular level, the policy consideration of cost-effectiveness" (*Idem*, p.227)). Nordmann concludes as a recommendation for the ethicist to disentangle the significances and to strive to a "productive disillusionment." But it seems to us that a third model would insist mostly on the process' dynamic. It falls to the actors themselves to experience the base values of their narratives along with testing and trying their contingency and their inherent conflictuality.

It is this way that the EGAIS project's recommendations join within an interpretative model. As it is explained elsewhere in this book[2], EGAIS pleads for a 'comprehensive proceduralism' offering to the actors the opportunity to test and try and to clarify their axiological commitment regarding norms, and especially the meaning that they attribute to them in terms of values. In doing so, the actors are requested to question and reconstruct their context (in Marc Maesschalck's sense) even if a total awareness of this latter is obviously impossible. EGAIS seeks to provoke the destabilization of the contexts and the awareness by the actors of the relativity of their moral position. But in this field, EGAIS goes beyond just argumentation and considers the possibility of a multiplicity of practices including among others the narration, the translation, or even the experimentation. The objective is to bring every actor, through a battery of personalizable means, to the apprehension of ethical norms from a point of view different to his own, and to try and test

the contextual thickness of his own commitment to the norms. It is at this price that an authentic justification of the ethical norms seems possible, as well as their application to a plurality of contexts.

In the end, the interpretative model is thus only the continuation of the first two ones on another mode. The first model considered the application of nearly self-justified ethical norms to new contexts, while resolving as it could the possible conflicts among these norms. The second one demonstrates that the choice of norms rests in narratives which are advisable to investigate, identify and reveal to the actors in order to improve the argumentation on the norms and to smooth out the oppositions. The third model assumes that only the interpretation by the values in play can guarantee an authentic awareness of the narratives, and the ability of the actors to consider other systems of reference, and coming, other norms than theirs. The selection of norms becomes from then on a dynamic process conducted by the actors themselves in a constant interaction which guarantees at the same time their applicability. The sad spirits will undoubtedly be able to see in this line of thought a long fall "from logos to pathos" as Gottweis' expression goes (Gottweis, 2005). Nevertheless, the attention given to the subjective investment of norms by the actors does not pave the way for moral relativism. As it is shown by Schummer in his conclusions of Schummer (2006) ("Cultural diversity without ethical relativism," p.228), this criticism is itself mostly dependent on particular values and biased interpretations regarding what human beings and their ethics are. To the contrary, this attention to the values vouches for the evolutionary possibilities of the actors, and in the first place, for their learning. Only this way of thinking the justification of norms seems to us moreover to allow apprehending what Swierstra and others (Swierstra, Stemerding, & Boenink, 2009; Lucivero, Swierstra, & Boenink, 2011, p.9) call the "techno-moral change," which means the co-evolution of technology and the surrounding moral standards, which, for instance, will undoubtedly show us with a different perspective the ethical challenges of human enhancement in a few decades.

Which Implementations for an Interpretative Model of Ethical Governance?

It is hard to find nowadays concrete examples of the implementation of such an ethical governance model. As the EGAIS project details elsewhere[3], there is maybe not an only ideal tool of governance likely to embody to their best the recommendations of the interpretative model. In this field, there is no "one best way," and the same governance tools are likely to be used differently. This way for instance, on the occasion of the discussions implemented by the Deepen project, some interpretative advances are realized here and there (even if the simple description of the narratives which is drawn from them tends undoubtedly to hide it). Other governance tools such as for instance the writing of scenarios can also undoubtedly get, at certain levels, in the standards of the interpretative model. It is important in that sense to go beyond the simple phase of the identification of the 'values' carried by the actors participating to those scenarios (which pertain as it is to the 'inquiry model'), and to turn those scenarios into learning moments in which the actors test and try as well as discuss the nature and depth of their axiological commitment. This is, it seems to us, towards which Laurent points. In Laurent (2010), he describes different scenarios imagined by a project called *Nanofutures* within the Center for Nanotechnology and Society of the Arizona State University. We find in it for instance the following scenarios: "'Living with a chip in the brain,' in which a brain chip provides information into the user's brain during his sleep. 'Automated surveillance of water treatment systems,'" in which an ultrafast sequencing technology is used to analyze the DNA fragments available in waste water allowing this way a more acute control of populations. "Disease detector," in which a device measures the protein levels of an individual and detects abnormal levels before the disease's symptoms arise'" (*Idem*, p.140). Laurent considers the way of apprehending these exercises as unique experiences to explore the social and technological identities, which disturb, transform and mold, in one same move, the systems of references of the actors, and the technological possibilities of the field (*Idem*, p.142). More generally, the entire set of governance practices associated with the so-called "Real-time Technology Assessment" carries undoubtedly characteristics which are likely to realize some aspects of the interpretative model. We can think of, for instance, the "embedded humanists'" practice in which an ethicist is integrated to a research laboratory somehow like the other researchers. Once again, if his practice is limited to observing, describing, analyzing and revealing some systems of latent values, the interpretative level is not achieved. But in certain cases the embedded humanist can also participate to the collective transformation, try and test with the others the systems of norms through the thickness of values, and finally, be embarked in an adventure in which the contexts are collectively redrawn. Obviously, his knowledge and the inherent purpose of his 'mission' always distinguish him and make him a piece apart from the process. But the important thing is that his mission be undertaken without *a priori* expectations, with flexibility, and with great care regarding the choices of the levers which are the most adapted to the individual and collective questioning (for some examples, see Laurent (2010), pp.138-139).

AN EXAMPLE: THE ISSUE OF PRIVACY IN ETHICAL GOVERNANCE OF NANOTECHNOLOGY

Privacy issues belong to nanotechnology's ethical challenges which are set forward the most in the literature. As we open the doors to smaller and smaller devices able to collect and share information as well as able to more efficiently store and network that information, nanotechnology can only mechanically expose society to a growing risk of excesses and corrupt uses of individuals' personal data. In France, the Commission Nationale de l'Informatique et des Libertés (CNIL) (the National Commission for Computing and Civil Liberties) greatly insisted, among others, on the irreversibility risk related to the growing miniaturization – and therefore to the growing undetectable aspect – of nano-technological devices (Türk, 2011). However, and despite the scope and uniqueness of the privacy issues raised by nano-technologies, it seems possible to recognize with regards to their ethical governance the same main tendencies as identified previously along with the resort to the different 'models' described above.

Thus, for certain authors (model 1), the privacy issues raised by nanotechnology are only joining the lineage of older and more classic concerns, such as those related to the growing data exchanges through the internet, the tracking and identification possibilities inherent to the use of smartphones, the more and more generalized resort to RFID mechanisms in the industrial and mass-market distribution fields, etc. In substance, these issues do not come under another ethical analysis, and can be tackled from the same theoretical framework as the previous ones. This theoretical framework is often the one of a system of principles inherited, for instance, from bioethics. It is this way that, for instance, Ebbesen, Andersen, and Besenbacher (2006) analyzes and studies the privacy issues by the yardstick of the fundamental principle of the respect of autonomy, which is another name for the right to self-determination. Indeed they understand privacy as the right to authorize or to decline access, which is directly justified by the right of autonomous choice. In that respect, the right to privacy benefits from the same justification as the right to give an informed consent (see *Idem*, p.460, which refers to Beauchamp and Childress). Ebbesen, Andersen, and Besenbacher feel nonetheless the need to complete the principle of the

respect of autonomy by a principle of respect of integrity, allowing among other things to ensure a respect for privacy for the people unable to exercise their autonomy (such as kids, patients under the influence of drugs or medication, etc.). Integrity is then defined as the sphere of experiences, information, etc. of an individual within which no one can breach (*Idem*, p.460). We could therefore, it seems, relate it to the concept of intimacy.

Still within this first model, other actors, while maintaining the reference to predefined principles' systems, insist on the inherent contradictions to this organization. Allhoff for instance analyzes the challenges of privacy in terms of the conflict between the two competing values of privacy and security. "As privacy increases, security decreases, and vice versa" (Allhoff, 2007, p.19). For Allhoff, the issues raised by nanotechnology require trade-offs and arbitration which in no way differ from those required by, for instance, the implementation of the United States Patriot Act. Voted and enacted after the September 11th 2001 attacks, this latter gives the federal government the authority to intercept wire, oral, and electronic communications relating to terrorism, particularly *via* certain "enhanced surveillance procedures" (*Idem*, p.19). The ethicist is then required for arbitrating the priority conflict between privacy and surveillance, and Allhoff seems to consider that this conflict can only be resolved through utilitarian terms with regards to the benefit probabilities to expect from the agreed infringements to the principle of respect for individual rights (*Idem*, p.20). In the same vein, other authors bring the privacy issues to arbitrations between the respect of individual data and efficiency, like in the case of nano-medicine where the possibilities for early diagnosis promised by nanotechnology will only be realized at the price of a systematic crossing of a multitude of data pieces collected on the individuals, which obviously exposes to the risk of information leaks or to corrupt uses of these information (see Toumey (2007), p.218).

In the end, the approach in terms of *a priori* principles such as the first model suggests it can seem limited. For instance, reducing 'privacy' to 'integrity' can seem very poor rescue in the resolution of concrete issues, and does not help filling the concept of privacy with meaning at all. However, if it is taken through its deductive angle, the first model can also lead to the promotion of new rights. This way for instance, the European Group on Ethics in Science and new technologies (EGE) proposes to extend the "habeas corpus" in a "habeas data," a sort of inalienable right of scrutiny of the individual on the broadcasting and use of its personal data no matter what the type of the data is (see EGE (2005)). One can also notice the tendency to create *ad hoc* rights, a way for the states to counter at the root new threats to privacy carried through new devices, often as a reaction to original excesses on the subject. Toumey thus points out how "the Driver's Personal Privacy Act arose after a stalker obtained someone's personal information from a data base of driver's licenses," as well as how "the Video Privacy Act constituted a reaction to a case of publicizing someone's record of renting movies" (Toumey, 2007, p.215). The approach in terms of first model will undoubtedly inspire rights of this type for the users of nanotechnology, in the same way as it inspires all the *value sensitive design* initiatives applied to nanotechnology, such as for instance the promotion of "privacy friendly" devices.

In the framework of a second model (model 2), inquiries can be conducted on privacy issues in order to collect information on the important challenges and on the principles enabling to face them. The oppositions between the principles mentioned above can particularly, through this, be referred to deeper oppositions of values, cultures, narrative structures, etc. In all cases, one will seek through this knowledge to 'fuel the debate', and particularly to improve the argumentation on these topics. First of all, the inquiry can be diachronic. Chris Toumey (*Idem*) for instance shows how the privacy issue in nanotechnology

can only be addressed through the reference of certain pre-existing patterns regarding the relationships between technology and privacy, since the Post Office Act enacted by British colonial authorities in 1710 which marks the origin of privacy policy in the U.S., until the Patriot Act of 2001 which provides the American president with quasi-unlimited powers in the fight against terrorism. Then, the inquiry can aim at taking a hold of more contemporary elements through diverse ways. Schummer (2006) for instance points out that the concept of individual privacy itself comes across big cultural variations. In Germany one hides one's salary while in Scandinavia, it is almost public information. One is very tolerant of nudity on German and Scandinavian beaches, but very restive about it in England. On the other hand, England is the first country which generalized the use of surveillance cameras, which remains considered a violation of privacy rights in a lot of European countries. In short, it is therefore possible that the challenges related to privacy for nanotechnology will be perceived differently depending on the country. Within the framework of this second model, one will be led to consider that the conflict between the competing values of security and freedom can be settled differently depending on the weight conceded to these values by the respective societies (Schummer, 2006, p.222), and no more uniquely by the utilitarian appreciations of ethical experts like in the first conservative model. Besides, one can also direct the inquiry on the 'privacy' practices, and on the positions regarding privacy identified within the society. In the absence of detailed data on nanotechnology, tendencies demonstrated with regards to other fields can prove to be informative. Thus one will be able to appreciate how much, paradoxically, "privacy is no longer a social norm," as the founder of Facebook Mark Zuckerberg declared and stated in January 2010. Hauptman, Sharan & Soffer (2011) provides for instance interesting data regarding the dedicated usage of teenagers for social networks. The conclusions of their investigation hold in the following six observations: "The concept and perception of privacy is important to adolescents but is transforming"; "Most adolescents (88%) have and use Social Network Sites (SNS)"; Many users have little knowledge or even misconceptions about visibility and privacy policy"; "Students' privacy settings on SNS are affected by their lifestyle"; "Students using SNS are aware of potential risks like privacy intrusions or misuse of personal data but they are not concerned about privacy invasion and data abuse"; "Adolescents are willing to use new technologies but they are balancing between the advantages of using new technologies and protecting their privacy/personal data" (*Idem*, p.143). Finally, in an interesting way, the inquiry can also attempt at revealing the 'narratives' or the thought schemes in which the appreciations and position takings anchor themselves on the topic of privacy. For instance Toumey (2007) identifies three different kinds of relations between the individual and the state in the sphere of privacy-and-nanotech which, it seems to us, function as 'grand visions' or large normative resources usually unquestioned. One of them is the famous Panopticon (based on the concept of J.Bentham which was revisited by Foucault), in which, "the state disciplines the behavior of the individual by making it possible to observe one's behavior while also making it uncertain whether it is actually observing at any given time." The second one is what Toumey calls the "Big Brother state," in which "the state observes some behaviour constantly, and punishes those who disobey its rules." The third one is the perspective known under the term "underveillance" (meaning observation from below, whereas surveillance is observation from above), and in which "the tools of observation are used to resist state surveillance and other state behaviour" (*Idem*, p.219). We can think that, for individuals joining these ways of thinking, the connection to the privacy norm will be largely influenced by these images, which will feed and partially determine their argumentative efforts.

A third model (model 3), such as the one advocated by the EGAIS project, would then aim at deeper interpretations and experiments of the terms and components of these narratives. It is hard to provide realized examples of it today. But, it would imply, by all means possible, to make the actors experience the significance by which they invest this privacy concept, and the fabric of values through which they apprehend it. Toumey for instance insists on two crucial points, which he borrows respectively from M. Mehta, and from J. van den Hoven and P. Vermaas (see Toumey, 2007, p.220). The first one belongs to the natural links between the concepts of trust and privacy. "A society that depends on trust needs to maintain privacy" (*Idem*, p.220). How, from then on, do individuals invest, in the most intimate manner, the privacy norm with regards to their basic conceptions of the value of trust? The second one is the observation that possessing data about someone is not the same as knowing the person subjectively (*Idem*). How, from then on, do individuals interpret the privacy norm with regards to the value they grant to the discovery of and knowing of others, and to the terms of such knowing? In the eyes of EGAIS, only the exploration and experience of the answers brought to this kind of questions would undoubtedly be likely to truly establish the privacy norm and to allow its effective application.

CONCLUSION

In the preceding pages, we have shown that the reflections developed within EGAIS were relevant regarding the general trends of the reflection about ethical governance of nanotechnology. EGAIS obviously does not intend to revolutionize this governance. Facing the already very big ensemble of tools available at the European level to frame the development of nanotechnology, EGAIS doesn't suggest really new tools. The emphasis of EGAIS is on a more interpretive use of the existing tools, which involves the reconstruction of the context of the actors involved in the governance process, and a consideration of the subjective values on which can be built and embodied the ethical norms. We have tried to show how some aspects of this requirement were already expressed in several schools of thought and in several projects that have taken nanotechnology as a case study. The ambition of EGAIS is to help unite these impulses, developing the overall concept of 'comprehensive proceduralism'[4] as a framework for covering them. We have outlined the possible directions that could take reflections consistent with our promoted 'interpretative model' of ethical governance of nanotechnology in the much discussed field of the privacy issues raised by nanotechnology. No doubt that, in the coming years, the need for ethical norms adapted to the new realities of the development of nanotechnology will become more and more pressing, and we hope that the principles put forward by EGAIS will be able to be of great help for the task of establishing them.

REFERENCES

Allhoff, F. (2007). On the autonomy and justification of nanoethics. In Allhoff, F., & Lin, P. (Eds.), *Nanotechnology and society: Current and emerging ethical issues* (pp. 3–38). Dordrecht: Springer. doi:10.1007/s11569-007-0018-3.

Allhoff, F., & Lin, P. (2006). What's so special about nanoethics. *The International Journal of Applied Philosophy, 20,* 179–190. doi:10.5840/ijap200620213.

Coenen, C. (2010). Deliberating visions: The case of human enhancement in the discourse on nanotechnology and convergence. *Sociology of the Sciences Yearbook, 27,* 73–87. doi:10.1007/978-90-481-2834-1_5.

Davies, S., Macnaghten, P., & Kearnes, M. (Eds.). (2009). *Reconfiguring responsibility: Lessons for public policy (Part 1 of the report on deepening debate on nanotechnology)*. Durham: Durham University.

Drexler, K. E. (1986). *Engines of creation. The coming era of nanotechnology*. New York: Anchor Books.

Dupuy, J. P. (2007). Some pitfalls in the philosophical foundations of nanoethics. *The Journal of Medicine and Philosophy, 32*(3), 237–261. doi:10.1080/03605310701396992 PMID:17613704.

Dupuy, J. P. (2010). The narratology of lay ethics. *NanoEthics, 4*(2), 153–170. doi:10.1007/s11569-010-0097-4.

Dupuy, J. P., & Grinbaum, A. (2006). Living with uncertainty: Toward the ongoing normative assessment of nanotechnology. In *Schummer* (pp. 287–314). Baird. doi:10.1142/9789812773975_0014.

Ebbesen, M., Andersen, S., & Besenbacher, F. (2006). Ethics in nanotechnology: Starting from scratch? *Bulletin of Science, Technology & Society, 26*(6), 451–462. doi:10.1177/0270467606295003.

European Group on Ethics in Science and new technologies (EGE). (2005). *Ethical Aspects of ICT implants in the human body. Opinion of the European Group on Ethics in Science and new technologies to the European Commission*. Retrieved January 19, 2012, from http://ec.europa.eu/bepa/european-group-ethics/docs/avis20_en.pdf

Ferrari, A., & Nordmann, A. (Eds.). (2009). *Reconfiguring responsibility: Lessons for nanoethics (part 2 of the report on deepening debate on nanotechnology)*. Durham: Durham University.

Ferry, J. M. (1996). *L'Éthique reconstructive*. Paris: Éditions du Cerf, Collection « Humanités. »

Fiedeler, U., Grunwald, A., & Coenen, C. (2005). *Vision assessment in the field of nanotechnology - A first approach.*

Gottweis, H. (2005). Regulating genomics in the 21st century: From logos to pathos? *Trends in Biotechnology, 23*(3), 118–121. doi:10.1016/j.tibtech.2005.01.002 PMID:15734553.

Grin, J., & Grunwald, A. (2000). *Vision assessment: Shaping technology in 21st century society towards a repertoire for technology assessment.* Springer. doi:10.1007/978-3-642-59702-2.

Grunwald, A. (2000). Against over-estimating the role of ethics in technology development. *Science and Engineering Ethics, 6*, 181–196. doi:10.1007/s11948-000-0046-7 PMID:11273446.

Grunwald, A. (2001). The application of ethics to engineering and the engineer's moral responsibility: Perspectives for a research agenda. *Science and Engineering Ethics, 7*, 415–428. doi:10.1007/s11948-001-0063-1 PMID:11506427.

Grunwald, A. (2004). *Vision assessment as a new element of the FTA toolbox.* Retrieved January 20, 2012, from http://forera.jrc.ec.europa.eu/fta/papers/Session%204%20What's%20the%20Use/Vision%20Assessment%20as%20a%20new%20element%20of%20the%20FTA%20toolbox.pdf

Grunwald, A. (2005). Nanotechnology - a new field of ethical inquiry. *Science and Engineering Ethics, 11*, 187–201. doi:10.1007/s11948-005-0041-0 PMID:15915859.

Hauptman, A., Sharan, Y., & Soffer, T. (2011). Privacy perception in the ICT era and beyond. In von Schomberg (2011) (pp. 133-147).

Heller, A. (2006). European master-narratives about freedom. In Delanty, G. (Ed.), *Handbook of European social theory* (pp. 257–265). London: Routledge.

Holm, S. (2005). *Does nanotechnology require a new "nanoethics"?* Cardiff Centre for Ethics, Law and Society. Retrieved December 20, 2011, from http://www.ccels.cf.ac.uk/archives/issues/2005/

Kahan, D. M., Slovic, P., Braman, D., Gastil, J., & Cohen, G. L. (2007). Affect, values, and nanotechnology risk perceptions: An experimental investigation. *GWU Legal Studies Research Paper No.261; Yale Law School, Public Law Working Paper No.155; GWU Law School Public Law Research Paper No.261; 2nd Annual Conference on Empirical Legal Studies Paper.* Retrieved July 19, 2012, from http://ssrn.com/abstract=968652

Kaiser, M., Kurath, M., Maasen, S., & Rehmann-Sutter, C. (2010). *Governing future technologies.* Springer. doi:10.1007/978-90-481-2834-1.

Karinen, R., & Guston, D. H. (2010). Toward anticipatory governance: The experience with nanotechnology. In *Kaiser* (pp. 217–232). Kurath, Maasen, & Rehmann-Sutter.

Laurent, B. (2010). *Les politiques des nanotechnologies.* Editions Charles Léopold Mayer.

Lucivero, F., Swierstra, T., & Boenink, M. (2011). Assessing expectations: Towards a toolbox for an ethics of emerging technologies. *NanoEthics, 5*(2), 129–141. doi:10.1007/s11569-011-0119-x PMID:21957435.

MacDonald, C. (2004). Nanotech is novel; the ethical issues are not. *Scientist (Philadelphia, Pa.), 18*(8).

Macnaghten, P. (2010). Researching technoscientific concerns in the making: Narrative structures, public responses, and emerging nanotechnologies. *Environment & Planning A, 42*(1), 23–37. doi:10.1068/a41349.

Macnaghten, P., Davies, S., & Kearnes, M. (2010). Narrative and public engagement: Some findings from the DEEPEN project. In von Schomberg & Davies (2010) (pp. 13-30).

Mordini, E. (2007). The narrative dimension of nanotechnology. *Nanotechnology Perceptions*, *3*, 15–24.

Nordmann, A. (2007a). If and then: A critique of speculative nanoethics. *NanoEthics*, *1*(1), 31–46. doi:10.1007/s11569-007-0007-6.

Nordmann, A. (2007b). Knots and strands: An argument for productive disillusionment. *The Journal of Medicine and Philosophy*, *32*, 217–236. doi:10.1080/03605310701396976 PMID:17613703.

Nordmann, A. (2011). *Between conversation and experimentation*. Lecture given on March 30, 2011, during the Second Workshop of the EGAIS Project, Brussels.

Nordmann, A., & Macnaghten, P. (2010). Engaging narratives and the limits of lay ethics: Introduction. *NanoEthics*, *4*, 133–140. doi:10.1007/s11569-010-0095-6.

Nordmann, A., & Rip, A. (2009). Mind the gap revisited. *Nature of Nanotechnology, 4*(5) *(Nature Publishing Group)*, 273-274.

Nordmann, A., & Schwarz, A. (2010). Lure of the "yes": The seductive power of technoscience. In *Kaiser* (pp. 255–277). Kurath, Maasen, & Rehmann-Sutter.

Pavlopoulos, M., Grinbaum, A., & Bontems, V. (2010). Toolkit for ethical reflection and communication. *Observatory Nano Project, CEA, June 2010*. Retrieved January 09, 2012, from http://www.observatorynano.eu/project/document/1598/

Retrieved January 20, 2012, from http://www.itas.fzk.de/deu/lit/2005/fiua05a_abstracte.pdf

Rip, A., & Shelley-Egan, C. (2010). Positions and responsabilities in the 'real' world of nanotechnology. In von Schomberg & Davies (2010) (pp. 31-38).

Schummer, J. (2006). Cultural diversity in nanotechnology ethics. *Interdisciplinary Science Reviews*, *31*(3), 217–230. doi:10.1179/030801806X113757.

Schummer, J., & Baird, D. (Eds.). (2006). Nanotechnologies challenges: Implications for philosophy, ethics, and society. World Scientific Publishing Co. Pte. Ltd.

Swierstra, T., & Rip, A. (2007). Nano-ethics as NEST-ethics: Patterns of moral argumentation about new and emerging science and technology. *NanoEthics*, *1*, 3–20. doi:10.1007/s11569-007-0005-8.

Swierstra, T., Stemerding, D., & Boenink, M. (2009). Exploring techno-moral change: The case of the obesity pill. In Sollie, P., & Düwell, M. (Eds.), *Evaluating new technologies (Vol. 3*, pp. 119–138). Springer Netherlands. doi:10.1007/978-90-481-2229-5_9.

Toumey, C. (2007). Privacy in the shadow of nanotechnology. *NanoEthics*, *1*, 211–222. doi:10.1007/s11569-007-0023-6.

Türk, A. (2011). *La vie privée en péril*. Paris: Odile Jacob.

van de Poel, I. (2008). How should we do nano-ethics? A network approach for discerning ethical issues in nanotechnology. *NanoEthics*, *2*, 25–38. doi:10.1007/s11569-008-0026-y.

von Schomberg, R. (Ed.). (2011). *Towards responsible research and innovation in the information and communication technologies and security technologies fields*. European Union.

von Schomberg, R., & Davies, S. (2010). *Understanding public debate on nanotechnologies, a report from the European Commission Services*. European Commission Services, European Union. Retrieved January 09, 2012, from http://ec.europa.eu/research/science-society/document_library/pdf_06/understanding-public-debate-on-nanotechnologies_en.pdf

Wickson, F., Grieger, K., & Baun, A. (2010). Nature and nanotechnology: Science, ideology and policy. *International Journal of Emerging Technologies and Society*, 8(1), 5–23.

ENDNOTES

[1] EGAIS was a Project funded by the European Commission under the Seventh Framework Programme (Science in Society). The acronym EGAIS stands for 'Ethical GovernAnce of emergIng technologiesS'. The project lasted from May 2009 to February 2012 and included 5 partners from 5 European Member States. Information about EGAIS can be found at http://www.egais-project.eu/.

[2] See in this book the article of Sylvain Lavelle on *Transformation of Proceduralism*.

[3] See for example the EGAIS Policy Brief on EGAIS Website (http://www.egais-project.eu).

[4] See in this book the article of Sylvain Lavelle on *Transformation of Proceduralism*.

Section 4
Paradigms of Governance

Section 4: Paradigms of Governance

The notion of governance is often used to designate the way by which an institution or an organisation (a state, a company, etc.) is ruled in a more or less democratic and inclusive way. In the political sense, the procedure and the process of democratic inclusion of the citizens or the 'civil society' into the framing and the making of collective actions, rules or decisions offers quite a contrast with the more traditional way of governing. It is often assumed that governance actually means democratic governance, whereas this is obviously not always the case if one considers its use in the management of firms, or the finance business, for instance, where it can appear as a kind of "window-dressing." It is then valuable in order to avoid some misleading confusions, but also in order to assess the relevance of models, to present the various *paradigms of governance*, or frames of reference.

The first type of governance paradigms lies upon the opposition between the *democratic* and the *non-democratic*, namely, as will be shown and defined, the *technocratic* (skilled-based power), the *ethocratic* (virtue-based power), and the *epistocratic* (wisdom-based power). The point in this opposition is that, contrary to the democratic paradigm, the non-democratic ones assume that the condition for social rules or decisions to be valid is their reflecting, discussing, and making by an elite of experts, virtuous or wise individuals or groups. There is no doubt in these paradigms a basic distrust as to the ability of 'the people' to take into their own charge matters of public affairs, and then to elaborate the appropriate standards and norms that account for the regulation of actions and conducts. The re-construction of these four paradigms (the democratic and the non-democratic) can be illuminating as regards the interpretation of the actual expert and law-driven trends in the ethical governance of technology. It appears, indeed, that the paradigms of technocracy as well as that of ethocracy still operate in the design of governance settings aimed at regulating research and innovation projects.

The second type of governance paradigm is more focused on the *European framework* in which the consideration of ethics is both essential as regards the expression of legitimate concerns and non-essential as regards the priority of business interests. In this European paradigm, governance—especially when related to technology development and technology policies—can be taken as a complex structure including a number of actors not necessarily with equal power structures but who can only act in interdependence and interaction with each other. Hence, the reciprocal relationship among actors results from an on-going interaction and learning among actors in order to reach mutually acceptable and useful end-results. The governance of technology is to be considered as a public policy concept at national, European or global levels and involves the "public" to take part in the technology development process from the beginning (design stage) until the market (diffusion) stage. However, it appears that the ethical governance within technology development projects is often treated from a non-reflexive (or a limited reflexive) stance towards norm construction.

The third type of governance paradigm is the option of the *Commons*, as opposed to public or private property as well as to general and particular interests. The *Governance of the Commons* can be seen, from the legal perspective, and possibly further, as an alternative solution for managing future projects on common property *in common*. The case of Creative Commons is a good example for describing how a new way to govern Commons has been invented, without much consideration of governance at the start, but with a logic of collective action arising. The notion of *commons patrimony* turns out to be helpful in order to analyze whether a Commons procedure and process necessitates a new type of governance, namely a "patrimonial governance." The idea of the Commons and the correlated notions of common patrimony and of patrimonial governance constitutes no doubt a promising way for the future of cooperative research, innovation, and projects.

Chapter 9
Paradigms of Governance:
From Technocracy to Democracy

Sylvain Lavelle
Center for Ethics, Technology and Society (CETS), ICAM Paris-Sénart, France

ABSTRACT

The elaboration of some paradigms of governance lies upon the opposition between the democratic and the non-democratic, namely, as will be shown and defined, the technocratic (skilled-based power), the ethocratic (virtue-based power), and the epistocratic (wisdom-based power). The point in this opposition is that, contrary to the democratic paradigm, the non-democratic ones assume that the condition for social rules or decisions to be valid is their reflecting, discussing and making by an elite of experts, virtuous or wise individuals or groups. There is no doubt in these paradigms a basic distrust as to the ability of the people to take in charge the public affairs and then to elaborate the appropriate standards and norms accounting for the regulation of actions and conducts. The re-construction of these four paradigms (the democratic and the non-democratic) can be illuminating as regards the interpretation of the actual expert and law-driven trends in the ethical governance of technology. It appears, indeed, that the paradigms of technocracy as well as that of ethocracy still operate in the design of governance settings aimed at regulating research and innovation projects.

INTRODUCTION

The notion of governance is often used in politics to designate the dynamics of inclusion of the citizens and the society within the political processes of decision-framing and decision-making.

DOI: 10.4018/978-1-4666-3670-5.ch009

The word 'governance,' however, which owns a common origin with the word 'government' (from the ancient Greek *kubernan*, the tiller) does not substitute the traditional nation-state government. It is rather an alternative regime applicable to a wide range of activities, institutions and organizations, as suggested for instance by Stoker's propositions on governance[1]:

- Governance concerns a range of organizations and actors, not all of which belong to the government sphere.
- It modifies the respective roles and responsibilities of public and private actors as established in traditional paradigms of policy making.
- It involves interdependence between organizations and actors engaged into collective action in contexts in which none of them has the necessary resources and knowledge to tackle the issue alone.
- It involves autonomous networks of actors.
- A key principle is that actions can be pursued without necessarily having the power or the authority of the State.

The basic idea in the governance theory and practice is to concentrate more upon the way the rulers and the citizens share and exercise their political power than upon the sole institutions and organizations of the government[2]. It does not mean that the analyst should have no interest in the procedures that stand behind the processes themselves on the ground that one should pay more attention to the informality than to the formality of things. On the contrary, one of the main stakes in the so-called 'reflexive governance' is precisely to examine the relationship between the processes and the procedures of inclusion of the 'outsiders'. Some procedures indeed are more likely to produce such or such governance process and to emphasize such or such aspects ('technical,' 'ethical,' 'epistemic,' etc.). It is then argued that we need a kind of mutual balance and adjustment of processes and procedures according to the means and the ends of the governance dynamics.

The overall trend in the evolution of governance is to substitute some democratic forms and experiments to some non (or less) democratic ones, the more well-known being the technocratic. It is often implicitly assumed that governance actually means *democratic* governance, while it is obviously not always the case if one considers its use in the management of firms, or the finance business, for instance. Thus, governance can also be viewed merely as a diversion for the citizens and the society aimed at keeping the power in the hands of its owners and at hiding their actual strategies and tactics in the course of their affairs. It is then valuable for the sake of self-judgment to present in a synoptic historical-philosophical fashion the various *paradigms of governance*, namely, their origin, their concept, their rationality, as well as their limit. The notion of paradigm can mean many things since Plato, but here I refer mainly to Kuhn's meaning, that is, the idea of a *frame of reference* including theories, practices and techniques[3].

Hence a set of questions: (1) What origins, concepts, rationalities and limits can one identify in the anti-democratic paradigms (technocratic, ethocratic, epistocratic) as compared to the democratic one? (2) Is there something like an ethical expertise in the same sense as the technical or the epistemic expertise that are perhaps justifiably more legitimate? (3) To what extent do the democratic options of participation and deliberation form a coherent paradigm? (4) Can the technical, ethical or epistemic expertise of the 'skilled,' 'the virtuous' or the 'wise' be shared or even replaced by that of the society?

I will now present in more detail the four main paradigms that we identified in our research: the *technocratic-instrumental*, the *ethocratic-normative*, the *epistemic-cognitive*, and finally, the *democratic-inclusive*[4]. The presentation of these paradigms enables a better understanding of the kind of anti-democratic models and trends the supporters of the democratic options(s) had and still have to struggle with. It can also shed some light on what remains attractive in anti-democratic thought and on why there remains some kind of 'aristocratic' resistance and reluctance to putting the power of the experts or the rulers into the hands of the people. This is particularly the case in the ethical governance of emerging technologies, where the technocratic and the ethocratic paradigms remain quite influential, which means that the technical and the ethical *expertise* tend to remain at the center of the play.

THE TECHNOCRATIC-INSTRUMENTAL PARADIGM

- **The Technocratic - Instrumental Paradigm:** Can be defined as the combination of a technical expertise provided by a restricted community and of an instrumental power of technical skills in the determination of social rules and choices.

Concepts of Technocracy

The term technocracy emerged supposedly at the end of the nineteenth century and refers to a science and skill-based form of power. It is separate from other long identified forms of powers, such as democracy (power of the people), autocracy (power of one man), plutocracy (power of wealth), etc. In fact, the concept of technocracy originates in the doctrines of the French philosophers Saint-Simon and Comte.

Saint-Simon, living at the time of the French Revolution, is the founder of a doctrine that gives the industry and its representatives (engineers, bankers, etc..) a major role in the project of society transformation. The view of Saint-Simon is that of a society organized according to the model of a factory, in which the happiness of individuals coincides with the satisfaction of their physical and moral needs. In this model, the government is fully identified with the administration of material things in the hands of specialists in Arts and Crafts. The ethics of technocracy as elaborated by Saint-Simon is quite close to that of utilitarianism, with certainly a more positivist orientation: "One cannot give to morals any others motives than tangible, ascertained and present interests."

This 'technocratic' doctrine, very influential around the 1830s, is related to that of Positivism as developed by Auguste Comte, although for him, the reform of society requires first an intellectual mutation. For Comte, political progress depends on scientific progress, and after the stages of religion and metaphysics ages, positivism believes in the advent of a scientific age. The new era is based on the universal application of the experimental method, which can be applied to social phenomena, through the creation a new scientific discipline that is called *sociology* by Comte. The spirit of positivism lies in the question 'How' that substitutes the question 'Why,' as well as in the link between science and action: 'Science, hence prediction; prediction, hence action'. It is noticeable that scientific positivism gave rise to several developments: logical positivism (Russell, Ayer, Carnap) as well as legal positivism (Kelsen); the form of technocracy could be said in that respect a 'technical positivism'.

The concept of technocracy is examined more than one century after, notably by the Swiss political scientist Jean Meynaud. The latter views technocracy very broadly as an 'allocation to a community of technicians of a key place in the conduct of human affairs' (Meynaud, 1960). Technocracy can be characterized by the possession of a particular skill, whether a strictly technical expertise, or rather, a managerial one. It is also a particular mode of conduct, reluctant to political compromises and targeting the sole efficiency. Of course, this idealized picture neglects the underlying implicit value choices in the decisions made by the technocrats (Maynaud, 1964). Technocracy can be said to be a 'politics of expertise,' as stated by Franck Fisher (1990), which entails a form of participation of interested groups.

The concept of technocracy particularly inspired the creation in the United States of the *Technocracy Incorporation*[5] in 1918. The purpose of this organization of 'social consultants' led by Howard Scott, that played a significant role between the two wars, was a social renewal based on widespread use of technology. He believed in the specificity of social development in America, as well as in the impossibility of increasing the efficiency of human activities without systematically replacing the technology at work[6]. He then developed within the Technocracy Incorporation an original concept of technocracy, distinct from those of Veblen ('Soviet of Technicians') or Taylor ('Scientific Management'). Technocracy Incorporation (Tech. Inc.) wanted to promote a new form of government, regardless of philosophical values, and only concerned with factual realities (including energy realities).

The supporters of this doctrine also dreamed of replacing the slow human deliberations by a new type of scientific decisions, as fast as the energy flows they must manage. As Scott stated:

It is well to realize here and now that Technocracy, like science, has no truth ; truth is a philosophical absolute, while in Technocracy all things are relative… We are concerned with the consumption and control of energy and their resultant production and consumption, which are all measurable, and have nothing whatsoever with truth or philosophical values. (Scott, 1965)

But later, Scott mentions the 1934's declaration (Scott, 1933), stating that "Technocracy speaks the language of science and recognize no authority but facts…We see science banishing waste, unemployment, hunger and insecurity of income forever."[7]

Instrumental Rationality and Procedural Technics

The instrumental rationality, as a field exercise of reason, is only concerned with the effectiveness of the action. It is characterized in Kantian philosophy by the concept of rational means and *technical imperative (hypothetical)*: *if I want to do X, then I must use the means Y or Z* [8]. The notion of technical imperative suggests that, in the course of action, there is no moral valuing of the ends, but only a technical appraisal of the efficiency of the means.

It also benefits from a sociological treatment in Max Weber's work who made an interpretation of modernity as a historical phenomenon of rationalization of the world and life. Weber made a distinction between the *end-rationality* (*Zweckrationalität*) and the *value-rationality* (*Wertrationalität*), the instrumental rationality being of the first kind[9]. The rationalization of the world and life is then a historical process in which the instrumental rationality, that of efficient means, is increasingly deployed throughout the modern societies. This process of rationalization is the key factor in the sharpening of the historical development of modern societies, characterized by the rise of experimental sciences, market economy, bureaucratic state and formal law.

Instrumental rationality is also related to technocracy in the concept of *social engineering* as developed by Popper[10]. A social engineer is precisely a technician or rather a technologist in charge of assessing the efficiency of means, regardless of the value of ends. This social engineering, as a technical reformism, is also called *piecemeal social engineering*, and is distinguished from socialism or communism, that are viewed as forms of total utopias. The social engineering aims at differentiating, with a supposedly absolute neutrality, what can be done from what cannot, what is possible, and what is impossible. It can establish, among the available means, whether they are compatible or feasible with an end, but not if the end itself is valuable. It that respect, Popper's social engineering, in the line of the technocratic tradition, can be said a form of *procedural technics*.

Critique of the Technocratic-Instrumental Paradigm

The main critique of this technocratic-instrumental paradigm is that by the defenders of a *technical democracy*, stating that ethical stakes, such as the good, or the just, are not dependent upon a procedure, nor an expertise.

Horkheimer and the Frankfurt School first developed a critique of instrumental rationality that was focused on the dangers of its historical development[11]. Then, Habermas added to the concept of *instrumental rationality* (effective action on the natural world) with that of *strategic rationality* (effective action on the human world). He thus endeavored to prove that these two rationalities suffer from a lack of normative legitimacy that only *communicative rationality* (discursive action-oriented understanding) can implement or restore[12].

In his early critique of instrumental technocracy, Habermas endorsed the demarcation made by Max Weber between domination and administrative leadership. According to Weber's decisionist model, administrative action leads to rationalization in the choice of means. Political action itself remains mostly concerned with values, beliefs and goals, but decoupled from any rational basis. Habermas considers however that in an advanced phase of the system, this model is outdated and replaced by the technocratic model, as first described by Bacon and St. Simon. This leads to a reversed relationship between technicians and politicians, in which policy-making is reduced to a body of rational administration, while decision-making loses all autonomy, as a full rational choice under uncertainty.

However, Habermas points out the shortcomings of this new model. On the one hand, it presupposes the ability to escape from the constraints of technical progress, which is autonomous only if one ignores the social interests at work. On the other hand, it assumes that the same rationale applies to technical and practical issues, while issues related to the practice are still there and pose a particular problem with the decision. Habermas identifies three models ultimately enrolling in the instrumental paradigm: *decisionist* (Weber), *technocratic* (Bacon, Saint-Simon) and *pragmatic* models (Dewey). But only the pragmatic model of

Dewey is likely to take into account the relationship to the political form of democracy[13].

The stream of *technical democracy*, as embodied by Callon and Latour in France, is also very critical to the shortcomings of social engineering[14]. Technical democracy states that it is possible, through a set of dialogical settings, such as 'hybrid forums' to make technological processes more democratic. Technical democracy is directly connected to the *Actor-Network Theory* (ANT), and to the translation paradigm, and it seeks to enable a democratization of technology, if not of a 'democratization of democracy'. This political option on democracy is, so to speak, the opposite to that of technocracy and social engineering, in Popper's sense. Even if they don't criticize directly Popper's view on social engineering, Latour and Callon offer a radical alternative way to technocracy in supporting a technical democracy. The core of their critique is the so-called 'neutrality' of technical procedures that relies on the myth of the fact-value dichotomy.

THE ETHOCRATIC-NORMATIVE PARADIGM

- **The Ethocratic-Normative Paradigm:** Can be defined as the combination of an ethical expertise provided by a restricted community and of a normative power of moral will in the determination of social rules and choices.

Concepts of Ethocracy

The term ethocracy (*Ethocratie* in French) was created by the Baron d'Holbach, who conceived it as government based on morals. It meant for him 'a union between morality and politics, the idea of legislation consistent with the virtue that could also be advantageous to the rulers, subjects, nations, families, to every citizen'[15]. The ambition of d'Holbach through the concept of ethocracy was to promote a form of government and law built on the respect and the imposition of a number of moral injunctions and obligations. Their implementation was to provide him access to the common good, the various ills of society being due to their negligence.

D'Holbach starts out on the basis of observation of moral ignorance of many men and of divorce between the interests of princes and those of their subjects. The reason, however, as well as the experience leads him to make a plea for a reform of morals, while remaining skeptical about the ability of reason alone to correct bad habits long ingrained. D'Holbach believed in the effectiveness of moral practice reinforced by the moral law, in the virtue of examples shown by sovereigns. D'Holbach set up a long list of legal incantations on all sectors of society: public and private, administrative and family, civil and military. Nevertheless, he was little concerned on the practical possibilities of establishing such guidelines. In this respect, Holbach was a good example of the typical optimism of the Enlightenment, but seemed to give the norms an inherently effective and binding force.

The concept of ethocracy can be taken also in the sense of 'ethical expertise,' as somewhat the equivalent in the field of ethics to the technical expertise in the field of technics. Thus, the concept of ethocracy has known further development in the business of ethicists or moral philosophers dealing with applied ethical issues. These experts in ethics provide guidance through advice and recommendations to various public authorities, hence the link of ethics to power. But the very possibility of a 'moral expertise' divides philosophers, perhaps since the origins of philosophy, since the time of Plato's *Meno* ('can virtue be a matter of teaching?'). Ryle suggests that the common man is just as well, or even better, equipped than the moral philosopher to rule on matters of ethics[16]. However, a philosopher like Peter Singer argues that a morally good man could go that way only

in the context of a society with a perfect moral code, which would be fully reliable. For Singer, the moral expertise is possible and useful to identify the good action that requires thought, skills, and information. In short, a form of work, the possession and exercise of which confer particular legitimacy to the expert in ethics[17].

Jonathan Moreno proposed three criteria in his characterization of ethical expertise in the field of health: (1) knowledge of general principles and moral theory (2) analytical skills such as discernment and insight (3) strength of will in order to avoid the easiest the way out[18]. Moreno also introduced several distinctions: on the one hand, between the *moral expert*, who is expected to take positions, and the *moral philosopher*, who should instead focus on an enlightening review; on the other hand, between *ethics committees* and *ethics consultants*. The functions of the first traditionally includes the review of cases, policy advice, staff training, although they may also evolve according to circumstances of unusual tasks, such as resource allocation in health United States. The ethics consultants are instead on real 'business ethics,' the action spectrum is more diverse and more involving[19].

Importantly, a historical example of ethocracy can be found in the National Consultative Ethics Committee (CCNE) in France. It was created in 1983 under the presidency of François Mitterrand, and it was the first national ethics committee in the world. The 6 August 2004 Act gives the mission 'to provide advice on ethical and societal issues raised by advances in knowledge in the fields of biology, medicine and health'. Its statutes are clearly in the perspective of applied ethics, more than in that of a moral philosophy divorced from issues of the time. The Council has an advisory mission, which can be understood in a passive sense (the response to a demand from the power) or active one (the advice to power). Its composition of 40 members is strictly defined by law: 15 persons from the area of biomedical research, 19 selected 'for their competence and their interest in ethical issues, and 5 belonging to the main spiritual families'.

Sometimes accused of being a form of 'ethocracy,' the CCNE is aware of the risk to embody a forfeiture of democratic thinking in favour of legal ethics experts. It strives to involve citizens in their thoughts at various annual meetings (Yearly Conference of Ethics, Citizen Meetings, etc…). It remains significant that the Committee can be attended only by some institutional actors who have been identified, and the question remains open whether this committee is in tune with the rest of society.

Normative Rationality and Procedural Ethics

In Kant's doctrine, duty is an objective constraint of action for the individual that is shown to be a *moral imperative (categorical)*: *X ought to do Y*. It manifests itself as a moral law, the criterion being the possibility of a universalization of the particular rule of action in order to make it a universal rule[20]. This possibility of universalization itself requires a judgment of the person, free from confrontation with judgments of others. The moral doctrine of Kant is also intentionalist, in the sense that the criterion of a good deed is the only good will, regardless of the consequences of the action. Nevertheless, for Kant as for the latest Ryle, morality is not an expertise or any special competence. It is rather for a person to have concern for people regardless of his own qualities of intelligence or judgment.

This deontological doctrine is illustrated in particular in the discourse ethics as developed by Habermas, who takes over the Kantian moral heritage. The ethics of discussion is procedural, formal, cognitive, both consequential and intentional, while claiming to escape the reproach of formalism addressed by Hegel to Kant. It is based on two principles[21]:

- **Principle U (Universalization):** All affected can accept the consequences and the side effects that (the norm's) general observance can be anticipated to have for the satisfaction of everyone's interests, and the consequences are preferred to those of known alternative possibilities for regulation.
- **Principle D (Discussion):** Only those norms can claim to be valid that meet (or could meet) with the approval of all affected in their capacity as participants in a practical discourse.

According to Habermas, adopting a moral perspective requires therefore, ideally, the participation to a reasoned discussion that is free of any form of coercion. This discussion should ground an emerging agreement on a standard, which all participants are willing to accept in considering all the consequences of the norm application. Habermas claims, however, that he refers to this procedure as a reconstruction of the moral perspective in general, that attempts to distinguish carefully any substantial bias on particular moral theories.

In appearance, the discourse ethics in Habermas' seems to function as the general framework for an *ethical democracy*, since it seeks to include all the interested people within the practical discussion. But, in fact, the formal procedure of the discussion does not inform in any way on the limits of the community of discussion: is it a small group, or can it be a bigger group, such as a city, or even a nation? It seems that a formal procedure assumes this problem not to be essential to the agreements on moral rules that are supposedly enforced by the sole power of a satisfying discussion procedure. However, from the point of view of an ethical democracy, the question of who is a member of the community of discussion is essential, insofar as a restricted community of members will be in a position to decide for the non members. There is a real danger that, in any way, a smaller community of people would stand as 'experts' of the moral duties in the absence of other people who, for some obvious reasons of time, skill or will, could not actually be members of the community of discussion.

Critique of the Ethocratic-Normative Paradigm

The main critique of the ethocratic-normative paradigm is that given by the defenders of an *ethical democracy*, stating that ethical stakes, such as the good, or the justice, are not dependent upon a procedure, nor an expertise. An ethical democracy is not so much the problem of the ethical basis of democracy, as suggested by Jose Ciprut[22], but rather that of a the procedures of participatory ethics[23].

The procedural ethics of Habermas is exposed to a series of critiques that questions its ability to guarantee an appropriate moral content of moral rules. First, some argue that certain moral norms (e.g., the condemnation of the senseless murder) seem too compelling or obvious to be made dependent on a process of discussion. One may also argue that the principle of discussion, in its very formulation, presupposes a prior establishment of certain standards, such as freedom and equality for example. It may also seem desirable to assess a reasoned discussion, leading to agreement that is clearly wrong (about the legitimacy of racial crime, for example), on the basis of some external standard. Thus, the specific commitment to rational and reasoned discussion seems to assume a set of quasi-substantial criteria (Alexy) that are not included in the construction claimed to be procedural and formal by Habermas.

Secondly, the 'agreement of all concerned' in the ethics of discussion may be possible within the reach of limited communities (family), but appear structurally impossible or utopian to larger communities (nation) or virtual ones (future generations). It may be affected by all the asymmetries of situation between partners involved in a discussion (between adults and children, experts and laymen, doctors and patients, etc.) In such cases, the practice of discussion requires representation of the interests of others, through a monological deliberation, rather than a dialogic process. Thirdly, the legitimacy of human interests is in fact disqualified, in favor of fair arguments, while a more open view on virtue and interest would enable some more realistic forms of consensus, or compromises, 'agreement on disagreement'. More generally, the ethics of discussion seems to ignore cases where the incommensurability of different positions actually makes a consensus impossible. This pretty naïve concept of communication is rather blind to certain realities of daily life, such as domination, violence, indifference, misunderstanding, etc…[24].

According to Maesschalck, Habermas gives the medium of discussion the function of a producing a temporary equilibrium, a procedural balance, allowing common sense to rebuild. The mutual understanding that structures lifeworlds is a fundamental premise of these ethics, but is required for a background that is 'de-contextualized'. This 'de-contextualization' of the background is necessary so that potential partners of standards development procedures can accept (or not) claims that they could share, despite their different situations. It is then not only the knowledge of experts that is important, but the expertise of the situation that develops each of his or her own experiences and that, through a proper argument, can challenge the regulatory bodies of the system so they adapt their management[25].

These are also depending in Maesschalck and Lenoble' notion of incompleteness of the 'scheme of application of discourse ethics of Habermas. The theory of discourse in ethics is the idea of universal presuppositions of communication that guide the decision on the conditions of a basic consensus. However, the pattern of application is less clear about the conditions for achieving empirical and sociopolitical forms likely to experience this ideal scheme. Habermas is quite confused about it, either on the tools used, the theory of society inherited from the Frankfurt School, or the conditions of satisfaction from this ideal of the discussion. We do not know specifically how public space is likely to produce agreements, and especially to grow and attract more and more people in its ability to produce agreements. However, the empirical question is not how Habermas transforms individual skills, but how to mobilize these skills in the context of social change. This is called an adjustment of problem behaviors, which may be particularly focusing on the policy of new objects, such as citizens' interest in issues of European governance[26].

THE EPISTOCRATIC-COGNITIVE PARADIGM

- **The Epistocratic-Cognitive Paradigm:** Can be defined as the combination of an epistemic expertise provided by a restricted community and the cognitive power of 'scientific' knowledge in the determination of social rules and choices.

Concepts of Epistocracy

The term epistocracy (or epistemocracy, from the Greek *episteme*, knowledge, and *kratos*, power) is newly established and remains of quite little use, although it refers to a conception of power that is ancient. We can define *epistocracy* as a form of government in which an elite of scholars or wise men is held to be the most likely to make the right decisions for the city. We find in Plato's *Republic* the origin of the concept of epistocracy, then criticized by Aristotle, but it is also found later in the principle of representative government and plural voting in Mill's view.

The epistocracy then appoints a political system in which an elite destined to govern the city is distinguished by their knowledge of the idea of justice. The tendency of the elite to produce correct decisions does not follow clearly, however, from the mere possession of a social science or a higher wisdom. We can thus assume that there is a small number of morally well-informed citizens who possess a high degree of practical wisdom, a 'science of the good,' which happens to be relevant for political affairs. This group of people is then supposed to know better than anyone among citizens outside the group what should be done in terms of policy. It therefore seems reasonable to conclude that the state should be headed by an elite of knowledgeable rulers, and that the people should conduct themselves in accordance with the guidelines proposed by this elite [27].

Aristotle also noted that rules imposed by the elite are likely to lead to correct political decisions more often than to follow directions from anyone else. However, decisions would be better if they were accompanied by a discussion with other groups, for example, the next most knowledgeable or the wisest of the society. The idea is that expanding the group to other members, although it will lower the average level, is generally always a better option. The reason is that cooperative work and 'collective intelligence' produces results that are better than the performance of a single individual or group in particular. It is on this basis that Aristotle rejected the argument that some individuals considered more knowledgeable or wiser should be leaders, because they know best what needs to be done. Yet, Aristotle believes that this argument is not sufficient to exclude the possibility that a person of science or wisdom clearly superior to others is of no benefit to the people who consult. He admits that if a person has a science or a greater wisdom than others, then it should lead them, and if others have the presumption to direct that person, their supposed authority is invalid or ridiculous. Consequently, the epistemic value of a wider group of officers is not a phenomenon general enough to destroy the legitimacy of the epistocracy, even in Aristotle's conception. Nevertheless, epistocracy is also challenged in the ancient world by the alternative way of epistemic democracy[28.]

Mill first supported an illiberal elitism, that is, the view that an intellectual and cultural elite should constitute an estate of the society – a church or caste with formal power. He then rejected this view on the grounds that it did not foster individual autonomy, which was incoherent with his attachment to liberty. But he maintained a liberal elitism, according to which an intellectual elite must exert influence through recognition of their authority in their sphere. The epistemic value of the discussion does not necessarily imply recognition of an equal position of participants in the final decision[29]. The idea of Mill, who advocated a system of plural voting, is that everyone should have at least one vote for several reasons. First, it is important to prevent a group or class from controlling the political process without having to give reasons for it. This is the problem of class legislation: as the largest class is the one with the lowest education level and social rank, this deficiency could be addressed by giving those with the best abilities a plural voting. Secondly, it is important to avoid giving equal influence to each person regardless of their merit, intelligence, etc. Mill considers that political institutions should embody the idea that there are opinions that are better than others, even if he does not claim that this is a way to produce better decisions.

Mill admits that deliberation is best if participants are numerous, which argues in favor of an extended vote to all citizens, or at least many of them. However, this does not imply that the subset formed by the most learned individuals should not have more votes (two, or three, for example). It is in fact a synthesis of ideas from Plato and Aristotle, who emphasized the merit of the class represented by the more educated people, not the wisest. It is indeed the test of wisdom that is meant to give legitimacy to the exercise of political leadership, but which poses many serious problems of identification (who is 'the wisest' in a society?). It would require that the wisest, as leaders, can, in their attempt to justify decisions taken, recognize their wisdom by the ruled. Mill's idea may at first seem relevant in that it relies on a simple intuitive idea: good education promotes the ability of citizens to lead wisely[30].

Cognitive Rationality and Procedural Epistemics

David Estlund opposes epistocracy and defends an *epistemic proceduralism*, arguing that we cannot let decisions be made by the 'knowers' of the good. First, he admits as generally acceptable the proposition that a population is most qualified to run if more people are better educated. Second, he admits as generally unacceptable that the subset of well-educated tend to contribute with more wisdom in the development of good rules. Indeed, even though there are truths, or people who are better at knowing them than others, there can be disagreement about who those are, even among the qualified. He then claims, on the basis of his earlier distinction between authority and legitimacy, that democracy exercises legitimate authority in virtue of possessing a modest epistemic power. Its decisions are the product of procedures that tend to produce just laws at a better chance than rate, and better than any other type of government that is justifiable within the terms of public reason.

Thus, epistemic proceduralism is justified by the tendency of democracy to lead to correct decisions more likely than random procedures or other competing procedures. This means that, instead of an ideal speech situation and a deliberative procedure, such as in Habermas, Estlund values the real speech situation and a real-world democratic procedure. According to him, it is likely to be close enough to the ideal one in its epistemic payoff, since it will tend to lead to decisions that minimize 'primary bads' (famine, genocide, etc). In epistemic proceduralism, specific political decisions in a democracy, whether correct or incorrect, are legitimate because they are the outcomes of a democratic procedure. That procedure itself is legitimate because it is likely to lead to correct decisions, that is to say, to qualifiedly acceptable decisions. The form of epistocracy is ruled out by the *Qualified Acceptability Requirement*, stating that the necessary condition on the legitimate exercise of political power is that it be justified in terms acceptable to all qualified points of view.

The concept of epistocracy is still defended by some philosophers who doubt the democratic performance of the mass of citizens in making the correct decisions[31]. On the one hand, Forcehimes argues that deliberative democracy cannot be justified by pure proceduralism but requires an epistemic component. Deliberative democracy needs an epistemic element because without it, deliberation lacks friction - i.e., it is rendered pointless. The obligations of legitimate decision-making provide a justification for the epistemic component of deliberative democracy. To have political legitimacy as a decision-maker, one must engage in certain epistemic practices: the exchange of reasons, arguments, and evidence. However, this epistemic justification seems to entail something like a Platonic epistocracy in which political authority would be based on epistemic ability or virtue. Yet, this is epistemically problematic, namely, because it diminishes the truth-tracking abilities of deliberation by limiting the range of perspectives within the deliberative process. Dougherty replies that an epistocracy is preferable to both Forcehimes's epistemically justified democracy and pure deliberative democracy. Indeed, elitism and exclusion, which as necessary components of epistocracy, are not prima facie negative qualities and, in fact, are beneficial within an epistocracy. Against Aristotle's preliminary objection to epistocracy, it can be argued that the combined knowledge of the masses could not be greater than that of the epistocrats.

Critique of the Epistocratic-Cognitive Paradigm

The main critique of this paradigm is the *epistemic democracy* challenging the possibility for experts to know the best outcome of rules and choices. Indeed, some deny that formal education can foster good leadership skills, while others purport to establish the right to equal treatment, which involves logically an equal vote. Others believe that the privilege granted to a minority of elected officials is an unwarranted insult or moral damage to the esteem of those to whom this privilege is denied.

The idea of an epistemic democracy is not new, since it could be traced back to classical Athens, as seen before, and was also taken over some years ago by Joshua Cohen[32]. The epistemic democracy is often identified with the renewed interpretation of the *Condorcet's Jury Theorem*, coming from the French philosopher Condorcet. This theorem states that if voters (1) face two options (2) vote independently of one another (3) vote their judgement of what the right solution to the problem should be (4) have, on average, a greater than 50% probability of being right, then, as the number of people approaches infinity, the probability that the majority vote will yield the right answer approaches. This is somehow a rational justification of what Surowiecki calls the 'wisdom of crowds,' or, as stated by him, 'Why the Many are Smarter than the Few and How Collective Wisdom Shapes Business, Economies, Societies, and Nations' (Surowiechi, 2004).

In fact, according to Elizabeth Anderson (2006), the epistemic powers of democratic institutions can be assessed through three epistemic models of democracy: the Condorcet Jury Theorem, the Diversity Trumps Ability Theorem, and Dewey's experimentalist model. Dewey's model according to her is superior to the others in its ability to model the epistemic functions of three constitutive features of democracy. These features are the following: the epistemic diversity of participants, the interaction of voting with discussion, and feedback mechanisms such as periodic elections and protests. It views democracy as an institution for pooling widely distributed information about problems and policies of public interest by engaging the participation of epistemically diverse 'knowers.' Democratic norms of free discourse, dissent, feedback, and accountability function to ensure collective, experimentally based learning from the diverse experiences of different knowers. In addition to that, Anderson (2008) criticizes the very idea of grounding a social rule or choice on the sole basis of knowledge. It appears, indeed, that the epistemic justifications of democracy neglect some other non-epistemic justifications, such as values of equality and collective autonomy.

THE DEMOCRATIC-INCLUSIVE PARADIGM

Def: The *democratic-inclusive paradigm* can be defined as the procedural combination of a democratic participation allowed to a community of citizens and of the inclusive power of political opening to society in the determination of social rules and options.

Concepts of Democracy

It is now well known that democracy is not a single political form, but gathers a variety of models, ranging from representation to participation. David Held (1996) identifies several of them, such as: classical democracy (participatory, direct), protective republicanism, protective democracy, direct democracy, competitive-elitist democracy, pluralist democracy, legal democracy, autonomous democracy, participatory democracy, deliberative democracy, cosmopolitan democracy (see Appendix II).

Democracy is a general ancient concept that requires a set of criteria to be satisfied in order for a political institution to be said to be inclusive. A set of criteria was given by Dahl (1998):

- **Effective Participation:** Before a policy is adopted by the association, all the members must have equal and effective opportunities for making their views known to the other members as to what the policy should be.
- **Voting Equality:** When the moment arrives at which the decision about policy will finally be made, every member must have an equal and effective opportunity to vote, and all votes must be counted as equal.
- **Enlightened Understanding:** Within reasonable limits as to time, each member must have equal and effective opportunities for learning about the relevant alternative policies and their likely consequences.
- **Control of the Agenda:** The members must have the exclusive opportunity to decide how and, if they choose, what matters are to be placed on the agenda. Thus the democratic process required by the three preceding criteria is never closed. The policies of the association are always open to change by the members, if they so choose.
- **Inclusion of Adults:** All, or at any rate most, adult permanent residents should have the full rights of citizens that are implied by the first four criteria.

The request for democracy has been particularly strong over the years as regards, for instance, the development and application of technological projects. It gave rise to a set of democratic tools for public deliberation and participation, such as focus groups, participatory assessment and technology assessment. In technological assessment, one can distinguish several tools:

Constructive Technology Assessment: This approach aims to identify the processes by which technology development makes assumptions about how it will be used. It particularly looks at confronting the Collingridge Dilemma (it is too early to make predictions about the consequences of technological development in its early stages, yet once the technology has been developed it is too late to change it). It aims to introduce participation methods 'upstream' of the development process (Fisher, Mahajan, & Mitcham, 2006). According to Heiskanen (2005), the requirements for CTA are the following:

- Inclusion of all interested parties.
- Beginning at the early stages of technology development, and continuing throughout.
- The ability for alternatives to be discussed and explored, in case of undesirable social impact.
- The ability to directly affect the technological development process.

Participatory Technology Assessment: This approach was developed in Denmark to enhance the democratic opportunities for the public to influence technology design and policy. It allows citizens to express opinions to boards which allows for their needs to be incorporated into development and policy. It uses consensus conferences, scenario workshops, focus groups, etc. to allow for inclusion of all opinions and ideas on particular technological developments or policies. Its aims generally are to:

- Enhance the knowledge and values base of policy-making.
- Open up opportunities for conflict resolution and achieving the public good.
- Foster the motivation of those involved and initiate a process of social learning.
- Provide economic actors with a better understanding of consumer and stakeholder concerns, and in doing.
- Improve the accountability and legitimacy of socio-technological decisions (Abels, 2006).

Real-Time Technology Assessment: RTA assesses the project at all stages of the development process. It involves a series of analogical case studies which look at previous, similar technologies in order to establish the likely future societal issues with the technology being developed. It also looks at key research and development trends, stakeholders, and institutional arrangements that might affect the technology's resources and capabilities. It also monitors the stakeholders for changes in their perceptions, knowledge, and attitudes that could affect the social impact of the technology. It then theoretically involves an ongoing assessment of the impact through analysis and participatory approaches that allow for it to develop an evolving image of the impacts throughout a project's life.

Inclusive Rationality and Procedural Politics

It is noticeable that, considering the well-known limits of representative democracy, sometimes identified with the 'crisis of democracy,' or even the 'crisis of politics,' the development of alternative models refers to those of *inclusive democracy*, especially in their deliberative and participatory forms.

For Young, for instance, "democratic political movements and designers of democratic processes [should] promote greater inclusion in decision-making processes as a means of promoting more just outcomes" (Young, 2000). Dahl states that:

Full inclusion: The citizen body in a democratically governed state must include all persons subject to the laws of that state except transients and persons proved to be incapable of caring for themselves. (Idem, pp. 76-78)

The democratic inclusion of citizens is certainly at the same time a democratic request and a democratic trend in political theory and practice. But it overlaps in many occasions with the procedural tendency of political thought, usually grounded upon a set of basic discussion rules, regardless of the discussion contents expressing the partners' specific cultural preferences.

Proceduralism in politics was originally involved in US legal-related governance systems in the 1960's and 70's, where it evolved from economic approaches that were critical of judicial activism and interest group politics and their effect on public policy (the Law and Economics movement, and the neo-Kantian approach to democracy) (Lenoble & Maesschalck, 2003, pp. 16-18.). Although these evolutions were not participatory, they laid the groundwork for the introduction of the procedural movements. In the 1980's, a movement started toward a procedural approach to action in governance, based around two theories of rationality: economic theory of efficiency, and deliberative democracy. The economic theory of efficiency is not a democratic participatory approach (and not in the scope of this work), so we will concentrate on the deliberative democracy theory, which grew out of the civic republican movement (*Idem*, p. 29).

Proceduralism appeared out of this movement as a solution to the problem of cultural and social pluralism. Pluralism is not so much a contingent fact as it is a permanent trait of modern democracies (Rawls, 1996). The emergence of cultural differences in societies (secular, multi-cultural, less authoritarian) has meant that a new method of discussion and cooperation has evolved to deal with these sometimes disparate communities. The inclusion of these different communities in the procedural method opens up the democratic and participative opportunities that characterize this paradigm. This method, based on the notion of *procedure*, requires that there be, at the very least, an agreement on the *way* to deal with problems, even if there not agreement on the content of the solutions. This appears to be something of an answer to the 'polytheism of values' identified by Weber (1919/1946), since a society, in order to function effectively, requires the establishment of a multiplicity of moral agreements on rules,

norms, and, if possible, values. But if the society's members cannot agree on the content or substance of values, especially with the heterogeneity of worldviews, they can, at least, agree on a fair procedure that can make agreement possible. The main aspect of proceduralism is the insistence on the non-substantive approaches to conflict resolution between the members of a society.

There are several approaches to proceduralism that have come out of this idea, each of which will be explained: the proceduralisms of Rawls, Latour, and Habermas[33]. These descriptions and analyses will not be exhaustive, but will instead give the reader a general overview and the salient points relevant to our discussion in this context.

Rawls' Procedural Approach

According to Rawls in his original position on proceduralism, the ideal situation for a participatory, procedural approach is one where the members of a society operate under a 'veil of ignorance' concerning their future position within the society (rich/poor, ruler/ruled, etc.). From behind such a veil, it is only then that the participants can honestly and seriously consider the morality of an issue, because 'no one knows his place in society, his class position or social status; nor does he know his fortune in the distribution of natural assets and abilities, his intelligence and strength, and the like'[34]. Rawls' approach to ethics concentrates on the idea of justice as fairness, end the veil of ignorance should allow for one to construct a set of principles for the fair treatment and to the advantage of all people involved. In this he displays some similarities with traditional Kantian deontology, but he also brings in elements of utilitarian theory. However, this notion is one of the weaker points of his procedural approach, since it ultimately functions as a very artificial basis for the whole system. Later, in *Political Liberalism* (1996), Rawls rethinks his position, and presents the idea of the 'overlapping consensus.' This appears to be a more realistic account on the process by which, in adjusting rules to cases, a society's members can come to an agreement about the principles of justice. However, such a consensus would be reached by, in part, avoiding some of the more fundamental and important arguments in the philosophies of the different members. Thus, although seemingly more realistic, there are some major questions that need to be raised about the background conditions that might enable such a consensus agreement.

Latour's 'Weak' Proceduralism

Latour's aim was to create a new humanism, so that scientific development became a way of life for society, bringing with it a democratic shift from laboratory work to becoming as embedded as a political norm (Latour, 1987). Latour's 'Actor-Network Theory' comprises a chain of both humans and non-humans involved in achieving the same objective, in a society constructed on the basis of production due to scientific organization. This approach uses methods set out by finalised normative representations, borrowed from intuitive pre-understandings of the workings of argumentation. It requires a multiplicity of actors involved in the decision-making process, and requires control of the variables introduced using different adaptation mechanisms. These adaptation mechanisms range from studying the connections in the network, to adaptive learning, the application of an inferential calculation that reinforces behaviour through retrospective adjustment.

Latour's approach, however, is purely conventional. It does not distinguish the rules of discourse, or the conventions for the institutionalization of discourses. Therefore it does not provide for the transformation of the actors or the bureaucratic arrangements that are supposed to aggregate the competencies of the institutionalized bodies of expertise. These are seen, instead, as necessary effects of the general process of recomposition required by the multiplication of actors. As well

as the problems of transformation, Latour's proceduralism suffers from three presuppositions: (1) A *pragmatic* presupposition: the principle of universalisation of discourse ethics is interpreted as being a requirement to extend the circle of actors involved, rather than the theoretical constraints of rational argument, before the learning and recomposition processes can take place. (2) A *semantic* presupposition: speech and collective action are linked which leads to the environment being seen as superior to the rationalisation of the social debate. (3) An *adaptation* presupposition: Latour presupposes that the learning and recomposition processes will be enough for the adaptation of the network to occur.

Habermas' 'Strong' Proceduralism

Habermas' (1981;1984) proposed solution to the problem of discourse is the creation of an ideal speech situation, in which the members of the society can discuss and agree on a normative statement solely on the basis of arguments, without any constraint, in an inclusive manner. It is not the normative statement's content that is important, but the method that the participants use to approve the statement that validates the statement. This requires a consideration and approval process, which is a consensus-oriented procedure of argumentation and justification that can avoid the manipulations inherent to the strategic actions.

In this philosophical construction, resulting from the idea of communicative activity, the ontological 'Is' and the deontological 'Ought' significance of communication are supposed to merge. In this way, the theory of communicative action (*pragmatics*) is articulated with ethics in the form of a discourse ethics. The coupling of the process of social evolution to a process of legitimization allows for individual capabilities to be able to determine common interests and for the implementation of a learning process that he deems necessary for social change. The learning process he implements requires 'authentic dialogue,' involving each speaker legitimately representing their claimed interest group, speaking sincerely, and making comprehensible and accurate statements to the rest of the participants. Yet, Habermas has always stated that the ideal speech situation is a counterfactual that can be taken as a normative reference for the evaluation of actual situations; that is, a description of what goes on in actual discussion. However, his theory does not make any strong statements about whether this procedure of discussion can help to solve the problem of the background conditions on the basis of which a consensus can be reached.

Habermas' approach is ambitious: on the one hand, it attempts to move beyond a functionalist framework and toward social transformations so that the reflexive involvement of the actors in their own development is ensured. On the other hand, he also refuses to sacrifice the intentionalist nature of such a framework, because it wishes to retain the references to rules of action and principles for the selection of behaviour. Habermasian strong proceduralism, much like in Rawls' approach, requires the actors to be reflexively involved in constructing their own cognitive capacities within the framework of institutional arrangements that demand these objectives be fulfilled. Latour's approach, on the other hand, has a necessity for the satisfaction of the deontological requirement for actors' roles to be interchangeable, but relies on the actors' capability to adjust through social adaptation. Weak proceduralism also favours reflexivity of routine adjustment using mechanisms that rely on the recurrence of sub-system equilibrium. On the contrary, strong proceduralism requires an adaptive learning within regulation mechanisms in order to determine the interest of different involved parties and to allow for the emergence of the largest possible group of participants that could have the capacity for reflexive cooperation. In the former, the adjustments made are functional; in the latter, the framework itself is involved so that retrospective consideration of adjustments to behaviour can be made.

Critique of the Democratic-Inclusive Paradigm

Behind the Democratic Paradigm is the idea of a participatory approach with stakeholder inclusion key to the accomplishment of both the participation and the democracy. By involving the public in decisions on policy and technological development approaches, those using models and tools under the Democratic Paradigm allow for a much more inclusive, bottom-up construction process than those of the other Paradigms. This can lead to more acceptance of a technology in society and more legitimacy for technology-related policy, given the structured approach to discussion and consensus that is open to all interested stakeholders. But, as suggested by Arrhenius (2005), it raises the classical 'Boundary Problem' concerning the criteria of citizen's inclusion in the decision-making.

However, it does not come without its problems and limitations. Although there is the greater level of social acceptance in the outcomes of the approaches under this paradigm, the fact that something is socially accepted does not mean that it is necessarily ethically acceptable. Societies are fully capable of creating extremely ethically unacceptable policies and actions. The neutralization and formalization of the approaches used under this paradigm (the consensus and compromise processes) are also highly problematic, since they lead to the neutralization and formalization of the relationship to the norm. Each participant might have a personal idea about what 'privacy' entails, yet this is highly likely to be quite different from other participants' ideas of the abstract norm of privacy. In order to make some sort of decision, a consensus needs to be reached, and this can only occur through negotiation and homogenizing the differing views, which may not only remove some of the fundamentally important values within the norm, but may remove the ethical nature of the norm (if there was one to begin with).

Proceduralism itself is problematic, insofar as it suffers from the *intentionalist*, *mentalist*, and *schematising* presuppositions. The participants in the procedural approach are presupposed to be capable of taking part in a meaningful manner, and when a result is decided on, there is a presupposition that this decision is enough to change the trajectory of the project; the participants come with a particular framing which influences their participation in the procedural approach; and they also come with a particular set of internal (or external) rules that they feel must be followed (this could be a set of societal or other organizational 'rules,' or their own personal deontology). Not only that, but the procedural approaches presuppose that the participants will agree on the ways that the discussion will unfold, and a whole series of cultural discursive norms that govern the way people interact with each other. This undermines the freedom extolled by the proponents of proceduralism; it is praised as a neutral solution to the problem of pluralism, but it presupposes a set of anthropological background conditions that ruins the pretention of it ever being a neutral option.

The participatory, democratic approaches presented here, although improving societal impact on technological policy and development, can, in fact, get caught up in this enthusiasm for participation and forget the ethical dimension. Ethics, in order to retain its authority, needs to impose a certain set of constraints on decisions. However, under these approaches, it is often the case that the issues are reduced to the problem of social acceptance. This can be seen in the development of legislation based on social pressure to enact rules based on a particular norm. But 'ethics and law cannot easily be reduced to each other' (Felt & Wynne, 2007), and ethics needs to impose its imperative lest it be lost in a reduction to contextual constraints.

The approaches that fall under the Democratic Paradigm are certainly a step forward in terms of inclusive governance. But they are ultimately held back by a restriction of framing and presuppositions about how the approaches will work and how the outcomes of such approaches will impact the technological development or policy. They also by

no means imply that any ethics will be involved, since these approaches could also be used for (for example) economic, scientific, or cultural norms and values, without any inclusion of an account of ethical norms.

CONCLUSION

The elaboration of the four main paradigms of governance, ranging from the technocratic to the democratic, is hopefully illuminating as to the kind of expectations that one can reasonably have in terms of governance procedures. One can recall the definitions of these paradigms which are divided into the 'Non-Democratic' on the one hand, and the 'Democratic' on the other hand:

- **The Technocratic-Instrumental Paradigm:** Is the combination of a technical expertise provided by a restricted community and of an instrumental power of technical skills in the determination of social rules and choices.
- **The Ethocratic-Normative Paradigm:** Is the combination of an ethical expertise provided by a restricted community and of a normative power of moral will in the determination of social rules and choices.
- **The Epistocratic-Cognitive Paradigm:** Is the combination of an epistemic expertise provided by a restricted community and of a cognitive power of scientific knowledge in the determination of social rules and choices.
- **The Democratic-Inclusive Paradigm:** Is the combination of a democratic participation allowed to a community of citizens and of an inclusive power of political opening to society in the determination of social rules and choices.

If we use these paradigms as a basis for evaluating the actual governance procedures, then we can identify some limits as regards the *normative* and the *contextual* aspects of each of these paradigms. The non-democratic paradigms (technocratic, ethocratic, epistocratic) can hardly satisfy the normative and contextual criteria of non-reduction of the significance and the relevance of legal, moral or social rules for the 'outsiders'. Nevertheless, it appears that even in the democratic paradigm, no procedure of inclusion of citizens, or stakeholders, would guarantee that the normative and the contextual criteria are satisfied. There can also be some kind of reduction of the actors' contexts that preclude the significance of norms to be fully scrutinized, discussed and experienced.

This means that the democratic approach in the governance procedures is certainly a necessary condition to avoid the pitfalls of the non-democratic models and settings and for the community of assessors to be enlarged to that of the citizens. But it is certainly not a sufficient condition for the normative and contextual demands to be fully satisfied, if no procedure enables participants to question the context-based value-significance of a set of norms.

REFERENCES

Abels, G. (2006, Jul 4). Forms and functions of participatory technology assessment – Or: Why should we be more skeptical about public participation? *Participatory Approaches in Science and Technology Conference* (pp. 1-12).

Adair, D. (1967). The technocrats: 1919-1967. A case study of conflict and change in social movements. (Thesis). Simon Fraser University.

Anderson, E. (2006). The epistemology of democracy. *Episteme*, *3*(1-2), 8–22. doi:10.3366/epi.2006.3.1-2.8.

Anderson, E. (2008). An epistemic defense of democracy: David Estlund's democratic authority. *Episteme*, *5*(1), 129–139. doi:10.3366/E1742360008000270.

Arrhenius, G. (2005). Vem bör ha rösträtt? Det demokratiska avgränsningsproblemet. In Tidskriften för politisk filosofi (vol. 2).

Assoun, P. (1990). *L'Ecole de Francfort*. Presses Universitaires de France.

Ayer, A. (1959). *On the analysis of moral judgement. Philosophical Essays*. London: Mac Millan.

Berten, A. (1993). Habermas critique de Rawls. La position originelle du point de vue de la pragmatique universelle. Retrieved from www. uclouvain. be/cps/ucl/doc /.../ DOCH _006_ (Berten).pdf

Brooks, T. (2008). Is Plato's political philosophy anti-democratic? In Kofmel, E. (Ed.), *Antidemocratic thought*.

Burris, B. H. (1993). *Technocracy at work*. New York: State University of New York Press.

Cahill, C., Sultana, F., & Pain, R. (2007). *Participatory ethics: Politics, practices, institutions*. ACME Editorial Collective.

Callon, Lascoumes, & Barthe. (2001). Agir dans un monde incertain. Seuil.

Callon, M. (1998). Des différentes formes de démocratie technique. *Les cahiers de la sécurité intérieure*, 38, 37-54.

Ciprut, J. V. (2008). Prisoners of our dilemmas. In Ciprut, J. V. (Ed.), *Ethics, politics and democracy. From primordial principles to prospective practices* (pp. 17–20). MIT Press.

Cohen, J. (1986). An epistemic conception of democracy. *Ethics*, *97*(1), 26–38. doi:10.1086/292815.

D'Holbach, P. H. T. (1776, 2008). *Ethocratie. Ou le gouvernement fondé sur la morale*. Coda.

Dahl, R. (1998). *On democracy*. Yale University Press.

Dosi, G. (1982). Technological paradigms and technological trajectories: A suggested interpretation of the determinants and directions of technical change. *Research Policy*, *11*, 147–162. doi:10.1016/0048-7333(82)90016-6.

Dougherty, M. (2010). Platonic epistocracy. A response to Andrew Forcehimes. Deliberative democracy with a spine. idem, pp. 79-83.

Felt, U., & Wynne, B. (2007). *Taking European knowledge society seriously. Expert Group on Science and Governance*. Luxembourg: Office for Official Publications of the European Communities.

Fisher, E., Mahajan, R., & Mitcham, C. (2006). Midstream modulation of technology: Governance from within. *Bulletin of Science, Technology & Society*, *26*(6), 485–496. doi:10.1177/0270467606295402.

Fisher, F. (1990). *Technocracy and the politics of expertise*. Newberry Park, CA: Sage.

Forcehimes, A. (2010). Deliberative democracy with a spine. Epistemic agency as political authority. *Dialogue*, *52*(2-3), 69–78.

Freeman, C., & Louçã, F. (2002). *As time goes by: From the industrial revolutions to the information revolution*. New York: Oxford University Press. doi:10.1093/0199251053.001.0001.

Gordon, J., & Finlayson, F. F. (2010). *Habermas and Rawls. Disputing the political*. Routledge.

Goujon, P., & Dedeurwaerdere, T. (2009). Taking precaution beyond expert rule. Institutional design for collaborative governance. The genetically modified organisms. In *Proceedings of the ICT that Makes a Difference Conference*. Brussels, November 2009.

Haber, S. (2001). *Habermas*. Presses Pocket.

Habermas, J. (1968). *Technik und Wissenschaft als 'Ideologie*. Frankfurt am Main: Suhrkamp.

Habermas, J. (1981/1984). The theory of communicative action. Cambridge. *Polity*.

Habermas, J. (1983). *Moral Bewusstsein unf Kommunikativ Handeln*. Suhrkamp Verlag.

Habermas, J. (1991). *Erläuterungen zur Diskursethik*. MIT Press.

Habermas, J. (1984-1987). The theory of communicative action. Cambridge.

Habermas, J., & Rawls, J. (2005). *Débat sur la justice politique*. Cerf.

Hamilton, A. (2008). John Stuart Mill's Elitism. In Kofmel, E. (Ed.), *Anti-democratic thought*.

Heiskanen, E. (2005). Taming the golem - an experiment in participatory and constructive technology assessment. *Science Studies*, *18*(1), 52–74.

Held, D. (1996). *Models of democracy*. Stanford University Press.

Horkheimer, M. (1967). *Zur Kritik der Instrumentellen Vernunft*. Frankfurt am Main.

Horkheimer, M., & Adorno, T. (2007). *Dialectic of enlightenment*. Stanford University Press.

Kalberg, S. (1980). Max Weber types of rationality. *American Journal of Sociology*, *85*(5), 1145–1179. doi:10.1086/227128.

Kant, I. (2009). *Groundwork of the metaphysics of morals*. Harper Perennial Modern Classics.

Kuhn, T. (1962). *The structure of scientific revolutions*. University of Chicago Press.

Kuhse, P. (2006). *Bioethics: An anthology*. Blackwell Publishing.

Latour, B. (1987). *Science in action: How to follow scientists and engineers through society*. Milton Keynes, UK: Open University Press.

Latour, B. (1988). *Science in action*. Harvard University Press.

Lenoble, J., & Maesschalck, M. (2003). *Toward a theory of governance: The action of norms*. Kluwer Law International.

Lenoble, J., & Maesschalck, M. (2006). Beyond neo-institutionalist and pragmatic approaches to governance. REFGOV, FP6.

Lenoble, J., & Maesschalck, M. (2009). *L'action des normes*. Presses de l'Université de Sherbrooke.

Maesschalck, M. (1981). *Normes et constextes*. Olms.

Meynaud, J. (1960). *Technocratie et politique*. Lausanne.

Meynaud, J. (1960). Qu'est-ce que la technocratie? *Revue économique, 11*(4), 500.

Meynaud, J. (1964). *La technocratie. Mythe ou réalité?* Paris: Payot.

Mill, J. S. (1973). On liberty and considerations on representative government. Lodon: Everyman.

Moreno, J. (1998). Ethics by committee: The moral authority of consensus. *The Journal of Medicine and Philosophy, 13*(4), 411–432. doi:10.1093/jmp/13.4.411 PMID:3246580.

Moreno, J. (2006). Ethics consultation as moral engagement. In Kuhse, H., & Singer, P. (Eds.), *A compagnion ot Bioethics*. Blackwell Publishing.

Moreno, J. (2009). Ethics comittee and ethics consultants. In Kuhse, H., & Singer, P. (Eds.), *A compagnion ot Bioethics* (pp. 573–584). Blackwell Publishing. doi:10.1002/9781444307818.ch48.

Ober, J. (2009). *Epistemic democracy in classical Athens*. Princeton/Stanford Working Papers.

Outhwaite, W. (2009). *Habermas*. Polity Press.

Perez, C. (2004). Technological revolutions, paradigm shifts and socio-institutional change. In Reinert, E. (Ed.), *Globalization, economic development and inequality, an alternative perspective* (pp. 217–242). Cheltenham, UK: Edward Elgar.

Popper, K. (1988). *The poverty of historicism*. Routledge.

Rawls, J. (1971). *A theory of justice*. Cambridge, MA: Belknap Press.

Rawls, J. (1996). *Political liberalism*. New York: Columbia University Press.

Ruol, M. (2000). De la neutralisation au recoupement. J. Rawls face au défi de la démocratie plurielle. *Revue Philosophique de Louvain*.

Scott, H. (1933). *Science versus Chaos*. Retrieved from http://www.archive.org/details/TechnocracyHowardScott ScienceVs.Chaos-June1933

Scott, H. (1965, 2008) History and purpose of technocracy. *The North American Technate* (TNAT). Retrieved from http://www.archive.org/details/HistoryAndPurposeOfTechnocracy.howardScott

Singer, P. (2006). Moral experts. In Selinger, E., & Crease, R. P. (Eds.), *The philosophy of expertise* (pp. 188–189). Columbia University Press.

Stoker, G. (1998, March). Governance as a theory: Five propositions. *International Social Science Journal, 155*, 17–28. doi:10.1111/1468-2451.00106.

Surowiecki, J. (2004). *The wisdom of crowds*. Anchor.

Weber, M. (1919/1946). Science as a vocation. In Weber, M., Gerth, H. H., & Wright Mills, C. (Eds.), *From Max Weber: Essays in Sociology*. Oxford: Oxford University Press.

Young, I. M. (2000). *Inclusion and democracy*. Oxford University Press.

ENDNOTES

[1] See Stoker (1998).

[2] As Rosenau says, 'governance is a more encompassing phenomenon than government. It embraces governmental institutions, but it also subsumes informal, non-governmental mechanisms whereby those persons and organizations within its purview move ahead, satisfy their needs, and fulfill their wants' J.N. Rosenau (1992) 'Governance, order and change in world politics,' in Rosenau, J.N., Czempiel, E.O. *Governance without government* (pp. 1-29). Cambridge University Press.

[3] Our understanding of the notion of 'paradigm' stems from Kuhn's (1962) concept of 'scientific paradigms,' in which we consider the concept first as a typical matrix gathering a coherent set of theories and practices (e.g. values, beliefs, facts, knowledge); then as a tool for structuring the reasoning, interpretation and the behavior of people; and finally as a cultural and social frame of reference one can identify as different from another competing paradigm. Further, within the context of emerging technologies, the notion of techno-economic paradigm can be utilised as a pattern used in understanding the relation between technological innovations, society and economic growth. Dosi defined a 'technological paradigm' referring back again to Kuhn's concept as 'a model and a pattern of solution of selected technological problems, based on selected principles derived from natural sciences and on selected material technologies' (1982, p.152, original emphasis).

[4] For more details on the paradigms of governance, the reader can see the EGAIS deliverable 3.1.

[5] For the history of the Technocratic Movement, see David Adair (1967) *The Techno-crats: 1919-1967. A case study of conflict and change in social movements*. (Thesis). Simon Fraser University. See also Beverly H. Burris (1993) *Technocracy at work*. New York: State University of New York Press.

[6] See Scott (1965, 2008).

[7] Idem.

[8] See Kant (2009).

[9] For an interpretation of Weber's types of rationality see Stephen Kalberg (1980) 'Max Weber Types of Rationality,' *The American Journal of Sociology*, Vol. 85, N° 5, pp. 1145-1179, as well as Habermas in the *Theory of Communicative Action*, Book I.

[10] See Popper (1988).

[11] See Horkheimer (1967) and Assoun (1990).

[12] See Habermas (1984-1987).

[13] See Habermas (1968).

[14] See Latour (1988) and Callon, Lascoumes, & Barthe (2009).

[15] See D'Holbach (1776, 2008).

[16] See Ayer (1959).

[17] See Singer (2006).

[18] See Moreno (2006).

[19] See Moreno (2009).

[20] See Kant (2009).

[21] See Habermas (1987).

[22] See Ciprut (2008).

[23] See Caitlin Cahill, Farhana Sultana, Rachel Pain (2007) 'Participatory Ethics: Politics, Practices, Institutions,' *ACME Editorial Collective*.

[24] For a summary of all these critiques, see Stéphane Haber (2001) *Habermas*, Presses Pocket and William Outhwaite (2009) *Habermas*, Polity Press.

[25] See Maesschalck (1981).

[26] See Lenoble, & Maesschalck (2003).

[27] See Brooks (2008).

[28] See Ober (2009).

[29] See Mill (1973).

[30] See Hamilton (2008).

31 See the debate between Forcehimes (2010) and Dougherty (2010).

32 See Cohen (1986).

33 For the comparison between Rawls and Habermas, see *Debate on political Justice*, and James Gordon Finlayson, Fabian Freyenhagen (2010) *Habermas and Rawls. Disputing the Political*, Introduction, Routledge. Also Muriel Ruol (2000) 'De la neutralisation au recoupement. J. Rawls face au défi de la démocratie plurielle,' *Revue philosophique de Louvain*. Also, André Berten (1993) 'Habermas critique de Rawls. La position originelle du point de vue de la pragmatique universelle'.

34 See Rawls (1971).

APPENDIX 1

Paradigms of Governance and Normative-Contextual Perspectives

Table 1. Contextual perspectives

	Technocratic-Instrumental Paradigm	Ethocratic-Normative Paradigm	Epistocratic-Cognitive Paradigm	Democratic-Inclusive Paradigm
Idea	Optimalityof means	Justification of norms	Selection of the best	Participation of society
Stake	Efficiency of decisions	Legitimacy of rules	Wisdom of choices	Inclusion of citizens
Form	Technical positivism	Ethical universalism	Epistemic elitism	Political neutralism
Example	Saint-Simon, Comte, Popper	D'Holbach, Kant	Plato, Mill	Rawls, Habermas,Latour
Limit	Axiological appraisal of choices	Empirical context- testing	Hierarchic determination of wisdom	Hermeneutical relationship to traditions
Norm-related problem	Factual 'devaluation' of norms, unethical reduction	De-contextualised abstraction of norms, no effective impact	Cognitive embededness of norms, epistemic closure	Neutralization, formalization of norms' contents

APPENDIX 2

Patterns of Deliberative and Participatory Democracy

Table 2 shows the comparative characteristics of the two models based on the principles of deliberation and participation as elaborated by Held (1996).

Table 2. Deliberative vs. participatory democracy

Deliberative democracy	Participatory democracy
Principle(s) of justification The terms and conditions of political association proceed through the free and reasoned assent of its citizens. The 'mutual justifiability' of political decisions is the legitimate basis for seeking solutions to collective problems. *Key features* Deliberative polls, deliberative days, citizen juries E-government initiatives from full on-line E-democracy programmes including on-line public for a group analysis and generation of policy proposals. Deliberation across public life, from micro to transnational settings. New uses of referenda tied to deliberative polls. *Movement*: From renewing representative democracy to radical, deliberative participatory democracy. *General conditions* Value pluralism. Strong civic education programme. Public culture and institutions supporting the development of 'refined' and 'reflective' preferences. Public funding of deliberative bodies and practices, and of secondary associations which supports them.	*Principle(s) of justification* An equal right to liberty and self-development can only be achieved in a 'participatory society,' a society which fosters a sense of political efficacy, nurtures a concern for collective problems and contributes to the formation a knowledgeable citizenry capable of taking a sustained interest in the governing process. *Key features* Direct participation of citizens in the regulation of the key institutions of society, including the workplace and local community. Reorganization of the party system by making party officials directly accountable to membership. Operation of 'participation parties' in a parliamentary or congressional structure. Maintenance of an open institutional system to ensure the possibility of experimentation with political forms. *General conditions* Direct amelioration of the poor resource base of many social groups through redistribution of material resources. Minimization (eradication, if possible) of unaccountable bureaucratic power in public and private life. An open information system to ensure informed decisions. Re-examination of childcare provision so that women as well can take up the opportunity to participate in public life.

Chapter 10

Governance in Technology Development

Aygen Kurt
Middlesex University, UK & London School of Economics (LSE), UK

Penny Duquenoy
Middlesex University, UK

ABSTRACT

With an increasing focus on the inclusion of considering the ethical and social impact of technology developments resulting from research in the European Union, and elsewhere, comes a need for a more effective process in technology development. Current ethics governance processes do not go far enough in enabling these considerations to be embedded in European Union research projects in a way that engages participants in technology development projects. Such a lack of engagement not only creates a distance between the technology developers and ethics (and ethics experts) but also undermines the legitimacy of decisions on ethical issues and outcomes, which in turn has an impact on the resulting innovation and its role in benefitting individuals and society. This chapter discusses these issues in the context of empirical work, founded on a theoretical base, undertaken as part of the EGAIS (Ethical Governance of Emerging Technologies) EU co-funded FP7 project.

INTRODUCTION

Governance of technology development is a central focus of the European Commission's (EC) funding strategies and the main economic development programme of the European Union. According

DOI: 10.4018/978-1-4666-3670-5.ch010

to the EC's 2002 Science and Society Action Plan, "if citizens and civil society are to become partners in the debate on science, technology and innovation in general and on the creation of the European Research Area in particular it is not enough to simply keep them informed" (European Commission 2002, p.17). This statement convincingly reflects that governance of technology is to be considered as a public policy concept at national, European or global levels[1] and particularly includes the "public" in technology development processes through to the market (diffusion) stage of technological innovations.

Different approaches to governance to solve specific systemic and coordination problems in areas such as economics, health and technical developments have received increasing attention over the last decade. Governance, according to Edler (2010) is "an ill-defined, amorphous concept (analytically and empirically)" but one which involves "some form of cooperation, persuasion and reflection". In recent years governance has been understood to refer to "a basically non-hierarchical mode of governing, where non-state, private corporate actors (formal organizations) participate in the formulation and implementation of public policy" (Mayntz, 2003, p.1) and further "By definition, governance refers to the solution of collective problems and the production of public welfare." (Ibid, p.7)

However, when it comes to understanding the contextual and horizontal differences or applications which certain governance structures could entail, a more focussed definition is needed. We take governance – especially when related to technology development and technology policies – as a complex structure including a number of actors not necessarily with equal power structures but who can only act in interdependence and interaction with each other. As Jessop puts it, governance is:

The reflexive self-organisation of independent actors involved in complex relations of reciprocal interdependence, with such self-organisation being based on continuing dialogue and resource-sharing to develop mutually beneficial joint projects and to manage the contradictions and dilemmas inevitably involved in such situations (Jessop, 2003, p.1).

Hence the reciprocal relationship among actors results from an on-going interaction and learning among actors in order to reach mutually acceptable and useful end-results. A significant question that arises in this context is in the realm of democratic processes and how democratically motivated decision-making processes should drive governance of technology development especially in relation to emerging technologies[2]. In order to understand such processes and establish well-functioning governance structures for technology development, we need to bring "ethics" into consideration, realise the uses and utilise the applications of ethical governance mechanisms before, during and after a technology is developed.

The objective of this chapter is to present a snapshot of governance tools in technology development projects funded by the European Commission and demonstrate their link to some governance models. The snapshot is taken from a number of projects analysed as part of the empirical work of the Ethical Governance of Emerging Technologies (EGAIS) project[3], based on a theoretical framework developed in the first stage of the project[4], and provides an overview of such approaches and their meanings. The following section sets the context for the empirical study, reports on the key findings and discusses their implications.

ETHICS, GOVERNANCE AND REFLEXIVITY IN EU-FUNDED TECHNOLOGY DEVELOPMENT PROJECTS

Integrating ethical considerations within EU co-funded technology development projects necessitates bridging a disciplinary gap (humanities and computer science), as well as introducing sets of values to the project that may challenge original technical or political objectives[5]. Various ethics governance mechanisms have been, and are still, deployed such as ethics check lists, external and internal ethics advisory groups. The consideration of ethical aspects of a technology project (whether EU co-funded or otherwise) is generally seen as the responsibility of the 'ethics expert', and the application of the recommendations of the 'ethics expert' is set in the domain of the technology development team, both having particular domain language and motivations. This arrangement gives rise to a divide between the technical community and the ethical community in integrating ethical governance into technological development. Ethics governance is often used to respond to ad hoc problems resulting from artificially designed contexts, and usually follows a categorical and consequentialist model. Overall, most approaches to governance in projects tend to undermine the role of the external context within which the technology development is situated, which also means that the external context within which the norm is to be constructed and operated is also undermined. There is a need to bring ethical reflexivity into the heart of the project, which should include reflexivity by all those involved in it, so that the results of reflection can more easily be implemented (i.e. putting the norm into action, or as referred to in the project the "action of the norm").

The investigation of how effective ethical norms are, and under what conditions they emerge and become implemented, formed a key aspect of the EGAIS project. In this respect special emphasis is placed on "the action of a norm"[6]in addressing the conditions for the effectiveness of ethical reflexivity within technological development. Therefore, understanding the conditions of the construction, and just as important, the implementation of the ethical norm in a specific context, is crucial if effectively embedding ethical norms in technology development is to take place.

One of the main constraints on effective reflexivity (that results in a norm being implemented) lays in the framing of a concept, in this case the framing of a project and its technological application. Conceptions of the world are related to specific cognitive framings - whether technical, political or cultural – therefore any cognitive framing carries its own constraints. This is referred to as "cognitive closure". In terms of research projects,

technical and scientific value systems may override moral reason and undermine the other external framings that are relevant to the construction of the norm when technology development is the case. In this case, making a distinction between ethical and Science and Technology communities needs to be questioned, because genuine reflexivity is likely to be suppressed by a technological rationality that may be imposing its own value system, which will undermine (or exclude even) alternative framings which may be relevant to constructing the norm. The 'opening up' of such framings therefore becomes significant in order to achieve a genuine ethical reflexivity. This also applies when ethical governance becomes sectorial ethics (i.e. the community of expertise)and is led by specified experts who have their own particular framing, and do not go beyond the boundaries of their own cognitive framing (leading once again to cognitive closure). What is needed to 'open-up' this cognitive closure is:

- A reflexive governance process that allows learning among the actors (which would eventually lead to capacity building for those actors).
- The destabilising of any prior framing.
- An ethical governance process that integrates a systemic continuous learning throughout the technology development.
- An investigation into the conditions for effective insertion of an ethical norm into its appropriate context from the beginning of the technology development.

The EGAIS approach emphasises going beyond considering ethics as restricted to a categorical field, which means that ethical governance must be about more than providing the elements to justify ethical decision-making – ethical governance needs to concern itself with the construction of the context (social, political, economic and technical)in relation to the technology. Following this line of enquiry the focus of our initial investigations was EU co-funded technology development projects within the Ambient Intelligence (AmI) realm. The concept of ambient intelligence combines intelligent technology with ubiquitous computing operating in the background, that is "an environment that is capable of recognising and responding to the presence of different individuals in a seamless, unobtrusive and often invisible way"[7]envisaged as facilitating everyday actions in the home, at work, in healthcare, and so on. With the vision of these technologies being embedded into day-to-day lives explicitly or implicitly, these projects seemed to offer a good starting place for assessing ethics governance[8].

By focusing on a number of AmI projects funded by the EU (mainly under the Framework Programme 6) we intended to investigate and address ethical governance patterns in the projects to see whether any governance tools were used and if so, whether they would show any similar patterns. The empirical process is described next.

EMPIRICAL INQUIRY ON EU-FUNDED AMBIENT INTELLIGENCE PROJECTS[9]

Our sample target was the coordinators/leaders of the EU-funded AmI projects, with the aim of retrieving information from project leaders about their project's engagement with ethics (if applicable) and the motivations behind any such engagement. A questionnaire was designed with simple (but not simplistic) questions, and additional interviews planned with project leaders to get a more in-depth understanding of their responses to the questionnaire.

The questionnaire was developed in accordance with a number of analytical parameters informed by our theoretical standpoint[10].

- **Ethical Issue Identification and Specification:** To identify whether the projects under investigation had any ethi-

cal issues that had to be addressed, and if they did, how they determined what ethical issues to look for.

- **Ethical Approach:** To find out whether any theoretical approaches were used to address the ethical issues, and what theories, approaches and principles were utilised.

- **Reflexivity:** To gain an insight into the types of reflexivity within the project, and at what stages (since the beginning, throughout the project or at the end). This parameter was the foremost element to explore the type of "ethical" questions project leaders might have asked, and if those questions were solely concerned with determining an ethical issue, or concerned with going further by asking a second-order question such as how the actors might reflect on the process of verifying ethical issues.

- **Governance Arrangements:** To find out the institutional arrangements (if any) in operation within the project to deal with ethical issues.

- **Implementation:** To discover the implementation process and tools of ethical governance, and the degree to which they were effective.

The questionnaires and the follow-up interviews were categorised according to the parameters outlined above[11]. The Table 1 summarises the questionnaire guide used to produce the questions we sent to project leaders.

Data Collection

The data collection strategy included compilation of primary and secondary data. Our starting point was to identify ethical governance structures in a number of AmI projects funded and completed under the EU 6th Framework Programme. Of interest were projects that had included ethical governance as well as projects which had not but

Table 1. Questionnaire guide

Ethical issue determination	What are the ethical issues (if any)? Has the project been to the EU review?
Ethical approach	What ethical theories, tools, approaches are used?
Reflexivity	Was there a first-order and/or second-order reflexivity implemented?
Governance arrangements	What institutional arrangements were implemented and what governance approaches were used?
Implementation	What tools were used to incorporate ethical values into the development of the project? Were they effective?

which appeared to have an ethical aspect within them. The intention was to understand ethical governance and reflexivity issues that had taken place or that could have been integrated in the projects, and to analyse the strategies used to address ethical issues, to consider their limits and to propose a strategy for overcoming those limits.

Three data collection methods were used: (1) initial questionnaire to project leaders, (2) follow-up interviews with some project leaders (3) project documentation.

Our normative approach to data collection involved a questionnaire aimed at project coordinators only. This was because we needed to understand where, how, and to what extent ethical issues within the project were tackled from the project leaders' perspectives (i.e. to gain insight into the projects leaders' understanding of the issues). The questionnaire included both closed and open-ended questions.

We used a qualitative interviewing technique with a small number of selected projects as a follow-up strategy after the questionnaires was seen necessary to be able to conduct an in-depth analysis. The research participants, who had responded positively to our initial email request regarding the questionnaire, were asked if they would agree to a follow-up interview. Not all participants did (which was an expected result). It was not necessary, in terms of our research sample, for

all questionnaire respondents to participate in an interview. The interviews enabled clarification of the questionnaire responses and a deeper enquiry into the project leaders' understanding of the place of ethics and governance in their project.

Finally, our secondary information collection stage focused on analysing publicly available documents, e.g. deliverables, project brochures or newsletters, web sites, to verify the match (or mis-match) with the responses gathered from the project leaders.

The research population included a total of 73 completed AmI projects identified from EU CORDIS project database and the possible respondents were all contacted via e-mail. We received 19 questionnaires, giving a response rate of 26%.

Each project leader was sent an explanatory email initially and a following email with the questionnaire attached to those who responded positively (See Table 2). The questionnaire included an introductory note that assured confidentiality and anonymity of the respondent.

The field of application of the AmI projects in the study ranged from: system integrations to small devices; establishment of communication networks among devices; micro-systems in medicine to biometric data detection and robotics.

Outcomes from the responses to the questionnaire revealed that in seven projects ethical issues were integrated at the initiation of the project; five projects had been subject to EU ethical review; ten projects had also considered, or engaged with, social issues; and three respondents were unsure. Among the most frequently mentioned ethical perceptions were: technology's impact on social life, privacy of individual data and informed consent where relevant.

The ways in which any ethical and/or institutional arrangements resulted in concrete outcomes were also at the kernel of our study. Those projects engaged with the EU ethical review process developed technology applications according to the recommendations. For instance, one of the projects, *Project L*, selected software and hardware

Table 2. Profile of research population and sample used in this deliverable

Introductory email asking for collaboration sent to	=	73 AmI projects
Positive responses to the first email	=	30
Questionnaires received	=	19
Response rate (project #)	=	26%

according to the review. On the other hand, those which relied on ethical protocols from the beginning, for example, *Project N*, did not encounter any concrete effects as a result of these arrangements. Creating a profile of the ethical and social issues present in projects from the project leaders' perspectives gave the project leaders' interpretation of these aspects at the initiation of their projects. However, the interpretation of the researchers, based on the project documentation (publicly available materials such as deliverables, websites, and publications) suggested a different picture. The differences could be seen where projects had indicated 'no ethical/social issues' according to the project leaders but which demonstrate a different interpretation in the project documentation (shown as highlighted sections in Table 3 below). The differences could be attributed to a growing awareness and realisation of social and ethical aspects of the projects during the research process.

The Table 3 presents the social and ethical aspects of the projects as understood by the project leaders (extracted from the questionnaires) and as understood by the EGAIS partners (researchers) from the public documentation.

Only five projects had been submitted for EU ethical review, of these, information about the outcome was only available for four. *Project L*, for instance went through an ethical review within the first six months and the project was approved with some recommendations including an ethical audit of the project. *Project R* covered the issues comprehensively to the extent that the review was sufficient and no ethical audit was

recommended. However, for *Project M*, the EU suggested enhancement of the ethical assessment approach and production of ethical guidelines for general use. Ethical audit was recommended as a follow-up strategy. The outcome of the review for *Project N* was the use of an ethical protocol for field work.

Governance Tools and Implementation

It is important here to emphasise the distinction between governance tools and approaches, since there can be some confusion over whether using a particular tool is sufficient to qualify as a full governance procedure for a project. However, it is often the case that projects do not use explicitly use governance, yet the ways the tools are used and incorporated into project design can determine what particular governance approach was used.

When the project leaders were asked how they used any strategy for addressing the social or ethical aspects of their projects; three projects had a separate work package or deliverable devoted to the exploration and definition of social and ethical aspects in relation to their projects (*Projects L, M and R*). One of these projects (*Project R*) also went further by integrating some legal and regulatory aspects of the technological applications they were operating. *Project N* produced a final report on ethical issues. Nine projects did not employ any governance arrangements for ethical issues (See Table 4).

On the table above we can see that among the listed strategies for integrating ethics or social aspects, technology assessment was the most used strategy by seven projects out of the total. As the respondent for *Project L* stated, "[...] without technology assessment it is unlikely that you can run a technology development research project". Focus groups and questionnaires followed technology assessment as each being the second most preferred strategy.

Table 3. Ethical aspects of the AmI projects

No.	PROJECT	Ethical/social issues (at the beginning of project) according to questionnaires	Ethical/social issues according to project documentation
1	A	No ethical issues, but to develop Internet as a better place for all.	human behaviour, user-centred technology development
2	B	No ethical/social issues.	user testing, user participation, technology relevance, user acceptance
3	C	No ethical issues. User-involvement.	user involvement in technology development, no ethical issue mentioned
4	D	No ethical /social issues.	information security
5	E	No ethical issues. User-acceptability, benefit for society,	social impact in daily life, benefit for society, no ethics mentioned
6	F	Yes to ethical / social issues but not explained.	Environmental and individual safety, security, quality of service
7	G	Unobtrusive private data, user-acceptance not acceptability.	Surveillance, profiling, identity, autonomy, privacy, confidentiality.
8	H	No ethical/social issues.	Trust and information security, confidentiality.
9	I	No ethical/social issues.	Video surveillance, monitoring.
10	J	Seeking ethical approval, no social.	Quality of service, informed consent, data handling, social / medical policy
11	K	No ethical issues. Gender and technology.	Human behaviour, cultural contexts, learning.
12	L	Data privacy, informed consent, vulnerability of the user group, information quality. No social issues.	Social exclusion, social and technological risk, data security, monitoring, confidentiality, information quality.
13	M	User-acceptance, user-care.	Privacy, autonomy, dignity, reliability, e-inclusion, benefit for society
14	N	No ethical issues. Social issues not explained.	User acceptability, informed consent, data confidentiality, video surveillance
15	O	No ethical issues. Organisational and social aspects of technology.	Social mobility, quality service, information security (sharing), organisational and social aspects of work
16	P	No ethical/social issues.	Privacy
17	Q	No ethical/social issues.	Data security and confidentiality, freedom of choice, reliability, obligation of information
18	R	Technological risk, data security, monitoring.	Public understanding of science, e-inclusion, monitoring, surveillance, data handling, quality of informed consent, data management.
19	S	No ethical issues. Social are user safety, privacy.	Privacy, security, data handling.
Projects without a questionnaire			
20	T	N/A	None are apparent.
21	U	N/A	User acceptability, security
22	V	N/A	User-centred space, data security, inter-operationability, privacy, data handling.
23	W	N/A	Dignity, body implants, reliability and quality of technology, user acceptability.

Table 4. Ethical governance arrangements used in the European projects in this research

Methods:	A	B	C	D	E	F	G	H	I	J	K	L	M	N	O	P	Q	R	S	Total
Ethical Committees							X			X			X	X				X		5
Expert Panels											X		X	X				X		4
Focus Groups								X		X	X	X						X		5
Hybrid Panels																				0
Participatory Assessment											X	X						X		3
Technology Assessment	X						X			X	X	X	X					X		7
Value-Sensitive Design													X							1
Delphi Studies	X																			1
Questionnaires							X			X	X		X					X	X	6
Other…													X					X	X	3
None		X	X	X	X	X			X						X	X	X			9

Project leaders were asked how they reflected upon ethical issues during the operation of the projects (i.e. as and when they had arisen). Ten respondents either gave no explanation or responded negatively to this query. However, *for Projects L, M and R,* a continuous involvement with stakeholders and end-users seemed to have a positive impact on the trajectory of the projects in terms of improvement of understanding the needs and limits of users and also the limitations of the technology being developed. For this reason, the reflexivity measures of these projects are highlighted in Table 5. These projects had the most clearly defined reflexivity measures according to the questionnaire in terms of application, utilisation and implementation of ethical tools and grasping the problems technology may solve or produce. Three projects (*Projects G, N and P*) on the other hand mainly relied on a management team or an expert board (not included in Table 5, as 'reflexivity' on the part of the project participants, and therefore 'embedded' had not taken place).

The governance arrangements and the governance approaches are linked to reflexivity measures. Here we intended to ascertain the institutional arrangements used for the implementation of ethical issues in the projects. The data retrieved for the ethical governance approaches is presented in a different format below. Here we can see, from the project leaders' perspectives, the degree to which these arrangements were effective.

Results as to the effectiveness of these different governance arrangements (see Figure 1) indicate that technology assessment (TA) was the most effective tool used in the projects. Half of the projects that used ethical committees considered it as an effective tool, whereas only one project for each tool, i.e. focus groups and expert panels, suggested they were somewhat effective.

The figure interestingly shows that the highest number was attributed to TA. It is also interesting to note that the indicators "not effective" and "not very effective" do not appear on the Figure (i.e.

Table 5. Profile of reflexivity measures for Projects L, R and M

			Project L (beginning)	**Project L (during)**
Reflexivity	Ethical problem identification		Yes	Yes
	Validation	Arrangements	EU ethical review	Technology and participatory assessment
		Tools	None	Focus groups
		Framing	EU approach, technological and social	Technological, social and eco-nomical
			Project R (beginning)	**Project R (during)**
Reflexivity	Ethical problem identification		Yes	Yes
	Validation	Arrangements	Ethical supervisory committee	Expert panels
		Governance Tools	Technology assessment (TA), expert views	Participatory assessment, focus groups, questionnaire
		Framing	Medical, technological, social, legal	Medical, technological, social, legal
			Project M (beginning)	**Project M (during)**
Reflexivity	Ethical problem identification		Yes	Yes
	Validation	Arrangements	Ethical advisory board	Ethical audit
		Governance Tools	Ethical review, expert panels, TA	Ethical manual, TA
		Framing	Utilitarianism, consequentialism	Technological, social, utilitarian-ism.

no project considered the tools they had chosen were not effective).

Focus groups and questionnaires were each the second most attributed tool used for addressing ethical issues in the projects. None of the projects used hybrid panels.

More than half of the sample also expressed their key opinions in form of statements very briefly in relation to ethical governance. *Project H*, for instance suggests that because of the different cultural contexts and different perceptions of technology arising from these cultures, one should not take anything for granted about ethics. On the other hand, *Project M*, which used ethical committees and advisory boards, and had to produce an ethical manual as recommended by the EU review, acknowledges that concrete ethical guidelines are not easy to define.

In terms of the distance between ethical and technological experts, *Projects G and Q* thought that having an ethical expert in the consortium would help, and that the task of social and ethical issues should be ethical experts' task.

Projects L and R, nonetheless, emphasised the significance of involving users and relevant stakeholders as early as possible within the project, and people [as users] should be asked what they want as often as possible (for example, without the pre-assumptions that people from vulnerable groups might have different needs from others).

Taking the governance tools one by one, we can categorise them in terms of their usage and implementation purpose. The categorisation shows the majority of the tools were used to gain feedback from users to enable social acceptance of the technology; some as part of the project's management structure; and a small number were

Figure 1. The effectiveness of the tools used to address ethical issues in the projects

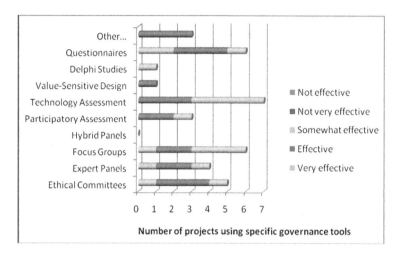

used in the form of expert feedback as an external input. Table 6 gives explanations of the tools used.

Governance tools are the mechanisms by which governance approaches are made manifest. In most cases technology development projects do not use governance approaches explicitly but employ some of the tools which allow the analysis of the governance approaches embedded within the project.

In the context of the focus of the EGAIS project (i.e. technology development projects) governance tools that emerged are methods such as technology assessment, focus groups, Delphi technique, and value-sensitive design. Although these tools are not sufficient on their own to form governance approaches they can be used to explore approaches to ethical governance in addressing ethical issues in technology development.

The data obtained from the selected number of projects revealed that the majority of projects and the tools used reflect the characteristics of a standard model of governance (discussed in the following section), with experts playing a central role and a tendency to de-contextualising the relationship between the norms and context. That is to say the project members, or those who are responsible for ethical issues, are aware of the ethical implications of the technology being developed but there is no solid imposition of ethical governance.

The second pattern is that the type of governance tools can be categorized under the revised-standard model of ethical governance, the majority of which have a relationship to the norm where the context is restricted. The stakeholders are involved, but since they do not participate in defining the context themselves, tools involving stakeholders are justified for the purpose of utilizing them as an information source for decision-making.

On the other hand, a small number of projects can be grouped under the consultation model, because the level and nature of risk perception differs between the public and the experts. Tools are mainly used for consultation purposes.

Finally, there are no projects with governance tools that fall into the co-construction model. From the ways of which the tools are used in selected projects and the reasons behind them, we would argue that the ethical and technical expertise are separated; the stakeholders, particularly the users or the public are involved with the project process only for consultation purposes about the technology's design or usage, which ultimately aims to lead to the social acceptance of technology. We have seen that although the users are included in the technology development process as a feed-back

Table 6. Governance tools used in the projects

Governance TOOLS	Justification
Tools used for Stakeholders' Feedback / Social Acceptability	
Delphi	Although this tool was used with the purpose of assessing the stakeholder view, the problem is not defined by the stakeholders and the decision making is left to those "who knows".
Focus Groups	Common purposes for the use of focus groups were consultation and help with decision-making, therefore it is in way involving the users. However the ethical issues and the problems are reduced to the problem of regulation and law. Hence, there is a sense of elitisim embedded in the decision making process, which brings in the more knowledgeable groups to the scene.
Questionnaires	Dominantly within the use of this tool, the ethical issues are part of the project but the project has no concrete imposition of ethical issues.
Expert Panels	Involving experts or expertise was limited to technological expertise in all the projects that used this tool.
Interview	This tool was used by only one project, and the key aim was "to facilitate the ethical and social acceptance of technology", but within the leadership and direction of the technical expertise and in the hands of those who were scientifically more knowledgeable.
Technology Assessment	This tool was mainly used for aiding decision-making which was under the authority of the technologist as the designer, and a small (elite) group of people as the follower of the procedures involved.
Partnership Networks	The usage of this tool did not focus on ethical issues although the project was aware of them, but aimed to involve stakeholders to address social impact.
Project Associated Partner Networks	The project consortium used this tool for consultation, however none of the stakeholders in partners' network consulted were part of the decision-making process, nor they had defined the context in relation to addressing the ethical issues.
Interaction of users	Same as above.
Participatory Design	Since all the ethical issues within the projects were reduced to legal formalities, and although the deliberative approach allowed some participation into design, project teams had the major influence on the end-results.
Social Acceptability	The idea centred on making the technology commercially viable through social acceptability, with the moral will embedded in it that the technology developed and its interaction with the human was seen crucial.
Participatory Assessment	Usage of this tool involved two key aspects; involving experts in assessing the technology's ethical issues through conferences, and involving stakeholders to give feed-back but mainly to realise the social acceptance of technology.
Value Sensitive Design	This was totally about informing the technical expertise.
Participatory Workshop	Although this tool was used with the purpose of collecting the stakeholder view, the problem is not defined by the stakeholders and the decison making is left to those "who knows".
Governance tools used within Project Management (2)	
General Assembly	As is seen, the technical and the ethical expertise both have equal weight in terms of usage of the tools within project management. In most cases the context is reduced to limiting the ethical issue determination and implementation to legal systems, or it is pre-defined even when the project partners are consulted. In one case the scientifically more knowledgeable group takes the lead in addressing the ethical issues, however, the principles referred to are far from being ethical norms, but legal norms.
Project Management Team	
Authority of Project Leader	
Project Coordination Committee	
Authority of Project Partners	
Dedicated Work-Package	

continued on following page

Table 6. Continued

Governance TOOLS	Justification
Governance tools used within Experts' Feedback (3)	
Ethical Committee	Relying on ethical expertise and not integrating the technical or the societal with the ethical is the major approach here. In most cases, the relationship between the norm and the context ignores explicit imposition of ethical values within the project or prefers to leave them within the realm of legal framing and/or the mainstream procedures existent in technology development.
Advisory Board	
Reference to ethical standards	
Ethical Review Board	

mechanism, the context has been pre-defined and those who already have some sort of 'high level knowledge' give the final decision.

This whole analytical process has demonstrated that norms are constructed under the influence of specific framings related to the context. The framing is connected to the experts' knowledge without considering the exploration and the construction of the context; and without taking into account the consideration of norm application; where the justification of the norm is perceived as sufficient for its application.

FROM GOVERNANCE TOOLS TO GOVERNANCE MODELS[12]

When we look at Lenoble and Maesschalck's comprehensive work (2006, 2003) explaining the theory and methods behind the notion of governance; we are able to identify two overarching perspectives towards governance. These approaches either represent participatory characteristics or procedural, efficiency perspectives. These two conceptions illustrate the two main trends that are the basis for almost all of the currently-used governance approaches[13].

These two conceptions are at opposite ends of the spectrum: the efficiency perspective allows for the effect of the norm, but without any legitimacy for the norm. This is because the characteristics of governance approaches with an efficiency focus are often an imposed, top-down system of expertise and forced implementation. On the other hand the

approaches with participatory characteristics has the legitimacy (through its participatory nature), but fails to effect change; thus it has no efficiency. The reasons for these characteristics are that the way this perspective functions is to have a participatory, bottom-up approach that, while taking in the views of all interested stakeholders, generally might have little effect in reality by reducing ethics to a social acceptance issue.

Four models of governance into these perspectives: the standard, revised standard, consultation, and co-construction models. These models were mostly developed in the context of risk assessment but they are also applicable outside of risk scenarios because of the conceptions they involve and the relationships they display between the norm and context[14].

Standard Model

The Standard Model presents a traditional top-down approach, which is based on the knowledge of experts. The knowledge of the experts determines the norms. Since the public is seen to have less knowledge or expertise on the subject matter; the risks perceived by the public is also seen to be subjective as it does not rely on expertise. However, experts are seen to have an objective approach to risk.

Expertise is considered to be generally independent of political or economic influence. The notion and practice of trust makes the model operate. The gap between the public and expert perceptions can be reduced through educating the less knowledgeable especially about ethical principles. It is a linear process of communication between the experts and the public about the risks involved; mainly processed from experts to public, not the other way around. Therefore, the model can be counted as part of the efficiency category as it entails a top-down approach in order to ensure that the risks are perceived efficiently resulting from the opinions of experts.

Consultation Model

With this governance model, the focus is on the perception of risk creating the difference between public and experts, not the knowledge levels or opinions of experts.

The public is consulted about risk through which they ask questions resulting from their perceptions of real risk (not the scientific risk) they do not have any explicit links to industry or businesses like some experts may do. According to this model, a risk taken voluntarily, more familiar, known and visible; and may be affecting a small number of people is more likely to be accepted. Risk is also communicated through a two-way mutual process between the public and experts.

In this model, trust and institutional legitimacy can be established through public participation. The public is engaged only in risk management but not in risk definition, as the public is still seen as less knowledgeable (and irrational). This model, is therefore is a participatory model but it uses the participation to gain efficiency as the public is still seen as an opponent not as part of the process and on the "same" side.

Revised Standard Model

In this model, the risk perception by public is normally so inadequate that public attitude towards risk is being influenced by the uncertainty created by media. This whole process eventually leads the legislative process of regulating risk to be trapped in a vicious circle. Therefore, risk management is

separated from risk assessment; and risk management is left to an independent expert body in order to prevent media influence and political pressure. Risk is also analysed, not measured in an abstract way; analysed from an economic perspective as well as with scenario building methods. Trust is built via competence and reputation.

In this context, the model is influenced by a technocratic vision and demonstrates a top-down approach as it reinforces independent scientific expertise. Thus, it fits with the *efficiency* category, because, like the Standard Model, it leaves public stakeholders out of the decision-making process.

Co-Construction Model

This model approaches the ways which technology development projects use experts in the process critically. This approach is an outcome of Latour's methodological path in which he proposes a critical view of the traditional notion of science and considers risk analysis from a pragmatic perspective.

The model takes both facts and values into account together as part of the analytical analysis in addition to considering the process a democratic matter. In this model we see a significant effort for establishing procedural methods appropriately, not only in terms of the discussions carried out within the model but in the construction of the model itself. It involves both policy-makers and stakeholders in the construction of the policies implemented (whether at governmental level or on a much smaller scale). The model is in the participatory category which has legitimacy but not much efficiency.

As the figure above illustrates, the four governance models are placed within the two main dimensions representing the ranges of possibilities for constructing a norm used in technology governance used in technology development projects. The efficiency measures represent the degree of delegation of ethical governance to experts and the degree of tendency towards efficiency. The participatory models, which are procedural at the same time, reflect the degree of participation of stakeholders in the process of norm construction and legitimization. Although they are traditionally involved in risk assessment methodologies, these models underlie all the approaches used in technology governance, whether explicitly or implicitly used.

CONCLUSION

In the EGAIS project we have argued that to effectively embed the consideration of ethical issues, and *implement* the resolution of these issues, processes need to be put in place to allow this to happen, and that these processes should be included in ethics governance mechanisms. The theme of this chapter has been to show the limitations of current ethics governance measures, using empirical data from EU co-funded (FP6) projects, and to explain how ethical norms – far from being imposed 'top-down' really need to emerge 'bottom-up' to be (1) relevant in context

and (2) to have legitimacy. Moreover, we argue that this bottom-up approach, which engages all involved in technology development projects (including the technical developers) brings about a 'real' embedding that has meaning – in contrast to a sort of 'subscription' to ethics to meet a higher authority's set of requirements. Indeed, if ethics in technology projects amounts to subservience to a higher authority (even if by an ethics expert) it is not a *real* ethics, as in a non-coerced action. Further, an ethics that is subscribed to for the purposes of meeting higher authority requirements falls into danger of (1) missing, or avoiding, the influence of the context in which the project is set (whether social, political, economic or technical) and (2) creating distance between science and innovation and the general public. Both of these outcomes are counter-productive in terms of initiatives put into place to overcome obstacles to 'social acceptance' and reduce negative impacts of technology development.

The conclusions from the empirical study undertaken in the project (EGAIS 2010a) and

Figure 2. Governance Compass. Source: EGAIS Project Deliverable 2.4, Interpretation of Ethical Behaviour, p.25

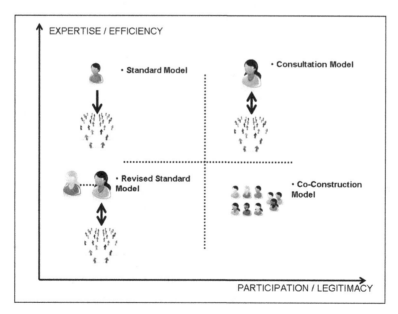

described in this chapter are that the arrangements in place for EU FP6 projects (and to an extent continued in EU Framework Programme 7) did not in themselves encourage project partners to consider ethical or social aspects of their projects either at the research design stage or during the development process. The results showed that in many projects project co-ordinators were not aware of potential issues (that researchers in EGAIS had noticed in publicly available project documentation). Some projects did however include what we have termed 'governance' tools (such as Technology Assessment, focus groups, etc.). Despite the limited facilitation of reflexivity enabled by such governance tools these projects at least demonstrated an interest, and willingness, to incorporate ethical and social aspects within their research development.

Opportunities for a more thoughtful reflection, under the types of mechanisms we have identified here (set out as the different characteristics of governance models in the previous section) need to be put in place so that ethical norms can emerge as grounded in the social and cultural communities for which the technology is destined. The four models of governance presented have their limitations and trade-offs (noted in Figure 2, and classified as expertise/efficiency versus participation/legitimacy) which present opportunities to explore different, or adapted, models of governance that more closely could meet the criteria needed to include participants in the information society in an ethical and democratic process – one which more closely reflects the societies that the European Union represents.

ACKNOWLEDGMENT

The research leading to these results received funding from the European Union's Seventh Framework Programme FP7/2007-2013 under grant agreement n° SIS8-CT-2009-230291.

REFERENCES

Ducatel, K., Bogdanowicz, M., Scapolo, F., Leijten, J., & Burgelman, J.-C. (2001). *Scenarios for ambient intelligence in 2010*. Seville: European Communities, IPTS. Retrieved November 27, 2012, from ftp://ftp.cordis.europa.eu/pub/ist/docs/istagscenarios2010.pdf

Edler, J. (2010, October). *Towards understanding the emerging governance marble cake in multi-level European R&D funding*. Paper presented at the Conference: Tentative Governance in Emerging Science and Technology: Actor Constellations, Institutional Arrangements & Strategies. University of Twente, Netherlands.

EGAIS. (2009). *Deliverable 2.1. Grid-based questionnaire development*. Retrieved November 27, 2012, from http://www.egais-project.eu

EGAIS. (2010a). *Deliverable 2.2. Empirical data collection*. Retrieved November 27, 2012, from http://www.egais-project.eu

EGAIS. (2010b). *Deliverable 2.4 Interpretation of ethical behaviour*. Retrieved November 27, 2012, from http://www.egais-project.eu

ETICA. (2010). *Deliverable 4.1 ICT ethics governance review*. Retrieved November 27, 2012, from http://ethics.ccsr.cse.dmu.ac.uk/etica

European Commission. (2002). *Science and society action plan*. Luxembourg: Office for Official Publications of the European Commission. Retrieved November 27, 2012, from http://ec.europa.eu/research/science-society/pdf/ss_ap_en.pdf

Goujon, P., & Lavelle, S. (2007). General introduction. In Goujon, P. et al. (Eds.), *The information society: Innovation, legitimacy, ethics and democracy (In honour of Professor Jacques Berleur s.j.)*. New York: IFIP, Springer. doi:10.1007/978-0-387-72381-5.

Jessop, B. (2003). *Governance and metagovernance: On reflexivity, requisite variety, and irony.* Lancaster: Department of Sociology, Lancaster University. Retrieved November 25, 2012, from http://www.comp.lancs.ac.uk/sociology/papers/ Jessop- Governance-and-Metagovernance.pdf

Lenoble, J., & Maesschalck, M. (2003). *Toward a theory of governance: The action of norms.* London: Kluwer.

Lenoble, J., & Maesschalck, M. (2006). Beyond neo-institutionalist and pragmatic approaches to governance, REFGOV, FP6.

Mayntz, R. (2003). *From government to governance: Political steering in modern societies.* Summer Academy on IPP: Wuerzburg, September 7-11, 2003. Retrieved November 25, 2012, from http://www.ioew.de/fileadmin/user_upload/do-kumente/Veranstaltungen/2003/SuA2Mayntz.pdf

Wright, D., Friedewald, M., Punie, Y., Gutwirth, S., & Vildjiounaite, E. (Eds.). (2010). *Safeguards in a world of ambient intelligence. The International Library of Ethics, Law and Technology* (*Vol. 1*). Netherlands: Springer.

ENDNOTES

[1] For example The United Nations General Assembly Resolution 56/183 (21 December 2001) in endorsing the first World Summit on the Information Society (WSIS) which was held in Geneva in 2003. The resulting "Declaration of Principles - Building the Information Society: a global challenge in the new Millennium" speaks of 'involving all stakeholders' which includes citizens. Arising from WSIS has been the Internet Governance Forum (IGF) 'a new forum for multi-stakeholder policy dialogue'. WISIS retrieved November 25, 2012 from http://www.itu.int/wsis/basic/about.html;Internet Governance Forum retrieved November 25, 2012 from http://www.intgovforum.org/cms/aboutigf

[2] Questions relating to democracy and the challenge for society vis-à-vis technology innovation are discussed in Goujon and Lavelle (2007). For example the authors note that any justification of social legitimacy of technologies being developed suffers from a "circular justification inside the technical ideology" (p. xvi). That is, questions are reduced to technical interpretations only, and therefore cannot be challenged (as the technical justifies itself, based on its own interpretation).

[3] The EGAIS project is a collaborative project (May 2009-March 2012) funded by the European Commission under the Seventh Framework Programme (Science and Society) taking an interdisciplinary approach (philosophy, organisation science, technology and information science) to integrating ethical considerations within technical development projects through governance mechanisms. http://www.egais-project.eu

[4] EGAIS project framework outlined in EGAIS (2009) Deliverable 2.1.

[5] See for example the discussion on the "status and of the responsibility of human sciences in technological project funded by European Commission" in MIAUCE (Multi modal Interaction Analysis and exploration of Users within a Controlled Environment. IST Call 5, FP6-2005-IST-5. Deliverable D5.1.2 Ethical, legal and social issues. pp.27-29. Retrieved November 28, 2012 from http://www.fundp.ac.be/pdf/publications/70144.pdf

[6] This concept and the theoretical perspective behind it (based on the study by Lenoble and Maesschalck, 2003) formed the core of the first stage of research in the EGAIS project, to provide a theoretical foundation to the research. A comprehensive account is

available in EGAIS (2009) Deliverable 2.1 retrieved November 27, 2012 from http://www.egais-project.eu

[7] Preface, ISTAG: Scenarios for Ambient Intelligence in 2010 (Ducatel et.al. 2001)

[8] For discussions on the ethical and social issues relevant to Ambient Intelligence see for example Wright et.al. (2010)

[9] The empirical work of the project was undertaken in EGAIS (2010a) Project Deliverable 2.2

[10] The theoretical work undertaken in EGAIS (2009) Project Deliverable 2.1 informed the Grid-based questionnaire development

[11] A template for analysis was produced by the authors in order to ensure synchronisation of data collection and analysis among all EGAIS project researchers. Since the partners in the EGAIS consortium come from inter-disciplinary backgrounds it was necessary to formulate a common interpretation of the data.

[12] This section is based on: (1) EGAIS (2009) Project *Deliverable 2.1. Grid based questionnaire development* (2) EGAIS (2010b) Project *Deliverable 2.4 Interpretation of Ethical Behaviour*; (3) ETICA (2010) Project *Deliverable 4.1. ICT ethics governance review* and (4) Lenoble and Maesschalck, (2003, 2006).

[13] As elaborated in the Introduction of this book and in the EGAIS project's framework outlined in EGAIS (2009) Deliverable 2.1.

[14] ETICA (2010) Project Deliverable 4.1 *ICT ethics governance review*; Lenoble and Maesschalck, (2003, 2006).

Chapter 11
Law and Governance:
The Genesis of the Commons

Danièle Bourcier
Centre d'Études et de Recherches de Sciences Administratives et Politiques (CERSA), France

ABSTRACT

The case of Creative Commons (CC) is a good example for describing how a new way to govern Commons has been invented. The Creative Commons (CC) Project was launched with no particular consideration of governance. Its primary aim was simply to share a common resource with common digital management. Several years on, the question of governance, as a logic of collective action, is coming to the fore. Between legicentrism and over-privatization, can both CC governance and governance by CC be seen as an alternative solution for managing future projects on common property in common?

INTRODUCTION

A lot of studies explore the relationship between commons, community and property[1]. Even before the current increase in digital commons Ostrom (1990) drew attention to the importance of governing commons.

Creative Commons is a project that enables the sharing of digital cultural goods through a range of licenses that can be applied to a work, according to the terms desired. It is based on copyright, but with a new rationale of patrimony. That potential for conceptual innovation has let to the success of this "singular legal object"[2], but has, however, also attracted criticism. In essence, this initiative is at the center of new tension between the logics of economics (commercial/non-commercial) and the law (exclusive ownership or shared use). How can we analyze these new institutional formats whose equivalents are seen in other contexts, and which require us to review certain legal concepts?

DOI: 10.4018/978-1-4666-3670-5.ch011

At this juncture, our focus is on questions of property, common goods, patrimony and governance.

We will firstly explain the difficulty of defining the goods and contracts concerned, then we will turn to *patrimony* with a view to renewing our approach to this notion. In point of fact, considering only the property regime (purely questions of property) means that some analyses have little relevance to the topic in hand. The question of patrimony goes beyond a simple distinction between public and private goods, for to administer is first to ensure management, that is, the best management possible in response to specific objectives. In other words, the notions of community and patrimony determine new choices of governance. This chapter looks at these choices.

When the market and the state were considered to have different roles, a balance could be found between individual and general interest. However, since the demarcations between market and state intervention are now becoming less clear, as in cases where the state distorts competition and innovation by strengthening regulation in favor of the market, new types of governance emerge.

PROPERTY RIGHTS AND GOVERNANCE

Goods relevant to CC are immaterial and/or digital goods[3], and may be private or "public." However, this category of "public goods" is difficult to clarify. For economists, a public good is characterized by non-rivalry and non-excludability. Water, for instance, is not a public good since appropriation of this resource may give rise to rivalry. In this, we see the first signs of possible confusion between economic and legal vocabulary: as far as lawyers are concerned, public informational goods are those subsidized by public funding, intended to be accessible to the public, while the law deals with private goods under the concepts

of property and private heritage. The question of governance arises when the goods are in a gray area between private property and public service.

Reconsidering Traditional Notions

In the case of copyright, some aspects of ownership have been extended to include intellectual property. Creative Commons and Science Commons have reintroduced the concept of commons. We will now return to these fundamental concepts to give a clearer appreciation of their development.

Property Rights

Ownership was defined in the Civil Code at the beginning of the nineteenth century, on the basis of the Roman law de Justinien (*plena in re potestas*)[4]:

Ownership is the right to enjoy and dispose of things in the most absolute manner, provided they are not used in a way prohibited by statutes or regulations (Civil Code Art. 544).

Some characteristics of property such as *exclusivity* are increasingly put into question. From a political standpoint, Angell (2009) stated that property may be considered not as "something owned" but as a government-sanctioned monopoly right which is legally enforced by courts.

However, over the last two centuries, this exclusivity has become limited, particularly where land property and areas covered by the rise in urban planning are concerned. There is now a distortion between fact and law and a growing number consider that a new definition would be desirable. It has been said that this change was the revenge of Greece on Rome, of Philosophy on Law. The Roman concept that justified ownership in relation to its source (family, dowry, inheritance) has been overtaken by a teleological concept that justifies ownership through its aim, its service, and its function. The first draft of the French Constitution in 1946 was thus written:

"Ownership cannot be exercised unless it is of social utility".

Some lawyers have attempted to bring together notions of general interest and public domain with a view to changing the current position. The use of a plot of land is forbidden if this is contrary to the designation of the land. Ownership will be removed if the use goes against collective interest.

The term "property" has been used in the field of intellectual property. Property is more than a metaphor: for Elkin-Koren (2006), "It constitutes an effective legal mechanism that allows exclusion." In this legal field, in fact, the changes have been the exact opposite: this right is used increasingly "in the most absolute manner." The CC project was one of the responses to this "insanely complex system" of copyright (Lessig, 2004).

Common Goods

Common goods are goods that have owners: thus they may not be appropriated by third parties, whereas *res nullius* (goods that belongs to no-one, such as wild game) and *res derelictae* (abandoned goods or waste) have no specific ownership. Common Goods are resources that are not divided into individual portions of property but rather are jointly held so that anyone may use them without special permission: for example, public streets, parks, waterways, and works in the public domain. According to the Oxford English Dictionary "*A commons is a legal regime in which multiple owners are each endowed with the privilege to use equally a resource, and no one has the right to exclude another.*" This means that no-one needs permission for access or use. Some rules may define particular rights *a priori*. No-one is owner because no exclusive right can determine if a resource can be let to the other.

A common is not only defined by its nature (water, beaches, or the theory of relativity) but also by its function in the community (Lessig, 2001). A policy choice decides on the organization. For Lessig, several factors may justify the creation

of common goods. Firstly, common goods imply certain values that would vanish if these goods were privatized. Secondly, some resources may be more *efficiently* used if they are held in common. Nuances have now been introduced into this notion.

When commons are referred to, there may be some confusion between the notions of a community of rights and common management.

For example, management of a common good may be undertaken as a result of a decision taken between a group of commoners and the owner. However, this management may also be handed over to a public person. In 1792 during the French revolution, the laws stated that every Municipality was the owner of several types of goods: commons goods ("bare and indeterminate land") in the stricter sense, and productive goods.

A common land (a common) can be a piece of land owned by one person but over which other people may exercise certain rights: this is considered a semicommons, which would in reality come under CC (Loren, 2007). Older texts use the word "common" to denote any such right but modern usage refers to particular rights of common and reserves the word "common" for the land over which the rights are exercised. By extension, the term "commons" has come to be applied to any resource to which a community has rights or access.

In reality, common land does not mean there is no owner. Rather, it signifies only that people other than the owner have rights: these are known as commoners. The owner may retain certain rights and any common rights abandoned by the commoners. The commoner may be interested in particular plots of land, but most of the time the rights of common are unconnected with ownership or tenure of land: these rights include pasture, *piscary*, *turbary*, *mast* and *pannage*, and *estovers* (small trees and fallen branches).

The term "commons" denotes a resource over which a group has access and rights of use, under certain conditions. This is used even more widely than the term "public domain." The first aspect is the *size* of the group that has *access* rights. Some would say it is a commons only if the whole community has access. The second aspect represents the extent of the restrictions on *use*. A commons may be restrictive. For example, some open source software gives the user the freedom to modify the software on condition that their own contributions will also be freely open to others. There is some discussion on what exactly should be considered as commons: all agree on water, but not on education, health, or the environment (Kiss, 1989). The issue of software, genes, and seeds are points of contention between those who wish these to be commons and others who want to extend patents. Developed societies tend to have a preference for private property over public interest. For John Sulston, Nobel Prize Winner for Medicine, the human genome was sequenced as an open project; consequently this type of collaboration and increase in knowledge is inalienably a common patrimony of humanity. The Global Earth Observing System of Systems (GEOSS) is a major international initiative which proposes that "all shared data, metadata and products will be made available with minimum time delay and at a minimum cost" to develop "coordinated, comprehensive and sustained earth observations and information" (Uhlir, 2009). Wikimedia Commons is free, unlike image and media banks that come under GNU licensing, where it is possible to use and modify the information. This lack of agreement on the size of the commons and the rights of use requires the involvement of the notions of public interest patrimony and human patrimony. This hypothesis will be explored later in this chapter.

According to Boyle (2008), "We have to invent the public domain before we can save It." CC0, the final licensing option in CC, is a way of putting goods into the public domain by waiving all copyrights and related contiguous rights such as moral rights (to some extent waivable), and database rights that protect the extraction, dissemination and reuse of data. In the same way, Science Commons launched the Protocol for Implementing Open Access Data, a method for ensuring that scientific databases may legally be integrated into each another. The protocol is built on the public domain status for data in many countries and provides legal certainty for data deposit and data reuse.

Governance of Cultural Goods

The "Governance of Goods" describes the types of management that regulate use of these goods. The management rules may be established by law, and complemented by individual contract for copyright, or in a form developed by the rights holders where common goods are concerned. Composite positions may be defined as a result of public policy or democratic principles (open access, for example) related to patrimony or the public domain.

The conditions for production of and the needs of the public with regard to informational or cultural goods have radically changed, to such an extent that property rights and particularly copyright, and also the size of the commons along with public domain status, have had to adapt to a movement towards *copyleft*. These changes have been necessary firstly because production of these goods, whether public or private, is increasingly the result of collective effort, and a new culture of collaborative work has emerged. This culture is linked to the new technological possibilities offered by Web 2.0, with its orientation toward user interactions via participatory portals. Users are active players and, consequently, authors. They feed content into databases, websites, blogs and wikis. Secondly, cultural goods are now more likely to be produced and disseminated by public bodies. Lastly, there is a public desire to use digital works for themselves, in order to modify and diversify the usages although digitalization of some holdings appears to have changed the conditions for access[5]. In this context, the mode of governance becomes capital, as suggested by Garrett Hardin (1968) in the (ideological) fable relating the Tragedy of the Commons: a community of sheep farmers who share a commons of pastureland may choose a dangerous type of management that leads to over exploitation of the resource and environmental deterioration. Garrett concludes that in order to overcome this paradox of the commons it will be necessary to reinstate appropriation by the state, or, even better, by the private sector, in the same way that in 1960 Ronald Coase advocated privatizing the environment so that it could be managed more effectively. Some believe that certain mechanisms restricting access to resources managed by the state must be extended. In fact, this is now the case for databases, as will be seen later. Nevertheless, there are other solutions. The two aspects of common property and free access must not be confused.

Private knowledge is governed by intellectual property law, with the complex restrictions of exclusivity denounced by believers in the opposite tragedy, anticommons. Public knowledge may be the result of a public service mission or material produced by public servants. Some public knowledge may be ordered by copyright law (abstracts of judicial cases) or in the "public domain" (Official Journals). The status of public knowledge is varied, as demonstrated under "legal information" in public sites constructed for museums and libraries. Public bodies often produce works and manage them as enclosures: these are made available to closed groups of users. For several years now these goods have become more accessible because of the Internet. However, opening public informational goods remains subject to exceptions. In addition, under European law, databases benefit from protection *sui generis* that prevents direct access to the content. Legal databases are an example of the tensions currently allied to the functions of information in a democratic society funded through circulation of content[6]. In the sphere of politics and public policy, environmentalism presents a *remarkable*

diversity of organizational forms and missions. We are at the beginnings of the replication of that *institutional diversity in the world of intangible property.*

MANAGING DIGITAL COMMONS: THE CREATIVE COMMONS PROJECT

Intellectual property rights are based on the regime of exclusive property and limited access to the resource, whether the property be *individual*, *collective* or *common*. Creative Commons has changed the given by immediately questioning how to best manage rights holders' freedoms in a digital commons, and how these management "rules" could be part of a minimal agreement between authors and the public.

In the Tragedy of the Commons, the commoners were invited into the area but no-one had defined the limits of the resource. For Garrett Hardin, only the right of the state (which regulates individual property and internalizes management on the part of each owner) is effective because common goods will inevitably be subject to anarchy of use until the moment they disappear. Two institutional modes of managing these knowledge pools have therefore been detailed: the market and the state. In the Internet world, the question has shifted from exclusion to access. Many resources now have different modes of access. The fact that digital goods are not rivalrous broadens the variety of agreements and types of cooperation, even though the cooperation is weak. These commons, considered by economists to be positive externalities, are facilitated by *social media* and Web 2.0 applications.

Mediated Communities: Are asking the State to develop all possible means to enable better circulation of information, for example to develop high speed broadband against outside company monopolies and to ensure arbitrage between differing collective preoccupations. However these communities believe there is another tragedy, that of *private goods*: *enclosure* models are closed in the short term. Although neither altruistic nor completely individualistic, communities of creators, such as Creative Commons, share a common vision and aim for equality of rights to the resource, with no limits to access, when the licensor waives his conditional rights.

The Objectives of Creative Commons

The aims of CC are thus to facilitate simple, legal sharing and reuse of creative, educational, and scientific content. To this end, it provides free legal and technical tools in order to:

Promote collaboration between content creators and users around the globe. Creative Commons supports a world in which people actively engage in — and don't just passively consume — cultural, educational, and scientific material around them and build a pool of creative, educational, and scientific content that can be freely and legally accessed, used, and remixed[7].

CC falls into the category of open information models. Any use that is not to the detriment of the common may be made. However not all the conditions are totally to the advantage of a community: non share-alike, non-derivative, non-commercial (Eechoud & Wal, 2008).

The notion of *anticommons* characterizes a resource that is not used for maximum benefit. Whereas a common good is a good that all can draw on, an anticommon is subject to vetoes on its use. The Nobel Prize winning economist J. Buchanan has shown that as too many people have the right of veto over patents, the resource has become underused. Thus private property rights combined with rights pertaining to databases result in some resources lying dormant. It is therefore necessary for the right attached to a good or a work to be returned to its original source. The costs

for use are too high. This exploitative monopoly has become a hindrance to the free circulation of information, works, and services.

Common Goods in Creative Commons

James Boyle's book puts forward the view that we are in the middle of a second enclosure movement. Things that were formerly considered to be common property, or "uncommodifiable," or completely outside the market, are now being covered by new, or newly extended, property rights.

Take the human genome, for example: supporters of enclosure have argued that the state was right to step in and extend the reach of property rights. However, opponents of enclosure have claimed that the human genome belongs to all, that it is literally the common heritage of humankind, and that it should not, and perhaps in some sense cannot, be owned. Here it is impossible to exclude economic factors from legal values. The point of property is that it may be destroyed, and also confiscated. When scientific progress is concerned, this method of managing the resource may be particularly unfavorable to the common good of health[8]. We have noted that "A common is a piece of land owned by one person but over which other people can exercise certain rights." A work created by the mind, like a piece of land, has an owner (the author) but other persons may also exercise some rights.

Moreover as the term "commons" has come to be applied to any resource to which a community has rights or access, it is agreed that all in the CC and more can access the piece of common. What exactly is *shared* in the CC project? At first sight it may be said that this is all works in the virtual pool. However, my hypothesis is that the Commons shared in CC is not simply limited to *private ordering* but that it applies to all the facilities and services put in common in this project.

The Size of the Commons

The knowledge pool is made up of all works that authors have placed under CC license. These multimedia works (books, reports, articles, images, photographs, music, films etc.) are produced by independent authors or producers; however, they may also be produced by public bodies. It must be borne in mind that all the works in question must be protected by copyright because the author chooses the type of license.

The commons includes rights over these productions because they are copyrighted, unless they are CC0 licensed or fall into the public domain[9]. What is put in common is open, free access to works: a non-exclusive general authorization to produce, distribute and communicate the work to the public, at no charge. This applies also to collective works and P2P networks. Other optional conditions apply to certain rights that the creator has relinquished: the possibility of altering the work, of using it commercially or non-commercially, and of circulating it under the same conditions.

A movement towards a community-oriented philosophy that views exclusion from content as undesirable has now been set in motion. Licensed content is accessible via search engines such as Yahoo and Google, and is thus completely open. The number of works online has now reached around 200 million, but as there is no central count of works, authors or users the number of commoners is unknown.

The larger the body of work online the greater will be the choice and the use of the commons. Bell & Parchomovsky (2009) explain this choice in terms of cost: "The CC movement was born out of a sense that in the information age the cost of excluding others from most informational works is too high." This means that the free commons (where there is no question of entry or exclusion) diminish "costs of transaction and governance," and are thus particularly efficient. There is there-fore no need to register in order to use the services of the commons: there are no controls and no sanctions regarding the use to which they are put.

Digital License Platform

CC offers open, free access to the platform providing the licenses[10]. A license is a contractual document that specifies what can and cannot be done with a work. It grants permission, and states restrictions: access, reuse and redistribution, with few or no restrictions. It is possible for example to print, to share, to publish, to make alterations and additions to it, to incorporate it in another piece of writing, and to use it as the basis for a work in another medium.

Six options, combinations of the different criteria, are available to authors. Once the author has chosen the license for a particular work, the platform produces three personalized documents: one in html format, one that is simplified, and, thirdly, a legal document. Currently there are more than fifty-two sets of national licenses that have ported the generic license in national legal systems.

The Creative Commons Tag

Creative Commons is more than simply a resource or a service: it is also an identity. CC upholds a certain number of objectives and values. Some conditions are optional (individual values) while others are common (collective values). Authors who deposit their work in CC commit to making it available at a minimal level of access, at the least.

The Australian Cutler report (Cutler, 2008) contains a strong recommendation endorsing the use of CC licenses for public information[11]. CC licenses can become the means for implementing Open Access policies. In particular, these increase the delivery of government services on line. This approach fosters participatory democracy with a two-way commitment between the state and the citizen.

The Governance of Creative Commons

An understanding of the way in which the Commons are governed is necessary. However, the choice of management method is not linked solely the concept of the "cost of governance." Overall governance of the project is under neither the control of the authors who provide their works *a priori* nor of those who will use them.

The resource management system has not eliminated authors' rights and none can currently predict whether it will move towards more, or less, open access. Demsetz' hypothetical model that postulates the development of a rights regime moving from Commons to exclusive rights is no more provable than the opposing model, wherein there is an ineluctable movement towards open access.

The CC empirical governance model has not been aimed at *good governance* or optimal management of the resource (by giving more rights to commoners or initiating more organized governance); but it has been led by the increase in goods, the development of collaborative work, and freedom of choice between those who wish to relinquish their rights and others who want to manage their patrimony. When this is compared to the *Open Source Movement,* their access is open in the same way as for CC, but the license is a mechanism that manages the future of the resource: use and adaptation are reserved to those who accept the terms of the Open Source License. In CC, externalization costs have been borne by CC but authors continue to manage their own goods individually, with no collective commitment to the future. In the following part, we describe the current method adopted to manage CC globally.

The Collaborative Mode of Developing the Licenses

As the rules on Intellectual Property Rights come under national jurisdictions – in spite of the existence of international law – an international organization has been set up to adapt CC licenses to every legal system. Creative Commons International was established to develop and coordinate worldwide national chapters. Questions common to several chapters may sometimes be discussed (e.g. moral rights). Vertical relationships are associated to questions that for the most part relate to legal and technical expertise. The different steps in the porting show some tension in the efforts to keep close to the (US) originals, in order to maintain the greatest degree of similarity possible between the licenses. This expertise belongs to the Commons.

Management by National Chapters

The national Project Lead team is a volunteer. This would constitute an expert in copyright law, often a law professor, a junior researcher or an artist generally associated with the institutions or a law practice. In most cases, university law faculties have volunteered. Governance is based on subsidiarity, and coordination.

These relationships are coordinated by a CC International (Berlin) authority, separate from headquarters (San Francisco). The chapters have tried to form regional groupings: these attempts have led to several initiatives in the Asia Pacific region and in Europe. The 2009 CC Asia Pacific Regional Conference, recognizing its "cultural and language diversity" proposed to adopt an action strategy in an Action Plan Statement. Their aim is to play a proactive role in expanding and building the CC Communities in Asia and the Pacific areas with the objective of administering a Common web portal to establish a "regional identity."

Discussions regarding the region of Europe are now in progress. These are expected to lead to similar regional organization. However, for the moment no decision has been made on the creation of these regional levels positioned between the national and global bodies.

No Central Committee and No Centralized Governance of Owners

To Richard Epstein an experiment such as open source (close to the CC project) must inevitably end in failure because it cannot scale up to meet its own success. Then Epstein asks for a "central committee" from which insiders will be unable to cash out – a mixture of communist and capitalist metaphors. *"All governance systems - including democracies and corporate boards - have problems. But so far as we can tell, those who are influential in the free software and open source governance communities feel that they are doing very well indeed.*[12]" (Financial Times, 2004)

In the 1980s and 1990s a large body of literature taught us that property-like governance mechanisms could and often do emerge in the absence of formal property rights. Robert Ellickson and Elinor Ostrom[13] described governance regimes to allocate resources and coordinate activities when property rights were nonexistent or ineffective. In this process, order, allocation, and coordination were not always synonymous with formal property rights. For CC, formal rights are managed by the legal systems and the licenses are self-managed. In the same vein, some interesting real-world situations – where in effect public resources emerge against a backdrop of private entitlements – have been described. "The upshot is the same: private re-engineering of the entitlement structure, in the interest of people getting things done" (Merges, 2004).

The Relationship between Commoners

CC commoners, whether authors or users, have a variety of relationships, as already described in the sociology of social networks. In effect a social network is shown as a chart of relationships and interactions between individuals, and may be based on geography, work organization, or simply on informal connections (Barnes, 1972). With CC, the network does not exist prior to its creation. The

very fact that the commons pool is not centralized (through a database) and that there is no relationship between authors and users – or a system of representation – makes it a network of informal relationships where the network is mainly used to create a global identity or to determine the "social capital" of individual participants.

In reality the project is based on the digital universe wherein the idea of territory is relatively weak. The system does exist at territorial level but only in relation to the transposition into a particular legal system. The organization of production and circulation is reduced to a minimum due to the externality of the platform and the transparence of procedures and documents. The social spread of informal relationships is described *a posteriori*. However the Legal Leads community has created a Thematic Network on the Public Domain at European regional level in response to a European invitation to tender (COMMUNIA)[14].

Governance, the Law and Regulation, in CC

The terms "Governance" and "Regulation" are often used together and sometimes interchangeably. However, there is a clear distinction: regulation pertains to the development of new forms of interaction between law and society, while governance aims to find new equilibriums between law and society (Bourcier, 2002) where a parallel form of State regulation is sought. We will argue that the CC project relates in part to both approaches. What is Creative Commons vis-à-vis the State? Is it self-governance? What is the underlying institutional question? For the majority of the problems relevant to Commons, the issues are typically of governance and institution.

First, Creative Commons is a complementary form of State regulation. However, public interest is replaced by the "civil society of Internet users" at the core of the freedom related to creation. This might be considered a new method of coordination between agents. Agents are Commoners as well as users. This method of governance is a manner of responding to the creator's offer and the needs of new creators and users.

Second, CC licenses constitute a learning process enabled by the exchange of works, information, and experiences. CC creates a new form of social governance by means of "soft self-regulation" providing a dynamic that is worth looking at from both a national and international point of view. This approach can be seen as a reaction to legal complexity. However, the CC instruments also define a new equilibrium between law (statute) and contract. The CC contract is proposed under conditions chosen by the contracting parties.

Here, the issue at stake is to find a balance between freedom and rules. Authors must be paid but at the same time they cannot have complete control over their works: in legal parlance, the rules are more those of liability than of property. In fact all the questions concern rivalry or non-rivalry; and in a world of non-rivalry and innovation, it is necessary to allow for openness, but also to make rules for the common pool. Common practice may be more effective than the State or the market. Management of property rights in a customary system or a social network is able to self-organize (Rose, 1986).

Finally the CC approach introduces a new logic moving from "management by regulation" to "management by coordination." It represents an incremental approach where breakthroughs are measured in real practice more than through normative texts (Bourcier, 2007).

In this context, large-scale centralized coordination is impossible: the process of creation must be modular, with units of differing sizes and complexities, each requiring slightly differing expertise, all of which can be added together to make a complete whole. However, the total enterprise will be much, much, greater than the sum of its parts. Governance processes may be assembled through local systems in a global network, by people with widely varying motivations and skills.

PATRIMONIAL GOVERNANCE FOR DIGITAL GOODS: REINVENTING THE NOTION

The CC project has thus created a common resource that can be shared, and to which all have access. In addition, CC has allowed those who wish to share, following certain conditions, to choose and to circulate their options; CC has also created open collaborative works in progress, an open collaborative Commons that will become a future Commons. Membership is not necessary, but there is a kind of moral "contract" based on an economy of exchange, of give and take. A CC community is developing in every country and participating in the dissemination of the project. The Commons is not one of ownership, but it creates a pool that may be accessed. The Commons draws up the conditions for use. As a result, these Commons, far from falling into decline, are likely to increase their spread. The question I will ask in this final section on CC is whether a new form of institution is coming into being, one with a will for preservation and a potential to create new goods that have the express characteristic of being accessible. The traditional notion of patrimony is becoming multidimensional. Successively, patrimony has been familial, genetic, cultural, national, in land now close to World Heritage status, and in Humanity. We will explore whether it is possible to go beyond the simple notion of common property, and to extend the concept to common heritage. Cultural heritage is not limited to the material, such as monuments and objects preserved over time: the notion also encompasses living expressions and traditions that countless groups and communities worldwide have inherited from their ancestors and now transmit to their descendants, in most cases orally. In the field of Creative Commons, there are also many works, especially of music, that we cannot protect other than by making them live.

The Notion of Patrimony

The notion of patrimony is evolving from the strict civilist notion (patrimonial rights) to that of Human rights: "Patrimony is the set of goods of a person, considered as forming an universality of rights"[15]. CC requires attributes from each of these concepts of patrimony.

For positivist lawyers, patrimony is an individualistic notion and can refer only to monetary values. Where copyright is concerned, patrimonial rights differ from moral rights. Patrimonial rights give the author "the possibility of living off his work." They confer the exclusive right to authorize third parties to use his creations, through agreement on transfers or licenses with these third parties. Here, the notion of exclusivity proper to the notion of property reappears. The author decides on the conditions for use and may instigate proceedings against imitation in the case of any non-authorized exploitation of his work. This concept could explain the origin of copyright laws. Moral rights, without monetary value, are outside of the concept of patrimoniality.

The notion of patrimony is currently changing dramatically (World Common Heritage) due to conflicts in patrimonial management. Formerly, patrimony was defined by its source and origin more than by its future, affectation, and use. However, the notion has evolved: it is now not enough to preserve heritage, it is necessary to exploit it, to display it in cultural inventories. The notion of common heritage may be a pathway between private and public heritage. Given the origin of the private conception of patrimony, public bodies are not comfortable with the notion of public patrimony. This notion of patrimony could re-qualify a resource. It allows for re-appropriating some goods and some rights. Indeed, patrimony is also the expression of the capacity for use (Declaration of Human Rights, 1789, Art. 1). It is inherent in personality. To be entitled to a patrimony means to hold a power to be included in a circulation of goods and to be registered in a relationship of exchange. For Ollagnon (1979), patrimonialization is a way of constituting a community between participants – not to solve conflict or earn money but to affirm identity. Patrimonialization can become an *operational* tool, with the aim of administering/governing (economic value) or managing (patrimonial value). Patrimony can be seen as an alternative to property rights: it is a way of finding other institutional solutions. In the long term, patrimonialization can take on a social identity.

The Objectives and Values of CC as Patrimony

CC responds to the main points we have described above.

Micoud[16] demonstrates that the role of patrimonalization (in environment) was firstly to conserve and to protect. Now the issue of governance is a major consideration where resources are concerned: "Patrimoniality is a way of building a community." The use of this notion can prevent exclusive appropriation, particularly in the case of scarce resources such as fish.

In the digital world, reuse is at the heart of common interest. The main advantage is to be able to access, sort, and consult intermediate pools. However, reuse is a new type of production. In fact, the issue is also not to create a new work for an individual but to build collaborative space. The new paradigm is to bridge the gap between production and creation. The term 'patrimonial economy more than commercial or patrimonial capitalism is debated'.

Governance of this patrimony has led to the development of new property rights: the right to access, the right to regulate, the right to dispose, the right to be transparent (right to expression, and related preferences)[17]. The right to property becomes more varied. This is based on an implicit patrimonial convention whose principal aims are laid out below.

To Create a Safe Global Commons

The project was not set up to help individuals and to assemble an aggregation of individual works. On the contrary, it was created to share something different, something new. It was more than simply a series of isolated actions. The ensuing result was the creation of a global "commons" of material that was open to all, provided the terms of the licenses were adhered to. Any contract must be in accordance with the rules selected by individuals, and must conform.

This patrimony is based not only on access to a work for private use but also on co-production. In this context CC is similar in some ways to the world of Free Software: it has followed on from the opportunity offered by file sharing applications. In fact, all the bywords used in free software development have their counterparts in the theory of democracy and open society. With open source the production process was transparent, and the result of that process was a "product" which outcompeted other products in the marketplace. In his noteworthy book on "distributed creativity" and the sharing economy, Yochai Benkler (2006) sets the idea of "peer production" alongside other mechanisms for market and political governance and puts forward powerful normative arguments regarding that future. Eric Von Hippel (2005) shows that innovation happens in more places than we have traditionally imagined, particularly in end-user communities. This reinforces the theme that "peer production" and "distributed creativity" is not something new, merely something that is given considerably more salience and reach by the Web. In addition, Jonathan Zittrain (2008) argues that "the main force of Creative Commons as a movement has not been in the Courts but in cultural mindshare" (p.225).

Other examples of commons-based, nonproprietary production exist all around us. The present teaches us about the potential of a new "hybrid economy" (Lessig), one where commercial

entities leverage value from sharing economies. That future will thus benefit both commerce and community.

To Preserve All Cultural Patrimony through Open Access

Offering open access to a complete knowledge pool, or placing it under CC license, may prevent "pirating" of works in the public domain for which individuals with few scruples request patents or copyright at the time of their digitalization. In India, this was the case for traditional yoga postures found in ancient texts. The Indian government has begun scanning the documents in order to store and preserve them in an encyclopedia, The *First traditional knowledge digital library*, which will be open to free access[18]. A second example concerns digital archives that disappear because they have been purchased.

Even the term "reuse" may be subject to different conditions, depending on individual, commercial or non-profit reuse. Where private authors are concerned, solely the author should be able to decide what is authorized. However, reuse also enables the content to be extended, so that works which are Creative Commons (Wikipedia, for example) become new universal pools that are added to voluntarily by users. Thus, Global Commons, part of the United States Library of Congress, and therefore repository of a body of work produced through public funding, is constituted of a group of individuals and institutions who wish to make their "treasures" available to the widest possible public. Similarly, Flickr contains photographs from all over the world and users are invited to expand on descriptions by adding tags. We are now in a period where services are exchanged and information is extended, where facts produced by a private author or by an authority can coexist in the same common work.

In this way we now see the question of licenses in a different light: they are not limited to accessibility, but now also concern making material available, reusing it, adding to it and even patrimonial preservation of digital content within the public domain. Producers of content are not determined by the public or private nature of works. Creators now have tools they can trust that, while respecting ownership of their rights, offer a public the potential of using and reusing their content *a priori* and without *intermediary*.

Collective Patrimony: Fundamental Rights and Public Debate

The justification for property rights to be included in patrimonial governance may also be supported by the change from a system where rights are reserved in the name of ownership of goods towards a system where these goods are preserved "in the name of a common patrimony and of universal access to knowledge and culture" (Bourcier, Dulong de Rosnay, 2004). For V. Hugo, the work as thought belongs to humankind[19]. Moreover, public space and public debate need to be developed, as promoted by Habermas. Over propriatarization and the enclosure of works could put an end to the circulation of ideas (Dussolier, 2005).

The Relationship between Commons and Publicly-Funded Goods

Exclusive rights, the prerogative of private property, have also invaded the non-commercial public sector, that pertain to the general interest and public service.

In fact, there are different approaches: one based on the notion of public service or general interest that dominates management of the state public domain, and one based on ownership of public assets. Thus the copyright of public persons has continually been reaffirmed by the French Council of State[20]. These two approaches are conflictual which would show that over-patrimonialization has penetrated all sectors of society. In France in fact a study[21] has been set up

to attempt to understand why informational goods subsidized by public persons are not explicitly open to access. States' lack of capacity to deal with environmental issues has now led lawyers to develop the concept of the "common heritage of mankind." The traditional status of *res communis* as applied to certain resources is not suitable (free access and free exploitation). However, the central question remains how can "mankind" manage this heritage? The Rio Convention on Biological Diversity (1992) omitted this concept, but recognized the creation of new copyrights that assist indigenous local communities. The subjective individualistic nature of patrimony is a significant impediment to the conceptualization of a new framework for cultural patrimony. The right recognized as pertaining to a community is not an individual right but an agreement for the exercise of a specific activity, legally attributed to the holder of the right: it is no longer a property right. In private law patrimony and property are closely bound. All goods in the patrimony are submitted to property rights. On the other hand Creative Commons creates a new link between individual works, commons deposits and collective governance. We are now in a culture of availability rather than authorization and a "mutual sharing of knowledge as the collective property of mankind"[22].

Reinventing Heritage

The categories of public, community and private must thus be examined and brought to the fore through the notion of patrimony. The traditional approach to good property governance was through familiar institutions: the government, the market, communal management, or through property rights. Some recommend that the state control the majority of natural resources in order to prevent their destruction; others advocate that privatizing these resources will resolve the problem. Nevertheless these solutions are not uniformly successful in enabling individuals to sustain long-term productive use of natural resources. "Communities of individuals have relied on institutions resembling neither the state nor the market to govern some resource systems with reasonable degrees of success over long period of times" (Ostrom, 1990). At present we do not have the relevant intellectual tools and models to integrate the governance and management of these issues. The notion of Market and State cannot contribute to this reinvention of models. Many communities (commons), groups, and entities need to be reconstituted around the notion of patrimony and of governing how they coordinate and cohabit. The concept of patrimony shows that in addition to the notion of *general values* there is a way of more effectively managing the resource and conforming to collective interest. We recognize a new trend in public policies: the state is founded on the notion of general interest and social contact. Today, new communities are able to help the State to find a new legitimacy through regulating the way that common goods and public goods can be managed by new players who are not private but patrimonial, particularly where cultural heritage is concerned[23].

From the Environment to Cultural Goods

Over the last forty years, the notion of patrimony, which first appeared in the context of the environment, has been a genuine, new institutional innovation. For economists it seemed a means of escaping market logic, of upsetting market rationality. It became fair to change the terms of the debate: the right of ownership confers the right to use but also the "right to destroy." It is in this area that debate on the environment can add to the general debate. In the digital world, this right to destroy is equivalent to *enclosing* cultural resources. However, the notion of patrimony highlights the necessity of preserving, adding to and transmission. It brings collective management solutions to the fore, since common goods require a communal approach to patrimony (Ollagnon,

1979). Thus in this situation there is collective appropriation with a multitude of rights holders and uses. This is known as "transappropriation" or "transpropriation" (Ost, 1995). In real terms this means that anyone may exercise a right of ownership but that maintaining the good quality of the resource depends on all players and managers involved. A changing perspective of patrimony consists of seeing this common good within a "system of circulation and exchange not only between successive owners but also between the sphere of being and having". Patrimony is something that can be exchanged: it remains synonymous with belonging and continuity, coinciding with different degrees of fluidity and transitivity. However, a management structure (non-private, non-state) must be created above and beyond these rights. Patrimonial mediation falls into this category. It is necessary to first identify then draw up patrimonial objectives. Next management procedures must be set out and conditions for access and control specified. The final stage is ritualization, that is, a phase of public legitimization. The notion of property is reductive: the notion of patrimony opens new perspectives for assigning wealth, and for its visibility and sustainability in the digital world.

CONCLUSION

The CC project oversees a facility for sharing, where works are open to free access because rights holders waive their exclusive rights. This is a *common* facility, but the works concerned are not considered common property: they may be used according to certain conditions. Owners do not waive their rights except in the CC0 license (no rights reserved). Other concepts such as *patrimonialization* must be looked into. However, in a situation where the right of (intellectual) property is still central, what are the fundamentals and identity of this type of patrimony? Certain points must be clarified to enable an understanding of how this form of organizing may establish new balances and renew social effectiveness. Where circulation, sharing and collaborative work in the digital world are concerned, the divisions between private and public goods appear somewhat difficult to operate in reality. Following our account of certain concepts used by Creative Commons, we have demonstrated that the concept of patrimony may be the *driver*, as seen in other domains such as the environment. Within this concept of patrimony works that carry the identity of their holder, a global identity, and above all, public rights, are selected and put in place. This imposes a type of governance known as patrimonial governance of the digital commons, which bases copyright on the freedom to receive and to communicate ideas, that is, on fundamental rights.

REFERENCES

Angell, I. (2009). All (intellectual) property is theft? *Presentation, Fifth Communia Workshop, Accessing, Using, Reusing Public Sector Content and Data*. London School of Economics.

Barnes, J. A. (1972). Social networks (pp. 1-29). An Addison-Wesley Module in Anthropology, Module 26.

Bell, A., & Parchomovsky, G. (2009). *The evolution of private and open access property. Theoretical Inquiries in Law, 10*(1). Retrieved May 3, 2010, from http://www.bepress.com/til/default/vol10/iss1/art4/

Benkler, Y. (2006). *The wealth of networks: How social production transforms markets and freedom.* New Haven, CT: Yale University Press.

Bourcier, D. (2002). *Is governance a new form of regulation? Balancing the roles of the state and civil society* (pp. 2). IWM Working Paper 6/2002, Vienna.

Bourcier, D. (2007). Creative commons: An observatory for governance and regulation through the internet. In Casanovas, P., Noriega, P., Bourcier, D., & Galindo, F. (Eds.), *Trends in legal knowledge the semantic web and the regulation of electronic social systems* (pp. 41–52). Firenze: EPAP.

Bourcier, D., & Dulong de Rosnay, M. (2004). Introduction. In Bourcier, D., & Dulong de Rosnay, M. (Eds.), *International Commons ou la Création en partage*. Paris: Romillat.

Boyle, J. (2008). *The public domain, enclosing the commons of the mind* (pp. xvi). New Haven: Yale University Press. Retrieved May 3, 2010, from http://www.thepublicdomain.org/download

Cutler, T. (2008). *Report on the review of the national innovation system*. Cutler & Company Pty Ltd. Retrieved from http://www.innovation.gov.au/Innovation/Policy/Documents/NISReport.pdf

Dussolier, S. (2005). *Droit d'auteur et protection des œuvres dans l'univers numérique* (p. 231). Bruxelles: Larcier.

Elkin-Koren, N. (2006). Creative commons: A skeptical view of a worthy pursuit. In P. B. Hugenholtz and L. Guibault (Eds.), *The future of the public domain*. Kluwer Law International. Retrieved May 5, 2010, from http://papers.ssrn.com/sol3/papers.cfm?abstract_id=885466

Epstein, R. A. (2004, October 24). Why open source is unsustainable. *Financial Times*. Retrieved November 27, 2012, from http://www.ft.com/cms/s/2/78d9812a-2386-11d9-aee5-00000e2511c8.html#axzz2DSFs33Q2

Hardin, G. (1968). The tragedy of the commons. *Science, 162*(3859), 1243–1248. doi:10.1126/science.162.3859.1243 PMID:5699198.

Kiss A. (1989). *L'écologie et la loi: Le statut juridique de l'environnement.* Paris, L'Harmattan.

Lehavi, A. (2009). How property can create, maintain or destroy community. *Theoretical Inquiries in Law, 10*(1). The Berkeley Electronic Press. Retrieved May 5, 2012, from http://www.bepress.com

Lessig, L. (2001). *The future of ideas, the fate of the commons in a connected world.* New York: Random House. Retrieved May 8, 2010, from http://www.the-future-of-ideas.com/download/

Lessig, L. (2004). Foreword. In Bourcier, D., & Dulong de Rosnay, M. (Eds.), *International commons at the digital age: La création en partage* (p. 7). Romillat.

Loren, L. (2007). Building a reliable semicommons of creative works: Enforcement of creative commons licences and limited abandonment of copyright. *George Mason Law Review, 14*(2), 271.

Merges, R. P. (2004). A new dynamism in the public domain. *The University of Chicago Law Review. University of Chicago. Law School, 71,* 183–203.

Ollagnon, H. (1979). Propositions pour une gestion patrimoniale des eaux souterraines: L'expérience de la nappe phréatique d'Alsace. *Bulletin interministériel pour la rationalisation des Choix budgétaires,* 36, pp. 33.

Ost, F. (1995). *La nature hors la loi* (p. 323). Paris: La Découverte.

Ostrom, E. (1990). *Governing the commons: The evolution of institutions for collective action.* Cambridge: Cambridge University Press. doi:10.1017/CBO9780511807763.

Rose, C. (1986). The comedy of the commons: Custom, commerce, and inherently public property. *The University of Chicago Law Review. University of Chicago. Law School, 53,* 742. doi:10.2307/1599583.

Stallman, R. (2003). The copyleft and its context. In *Proceedings of the Copyright, Copywrong.* Nantes.

Thoumsin, P. Y. (n.d.). *Creative commons le meillur des deux mondes?* Retrieved from http://creativecommons.org/licences/by-nc-nd/2.0/be

Uhlir, P. (2009). Global change in environmental data sharing: Implementation of the GEOSS data sharing principles. *Communia Workshop Proceedings*, Torino, July 2009.

van Eechoud, M., & van der Wal, B. (2008). *Creative commons licensing for public sector information opportunities and pitfalls.* IVIR, University of Amsterdam. Retrieved May 3, 2010, from http://www.ivir.nl

von Hippel, E. (2005). *Democratizing innovation.* Cambridge, MA: MIT Press.

Zittrain, J. (2008). *The future of the Internet — And how to stop it* (p. 225). New Haven: Yale University Press.

ENDNOTES

[1]   See for example Community and property. Special Issue. *Theoretical Inquiries in Law, 10*(1), January 2009. The Berkeley Electronic Press http://www.bepress.com

[2]   See Thoumsin (n.d.).

[3]   The CC licences can be also used for material goods (books for example).

[4]   *Institutes* 2,4,4.

[5]   An investigation into public library websites in France showed that one third of the sites carry NO legal indications and that a large number of indications were biased or totally illegal.
http://www.slideshare.net/calimaq/bibliothques-numriques-et-mentions-lgales-un-aperu-des-pratiques-en-france

6   For instance: The State of California publishes its laws on the Internet. Its copyright claims mean that people have to pay to download or print state laws and regulations. A user decided to digitally scan the 33,000 pages and put them online for free on the website The State claimed that this user needed its permission to put state laws online. This user wanted to provide open access to common public resources. If documents are in standardized formats companies can improve services (annotation of codes, wikis). The point is that government information is monopolized by the State and above all by exclusive vendors. This content is therefore not a commons. Lessig (2001) responds: "The essence in other words is that no one exercises the core of a property right with respect to these resources – the exclusive right to choose whether the resource is made available to others."

7   http://creativecommons.org/

8   Health is considered to be part of common goods or global public goods by the United Nations Development Program (in English) (UNDP).

9   CC0 is similar to the current allocation to the public domain; but in the case of CC0 the rights waiver will also have more force because there is a platform for reputable systems to develop the copyright status of content on the basis of who has done the certifying.

10   Under French jurisdiction: www. fr.creativecommons.org

11   Report on the Review of the National Innovation System: Venturous Australia. Recommendation 7.8 "Australian governments should adopt international standards of open publishing as far as possible. Material released for public information by Australian governments should be released under a creative commons licence." (p.170)

12   See Epstein, R. A. (2004, October 21).

13   See for example: Robert C. Ellickson, Order without Law: How Neighbors Settle Disputes (Cambridge, Mass.: Harvard University Press, 1991); Elinor Ostrom, Governing the Commons: The Evolution of Institutions for Collective Action (Cambridge: Cambridge University Press, 1990). Governing The Commons: The Evolution of Institutions for Collective Action, Cambridge University Press 1990.

14   www.communia-project.eu

15   Cf. a traditional book in French doctrine: Aubry & Rau, *Droit civil français*, t. IX, 6th Ed.

16   Patrimony in Environmental field has been broadly analyzed in: A. Micoud, "Redire ce qui nous relie?," C. Barrière, D. Barthelemy, M. Nieddu, F.D. Vivien (eds), *Réinventer le patrimoine, De la culture à l'économie, une nouvelle pensée du patrimoine ?* Paris, L'Harmattan, p. 81 and ff.

17   Recommendation at the National Meeting of Internet Players, Autrans, France, January 2009 (300 participants): For free access to public data: to change the legislation and practice in favour of access and reuse in order to develop a knowledge-based society – to indicate the conditions for use clearly in all public data financed by public budgets; these rights could be guaranteed by licences such as CC and Art Libre.

18   Where asanas are concerned, the UK daily www.telegraph.co.uk recently reported that there have been more than 130 yoga-related patents, 150 copyrights and 2,300 trademarks in the United States alone. And in the USA alone, the yoga business brings in $5.7 billion a year, according to Yoga Journal, including money spent on yoga classes and products. http://www.communia-project.eu/node/217 (July 10 2009).

19   V. Hugo, Discours d'ouverture du congrès littéraire international de 1878.

[20] Public data can thus be "Works of the Mind" which do not belong to the civil servant authors, but to the public person itself: CE, Ass. 10 juillet 1996, Sté Direct Mail Production, AJDA, 20 février 1997, p. 189, Note H. Maisl.

[21] http://www.bicoop.net/index.php/BI-COOP,_des_Biens_public_biens_communs

[22] See Stallman (2003).

[23] Public law historians have raised a new topic of discussion: that in Europe the role of Public Administration has been linked to a rationale of management since the beginning of the nineteenth century. "To govern is to apply law to manage *common* interests."

Section 5
Pragmatic Approaches to Governance

Section 5: Pragmatic Approaches to Governance

The various approaches to governance can be said to be *pragmatic* in the sense they must all satisfy criteria as well as a certain "quantity" and "quality" of *action*—especially in *communication* for some doctrines. In this respect, a pragmatic approach to governance contrasts some other approaches by articulating or submitting cognition to action, or theory to practice, or by regarding communication as an action that refers to some ethical/dialogical norms of interaction. In order to avoid confusions, one can emphasize that the word "pragmatic" here encompasses in a broad sense the two meanings, while they are sometimes distinguished when using the word 'pragmatist' for the first meaning and the word "pragmatic" for the second meaning. In any event, as far as doctrines are concerned, "pragma-*x*" always refers to the primacy of action or communicative action over speculation, hence it does not mean that one should encourage a kind of "realistic," "down to earth." or "business-like" attitude in life.

The relation of theory to practice and the question of dialogue are quite obvious in the *governance of a complex multinational and multicultural political set*, such as Europe, shifting from a harmonic to a governance model. In a pluralist system, a concerted effort is made to shift policymaking from simple community "harmonisation" in law to representative, pluralistic governance, respectful of difference. *Harmonisation* lawmaking seeks uniformity in law across the union in order to ameliorate trade and national-cultural differences, particularly in cases where national identity or culture cause (or are seen to cause) inequalities. *Governance*, in contrast, seeks to account for difference by taking into account plurality and seeking authentic, representative dialogues on a "thick" basis that explains not just the action, but its context as well. National identity and culture are important here in terms of policy, as the principle of subsidiarity means that national self-determination must be respected, not dominated. Ethics is immediately brought up as autonomy and human interests are live issues when we attempt to account for pluralism and wish to include difference in governance.

The pragmatic approach is also helpful to reflect *governance* as an *activity* and a *practice* that requires some evaluative or normative judgements, not all them moral, but rather "polynormative." The quality of the activity of governing is that of guidance in attaining some goal in some proper, orderly, favourable or good way. Hence, the very idea of governance contains an irreducible evaluative or normative element. The problem of *polynormative governance* comes from the fact that processes of governance are as multifarious as the practices that need governance. It can be said that moral judgments, although providing an important resource for critical judgments about governance practices, are not the only such resource. Moralists tend to think that moral discourse is the ultimate governance practice, but this is a serious mistake in thinking about governance that pragmatists will avoid. Current notions of democratic governance and in particular *deliberative* democratic governance can be re-construed as particular specifications of the general conception of normative governance of normative textures. This particular specification of the general conception is oriented to democratic legitimacy, i.e. a complex mode of validity. In a deliberative democracy, other modes of validity (e.g. moral and juridical legitimacy) have to be integrated into democratic legitimacy in complex ways.

The *pragmatist turn* in the theory of governance refers to recent work by a range of researchers who have developed a pragmatist-experimental approach to democracy that they term *democratic experimentalism*. These theories constitute a genuine turn in the theory of governance, for they introduce the key issue of the *self-capacitation* of the actors. They seek to reflect on how the actors can organize themselves to acquire new capacities and to learn new roles, which is a novelty compared to other paradigms in the theory of governance. The paradox in our societies is, though there are more and more opportunities for participation, the influence of citizens does not seem to have been really increased. This paradox of *participatory* democracy refers to an unsettled question in the deliberative paradigm, of which Habermas is the most famous representative. This is namely the question of the capacitation of the stakeholders to assume their discursive role within the deliberative programming of the society. It can be shown how democratic experimentalism makes this question of the capacitation of the actors a central issue of theory of governance, despite some proper limits of the theory.

Chapter 12
Governance Theory and Practice:
The Case of Europe

Stephen Rainey
Facultés Notre Dame de la Paix, FUNDP, Belgium

ABSTRACT

European political life involves a productive tension between liberalist and communitarian tendencies. This 'Libero-Communitarianism' in the EU is the backdrop to various governance policies and potentials. This chapter develops a broad analysis of the governance setting in Europe and draws out some key areas of potential problems. This is based in the Ethical Governance of Emerging Technologies (EGAIS) project findings, and mirrors some of the issues flagged as ethically important in the field of emerging technologies. That such issues permeate European research and approaches to governance is testimony to their centrality and to the influence of Libero-Communitarian interactions.

INTRODUCTION

Between the revisions of the Treaty of Rome in the single European Act of 1987 and the Lisbon Treaty's coming into force in 2009, there has been a concerted effort to shift policymaking in Europe from simple community 'harmonisation' in law to representative, pluralistic governance, respectful of difference.

Community harmonisation lawmaking seeks uniformity in law across the union in order to ameliorate trade and national-cultural differences,[1] particularly in cases where national identity or culture cause (or are seen to cause) inequalities. Governance, in contrast, seeks to account for difference by taking into account plurality and seeking authentic, representative dialogues on a

DOI: 10.4018/978-1-4666-3670-5.ch012

'thick' basis. Clifford Geertz made famous this term, one he adapted from Gilbert Ryle. A thick description of human actions is one that explains not just the action, but its context as well, so that the action becomes meaningful to an outside observer in terms amenable to the actor.[2]

National identity and culture are important here in terms of policy as the principle of subsidiarity[3] means that national self-determination must be respected, not dominated. Subsidiarity requires that law must find its ultimate source in nations, not supra-national authority. In other words, while the supra-national authority can make recommendations and even requirements for nations, it cannot actually enforce change at the national level:

Any national Parliament or any chamber of a national Parliament may, within six weeks from the date of transmission of a draft European legislative act, send to the Presidents of the European Parliament, the Council and the Commission a reasoned opinion stating why it considers that the draft in question does not comply with the principle of subsidiarity. It will be for each national Parliament or each chamber of a national Parliament to consult, where appropriate, regional parliaments with legislative powers.[4]

Ethics too is immediately brought up as autonomy and human interests are live issues when we attempt to account for pluralism – how different people conceive of their own possibilities as a function of their personal, historical, cultural or their national identity are elements that must come into play when the ambition to include difference in governance is voiced.

The analysis here will focus on these moments (policy, ethics) separately in order to illuminate where systematic governance approaches and ethics meet, how they meet and what the outcomes and limitations are.

BACKGROUND TO POLICY

The European Union has a difficult task. Across a diversity of nations, each home to a diversity of ethnic, linguistic, cultural and social differences, it seeks to bring some sort of unity in policy to permit the fair and free expression of identity, from a basis of universal human rights. This is evidenced in the adoption of the Charter of Fundamental Rights (CFR) in 2000 and its subsequent strengthening in the Lisbon Treaty. In fact, this introduces a considerable tension from the start as regard the intentions of the practice of governance in Europe. This tension comes in the form of competing views of freedom and its realisation by political means – political liberalism and communitarianism. We look now to each in turn in order to flesh out this tension, thereby exposing a challenge for governance that any prospective recommendations we offer for improving the ethical governance of emerging technologies must address. With regard to ethics, this is important as we will see that for liberalism, ethics is only ever a matter of politics among a well-described group of individuals. Communitarianism, on the other hand, can be seen as operating on a basis that ethics ought to ground politics toward shaping a group's arena for autonomy.

POLITICAL LIBERALISM

It is an ambition of political liberalism in general to found policy despite difference. The idea is that, given broad enough consideration, policymaking can accommodate all individuals despite their ethical differences as well-argued for policy will convince universally by dint of its reasonableness (Lund, 2000). The idea is founded upon a deep notion that fairness involves annexing public deliberation from value, as value is a private reaction to the world, whereas public deliberation has to represent more than a narrow set of values, or else risk ruling out *a priori* certain conceptions of the good and their pursuit in given ways of life. As a prominent liberal thinker, John Rawls stands as a good representative of this thread of thinking.[5]

Rawls uses the fact of value pluralism as a basis to reject the role of value in public decision-making at all. In order to reasonably represent a people who will be divided on many matters of import, it is centrally important not to rule out *a priori* any particular conception of the good life. To this end, Rawls develops an account of public reason that tries to limit itself to the reasonable core of any particular conception of the world. He supposes that in any given group, an 'overlapping consensus' will be possible on many areas of principle:

...the point of the idea of an overlapping consensus on a political conception is to show how, despite a diversity of doctrines, convergence on a political conception of justice may be achieved and social unity sustained in long-run equilibrium, that is, over time from one generation to the next. (Rawls, 1987, p.5)

And:

Justice as fairness aims at uncovering a public basis of justification on questions of political justice given the fact of reasonable pluralism. Since the justification is addressed to others, it proceeds from what is, or can be, held in common; and so we begin from shared fundamental ideas implicit in the public political culture in the hope of developing from them a political conception that can gain free and reasoned agreement in judgment, this agreement being stable in virtue of its gaining the support of an overlapping consensus of reasonable comprehensive doctrines. (Rawls, 1993, pp.100-101)

While the details of, say, how privacy ought to be protected might differ between individuals based upon their history and culture, nonetheless all will agree that privacy is an important issue.

Thus, Rawls supposes in overlapping consensus there to be a basis for political progress if public reasoning limits itself to general matters in this fashion.

Liberalism is the view that above all else, self-determination is the value worth most in human beings. The only genuine way to respect human beings, liberal individualists suppose, is to respect their self-determination. The liberalist state, therefore, should recommend no one way of life over another, either directly or via incentives. To do so would be an affront to human dignity, they suppose, and risk penalising those individuals who did not share the values of the state, which would be unjust. The major spectre the liberalist wishes to avoid is that of state perfectionism, wherein the government urges its citizens to adopt a particular position to the exclusion of others. Under such a scheme, tyranny is a possibility as the notion of freedom is curtailed via a narrow conception of human nature.

For liberalism, it could be seen that choices amount to the expression of values in that they are the outward face of preferences. Preferences, moreover, are irrational in a substantial sense, and so each is as fitting as the next. To stifle some, therefore, while encouraging others is to impose an ideal for living upon people despite themselves. Preferences, the liberalist could be seen to hold, are the manifestations of persons who are prior to any contingency. However, in the detail of liberalism there can be seen the more subtle notion of 'the good life.' The notion of the good life, or *eudamonai*, stems from Aristotle's moral, practical and political philosophy and connotes a sense of 'doing and living well,' or exercising virtue according to character[6].

The reason that choice is so important in liberalism is that it is the product of deliberation aimed at bringing about a good life. Deliberation is done carefully, moreover, as we know we could be wrong. The notion of 'revealed preference,' where an agent discovers after making a choice that they didn't want the outcome, can appear in the course of a life lived. But it is the deliberate aiming at a good life and the learning and fulfilment this can bring that for liberalism is a pervasive mark of living a life.

Values aren't merely expressed in choice, but are aimed at in a context of a hypothesised 'good life,' and approached with an open mind. John Rawls expresses this point as follows:

As free persons, citizens recognise one another as having the moral power to have a conception of the good. This means that they do not view themselves as inevitably tied to the pursuit of the particular conception of the good and its final ends which they espouse at any time. Instead, as citizens, they are regarded as, in general, capable of revising and changing this conception on rational grounds. Thus it is held to be permissible for citizens to stand apart from conceptions of the good and to survey and assess their various final ends. (Rawls, 1980, p. 544).

Put bluntly, the enforcement of a good way of life upon a citizenry amounts to an unacceptable paternalism that uproots the dignity of being human at all. In fact, given the conception Rawls describes, the perfectionist injunction would trivialise the enforced way of life – should I be forced to act in a way according to the dictates of the paternalistic state, I will merely simulate the actions for fear of penalties or to receive incentives. This means the state in fact promotes trivial action in the pursuit of ensuring citizens avoid trivial ways of life (Kymlicka, 1990, p.204).[7]

The way around this absurd outcome is for a neutral state – a state which acts to minimise constraint, and to maximise opportunity, without prescribing a way of being, nor proscribing in principle anything not harmful to the total aggregate freedom. This is 'negative freedom' in the sense that it is a freedom from external impediment. Only by being neutral in this sense does the government respect all equally.

COMMUNITY VALUES

Contrary to liberalism, the communitarian suggests that there are in fact more and less valuable (dignified, worthy) ways to live a life. There are good and there are bad life choices. Given this, it ought not to be outside a government's purview to recommend one way of life, or one subset of all possible ways of life, over others. In so doing, the government shows respect to human freedom in that it aids those who might otherwise make bad choices or be a victim of having few choices at all. To fail to do this, to enact the liberalist's idea, is in effect to abandon citizens to the possibility of their own bad judgement or to the arbitrary distributions of natural talent (see for instance Dworkin, 1981).

A key point the communitarian makes against liberalism is that the revisability of deliberatively chosen ways of life does not occur in a vacuum. Part of how we assign value and achieve personal fulfilment in ways of life is in the adoption of roles in society. Attachments with others in social groupings constitute at least to an extent our worth as persons. This is not directly in contrast with liberalism, but the communitarian position regards this 'social thesis' (Kymlicka, 1990, p.216) as having impacts politically. Effectively, with the realisation of the need for substantial social links in order to realise oneself meaningfully, the state neutrality thesis has to be revised, it is argued.

In order to have freedom of choice in anything like the measure required by liberalism, one assumes free access to the cultural resources offered by any particular social group. If my ability to deliberately conceive of the good life, evaluate and potentially revise it, is to be thought of as genuine it must assumed I have access to good information. But where nothing is endorsed by the state on a liberalist doctrine (e.g. tax breaks given to quality education institutions and news organisations) there is nothing to suppose that the trivial won't prevail.

The attractive, therefore the market-successful, may not be the valuable. There may not even be a range of options as potentially valuable but unattractive or suppressed ends wither. In a context where access to information is dictated by impersonal, market forces, freedom of choice for the liberalist could well be impeded. Since forces dictate availability anyway, shouldn't some form of governance body step in and direct the process in order to optimise circumstances? One communitarian way of putting this would be to say that freedom must always be 'situated' somehow, or else risk becoming nihilism:

...complete freedom would be a void in which nothing would be worth doing, nothing would deserve to count for anything. The self which has arrived at freedom by setting side all external obstacles and impingements is characterless, and hence without defined purpose (Taylor, 1979, p.157).

Given the characterisation of Liberalism as the attempt to maintain freedom via respecting individual values, and the communitarian notion as being centred on maintaining adequate choices in order that non-trivial freedom of choice can be offered, we can see this opposition in terms of where perfectionism ought to stem from. For the liberalist, civil society perfectionism is in order. State perfectionism seems the proper means for the communitarian. For the communitarian, the liberalist position has the untenable assumption that people, once in social groups, will tend to realise and perpetuate worthwhile institutions. This will be mediated in civil society. For the Liberalist, the communitarian holds the unjust position that a people should have recommended for them a way of living via state intervention. The heart of the issue is in the role of value. For the liberalist, value is private and must be excluded from public matters, as it serves to inform private deliberation on the good life. For the communitarian, values must be included in public matters so as to reflect the common good in public policy.

This is a huge debate in which the stake is the priority relation between the right and the good. We do not need to delve further into this topic here as we need only note the tension between individualist liberalism and communitarianism as it is this tension that manifests in the European picture – the tension between the individual self as autonomous or the situated self as autonomous – as in European governance we effectively see a kind of libero-communitarian synthesis.

LIBERO-COMMUNITARIANISM IN THE EU

This is a further element to the complexity in the European situation. Whilst it seems clear that political liberalism features significantly in European policy-making, there is also that 'core' of values as represented in the CFR and Lisbon. Essentially, there is a core of community values that must be adhered to in order to be 'European.' However, given subsidiarity and the proliferation of difference (national, historical, linguistic, cultural etc.) the pressing of the value agenda is highly problematic.

The liberalist tendency could be interpreted as effectively anaesthetising difference – the presumption is that once 'we' decide privacy (or anything else) is important we have a basis for action and can deal with the details later, purely through politics. But it's the details that are at stake, they aren't what need to be 'dealt with later,' rather they remain opaque between different parties on the basis of their different backgrounds (including values, ethics, history and much more besides). Such opacity is at the very least a troublesome basis for action, at the worst no basis at all.

The tension within the particular European problem is clear here when the core of values themselves, as represented by fundamental rights, is put into this mix. Being 'fundamental' doesn't mean these come with a ready-made interpretation. Rather, even the core values must be interpreted and dealt with according to national, historical, linguistic, cultural etc. differences. In fact, in a nutshell, this *is* pluralism, and can be taken as a fact between individuals as much as it can across the vast area and cultural diversity of the EU. 'The fact of pluralism,' (Lund, 1996) indeed, can be argued to be *the* governance problem.

In the European context, given the role of subsidiarity, this can have a profound effect upon the efficiency of a normative injunction. For instance, where a policy is thought of at the supranational level, it must always be conceived of in extremely formal terms in order that no hint of domination is implied. Take for instance this statement of the right to privacy:

Article 8: Right to respect for private and family life.

1. Everyone has the right to respect for his private and family life, his home and his correspondence." *(European Court of Human Rights, 2010, p.6).*

Just what this means or how it can be regulated is far from determined from this spare wording alone. Given pluralism, we have to assume there will be different and perhaps competing interpretations of what this means and what it prescribes. So it is that national units must interpret and regulate these notions themselves. This allows the possibility for ethics to be subverted in policy as contents for formally-worded requirements are delegated to the national level. But that national level itself can hardly be thought of as comprising a homogenous group; the nations of diverse Europe are themselves diverse nations.

Given this role for pluralism as 'the' problem, it's necessary to look at some current attempts to address it. These will be models of governance predicated on the fact of pluralism and aimed not at reconciling difference, but reconciling governance with the persistence of difference. This is a subtle and important distinction, and is one characteristic of European governance as opposed to that typical of the US, for instance. This difference was highlighted in the EGAIS workshop report (Deliverable 3.2) in summarising the input of Alfred Nordman shown in Table 1 (cited in Santuccio, A., & Valentino, F., 2011, p.17):

The difference in US and EU approach can be discerned and posited as resting on the European ambition to include 'from the start' a core of community values informing a liberalist structure.

One immediate consequence of a position such as this, as pointed out by R. Mayntz (2002), is that individual preferences cannot necessarily be satisfied maximally in a way that respects the fulfilment of common, public goods. This means that, where common values are in play, and where there is a plurality of perspectives on them, each individual perspective may be incompatible with every other, formally speaking. Thus, it is a task

Table 1. US governance vs European governance

US Approach	EU Approach
Substantive vision of a new frontier	Procedural norms
Technological *determinism*	*Construction* of technology and society
American *individualism*	Societal welfare
Subservience of social science and the humanities	*Leading role* of social science and humanities
Constative "this will happen"	Conditional "this might happen"
Technological innovation as a mean of realizing human potential	*Social innovation* as away to realize technological potential
Technology is a means of *overcoming limits* to produce a Second Creation	*Technology ingeniously adapts human desires to the limits of nature*

of governance to permit the articulation of these perspectives so that their nature, source, implications and compatibility can be authentically discussed. In fact, to pre-empt, we will see that current approaches to governance posit 'public dialogue' as the solution to such problems of plurality. We can suggest, given the evolution of the position taken in EGAIS, that this ought to be a starting point, not the end of the matter. This is to say, dialogical articulation of perspectives requires reflexivity and opening of framings that we know is typically missing from governance approaches. Thus, we can expect to have to do some supplementary work in order to flesh out just what is meant by 'public dialogue' in governance.

For instance, in general the role of the public in liberalism is assumed to be describable and used as a basis for the work of politics on establishing fairness etc. In communitarianism it is more likely to be thought of as being in need of construction, something that must be clearly formulated as a result of analysis of prior concepts. Rawls' position can be seen to be in the descriptivist camp, whereas someone such as Habermas can be thought of as falling into the latter camp (Cf. McCarthy, 1994).

We must now look at an important European governance structure, direct deliberative polyarchy (DDP), which can be seen as trying to reconcile this liberalist/communitarian tension. It has to be stressed that 'structure' here refers to *theoretical structure*, i.e. DDP is the analytic heart beating in implementations of governance such as the open method of coordination. We outline this theoretical structure in order to present clearly the elements in play in the theory and in order to facilitate the clear understanding of the implementations of it. This will represent a deep background to a governance approach with the reconciliation (or at least mediation) of liberalism and communitarianism as a problem. The analysis of this deep background, moreover, will facilitate the subsequent analyses of other governance approaches as well as providing a critical basis for discovering problems. As it happens, although important, this dissection

of DDP will be brief and so will add much of value with little exertion. This is due in part to the foregoing examination of liberalist and communitarian thinking, and so is a demonstration of the importance of discussing those themes.

DIRECT DELIBERATIVE POLYARCHY

Cohen and Sabel argue for DDP on the basis that is represents an innovative response to institutional failures:

…it is an attractive kind of radical, participatory democracy with problem-solving capacities useful under current conditions and unavailable to representative systems… directly deliberative polyarchy combines the advantages of local learning and self-government with the advantages (and discipline) of wider social learning and heightened political accountability that result when the outcomes of many concurrent experiments are pooled to permit public scrutiny of the effectiveness of strategies and leaders. (Cohen & Sabel, 1999)

From the start, then, we see the inter-mingling of liberalist deliberation with community inclusion, with an emphasis upon bearing in mind the views of 'the other.' Through participation, the story seems to go, we get values into the mix and so can perhaps marry the best features of each of liberalism and communitarianism – we can have values at play within a liberalist structure. This saves us from the excesses of a narrow, market fixation wherein 'rational choice' is felt to be a quantitative matter of objective reality (and which ignores bounded rationality at the very least). It also avoids ideological fixation, which would posit some set of values upon which it then becomes necessary to found decisions.

On the face of it, moreover, DDP can be represented very straightforwardly and can be thought of as the dynamics of a familiar conception of democracy.

Imagining Figure 1 in terms of an ongoing process we see that the people elect representatives who determine policy to treat issues that affect the people who elect, and so on.

Among the constitutional principles of DDP that characterise it is deliberation. This is highly important as it instantly problematises Figure 1. Collective decision-making is intended to proceed by way of free and informed dialogue between interlocutors of equal status. Proposals must be acknowledged, advanced and defended as reasons for collective actions. Thus, there are interrelations between the different parties. This introduces another dimension absent from the simple depiction of representative democracy (See Figure 2).

The perception and determination of issues between elements of the structure become part of the first-dimensional process. People, along with representatives perceive and determine issues that policy-treatments themselves feed into. For instance, the banning of wearing religious veils in France recently was aimed at treating one perceived issue, but it also became an issue in itself. A third dimension between policy and people emerges as policy feeds into issues, creating new ones.

The nodes themselves, moreover, cannot be thought of as black boxes but can be 'exploded' so as to reveal the elements upon which they run. Given the deliberative characterisation of DDP, influences can be spelt out that inform (or deform) deliberation. There is a further 'internal' dimension, in other words, besides those mentioned so far. 'People' includes internal relations between media, lobbying, tradition, culture etc. These affect ethical/social issue perception. 'Policy' includes economic components, lobbying, tradition etc. etc. These can affect the possible trajectories for policy (see Figure 3).

The influence of tradition, media and CSO activity are not exhaustive of influences upon people, but are samples of pervasive influence. The point is that one cannot think of deliberation or social or ethical issue perception or determination without bearing in mind a nebulous background of influences. We end up with a quite complex set of inter-related parties, in contrast to the initial diagram that seemed to sum of a democratic basis so neatly. It is at this moment that 'local' and 'wider social' learning are relevant. This ought to remind us of the point raised above, with reference to Mayntz, in that we see here dialogue brought into the suite of measures aimed at good governance, but we are immediately struck by the complex array of factors that influence it. The procedures associated with inclusive dialogue posited in DDP at this point, or in proceduralisms in general, need to permit learning in the following sense:

The contingent and constructed nature of models of reality mean that they are inherently unstable and must be continually open to modification. In other words, they must be continually open to the possibility of learning. As a consequence, the emphasis shifts away from improving information and action based on a dominant model, as in formal and substantive rationality, and towards

Figure 1.

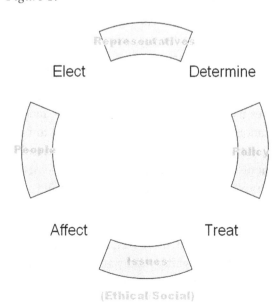

Figure 2.

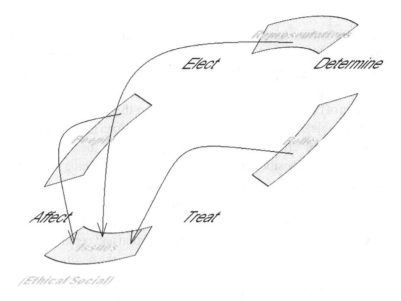

a concern with the adequacy of the procedures by which different models are exposed to each other, that is, confronted with their own contingency and encouraged into a posture of collective learning. In this way, what is universal is less the content of models than the procedures which develop this understanding of contingency and the need for learning. (Lebessis & Paterson, 1997, pp. 14)

The first sentence here sums up neatly the notion alluded to earlier, that the objectivising trend in governing must be resisted. Context is all important, especially where plurality among the governed is at stake, as it is not the resolution of one perspective with another that is at stake, but the more or less felicitous arrangement of

Figure 3. External factors affecting people's views (or values?)

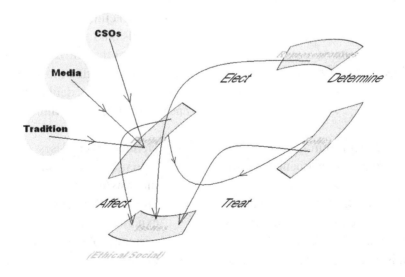

social groups *salva differentiae* ('maintaining the differences').

'Learning' in the sense it is used here, seems to be a means toward that end. The idea is that 'co-monitoring' of different groups' approaches to problem-solving can, via a process of critical iteration, lead to the adoption of best practices. This is clearly via bottom-up means and is based upon actual practices in terms of their effectiveness. There is therefore no question of imposition from above. Neither is there a paring off into relativistic groups (in terms of social action, for instance, such a paring off could be seen in the chance of 'hyper-regionalism.) (Cf. Gerstenberg, 1997).

A DIRECT, DELIBERATIVE EXAMPLE

The Electricity Regulatory Forum, or Florence Forum, was set up to discuss the creation of a true, European internal electricity market. Among the issues on its agenda is the cross-border trade of electricity, in particular the role and level of tariffs for cross-border electricity exchanges and how to manage inconsistencies regarding interconnection capacity between various nations.

The forum boasts a diverse makeup and meets regularly.

Participants include national regulatory authorities, Member State governments, the European Commission, transmission system operators, electricity traders, consumers, network users, and power exchanges. Since 1998 the Forum has meet once or twice a year. (cf. European Commission, 2011)

The minutes from the forum's meetings are open and available to all, emphasising its commitment to openness and transparency.

How is this related to DDP? When it comes to the taming of deliberative problems on difficult issues such as cross-border tarification, constructive participation is incentivised by the potential

for European Commission invocations of 'hard' legal means –competition law powers, antitrust, merger control, and state aid rules. Deliberation is thus more in the interests of all parties as intransigent participants would risk being worse off than under a compromise reached in the Forum.

From this it follows that EU experimentalist governance can rarely be understood as regulatory cooperation in the shadow of traditional public hierarchy. 'Bargaining in the shadow of the state'… assumes that public authorities can effectively regulate in case of deadlock, mismanagement, or policy failure. But both in the case of the public deliberation and penalty default mechanisms, the EU *induces the actors to provide rules for themselves*, presumably because it cannot itself provide defaults that would do the job as well or better. Not 'bargaining in the shadow of hierarchy,' but 'deliberating when hierarchy becomes a shadow.' (Sabel & Zeitlin, 2006)

Having thus induced participants to rule upon their own way forward, it remains for the results of the forum to go on to be interpreted by national and other bodies 'lower down,' and so further deliberative action is invoked. In other words, the process doesn't end in an act of legislation, or in the conclusion of a forum meeting. Instead, governance becomes an ongoing process involving a wide range of interested parties, driven by a deliberative engine. The workings of the networks required to facilitate this interconnection between individuals and statutory bodies themselves contribute to the horizontal and vertical checks characteristic of this mode of governance. The result is a dynamic, deliberative, accountability-ensuring whole. This whole is therefore:

- **Deliberative:** Uses argument to disentrench settled practices and redefine interests/preferences.
- **Directly-Deliberative:** Uses the concrete experience of actors' differing reactions to current problems to generate novel possibilities for consideration.

- **Polyarchic:** A system in which local units learn from, discipline, and set goals for each other *(Idem).*

With the key structural elements and motivations in mind, and this brief example, we have enough on DDP in order to pause the structural analysis in order to examine another example of how this has been instantiated in practice. Given this structural basis, and the complexities we have determined regarding the basis for DDP we can look now to the Open Method of Coordination (OMC), a key innovative governance tool deployed in EU circles with respect to social solidarity issues. The reason we turn to this now in determining the limits of current ethical governance approaches is owing to its influence and to its promise for providing ethical governance. To pre-empt somewhat, the approach seems to identify inclusivity with ethical governance. In short, we suppose this concept of inclusivity and its mediation in dialogue, to be *problems* not solutions. Thus, this is a key approach in need of some analysis.

OPEN METHOD OF COORDINATION

OMC is intended as a 'soft,' i.e. non-law-based, governance approach wherein community deliberations inform policy decisions nationally aimed at converging with policy goals set supra-nationally[8]. In terms of Europe, EU ministers first agree on framework goals. Second, member-states translate these guidelines into national and regional policies (including the input of nationals and experts). Third, the ministers agree on benchmarks and indicators, to measure and compare best practice within the EU and worldwide. Finally, through evaluation and monitoring, member-states' performances are assessed - relative to each other and to their declared goals. This is depicted in Figure 4. In this can be seen the inclusive, deliberative

thread from DDP, along with the fundamental adherence to subsidiarity.

So much for *what* it is, but the question might arise, 'why discuss the Open Method of Coordination in the context of a deliverable on the ethics of governance in research of emerging technologies?' In fact, there is a strong basis for this discussion. The Science in Society programme raises the insight that:

…while new developments can improve our quality of life and understanding of the world, scientists and policymakers may not always properly assess the potential risks or take full account of the public's concerns. Opportunities must be created for scientists and the general public to exchange views in a two-way dialogue of mutual respect and trust.[9]

We need governance rather than government – by hypothesis we are accepting the fact of plurality as a shorthand for the problems motivating governance.[10] This governance has to be based in dialogue to permit the sort of learning mentioned in the structure of DDP and with reference to Overdevest. The OMC scheme or model thus seems like a good candidate as a potential provider of the means to fulfil these ends. In the EU recently, moreover, OMC has played a prevalent role in social solidarity contexts:

In… social solidarity… the key innovation was the Open Method coordination (OMC), in which representatives of the Member States in consultation with the Commission set broad 'non-binding' goals and metrics, to be implemented through national action plans or strategies, and periodically revised following peer review of implementation experience. (Sabel & Zeitlin, 2007, p. 13)

Part of the reason that emerging technologies have the potential to raise ethical issues is the manner in which they can swiftly and pervasively alter ways of life. The problems seen in technology

Figure 4. The flow of OMC

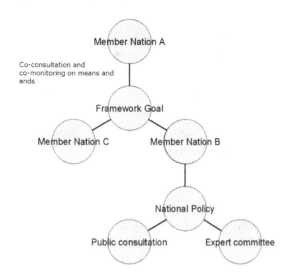

research centred upon a lack of awareness of this and a lack of thought regarding the contexts in which technological developments would appear. The Science in Society quotation above seems to seek what OMC claims to provide, structurally at least. The established role of OMC in social matters also serves to illustrate its potential in dealing with ethical governance of emerging technologies where social perspectives are the loci of ethical responses and concerns to ongoing research.

The problems of expert framings and de-contextualisation have loomed large throughout EGAIS' findings, not least owing to the gap between 'lay people' and experts in the technology field (EGAIS, 2010b, p.45). So, it is the case that emerging technology research presents social issues that need careful attention in order that they can be determined before they can be addressed. For EGAIS, moreover, such developments in technology research can present ethical issues that too need to be constructed to be understood – it is not the case, as we have established, that ethical issues can simply be *perceived*, owing to the role of context. Ethical issue determination involves constructing the perspective of the social actor and their context, which requires a means for this input to be made apparent. As an instance of

DDP, OMC claims to put the position of the social actor 'up front' and makes authentic deliberation a condition for governance.

Moreover, what EGAIS has discovered as missing in research projects under FP6 in general was reflexivity. OMC is a governance measure designed to incorporate such reflection, as demanded by a theoretical basis seeking to marry liberalism and communitarianism, itself to permit fairness under common values in the EU. Referring back to Figure 3, we can see how, in OMC, there is the attempt to mediate the complicated picture represented there. The integration of values and interests in a horizontally and vertically regulated dialogue, centred on authentic self-portrayals, is what the structure of DDP requires. This is also what facilitates the 'local' and 'wider social learning' that Cohen and Sabel (1999) flag as positive features of DDP. OMC in the case of social solidarity cases represents an attempt to instantiate this model. This is, structurally, a treatment of the issues EGAIS has determined to be present in research projects and more widely in other fields.

In other words, in OMC there seems to be the means to incorporate the elements missing from the ethical governance of emerging technologies – context construction and reflexivity[11]. If so, we can adopt or adapt the model in order to define for ourselves the parameters required for ethical governance *per se*. This in turn may be used to found a means of policy assessment.

We can make a brief summary now of some key features of governance discussed here before going on to determine shortcomings of this story do far.

- OMC is an instance of DDP.
- It is a governance approach felt legitimate owing to the deliberative inclusion of different stakeholders.
- Inclusion is thought to increase accountability through representing accurately views of people.

- Deliberation enables feedback and monitoring on policy content and practices.
  - Horizontally (across levels).
  - Vertically (between levels).

We will now scrutinise OMC with respect to ethics directly in order to begin developing a few problems with the approach.

OMC AND ETHICAL INTERFACE

In the diagram in Figure 5 shows a depiction of OMC with arrows indicating points where social or ethical issues might be predicted to arise.

For instance, where 'public consultation' is deployed there seems to be at least three questions that can immediately arise: *why* is the public being consulted and *how* is the public being consulted? Is this a symptom of the objectivising trend? These questions relate to the meaning of the practice being deployed in the course of governance – is it a popularity vote for a measure already enacted? Is it a measure designed to gauge public opinion so that a policy may be developed? Is it a truly consultative measure wherein the stakes include goals as well as means? Similar questions can be asked at each place indicated

with an arrow. At stake are the construction of context and the epistemological position of the actor in each case.

Why are those arrows relevant to ethics? The point is that one way of conceiving of ethics is the difference between what one can and what one ought to do. 'What one can do' is not simply a matter of operating within the laws of physics, moreover. The possibilities for action as they are conceived of by any given individual relate to matters of self-image, constraints of a legal and social nature, cultural norms and so on. Any given social actor thus has a complex set of norms and values that structure the world of possibilities for them. Should any governance approach fail to take this into account, it sells its governed short. Given that, structurally speaking, consultation can be more or less open, more or less authentic and can have any number of ends in sight, it seems clear that the mere adoption of inclusivity itself does not guarantee ethicity. This can be usefully thought of in terms of the now familiar concept of framing: if the structural elements of OMC are deployed according to a predetermined, perhaps unquestioned framing, then the governance approach as a whole will fail to reflect the complexity of the governed regarding the world of possibilities as it is seen by individuals.

Figure 5. Social and ethical issues predicted to arise

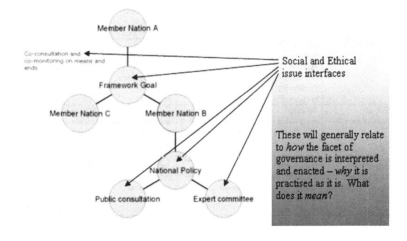

A brief example may serve to illustrate further. 'A shared energy grid' is a framework goal. There are presuppositions embedded in this that cannot be questioned as the goal is decided external to the system, which then merely operates on its basis. Interpretation of the goal at national level is open, as is the means with which to pursue it (subsidiarity). The means of selecting the means of pursuing the policy are not prescribed by OMC so within the open method there is room for arbitrariness, the pursuit of special interests, sectoral domination etc. Veto and lobby are the means of co-monitoring by which the worst case scenarios ought to be ruled out. But such concepts aren't exactly at the heart of open dialogue. In the pursuit of a shared energy grid Nation *A* might dissolve national monopolies by incentivising market behaviour in a power industry, while Nation *B* might nationalise a hitherto liberalised power market, stealing business from private hands, while greasing the wheels of such action with international political favours (to counteract veto-effects).

The upshot, then, can be mere powerplay, then. The aim should be ethical governance, governance centred upon the possibilities as perceived by a citizenry for that citizenry. The means ought to be ethical reflexivity in order that unquestioned framings be opened and recognition of the other's perspective be permitted. It seems the potential for something quite unlike this resides within the formal structure of OMC, and so we suggest that structure to be insufficient *in itself* to guarantee a role for ethics.

In the Table 2 we present some notional questions in the form of problematising elements for the stages of the process indicated above as being 'ethical interfaces.' The table is by no means an exhaustive analysis, but rather represents a point of reflective departure aimed at signposting that ethics is not inherent within the system merely by dint of the use of OMC:

Table 2. Questions for stages of ethical interfaces

Interface	Problematic	Stake	Negative Potential	Presuppositions
Consultation and co-monitoring on means and ends	Used for control of process given an agenda, or in an open-ended procedural manner?	Openness, representation	Power play, degradation of the process to complex technocracy	Mentalist, intentionalist, schematising
Framework goal	On what basis are these set? Open too?	Rationale of the procedure	The deformation of the open procedure to a mere system promoting arbitrary goals	Mentalist, intentionalist, schematising
National policy	Set to reflect national opinion? To respect co-monitoring pressures?	National accountability, possibilities as seen by the people	Mediating unaccountable imperatives resulting from external pressures	Mentalist, intentionalist, schematising
Public consultation	Seeking approval or co-construction?	Openness of dialogue, locus of citizens' possibilities	Social acceptance rather than ethical acceptability sought	Mentalist, intentionalist, schematising
Expert committee	Is expertise appealed to as another voice, or as a special voice?	Framing of the debate, cognitive closure	Triumph of sectoral interest over ethical imperatives	Mentalist, intentionalist, schematising

This suggestive device and the foregoing imagined scenario are supposed to hint that systematic approaches to governance require an ethical component. One is not inherent in the system itself, each being open to framing. Simply including people, for instance, guarantees nothing if the meaning of their engagement is not established; 'co-monitoring' and other such concepts are mere labels without substantial interpretation – without an answer to *why* and *how* they are deployed. The systematic approaches are limited to mere procedures, it would seem, and are in need of an external element should they be assessable as ethical.

CONCLUSION

- With DDP, liberal mechanisms are thought to be sufficient to operate with communitarian values.
    - This informs governance *per se* and can be seen lying behind the drift toward 'inclusion' evident in tech research.
    - As EGAIS has shown, this drift *merely* includes people (the absence of co-construction).
- OMC is intended to marry liberal self-determination with communitarian values.
- In the light of the portability findings we can transpose the limits from EGAIS' hypotheses into the OMC schema.
    - At each of the 4 levels of operating, the limitations concerning framing, presupposition, etc. will arise.
- It doesn't incorporate ethics and reflexivity.
- It relies upon the idea of the system itself manifesting ethics via good procedures.

As stated above, pluralism can be seen as *the* governance problem insofar as its investigation leads on to the various issues of 'norm,' 'value,'

'can' and 'ought' that characterise ethics in public decision-making. We have seen that the major structural manifestation of DDP as practised in OMC seek inclusion of the public in dialogue as a means to make governance responsible, to respect difference and to bring accountability to public decision. The problem of plurality, then, leads to the inclusion of the public and other stakeholders in moderated dialogues.

But, as stated, in the light of EGAIS own findings, mere inclusion is not a guarantee of reflexivity or of the implementation of ethical thinking. The revised standard and consultation models discovered throughout approaches to ethical governance in research projects each had a place for public consultation, for instance, but in a way inadequate to the task of producing genuinely ethical governance (EGAIS, 2010a, p.21).

REFERENCES

Aristotle. (1999). *Nicomachean Ethics* (2nd ed., T. Irwin, Trans.). Hackett.

Cohen, J., & Sabel, C. (1999). *Directly deliberative polyarchy*. Columbia Law School papers. Retrieved June 1, 2011, from http://www2.law.columbia.edu/sabel/papers/DDP.html

Dworkin, R. (1981). What is equality? Part 1: Equality of welfare. *Philosophy & Public Affairs*, *10*(3), 185–246.

EGAIS. (2010a). *Deliverable 2.4.: Interpretation of ethical behaviour*. Retrieved June 1, 2011, from http://www.egais-project.eu/sites/default/files/deliverables/EGAIS_D2.4_23092010.pdf

EGAIS. (2010b). *Deliverable 3.1.: Ethical governance models, paradigm recognition and interpretation*. Retrieved June 1, 2011, from http://www.egais-project.eu/sites/default/files/deliverables/EGAIS_D3.1_30092010-1.pdf

European Comission. (n.d.). *European Commission's Science in Society Portal.* Retrieved June 1, 2011, from http://ec.europa.eu/research/science-society/index.cfm?fuseaction=public.topic&id=1223&lang=1

European Commission. (2011). *Electricity market regulatory forum (Florence).* Retrieved July 1, 2011, from http://ec.europa.eu/energy/gas_electricity/electricity/forum_electricity_florence_en.htm

European Court of Human Rights. (2010). *European convention on human rights.* Strasbourg: European Council of Europe. Retrieved October 23, 2012 from http://www.echr.coe.int/nr/rdonlyres/d5cc24a7-dc13-4318-b457-5c9014916d7a/0/englishanglais.pdf

European Union. (2004). Protocol on the application of the principles of subsidiarity and proportionality. *Official Journal of the European Union, C310/207.* Retrieved June 1, 2011, from http://eur-lex.europa.eu/LexUriServ/LexUriServ.do?uri=OJ:C:2004:310:0207:0209:EN:PDF European Union. (2010). *Single European act.* Brussels: EU. Retrieved June 1, 2011, from http://europa.eu/legislation_summaries/institutional_affairs/treaties/treaties_singleact_en.htm

European Union. (2007). Treaty of Lisbon amending the Treaty on European Union and the Treaty establishing the European Community, signed at Lisbon, 13 December 2007. *Official Journal of the European Union, 50*(C306), 1-202. Retrieved June 1, 2011, from http://eur-lex.europa.eu/JOHtml.do?uri=OJ:C:2007:306:SOM:EN:HTML

Geertz, C. (1973). Description: Toward an interpretive theory of culture. In Geertz, C. (Ed.), *The interpretation of culture.* New York: Basic Books.

Gerstenberg, O. (1997). Law's polyarchy: A comment on Cohen and Sabel. *European Law Journal, 3*, 343–358. doi:10.1111/1468-0386.00035.

Jessop, B. (2003). *Governance and metagovernance: On reflexivity, requisite variety, and requisite irony.* Retrieved November 1, 2012, from http://www.lancs.ac.uk/fass/sociology//papers/jessop-governance-and-metagovernance.pdf

Kymlicka, W. (1990). *Contemporary political philosophy.* Oxford: Oxford University Press.

Lebessis, N., & Paterson, J. (1997). Evolution in governance: What lessons for the commission? A first assessment. *Forward studies unit, Working Paper 1997.* Brussels: European Commission. Retrieved November 1, 2012, from http://www.pedz.uni-mannheim.de/daten/edz-mr/pbs/00/evolution_in_governance.pdf

Lund, W. (1996). Egalitarian liberalism and the fact of pluralism. *Journal of Social Philosophy, 27*(3), 61–80. doi:10.1111/j.1467-9833.1996.tb00253.x.

Lund, W. (2000). Fatal attraction: 'Wilful liberalism' and the denial of public transparency. *Political Research Quarterly, 53*(2), 305–326.

Mayntz, R. (2002). Common goods and governance. In Heritier, A. (Ed.), *Common goods: Reinventing European and international governance* (pp. 15–27). New York, London: Rowman & Littlefield Publishers.

McCarthy, T. (1994). Kantian constructivism and reconstructivism: Rawls and Habermas in dialogue. *Ethics, 105*, 44–63. doi:10.1086/293678.

Rawls, J. (1980). Kantian constructivism in moral theory. *The Journal of Philosophy, 77*(9), 515–572. doi:10.2307/2025790.

Rawls, J. (1987). The idea of an overlapping consensus. *Oxford Journal of Legal Studies, 7*(1), 1–25. doi:10.1093/ojls/7.1.1.

Rawls, J. (1993). *Political liberalism.* New York: Columbia University Press.

Sabel, C., & Zeitlin, J. (2006). New architecture of experimentalist governance (Powerpoint presentation) at the *EU 6th Framework Research Programme's CONNEX Network of Excellence on European Governance Workshop* (Roskilde University). Retrieved June 3, 2011, from http://eucenter.wisc.edu/OMC/New%20OMC%20links/Learning%20from%20Difference%20-CONNEX%20presentation.ppt

Sabel, C., & Zeitlin, J. (2007). Learning from difference: The new architecture of experimentalist governance in the European Union. *Eurogov Papers, No. C-07-02*. Retrieved November 27, 2012, from http://edoc.vifapol.de/opus/volltexte/2011/2466/pdf/egp_connex_C_07_02.pdf

Santuccio, A., & Valentino, F. (2011). *EGAIS Project deliverable 3.2: Second workshop report.* Retrieved June 1, 2011, from http://www.egais-project.eu

Taylor, C. (1979). *Hegel and modern society.* Cambridge: Cambridge University Press. doi:10.1017/CBO9781139171489.

ENDNOTES

1. EUR-Lex http://eur-lex.europa.eu/LexUriServ/LexUriServ.do?uri=CELEX:12002E095:EN:HTML (accessed June 2011) 'Article 95 EC' is seen by many as the provision in European law for harmonization.

2. Cf. Geertz (1973), Chapter 1.

3. Cf. European Union (2004) 'protocol 30' of the European Community Treaty.

4. Article 6 of 'protocol 30,' just cited, which will be seen to be highly relevant in the context of the development of an ideal practical and an actually practicable piece of advice regarding ethical governance, something with which we must concern ourselves.

5. We cite Rawls as he is a liberalist *par excellence* in being a major figure in contemporary debates.

6. See Aristotle (1999).

7. This is a violation of the 'endorsement constraint.'

8. The status of OMC is evident from the scale of a conference held on its deployment by the commission in conjunction with the then Austrian presidency, retrieved July, 1, 2011, from http://ec.europa.eu/invest-in-research/pdf/download_en/programme_omc.pdf.

9. See European Comission (n.d.).

10. Governance is "the reflexive self-organisation of independent actors involved in complex relations of reciprocal interdependence" (Jessop, 2003. p.1), with more recent EU developments further qualifying the mode of co-ordination as democratic, participative, and pragmatic, and recommending support for collective action (Maesschalck 2007). Some commentators see governance as a way to describe "the changing nature and role of the state in advanced societies and the changing boundary between state and civil society" (Lyall, Papaioannou, & Smith, 2009), with an emphasis on the increasing importance of the involvement of stakeholders in policy research, as opposed to the traditional "top-down imposition" forms of policy formation and implementation.

11. We will see in due course that, following analysis, these issues carry with them further complications to do with reason and values that themselves must also be treated. For the purposes of this section, however, these terms can remain unanalysed without affecting the point.

Chapter 13
A Pragmatic Approach to Polynormative Governance

Matthias Kettner
Witten/Herdecke University, Germany

ABSTRACT

In the first and second part of the present article, the author provides a pragmatic reading of the very idea of governance. With the help of the late pragmatist Frederick Will's thoughts about the philosophic governance of norms, governance can be construed as a practice that is situated within other practices and whose aim is lending guidance to these practices. Since the point of establishing governance practices is to improve the targeted governed practices, governance is characterized by normativity, e.g. rationality assumptions, reflexivity and relativity to the general and particular significance of the governed practice. A schema is introduced for abductive inferences (as outlined by Charles Sanders Peirce) from observed defects in practices to expected improvements in governance practices. In response to the question, how governance itself is to be governed where it stands in further need of governance, I argue in the third section that there is an interesting problem of "polynormative" governance: Different forms of governance in different domains of practice may differ drastically in their advantages and disadvantages when compared from some particular evaluative point of view, and they will differ drastically across different evaluative points of view. The author argues that argumentative discourse, not in Michel Foucault's, but in Karl-Otto Apel's and Jürgen Habermas' sense of the term, is the governance practice of last resort for our giving and taking reasons. The relation of argumentative discourse to democracy (being the governance practice of last resort for political power) remains to be explored.

DOI: 10.4018/978-1-4666-3670-5.ch013

INTRODUCTION

On the Very Idea of Governance: Grammar, Practice, Reflexivity, and Relativity

There is more than a decade of debate and a vast literature about wider and narrower meanings of the term governance. The term governance, and the range of concepts covered by that term, has been used in a variety of ways with a range of different emphases (Pierre, 2000). One quite common reading of the term is that governance has primarily to do with the steering of actions of *public authorities* to shape their environment (Mayntz, 2003). Other authors (e.g., Pierre & Peters, 2000) see governance more as an *administrative learning process* of state-governments and nations to steer society in new ways. On this government-oriented reading, the concept of governance marks an emphasis on more *bottom-up* and *participatory* approaches to political decision-making (Kamarck & Nye, 2002) and on the development of *complex communicative networks* (Newman, 2001) of democratic debate and other forms of exchange within civil society. This is set against conceptions of control as top-down hierarchical power structures. In this vein, the term governance can come to mean a process of political communication in which both the governing instances and the people who are being governed negotiate a common way more or less on an equal footing (Bang, 2003). A third reading of governance with a view to business corporations (Mallin, 2003) and to public management emphasizes *control through contracts* (Donahue & Nye, 2002). Some writers use the semantics of governance in order to indicate alternative mechanisms of decision-making that arise where economic market forms falter.

Notwithstanding these differences most writers on governance agree that in some way governance is about collective decision making in various forms, on different levels, and in different arenas.

The semantics of governance, then, centers on formal and informal *rules and forms that guide collective decision-making*. The reference to rules conveys a sense of procedures that are expressed in institutional form and relatively stable over time, although not unchanging. Perhaps one reason for a growing interest in governance is a growing awareness that established institutional forms of governance appear insufficiently flexible and responsive in the face of increasingly complex and dynamic modern problems (e.g., climate change, developing the global internet, controlling unemployment and inflation) with which state and other agencies have to cope.

Governance, in distinction to related notions such as shaping, designing, steering, directing, and reigning, has a pronounced *dynamic connotation* of control-in-action. To characterize control processes as governance processes is often to point to their *reflexivity*, i.e. reacting to itself all along the way with the ever changing subject matter that is the object of governance[1].

Obviously, we are confronted with multifarious conceptual articulations of governance. Instead of attempting to integrate and combine this variety into one single conception at a more abstract level, I will try to return to basics with the following question: Is there something like a general depth-grammar underlying any particular project of governance on any level that one might want to distinguish?

Conceiving of governance merely as some process would not do. What specifically does it take for a process to count as a process of governance? Governance processes must be in principle intentional, though the intentionality of governance need not be the full-blown intentionality that we attribute to intentional action as performed by autonomous agents (Mele, 1995). Governance is an *activity*.

This activity has as its proper *object* or targets some *other activity*. And it has as its proper *subject* an actor doing the governing who is, and can be held to be, responsible for the activity of governing and its consequences for the proper object of governance.

It seems we cannot be said to govern natural processes as such. We cannot say that we govern, for instance, "the atmosphere." Properly speaking, we can govern e.g. emission rights of the atmosphere, traffic activities in the atmosphere, or the use of the atmosphere. Where governance is exercised, there must be some cultural form already in place and it is in this cultural form that governance gets a grip on its object. Consider this distinction: The laws of nature (if there are such things) determine all events in the universe, they do not govern them. In contrast, natural law (if there were such a thing) would govern all legitimate human affairs in all cultural worlds, it would not determine them.

The subject of governance, the governor, need not be an individual natural person. A collective form of agency (e.g., a government, a board, a council, a task force, a committee) is sufficient, provided that attributions of *rationality* (in a minimal sense, cf. Cherniak, 1986) and *accountability* can be channelled through that form and attributed to the individuals who are acting under that collective form of agency.

The *quality* of the activity of governing is, in the widest sense possible, that of guidance in attaining some goal in some proper, orderly, favourable or good way – as it were, in a ceteris paribus better way than ungoverned. The very idea of governance contains an irreducible *evaluative or normative* element.

This analysis of the depth-grammar of governance invites us to construe governance as a practice that is related in a specific way to other practices, a practice *within, or internal to,* practices. Note that it would be odd to construe governance as a meta-practice, i.e. a practice that has other practices as its object in the sense in which a so-called meta-language has other languages as its object. The activity of governing, for being a practice, is continuous with the practice that is its proper object. The target practice must already contain in itself some augmentable quality of the very governance that is provided to it, it is not a natural process completely devoid of governance.

In conceiving of governance as a practice that is situated within other practices, whose aim is lending guidance to these practices, the question arises whether the practice of governance itself stands in further need of governance. This question is superficially similar to the notorious problem of an (infinite?) hierarchy of meta-languages. But unlike to meta-languages, here an answer can be provided by looking at the reflexivity of governance: It does not make sense to search for an ultimate ungoverned governor, or absolute governance process, since any practice of governance GP is always relative to the limited affordances of some particular practice P, where P stands in need of governance only in as much as we think that within P things go comparatively worse than they should go and could go. The idea of "the best governance possible" is an incomplete thought and does not make sense. Governance is better or worse only in relation to the practices that are its objects. This leaves room for iterations of the governance relation: Given PG/P, if things within PG go comparatively worse than they should go and could go, an augmentation of governance is called for, now taking PG as its object: PG'/PG/P.

An Elementary Pragmatic Conception of Normative Governance

In this section, I introduce a general conception of governance as the governance of norms and of normative textures. This account brings in classical and recent resources of philosophical pragmatism. We can go one step in the direction of the pragmatic account of governance of normative textures by asking what it means to govern *a practice*. What is a practice? And what about a practice makes practices amenable to being governed by a practice *of governance*? Philosophical pragmatism provides an answer in terms of a realistic theory of norms. Practices are concrete compositions of norms. Norms are guides to what is a reasonable procedure; guides to appropriate ways of doing things, e.g. to appropriate ways of pursuing goals. Governance, then, is guidance in reference to norms. Within practices, the norms that compose the respective practices are what governance latches onto. Norms are the direct object of governance. It is by directly influencing *their norms* that a practice of governance indirectly influences the corresponding *governed practices*.

How are we to think about norms? I favour a realistic account of norms as social-psychological entities. As far as I see, the best such account has been worked out in contemporary philosophical pragmatism, in particular by Frederick L. Will (1988; 1993)[2]. According to Will, "[a] norm is not just a form of action. It is a form of appropriate action, of appropriate response; and this appropriateness is something that is determined in its relations with, among other things, other norms: by the role which this or that mode of action or response plays in the constitution of individual and communal life" (Will, 1988, p. 93).

Norms are clearly "collectivist" entities. This runs against the tide of methodological individualism which is the preferred style of theorizing not only in applied ethics but in many other branches of philosophy, theories of normativity and rationality included. The concept of a norm at least in the Anglo-Saxon universe of discourse tends to be biased towards, on the one hand, strongly socially sanctioned and formally elaborated norms (usually of the criminal law, cf. Ullmann-Margalit, 1977, pp.12-17) or, on the other hand, highly contingent social conventions and rules of etiquette. This overly narrow range of paradigms of norms tends to obscure the fact that any and every human practice is composed of norms of all kinds, or is, as I would prefer to say "normatively textured."

Viewed realistically, norms are intrinsically socio-psychological entities that interlock with each other and are rooted deeply in the practices of individuals and their communities. Embedded

in these practices, norms are principally open-textured. By "open-textured" I mean that norms are open to further definition and revision whenever and wherever serious anomalies in their existing forms in ongoing practices are encountered.

For a realistic doctrine of norms, it is important to recognize that norms defy an ontology of discrete atomistic entities and to acknowledge the corporate, or compositional, or - if you will - holistic nature of norms. In order to invent a distinctive term for the intrinsically polymorphic nature of norms when norms compose practices I propose to speak of *normative textures of practices* (Kettner, 1999). Of course, we can count norms and linguistically or otherwise symbolically represented normative contents one by one, just as we can count linguistically or otherwise symbolically represented propositions or beliefs about what is the case, or just as we can count numbers. However, we should be aware philosophically that what we single out as some determinate normative content is a symbolically stylized representation. A symbolically stylized representation is *abstracted* from the practice to which the norm pertains, a representation whose determinate identity conditions are tailored to the very purpose of sorting things out, whereas in order to grasp the very point of the norm it takes much more in the way of understanding the purposes of the practices in which some particular norm plays a role in conjunction with other norms that are equally part and parcel of the fabric of those practices.

Piecing together some of Will's acute observations[3] on normative change and the changing of normative textures, the realistic account of norms yields a realistic account of governance: We experience again and again how norms, or more or less extended components of normative textures for a variety of reasons, and in various degrees, falter and fail in praxis. However, such failures are not always merely "negative." Rather, they often provide stimuli and grounds for change. Bodies of norms enduringly - sometimes more and sometimes less - suffer from (1) incompleteness and (2) inconsistency. As to incompleteness, we encounter our normative repertories never as something complete: Occasions for decision, conviction, fixation of opinion and action arise for which there is no ready decision procedure.

As to inconsistency, inconsistency appears at various places in the repertory where contrary procedures seem to apply, yielding indecision which can be resolved only if we manage to revise components of the body of norms so that the debilitating clash is overcome. Further, as a result of the corporate character of these bodies of norms, if changes are generated in one set of norms these changes will have certain impacts upon other sets of norms are associated with the respective set because of the holistic nature of interlocking practices. Such resonance effects may sometimes include responsive changes and sometimes defensive dispositions to resist change. Always and everywhere, we are faced with having to *mutually accommodate* many different sets of norms. The stress that is generated among many different sets of norms, the abrasion and confusion that arises in their being jointly followed, as they should, provide motives for this accommodation: "What may appear in its local environment as merely negative, as the faltering, failure, or confusion in certain readily formalisable procedures, in a larger context may be revealed to play a most important positive role. The faltering, failing, and confusion may provide not only motivation for revision of these procedures, but also indications of the general location where revision is called for and the general direction the revision should take" (Will, 1988, p. 39).

A pragmatic conception of governance can now be formulated. Governance covers both the critical and the constructive activities in which we are engaged with the character, the establishment and the maintenance of norms, as well as with their dis-establishment and elimination. Governance of norms (following Will, 1993, p. 21) encompasses the development of norms, criticism of norms, refinement, reconstruction, reinforcement of norms, as well as the weakening and the elimination of norms. Governance comprises the entire range of activities, individual *and* collective, deliberate *or* without much conscious effort, of

- Fitting multiple norms to each other and to the circumstances in which they are followed.
- Generating new norms and reconstructing already extant norms once they have revealed serious insufficiencies in practice.
- Removing norms that have been rendered anachronistic because of substantial change in the very forms of life whose circumstances used to provide contextual fit for the norms.
- Broadening or narrowing norms in appropriate reaction to the surprises and vicissitudes that we permanently encounter in the practices whose workings are textured by those norms.

If normative governance has a rational quality, how can we model the rational quality of governance activities? A rational model of governance activities would have to model the kind of improvement that the gover*ning* practice GP is seeking, and it would have to model changes in normative textures of the corresponding gover*ned* practices P. Neither the logical forms of inductive reasoning nor that of deductive reasoning can adequately represent changes in concrete practices. In order to model the rational quality of governance processes it is necessary to enhance these forms by the logical form of abductive reasoning. According to the early American pragmatist Charles Sanders Peirce, abduction is a valid form of inference which permeates human experience. Abduction is the kind of reasoning that we employ when we are faced with explanatory gaps, or other gaps in our coherent understanding of things, in order to frame new hypotheses that are rationally warranted, would restore coherence, and would serve to close such gaps.

According to Peirce, the logical form of abductive reasoning can be captured in the following schema of premises and conclusions: First premise: A surprising fact is observed. Second premise: But if such-and-such were true, this fact would be a matter of course. Conclusion: Hence, there is reason to suspect that such-and-such is true (Fann, 1970; Kettner, 1991). Abduction can never be formulated as an algorithm. Abduction can only be represented as a rational constraint on creative invention, not as a calculation of creative invention.

Using the schema of abductive inference, the rational quality of governance activities can be modelled in the following way. First, some hypothesis has to be framed concerning an identifiable source of the problem that surprisingly occurs in some practice P composed of a certain set of norms:

1. Unexpected undesirable results are observed in the performance of some generally accepted practice P, where P is traditionally textured by the norms N_i, N_j ... N_z.
2. If some norm N_x (within the entire normative texture of P) were to be followed routinely, those undesirable results would be a matter of course.
3. Therefore, it is reasonable to devote and direct governance specifically to N_x within P.

In order to lay out and set up a promising governing practice GP concerning P, four assumptions have to be made and interpreted:

1. **Assumption:** P is important enough for us, all things considered, to want to continue P. Otherwise we might as well abandon P and try to establish instead an altogether different practice Q.
2. **Assumption:** It is not equally important for us to maintain N_x.

3. **Assumption:** We are sufficiently clear about the extent to which P allows for change in its normative composition while remaining recognizably a composition of that very practice, namely P.
4. **Assumption:** We are sufficiently clear about the constraints that limit the range of possible and acceptable surrogates (successors, alternatives) of N_x within P.

These assumptions prepare the ground for more reasoning. A range of variation in the normative composition of the governed practice has to be staked out, and we have to zero in on some norm within that range and change that norm:

1. Bearing constraints on sameness of P in mind, we explore how the set of norms that compose P can be extended. We find that P is recognizable as P not only in the received composition of norms $\{N_i, N_j ... N_z\}$ but also in a similar but amplified composition $\{N_a, ... N_d, N_i, N_j ... N_z\}$.
2. If some norm N* of those belonging to the set of possible amplifications $\{N_a, ... N_d\}$ could replace N_x in P and were to be followed routinely, then the unexpected undesirable results that we observed in P would probably not occur in P.
3. Therefore, it is reasonable to modify P so that N* succeeds N_x in P.

How to Govern Governance?

The preceding section introduced an elementary pragmatic conception of governance as a practice within other practices and attempted to model the rational quality of governance activities. In the third section, finally, I want to combine two of the points that were made in the first section. The first point that I want to take up again is (1) the *reflexivity* of governance practices, i.e. their reacting to their own activity all along the way with the ever changing object of governance. The second point is (2) that governance is itself *normative*, not only in the sense that norms are the object, but in an adverbial sense, i.e. that governing contains an irreducible evaluative or normative element in virtue of which governance practices themselves are open to critical judgment, for instance, critical judgments as to how well they reach their aims. Combining both points makes clear that the normative governance of a normatively textured practice is itself a normatively textured practice. This observation brings into focus the question of *full reflexivity*. How to govern governance? Can governance take care of itself?

Governance practices can be more or less reflexive, no doubt. But what would a fully re-flective mode of governance amount to? Is there a paradigm for such a mode?

I admit that the only practice that looks to me like a fully reflective governance practice is argumentative discourse, the term "discourse" taken not in the descriptive sense of Michel Foucault but in the normative sense elaborated by philosophers like Karl-Otto Apel and Jürgen Habermas. Argumentative discourse is a dialogical practice of evaluating determinate reasons as better or worse in relation to the purpose of justifying validity-claims with good reasons. In a certain sense, argumentative discourse serves as "the governance practice of last resort" for all of our practices of giving and taking reasons. Discourse, I would hold, is also the governance practice of last resort when governance practices of any kind come under critical scrutiny.

In utmost generality, the following consider-ation can be used in critical judgments of gover-nance practices:

Of two different governance practices GP_1/P and GP_2/P,

- One is better governance if and only if:
- One of the two different recompositions of P.

- That are effected by GP_1 and alternatively by GP_2.
- Remedies more "deficiencies" in P.
- And creates fewer new "deficiencies" in P.
- And in all associated practices P'… P'.
- Than the other recomposition.

Deficiencies in which respect(s)? By what kind(s) of standard(s)? This raises what I propose to call *the problem of polynormative governance*. Processes of governance are as multifarious as the practices that need governance. Different forms of governance in different domains of practice may differ drastically in their advantages and disadvantages when compared from some particular evaluative point of view, and they will differ drastically across different evaluative points of view. Take, for example, bureaucracy as a form of governance and think of what be-comes of academic life when *universities* are predominantly governed by bureaucracies. Then compare this rather nasty scenario with the effects of bureaucratic governance in a different setting, for instance, think of the indisputable advantages in the change from feudal practices of governing the affairs of *state-administration* to bureaucratic modes of governing these affairs.

The general point – that different forms of governance in different domains of practice may differ drastically when compared from different *evaluative* points of view – holds also for *moral* points of view. Moral points of view are a subset of evaluative points of view. As a consequence, moral judgments, although providing an important resource for critical judgments about governance practices, are not the only such resource. Moralists tend to think that moral discourse is the ultimate governance practice. This is a serious mistake in thinking about governance. Pragmatists will avoid that mistake.

CONCLUSION

Current notions of democratic governance, and in particular deliberative democratic governance (Kettner, 2007), can be reconstrued as particular specifications of the general conception of normative governance of normative textures. This particular specification of the general conception is oriented to democratic legitimacy, i.e. a complex mode of validity. In a (deliberative) democracy, other modes of validity, e.g. moral and juridical legitimacy, have to be integrated into democratic legitimacy in complex ways.

Within the confines of this essay, it is not possible to pursue this question further. The complex relation that holds between argumentative discourse (as the governance practice of last resort for all our practices of giving and taking reasons) and democracy (as the governance practice of last resort for all our practices of exercising political power) needs a thorough exploration.

REFERENCES

Bang, H. P. (Ed.). (2003). *Governance as social and political communication*. Manchester: Manchester University Press.

Cherniak, C. (1986). *Minimal rationality*. Cambridge, MA: MIT Press.

Donahue, J. D., & Nye, J. S. (Eds.). (2001). *Governance amid bigger, better markets*. Washington, DC: Brookings Institution Press.

Fann, K. T. (1970). *Peirce's theory of abduction*. The Hague: Martinus Nijhoff. doi:10.1007/978-94-010-3163-9.

Held, D. (1995). *Democracy and the global order: From the modern state to cosmopolitan governance*. Cambridge: Policy Press.

Kamarck, E. C., & Nye, J. S. (Eds.). (2002). Governance.com: Democracy in the information age. Washington, CD: Brookings Institution.

Kettner, M. (1991). Peirce's notion of abduction and psychoanalytic interpretation. In Epstein, P. S., & Litowitz, B. E. (Eds.), *Semiotic perspectives on clinical theory and practice: Medicine, neuropsychiatry and psychoanalysis* (pp. 163–180). New York: Mouton De Gruyter.

Kettner, M. (1999). Neue perspektiven der diskursethik. In Grunwald, A., & Saupe, S. (Eds.), *Ethik technischen handeln: Praktische relevanz und legitimation* (pp. 153–196). Heidelberg: Springer Verlag.

Kettner, M. (2007). Deliberative democracy: From rational discourse to public debate. In Goujon, P., Lavelle, S., Duquenoy, P., Kimppa, K., & Laurent, V. (Eds.), *The information society: Innovation, legitimacy, ethics and democracy, in honor of Professor Jacques Berleur* (pp. 55–66). Berlin: Springer. doi:10.1007/978-0-387-72381-5_7.

Mallin, C. (2003). *Corporate governance*. Oxford: Oxford University Press.

Mayntz, R. (2003). New challenges to governance theory. In Bang, H. (Ed.), *Governance as social and political communication* (pp. 27–40). Manchester: Manchester University Press.

Mele, A. (1995). *Autonomous agents: From self-control to autonomy*. Oxford: Oxford University Press.

Newman, J. (2001). *Modernising governance. New labour, policy and society*. London: Sage.

Pierre, J. (Ed.). (2000). *Debating governance*. Oxford: Oxford University Press.

Pierre, J., & Peters, G. (2000). *Governance, politics and the state*. Basingstoke: Palgrave.

Ullmann-Margalit, E. (1977). *The emergence of norms*. Oxford: Oxford University Press.

Westphal, K. R. (Ed.). (1998a). *Frederick L. Will's pragmatic realism*. Chicago: University of Illinois Press.

Westphal, K. R. (Ed.). (1998b). *Pragmatism, reason, and norms: A realistic assessment*. New York: Fordham University Press.

Will, F. L. (1988). *Beyond deductivism: Ampliative aspects of philosophical reflection*. London: Routledge.

Will, F. L. (1993). The philosophic governance of norms. *Jahrbuch für Recht und Ethik, 1*, 329–361.

ENDNOTES

[1] These introductory remarks have profited from instructive overviews of various debates about the conceptual content of the term governance. Relevant research-papers are publicly accessible on the website of The Institute On Governance (IOG). The IOG is an independent, Canadian, non-profit think tank founded in 1990 to promote better governance for public benefit (http://www. iog.ca/).

[2] For a collection of Will's articles and for a collection of philosophical essays on Will's pragmatism, see Westphal (1998a, 1998b).

[3] Cf. Will, 1988, especially pp. 39, 148, 157, 189.

Chapter 14
The "Pragmatist Turn" in Theory of Governance

Alain Loute
Université catholique de Louvain, Belgium

ABSTRACT

In this essay, the author focuses on what Jacques Lenoble and Marc Maesschalck call the "pragmatist turn" in the theory of governance. Speaking of pragmatist turn, they refer to recent work by a range of authors such as Charles Sabel, Joshua Cohen and Michael Dorf, who develop an experimental and pragmatist approach of democracy. The concept of "turn" may raise some perplexity. The author believes that we can speak of "turn" about these experimentalist theories because these theories introduce a key issue, what we may call the question of "self-capacitation of the actors." The author tries to show that this issue constitutes a novelty compared to the deliberative paradigm in the theory of governance. While the issue of collective learning is a black box in the deliberative paradigm, democratic experimentalism seeks to reflect on how the actors can organize themselves to acquire new capacities and to learn new roles. The author concludes in revealing the limits of this approach.

INTRODUCTION

In this essay, I would like to focus on some research results of the Centre for Philosophy of Law at the Catholic University of Louvain. In particular, I will focus on what Jacques Lenoble and Marc Maess-chalck call the "pragmatist turn" in the theory of governance. Speaking of pragmatist turn, they refer to recent work by a range of authors such as Charles Sabel, Joshua Cohen and Michael Dorf, who develop an experimental and pragmatist approach of democracy, what they call "democratic experimentalism" (Dorf & Sabel, 1998).

The concept of "turn" may raise some perplexity. It refers to an idea of novelty or overcoming,

DOI: 10.4018/978-1-4666-3670-5.ch014

an idea of transition from one paradigm to another. Can we really speak of "turn" about these experimentalist theories? Do they really permit to overcome other paradigms in the theory of governance or do they only constitute a deepening of these? Should we not rather regard such theories as a form of deepening of the deliberative paradigm?

Based on the work of Lenoble and Maesschalck (2010), my goal is to demonstrate that these theories constitute a real "turn" in theory of governance. The reason is that they introduce a key issue, what we may call the question of "self-capacitation of the actors." They seek to reflect on how the actors can organize themselves to acquire new capacities and to learn new roles. It seems to me that the issue of "self-capacitation" of the actors is a novelty compared to other paradigms in the theory of governance.

To demonstrate this thesis, it seems useful to begin with the diagnosis of a paradox in our societies. The paradox is that although there are more and more opportunities for participation in our society, the influence of citizens does not seem to have been really increased. In a second step, I will try to show that this paradox of participatory democracy refers first of all to an unsettled question in the deliberative paradigm, of which Habermas is the most famous representative, namely the question of the "capacitation" of the stakeholders to assume their discursive role within the deliberative programming of the society (Maesschalck & Loute, 2007). In a third step, I will show how "democratic experimentalism" makes this question of the "capacitation" of the actors a central issue of theory of governance. I will conclude in revealing the limits of this approach.

THE PARADOX OF THE PARTICIPATIVE DEMOCRACY REVIVAL

In the last few years, some reform practices that have taken place within our States, have revived the ideal of participation: recurring theme of participative democracy, deliberative practices, implication of the users in the evaluation of public services, etc. Some authors, like Blondiaux and Sintomer (2002), refer to the emergence of a "deliberative imperative." The European policy context is also strongly influenced by the theme of participative democracy. This is for instance illustrated with the White Paper on European Governance (European Commission, 2001) which highlights participation as one of the principles for a good governance. Other authors, like Pierre Rosanvallon (2008), show how much democratic legitimacy implies the necessity and prescription of proximity and reflexivity. Our societies have thus entered the era of "reflexive modernization" (Beck, 1986; Beck et al., 1994) which brings into question the strict divisions of the task of our representative societies and our societies founded on the power of experts (Callon et al., 2001). For us, although they constitute a new mode of participation, the paradox of these new practices initiated by politics is that they do not seem to induce a real growth of power of citizens in collective decisions. They do not seem, using an expression of Marcel Gauchet, to render power *appropriable* by the members of the political community (2002). These offers of participation touch only a small part of the population, when they are not "colonized" – using an expression from Habermas – by the lobbies or by administration discredited which seek to acquire some form of legitimacy[1]. The multiplication of deliberative spaces has instead had the effect of making possible new forms of opportunism and strengthening the domination of majority interest.

COLLECTIVE LEARNING: A BLACK BOX IN THE DELIBERATIVE PARADIGM

We believe that this "deficit" of the revival of participative democracy refers first of all to an unsettled question in the deliberative paradigm, one of the dominant paradigms which theorize participation, namely the question of the "capacitation" of the stakeholders to assume their discursive role within the deliberative programming of the society (Maesschalck & Loute, 2007). Deliberativism presupposes that the constitution of space of deliberation and the convocation of the actors suffice in themselves alone, to stimulate communicational competence, inciting participation in a "collective action" which is aimed at the discursive formation of the general interest. Following Lenoble and Maesschalck, this paradigm presupposes that "the aggregation of communicative competencies alone suffices to generate the adaptative capability needed to solve problems in the most satisfactory way possible from the point of view of group members' normative expectations (…); learning capacities are believed to be provided and activated simply by placing the various actors in dialogue" (Lenoble & Maesschalck, 2010, p. 142). This paradigm does not really problematize the issue of the acquisition of new skills and the learning of new roles by the actors.

The question of the learning of new roles by political actors is not really addressed. Such a question seems essential, however. If an author like Habermas (2006), following Arato and Cohen (1992), recognizes that the civil society has an essential role in democracy, he confines this position however to a simple peripheral role of informing the political system, while the political system itself remains the center of political life. This centralized and hierarchical "framing" of the communicative collective action raises different questions: What guarantees that the politicians will agree to be taught by the deliberations of civil society? Does not the political system run the risk of exploiting the public space which constitutes civil society, with the aim of legitimization or control civil society? (Blésin & Loute, 2011). Does not a real participative democracy imply the institution of a form of "polycentrism" (Ostrom, 1997), and a fragmentation of power that some define as "polyarchy"? (Cohen & Sabel, 1997).

Do not the actors of the political system need to learn new roles with reference to those fixed and specialized roles as we see in representative democracy?

In *The Postnational Constellation and the Future of Democracy*, Habermas (2001) addresses somewhat the issue of the transformation of the political actors, without developing a true reflection on the conditions of learning. In this paper, Habermas raises the question of the future of democracy in the context of globalization. For Habermas, globalization has affected both the functioning and the legitimacy of the democratic nation state. Facing the de-regulation provoked by globalization, he argues, a "re-regulation" (*Idem.*, p. 112) is needed if politics want to catch up with global markets: it is necessary to implement a "world domestic policy" (*Idem.*, p. 104), a global governance. By global governance, Habermas does not mean the government of a world state, which he considers both unlikely and undesirable but rather an interactive multi-level governance (national, international and global).

What particularly interests us in this paper is that Habermas is not limited to the question – normative – of the legitimacy of a postnational normative order. For Habermas, the implementation of a globalized public space of deliberation raises a second important question: "what are the conditions for a transformed self-understanding of global actors" (*Idem.*, p. 110)? He wonders what will drive the actors to act as partners of a cosmopolitan community and not like political actors who act only in order to get re-elected by their national voters. For him, such a transformation of the actors depends on the emergence of a "consciousness of a compulsory cosmopolitan solidarity" (*Idem.*, p. 112) in civil society which will lobby politicians. This answer leaves us dissatisfied. Is the sole pressure from civil society sufficient to induce political actors' learning?

Far from resolving the question of the transformation of political actors, highlighting the role of civil society raises a second question. How could a "cosmopolitan consciousness" emerge in civil society? Does Habermas presuppose, as Ulrich Beck (2001, p. 312; 2003, p. 22), that dramatization and mediatization of global risks (environmental, financial, etc.) lead almost automatically to a "consciousness of a compulsory cosmopolitan solidarity"? How can civil society actors become capable of deliberating and influencing the political power? In our view, Habermas does not really address the question of the capacitation of civil society actors. He merely states that such civil society's action is possible without really thinking the conditions of "capacitation" of civil society actors[2] (Habermas, 2003). Habermas does not really develop the issue of actor's learning.

CONTRIBUTION OF THE DEMOCRATIC EXPERIMENTALISM

Reading the work of Marc Maesschalck and Jacques Lenoble, it seems that we can consider that on this question of the "capacitation" of the actors, "democratic experimentalism" constitute a real "turn" in the theory of governance. The authors of this trend (i.e. C. Sabel, M. Dorf, J. Cohen) directly reflect on the conditions that enable the learning of new roles by actors and the learning of a new collective exercise of power. By "democratic experimentalism", these authors designate a public solving-problem system that combines a federal learning with the protection of the interests of federal jurisdictions and the rights of individuals (Dorf & Sabel, 1998, p. 288). According to such a governance model, at national level, the objectives, the contents of public policies must only be defined to a limited extent and remain vague and general to enable federal local unities to experiment contextual solutions. A double social learning process is therefore expected: the one which results from necessary choices to implement solutions by local experimentations and the one which results at federal

level from the evaluation and the comparison of the different local solutions.

We believe that this proposal draws the features of a new culture of governance. While in the deliberative paradigm, the attention was centered on the rules of rational debate within a community of discussion, these authors shift the attention towards that which makes possible a learning process through the confrontation between different groups of actors facing a common problem to solve. Thus they expect that these interactions will make possible a true "democratic experimentalism."

The pragmatist theory of participation leads to a fundamental shift in political philosophy. Following Maesschalck, it shifts the focus of the "intra-group learning" of the community of discussion, to a dynamic of "exo-group learning" (Maesschalck, 2008, p. 191). Between the level of individuals and that of the ideal community of discussion, pragmatism reinvests the level of inter-group relations in political philosophy. By group, there is no question of the community of the American communitarianism or of lobbies of the interest group theory. The pragmatist authors rather refer to a community of action, a "public" in the sense of John Dewey (1927). By "public," one does not need to understand an ideal community of deliberation, or in the sense meant by Arendt, a public space of apparition. The public space for Dewey is neither an idealistic space nor a space already constituted. The public is rather constituted experimentally through a process of collaborative and cooperative inquiry. Such a conception of the group enhances both individual freedoms – understood as the release of the personal potentiality of individuals through the association[3] – that democracy being experienced continuously through group's interactions.

The interest of democratic experimentalism does not seem to reproduce the aporia of the deliberativist model. It does not presuppose that the only constitution of space of deliberation and the convocation of the actors are sufficient to enable collective learning. For learning to occur, we must *act* on the institutional design of our democracies. We must organize ourselves to be taught by the course of action and by others. In addition to a decentralization of the management of public problems, different mechanisms such as benchmarking and comparative evaluation should be set up to provoke a collective learning.

For experimentalists, the benefits of democratic experimentalism are numerous. They expect from this experimentalism that it makes possible a destabilization of existing rules which moves the attention away towards other possible normative choices. In addition, the pragmatist's public sphere seems less ambiguous than the deliberativist's one. There is no question of a hierarchical and centralized political public sphere encompassing the deliberations of civil society. In democratic experimentalism, far from being centralized in the parliamentary forum, the public sphere is "organizationally dispersed" across all areas of local problem-solving (Cohen & Sabel, 1997, p. 337). These different "pieces of public space" are connected to each other by benchmarking procedures, which enrich the debate at the global level.

LIMITS OF THE DEMOCRATIC EXPERIMENTALISM

This experimentalist's approach of participative democracy brings to the fore some essential questions: how to make possible the realization of a social learning process which "capacitates" the actors and which enlarges their normative horizon? What organization of the public space can make possible such a democratic experimentalism? The theoretical proposition of democratic experimentalism remains, however, unsatisfactory. Democratic experimentalism does not really favor the elucidation of the conditions under which the social learning process would take place. It presumes that the existence of a common problem and the bringing into interaction of groups of actors by practical incentives like *benchmarking*

or comparative evaluation are sufficient to incite the groups of actors to position themselves in the public space and to cooperate with other groups.

On many occasions, Sabel and Cohen seem to think we can assume that all individuals agree on the urgent need to solve some common problems and to cooperate:

The problems of modern democracy arise quite apart from the clash of antagonistic interests or any guileful exploitation by individuals of blockages created by constitutional arrangements: they are (in the game-theoretic sense) problems of failed coordination, in which mutual gains are available, but different parties are unable to come to terms in a way that captures those gains (...) Put another way, we assume that for some substantial range of current problems, citizens agree sufficiently about the urgency of the problems and the broad desiderata on solutions that, had they the means to translate this general agreement into a more concrete, practical program, would improve their common situation, and possibly discover further arenas of cooperation. (Sabel & Cohen, 1997, p. 323)

They do not defend the idea that there is consensus on how to solve common problems. But they seem to be assuming that a common motivation to solve common problems is given. They write:

More immediately, we assume that citizens – despite conflicts of interests and political outlook – agree very broadly on priorities and goals, but cannot translate this preliminary agreement into solutions fitted to the diversity and volatility of their circumstances because of constitutional uniformity constraints. (Idem., p. 326)

Following Sabel and Cohen, mechanisms of benchmarking or comparative evaluation should allow citizens to translate this general agreement about the urgency of problems into concrete and innovative solutions. But, are such incentives sufficient to ensure confidence, cooperation and commitment to the joint objectives? An author such as Donald Schön has shown that defensives routines may constitute obstacles that a sole comparative evaluation process cannot dissolve (Cf. Lenoble & Maesschalck, 2010, pp. 179-189).

It also lacks a more comprehensive theory of learning at the institutional level. One has to realize that an approach such as democratic experimentalism "must not only rely on self-governance capacities of groups of stakeholders involved at local level, but also on the ability of the regulating power to guarantee equality of status and freedom within each experimentation and between experimentations themselves" [4].However, nothing is stated on what can lead the regulatory power to transform itself in its role as guarantor of general interest. However, nothing is stated on what can lead the regulatory power to transform itself in its role as guarantor of general interest. For example, some authors have shown that the Open Method of Coordination, inspired by democratic experimentalism (Sabel & Zeitlin, 2010), have not succeeded in transforming cultural preferences that still dominate the political decision of public governance in Europe (cf. De Schutter, 2007).

In this essay, I have focused on what Jacques Lenoble and Marc Maesschalck call the "pragmatist turn" in the theory of governance. I tried to show that the experimentalist and pragmatist approach of governance developed by authors such as Sabel introduce a key issue, what we may call the question of "self-capacitation of the actors." This issue constitutes a novelty compared to the deliberative paradigm in the theory of governance.

The theoretical proposition of democratic experimentalism remains, however, unsatisfactory. Democratic experimentalism does not really favor the elucidation of the conditions under which the social learning process would take place. In their proposal of a genetic approach to governance, Jacques Lenoble and Marc Maesschalck attempt to overcome these limits of the experimentalist

paradigm. However, they do not abandon the issue of the self-capacitation of the actors. They sought instead to pursue this issue. Despite its limitations, the experimental paradigm will have raised an issue that seems essential. Indeed, without a real actor's learning and without an institutional learning, the risk is great that decentralization will make possible the proliferation of opportunistic behavior.

REFERENCES

Arato, A., & Cohen, J. L. (1992). *Civil society and political theory*. Cambridge, MA: MIT Press.

Beck, U. (1986). *Risikogesellschaft*. Frankfurt am Main: Suhrkamp Verlag.

Beck, U. (2001). La dynamique politique de la société mondiale du risque. *IDDRI Working Papers No 01/2001*. Retrieved October 23, 2012, from http://www.iddri.org/Publications/La-dynamique-politique-de-la-societe-mondiale-du-risque

Beck, U. (2003). *Pouvoir et contre-pouvoir à l'ère de la mondialisation*. Paris: Aubier.

Beck, U., Giddens, A., & Lash, S. (1994). *Reflexive modernization, politics, tradition and aesthetics in the modern social order*. Palo Alto, SA: Stanford University Press.

Blesin, L., & Loute, A. (2011). Nouvelles vulnérabilités, nouvelles formes d'engagement, apport pour une critique sociale. In M. Maesschalck & A. Loute (Eds.), *Nouvelle critique sociale, Europe – Amérique Latine, Aller - Retour* (pp. 155-192). Monza: Polimetrica. Retrieved October 23, 2012, from http://www.polimetrica.com/form/form2850.php

Blondiaux, L., & Sintomer, Y. (2002). L'impératif délibératif. *Politix*, *15*(57), 17–35. doi:10.3406/polix.2002.1205.

Callon, M., Lascoumes, P., & Barthe, Y. (2001). *Agir dans un monde incertain: Essai sur la démocratie technique.* Paris: Seuil.

Cohen, J., & Sabel, C. (1997). Directly deliberative polyarchy. *European Law Journal, 3*(4), 313–342. doi:10.1111/1468-0386.00034.

De Schutter, O. (2007). *The role of collective learning in the establishment of the AFSJ in the EU.* The EU Centre of Excellence, University of Wisconsin. Retrieved October 20, 2012, from http://eucenter.wisc.edu/Conferences/GovNYDec06/Docs/DeSchutterPPT.pdf

Dewey, J. (1927). *The public and its problems.* New York: Holt.

Dorf, M., & Sabel, C. (1998). Constitution of democratic experimentalism. *Columbia Law Review, 98*(2), 267–473. doi:10.2307/1123411.

European Commission. (2001). *European governance: A white paper.* COM (2001) 428 Final. Brussels: European Commission. Retrieved November 1, 2012, from http://eur-lex.europa.eu/LexUriServ/site/en/com/2001/com2001_0428en01.pdf

Gauchet, M. (2002). Les tâches de la philosophie politique. *La Revue du MAUSS, 19,* 275–303. doi:10.3917/rdm.019.0275.

Habermas, J. (2001). The postnational constellation and the future of democracy. In Pensky, M. (Ed.), *The postnational constellation, political essays.* Cambridge, MA: MIT Press.

Habermas, J. (2006). *Between facts and norms, contributions to a discourse theory of law and democracy.* Cambridge, MA: The MIT Press.

Lenoble, J., & Maesschalck, M. (2010). *Democracy, law and governance.* Aldershot: Ashgate.

Lenoble, J. & Maesschalck, M. (2010). *Political pragmatism and social attention.*

Maesschalck, M. (2008). Normes de gouvernance et enrôlement des acteurs sociaux. *Multitudes, 4*(34), 182–194. doi:10.3917/mult.034.0182.

Maesschalck, M. (2008). Droit et "capacitation" des acteurs sociaux: La question de l'application des normes. *Dissensus, 1.* Retrieved October 12, 2012, from http://popups.ulg.ac.be/dissensus/document.php?id=231

Maesschalck, M., & Loute, A. (2007). Points forts et points faibles des nouvelles pratiques de réforme des Etats sociaux. In Schronen, D., & Urbé, R. (Eds.), *Sozialalmanach 2007* (pp. 191–203). Luxembourg: Caritas.

Ostrom, V. (1997). *The meaning of democracy and the vulnerability of democracies.* Ann Arbor, MI: University of Michigan Press.

Rosanvallon, P. (2008). *La légitimité démocratique: Impartialité, réflexivité, proximité.* Paris: Seuil.

Sabel, C., & Zeitlin, J. (Eds.). (2010). *Experimentalist governance in the European Union: Towards a new architecture.* Oxford: Oxford University Press.

ENDNOTES

[1] Following Marc Maesschalck, "Dans certains cas, les mécanismes délibératifs ont même rendu plus vulnérables les appareils de contrôle en favorisant l'opportunisme de nouveaux agents par leur option systématique pour la décentralisation et la multiplication des intervenants" (Maesschalck M., Normes de gouvernance et enrôlement des acteurs sociaux, *Multitudes 2008/4, n° 34,* p. 182).

[2] "The sociology of mass communication depicts the public sphere as infiltrated by administrative and social power and dominated by the mass media. If one places this image, diffuse though it might be, alongside the

above normative expectations, then one will be rather cautious in estimating the chances of civil society having an influence on the political system. To be sure, this estimate pertains only to a *public sphere at rest*. In periods of mobilization, the structures that actually support the authority of a critically engaged public begin to vibrate. The balance of power between civil society and the political system then shifts." (Habermas, J., *Between Facts and Norms, Contributions to a Discourse Theory of Law and Democracy*,

p. 379) "In the present context, of course, there can be no question of a conclusive empirical evaluation of the mutual influence that politics and public have on each other. For our purposes, it suffices to make it plausible that in a perceived crisis situation, that *actors in civil society* thus far neglected in our scenario *can* assume a surprisingly active and momentous role" (*Idem*, p. 380).

3 See Lenoble and Maesschalck (2010).

4 See Maesschalck (2008).

Section 6
Fields of Application (2): ICT's and Internet

Section 6: Fields of Application – ICTs and Internet

The idea of an ethical governance of emerging technologies can be examined also on the basis of some examples of Information and Communication Technologies (ICT), including the Internet. It is of no use to recall the impact of ICTs and Internet on the daily life, work, and leisure of hundreds of millions of people around the world. The emerging technology of Internet is to be counted among those that have produced a genuine historical revolution, perhaps more important than that brought about by Gutenberg and his printing machine. However, Internet is also known for giving rise to some ethical questions, for instance, those related to threats on privacy, child pornography, or terrorist organisations. These ethical issues require a certain framing not only of the questions themselves, but also of the governance devices that allow for asking those questions in an open relevant way.

One remark that can be made about Internet is the importance of the *architecture*, more than the technical knowledge itself, in the handling of governance issues on networks. A better understanding for choosing between different architectures gives a clearer view about constraints, opportunities, and governance orientation. The architectural choice as regards the Internet can be for instance: "What is a network?" or "What kind of problem do we focus on?" as suggested by the collective reflection on the future of Internet. It is now pretty clear, especially after Lessig's works ("Code is Law"), that one can embed some values inside the architecture, be it the values of freedom, or those of adaptation or innovation, or something less friendly. The possibility to run several architectures at the same time means that we will have to deal with several orientations leading to plenty of new governance issues. This makes things more complex and provides a less efficient way to improve the network and the networked society being governed, but perhaps it is a more sustainable option.

The Internet, like most of the information and communication technologies, is a *normative device* the characteristics of which is to make up a *computing ecosystem*. Nowadays, life in general and culture in particular are matters of technological inter-connectedness, which means that connecting computers equals producing culture. We should then consider that all of our activities, however culturally engaged, more or less relate to Information and Communication Technologies (ICTs). The Internet is a political, economic, and social device, and it can be also a *cultural device* if we view its industrial infrastructure and its coded architecture as a series of neutral and transparent means to reach ends or goals. Defining the Internet as a cultural ecosystem, and not as a device, is a way to apprehend the original pattern under which technique and culture are not just contiguous but truly continuous. It also implies that, contrary to what many people say about the dangers of the Internet, the ICTs essentially, and not accidentally reveal an atmosphere of both meaningfulness and meaninglessness. They are interrelated as we cope with or even struggle against the technical and informational constraints that affect our creativity.

The difficulty of any *ethical-normative reflexivity* upon a technological device can be usefully illustrated by the concrete example of an *e-government system development*. It can happen in such a project that between all the systems studied, the common denominators are things like extended delivery times, increased costs and non-functional solutions. On the technical side, testing, prototyping and analysing the systems only concentrates on the fields of *functionality*. On the ethical side, the only discussion is between a minority of specialists and civil rights activists and is mostly ignored in the procurement of the systems. The inability to handle ethical questions in relation to ICTs can be traced to a wider set of problems: partial or complete inability to handle problems related to technology. Therefore, blind faith in the capability of technology to solve every problem is mixed with ignorance or lack of knowledge about the problems technology brings forth itself. In the worst case, it is presumed that a developing technology is solving the problems fundamental to the technology itself. If one does not understand the descriptive reality, one cannot understand the *normative reality* either—thus understanding the ethical problems in the first place is impossible. Therefore, before designing or implementing technological solutions one must first acquire adequate technological understanding before one is capable of ethical reflexivity towards the technology at hand.

Chapter 15
Impact of Architecture on Governance:
Ipv6 and Internet Post-IP

Jean-Michel Cornu
Next Generation Internet Foundation (Fing), France

ABSTRACT

This paper presents the basics about network architecture and some of the current proposals for the future of the internet. There are two keys factors to understand the ongoing discussions: the definition of what is a Network usually depends of the industry you are coming from, while all of these kinds of networks are needed. The second key deals with two different kinds of values: the value of scarcity and the value of abundance: efficiency versus adaptability. This leads to new technology such as Cognitive Radio.

INTRODUCTION

To handle governance issues on networks, technical knowledge is less important than architecture. A better understanding for choosing between different architectures gives a clearer view about constraints, opportunities and governance orientation. Hopefully, this means that you do not have to be a technical expert to deal with governance. This paper presents the basics about network architecture and some of the current proposals for the future of the internet.

DOI: 10.4018/978-1-4666-3670-5.ch015

1ST ARCHITECTURAL CHOICE: WHAT IS A NETWORK?

When you ask an expert the question "what is a network?" the amount of various different replies that you may get is amazing. According to their culture and history, everybody gives a very different explanation.

Which Industry do you Come From?

Someone who comes from the media industry may see a network as one emitter broadcasting information to several receivers. A radio or TV network is based on this kind of architecture. This is also the case for the architecture of a meeting, when a speaker presents a speech to an audience. But ask the same question to someone from a telecom operator and he may answer that a network is a way to connect one point to another. After the connection, the two terminals can use the whole new route, and exchange information. This kind of architecture is the one used for the telephone network, but also the one you use, during the break at the meeting, when you call someone and say: "hey john, I've got something for you". In this case, you ask John to "unconnect" any other discussion and to pay full attention to you. The third way to describe a network is the one given to us by the Information Technology industry. Everyone can send info to anyone. If you use this architecture, you don't have to connect before having a conversation with someone, but each time you need to state to whom the information is being addressed. This is the kind of architecture the Internet has – at least at its early days. Each packet of information has a source address and a destination address, just like in the postal service. Getting back to our meeting example, this is rather like a working group: everyone sits round the table. If someone speaks to one of the participants, anyone can listen in, and get what he needs. This is also the architecture of roads. All cars (similar to packets of information given in the previous example) use the same road. You don't need to block the whole road for just one car, but this also means that if there are too many cars, there will be a traffic jam.

There is a fourth architecture for networks that can be represented by the Electronic industry: any terminal can send an alert to a supervisor. Home automation works this way: aggregating information from everywhere to a central point. In our meeting example, this would mean that various people report to the boss on what they have been doing recently and the result (See Figure 1).

"Intelligence is at the Edge of the Network"

There are four ways then to make something communicate with another (a technical terminal, a human, someone at a meeting or... a protein in a cell):

1.   One broadcaster sends information to everyone.
2.   First of all, the network establishes a connection between two terminal points.
3.   Everybody can communicate with each other, by simply indicating the destination each time.
4.   Any terminal point can send information to a centralized destination.

In most cases, the central network is just a neutral infrastructure. Only the second example – the telecom architecture – gives greater importance to the network by blocking and unblocking routes to reorganize itself and give a full path to connected terminals. This is why Lawrence Lessig says, "Intelligence is at the edge of the network".

The Push and Pull Strategies

There is one further aspect. In all of theses architectures, the information may be proposed by the sender (the right term is "push") or asked by the destination ("pull"). For example if a speaker makes a speech to an audience, we have two possibilities: the speaker gives the information he planned to provide (push) or answers a specific question from the audience (pull). Of course, one speech could use both ways. Furthermore, a meeting may have formal presentations, reports from participants, feedback times and breaks.

Figure 1.Four industries, four ways of understanding what a network is

Media industry

Telecommunication industry

ICT industry

Electronic Industry

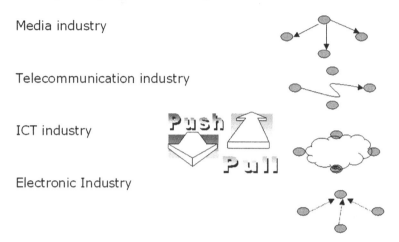

Choose the Right Strategy, Rather Than Always Using the Same One

The Internet Network, while based on "connection-less" and routing of information (i.e. Information Technologies like network), may use connected mode (i.e. telecom style network), especially in core networks where optimization is more important than adaptability. But whatever the technical architecture of internet is, the application running at the top of the network can use any kind of architecture. For example, the Web is broadcasting information from a server using the pull architecture (the user sends a request and the server answers this request), while live radio on the internet is a push-broadcasting architecture. Forums are like an Information Technology network –everyone can see what you write- with a pull orientation -you have to go on the forum to see what others have written. Mails may be seen as a push and connected architecture (the mail, whatever its size and the number of technical packets of information in it,, is sent as a whole to one of few persons). However, if you use an email list, the architecture becomes very different. It is much more like an Information Technology network: everyone on the list can read what you say. This means that an email list is much closer to a forum than to a private mail, but with a push architecture rather than a pull one: you don't need to go to the forum server, you get all the information directly pushed into your mailbox. It is interesting to note that while you may use the same application –a mail program- you can have at least two very different kinds of architecture: direct mail or a mailing list. These two architectures lead to different orientations: private mail or public expression. Most of the difficulties arise from the confusion between different architectures using the same network or application.

If we cannot see that there are various kinds of application architectures in the same network, there is a danger that we reduce the possibilities. Multiplying different architectures means multiplicity of economic models, of operational models, of governance models.

2ND ARCHITECTURAL CHOICE: WHAT KIND OF PROBLEM DO WE FOCUS ON?

Handling the Problem of Scarcity

Many of governance issues focus on problems of scarcity, mostly scarcity of resources. If bandwidths for a network are scarce, we have to decide which kind of application and which customer gets priority (this is called "Quality of Service"). If frequencies for wireless electromagnetic networks –such as radio, TV or wireless computer networks are scarce - we settle a regulation for the spectrum. Normally, we optimize managing scarcity by making predictions. We even need to predict the frequency of possible future network breakdowns so that we can sign a contract with penalties if there are more problems that planned. The "five nine" strategy for the telecom operators means that there should not be less than 99,999% availability for the network.

An Unforeseen Accident Just Before the Olympic Games…

Predicting scarcity however could add another problem. Many domains are hard to predict, some of them even fundamentally unforeseeable. We don't know in advance, for example, which submarine cable for transatlantic network will break? We try to solve this problem by having several cables so that we can switch from one path to another.

In September 2000, a few weeks before the Olympic Games in Sidney, the only Australian transcontinental cable, called Sea Me We" was broken a hundred kilometer off Singapore. This caused dramatic reduction for all transmissions planned for the games: most of the telephone and internet was slowed down by 60%. The problem was eventually solved, and the Olympic Games were saved. What a shame, when a second cable, called "Southern Cross," linking Australia to the USA was almost ready to be switched on but was not yet open for the games (it opened on November 15th, few weeks after the end of the games). Since then, there are about a hundred submarine cables breaks a year for various unforeseeable reasons such as micro-earthquakes or because fishing boat anchors catch them. These incidences are not spectacular enough though to be heard about in

the newspapers. We solved an unforeseeable problem with abundance: In this case, an abundance of possible routes for the traffic of information.

Internet and the Strategy of Abundance

The internet is very much oriented on having abundance to solve unforeseeable problems. It uses routing in a connection-less network (sending independent packets of information with a destination address so that they may be rerouted on one of the numerous other paths if there is a problem with one of them) rather than commutation (a route is established once and for all before sending information during the connection). While commutation is more efficient –when the route is established, there is no need to calculate a new one for each packet of information- it is less versatile than routing to any unforeseen situation– because all packets in a single connection go through the same path and are blocked if it is broken somewhere. The internet uses a "best effort" strategy: "I can't predict the rate of the traffic flow, but I will do my best…" This means that telephone or radio streams in the internet can't be sure that there will be no traffic jams – causing breaks in the speech or music stream- but we assume this will not last long most of the time. This kind of strategy leads to cheaper technologies, which are less efficient but can be adapted to any unforeseen situation.

Cognitive Radio and the Spectrum Becomes Abundant

A very good example of this kind of strategy by abundance is the recent so called "cognitive radio". If there is limited bandwidth for the electromagnetic spectrum we use for radio, TV and numerous other applications, there are two ways out. The first one, the currently most developed, reserves part of the bandwidth for a specific applications to make sure that the full domain is available (we can predict the rate of the traffic flow because there is only one radio, TV, telephone call or computer network on a specific frequency range. Most of the time, however, several of the frequencies that has been reserved is not used. It is just reserved just in case we need room for this application. Cognitive Radio uses a different option: first of all the transmitter listens to various frequencies. It then finds an available frequency and start transmitting on it with additional information so that the receiver can find it. If the usual occupant of this space starts transmitting, the cognitive radio transmitter just stops and drops to another place. With this kind of approach, what was known as "Spectrum scarcity" becomes "Spectrum abundance": there is plenty of room for many more wireless applications.

Predicting to Manage Scarcity or Abundance to Manage the Unforeseeable: The Janus Syndrome

Therefore, we have two different strategies (See Figure 2). The first one deals with scarcity. We reserve to make sure that we have the resources we need. We can then predict the performance of our system. This strategy focuses on performance and efficiency. The second strategy deals with the unforeseeable. We settle on abundance of choices to make sure that we reach a solution, whatever happens. This strategy focuses on adaptability.

These two strategies are not as independent as they appear: the solution used for one of them is the opposite of the problem we try to solve for the other. The first strategy tries to predict to manage scarcity, but we also encounter unpredictability. The second strategy uses abundance to manage the unforeseeable, but sometimes we get problems of scarcity.

Because we are not Janus, a Greek god with two faces looking in two different directions, we often solve this problem by forgetting one of the issues. This is a cognitive limitation of the way we use our intelligence (Cornu, 2011). Normally, we focus on scarcity, a well known problem since the birth of mankind. Sometimes, we even replace abundance by scarcity so that we have only to deal with the duo scarcity/prediction. This is the way we usually manage economy: better to burn tomatoes than having too many which may kill the offer/demand mechanism of price.

Where does the Value Come From?

We normally think that value comes from scarcity: the value of a specific raw material depends on its availability or shortage; any news has more value the fewer people know about it.

This is true, but the problem is that, in some ways, the opposite is also true: the value of a telephone or computer network depends on the amount of members. The more members a network has, the more I want to join it. If I am the only person in the world to have a telephone, I can do very little with it. This is known as the Metcalfe law: "the value of a telecommunications network is proportional to the square of the number of connected users of the system".

This is why when we talk about networks; there is a trend for concentration. The biggest, the best; "The winner takes all". We have seen this in the last few months, by the social network called Facebook, which has become the leader. Because there are so many more people on Facebook than on any other social network, you have to subscribe to it or you are out. This means that for a network, the value is on abundance, while for the rest, it is on scarcity.

Figure 2. Efficiency and adaptability are two opposing strategies

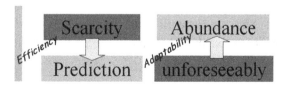

Likewise, the value may be placed on efficiency of industrial productivity (how we manage scarcity and then the price of raw materials, workers and funding), or on innovation, which is by nature unforeseeable.

Do We Need to Manage Scarcity or Abundance?

Which strategy should we choose? Scarcity or abundance? Prediction or the unforeseeable? Sometimes the answer is easy: when there is a shortage of resources but the situation is predictable we plan to make an efficient choice "a-priori". On the contrary when the situation is unpredictable but there is an abundance of possibilities we can adapt to whatever happens by making a choice "a-posteriori". Most of the time though, the situation is not so simple.

Let's have a look at how networks work. Technical networks designed by engineers exist, but also social networks and even biological networks. In a cell for example, you don't only need to know about all the genes of a DNA molecule (the "genome"). We very recently entered the "post-genomic" era where interrelations between genes or between proteins are at least as important as the genes themselves. In such a biology, based more and more on networks, we have discovered

that adaptability is just as important as efficiency. Robert E. Ulanowicz, former Professor of Theoretical Ecology at the University of Maryland's Chesapeake Biological Laboratory, demonstrated in 2008 (Ulanowicz, 2008) that in biological systems, sustainability is in relation with the amount of efficiency and resilience (which relate to some notions of adaptability). If the system is too efficient or if it is too resilient, it is not sustainable. There is an optimum, a "window of viability" which is situated approximately at 2/3rd of resilience and 1/3rd of efficiency: we need both efficiency and resilience with a little bit more of resilience (See Figure 3).

It seems that this result is not only specific to biology, it can be applied to any network – or in other words: this is a theorem of the science of complexity. Bernard Lietaer, one of the inventors of the ECU which then became the Euro, applied these rules to design a not too efficient currency (Lietaer &

Kennedy, 2008). When we apply this rule to technical domains, it leads to something very similar to the current internet: a bit of efficiency (the best effort) with a bit more adaptability. We have the same orientation in the development of free software: many people work on the same problem, which is not completely efficient, but produces good software which can be adapted to

Figure 3. Sustainability curve mapped between the two polarities of efficiency and resilience. Nature selects not for maximum efficiency, but for an optimal balance between these two requirements (Lietaer, Ulanowicz, &Goerner, 2009)

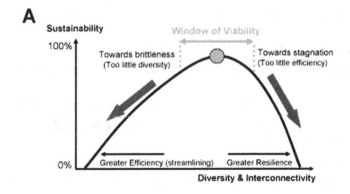

many different situations. As Eric Raymond says, the classical proprietary software is almost based on a "cathedral model" while the free software approach is based on the "bazaar model" (Raymond, 1999) (we should add "with a bit of cathedral model" to gain sustainability) (Cornu, 2001).

The Internet, a Sustainable Network?

The design of the current internet network is relatively old. The first experimentation started in 1969 with two, and then four computers connected. Forty years later there are 1.5 billions devices connected to the Internet and the first basic model is still running with very few adaptations. The current version of the internet protocol, Ipv4, was designed in 1981 (Internet Protocol, 2011) and is still the most widely deployed internet layer protocol. But sustainability has its limits and while the internet remains one of the most important and sustainable success stories of technical history, it has its limits. For example, the number of addresses for this "network of networks" is limited to 4.2 billion and the reservation of the range of addresses makes the picture even worse. China for example, has less than 60 million addresses for 384 millions internet users... Even if all the users are not online at the same time, this situation can only last two or three years longer at the current distribution rate (People's Daily Online, 2009).

Indeed, the current internet needs to evolve. Some people want to make the most of this opportunity to enhance the network, to make it even more efficient... But will the evolving internet have as much sustainability as the previous one?

Understanding the Present before Discussing the Future

The current internet situation is no longer the ideal vision of the network when it started. Several adaptations push the architecture towards more efficiency for some and towards more adaptability for others. For example, the number of optical fibers all around the world has dramatically increased including in redundant places.

Backbones of the Internet are Resilient

One of the most recent kinds of optical fiber used to carry internet traffic, called OC 3072, can transmit up to 160 Gbps (Gigabits per second): a hundred and sixty billions of elementary information in just one second! Even more, this is only for a specific color of the light used inside the fiber. It is even possible to use the same optical fiber with several beams, each of them using a different color. In we can carry up to 1022 colors beams using this technology called DWDM. Mix these two up-to-date technologies and you get, in theory, a single optical fiber which can carry more then a thousand times 160 Gbps. This is slightly more than the whole of 2008 internet traffic which was estimated at one hundred and twenty height thousand billions bits per second. Imagine a motorway big enough to carry the entire world's car traffic at one given time. In reality, and luckily this is unnecessary, because cars are distributed around different places on the planet. But internet motorways can almost carry internet's entire traffic ... So there are plenty of optical fibers (and also some satellite liaisons) which transmit the internet across long distances, and only one of these fibers (in theory, if it was absolutely up to date) could carry the whole internet traffic by itself. Such a non optimized, redundant network can adapt to the variety of sources and destinations for the information and to any unforeseen accident the information finds on its road. Long distance liaisons are very redundant and resilient most of the time (Cornu & Méadel, 2006).

Core Networks of the Internet are Efficient

Some other parts of the internet are called "core networks". These are where operators route most of the traffic to their own networks and also to other operator's networks, like a kind of marshalling yard with a huge traffic. This is why the core networks were optimized. They usually use commutation –a connection to establish a route at the beginning and then all packets of information from the same source and same destination use the same path. This part of the internet is usually efficient… and less redundant and adaptable.

We may have created a more efficient network with much fewer intercontinental cables (which could cause problems with unforeseen submarine earthquakes) or a more resilient network with routing of information not only aggregated in a few places but disseminated all around the network (but this would have been harder to manage). Theses decisions have been made, not because of a global view of the right balance between efficiency and resilience, but because of an unstable equilibrium between different cultures of the network, each pulling its own way, like in a Tug of War game.

SPECIFICATIONS FOR THE FUTURE OF INTERNET

We now have a basic understanding of network architectures. You don't have to be an Engineer with years of study behind you to get a good idea of the most important questions. Whilst technical aspects need a lot of background, a global view of the situation and a modicum of common sense is almost enough to understand architectural questions.

Can we use what we know to talk about internet governance and internet evolution? We already know that some aspects of the internet, have already got great sustainability, yet they must be modified in the short term (mainly the number of internet addresses). We could also use this opportunity to improve internet. But what does a "better internet" means? A more efficient or a more flexible network? From what we said before,

let's assume that we need both to have a really sustainable network. With these points in mind, let's try to understand the current propositions.

If we want the internet to evolve, we should also understand what the requirements are for the new coming network of networks. This is not such an easy task: In the past, internet was used mostly by humans to exchange information, but today, it is used in a lot more different ways. We can group them into three main categories.

Humans First!

The first sort of users are humans. Since the time when only four machines were connected to the network, there has been a huge evolution. Around one and half billion terminals are connected to the internet, and Internet has an impact on almost everyone on the planet. The rain season weather forecast in Africa is prepared across all Meteorological Organizations on the Continent, and coordinated by the African Center of Meteorological Application for Development (ACMAD). The internet is of most importance for such a distributed task. The results are sent to radio stations all around the continent. The final link of this information chain may use a "hand crank radio". Its user may not be directly connected to the internet, but the internet is an essential part for getting this vital information to him. Such a user is not usually counted as an internet user in statistics. I must admit that, while I'm counted as an internet user, the internet is not as important to me as it is for this user, for whom it is vital to know the long term weather forecast in order to cultivate his field.

The new internet is a real global network. But humans also want a huge diversity of applications running on it: Not only raw information exchange, but also audio (for Web radios and telephone), video… Multimedia is clearly a must for the future network. Another one is mobility. I no longer only connect to internet from computer placed on a single desk. Another requirement is that the network can support future innovative applications that do not yet exist... The capacity of adaptating, and making room for a huge amount of innovations that did not exist when the internet was created, is one of the reasons for its huge success.

There are More Objects than Humans

In 2002, the number of micro-controllers, these chips which make objects smarts, became greater than the number of human on this planet. In 2008, the number of RFID tags, another kind of chip which gives some kinds of communication abilities to objects, became also greater than the whole human population.. Today we are currently designing smart and communicating objects. There are already 1.5 billion of them. This is already the same number as humans with an internet connection at home or at work. In a few years, there will be much smarter, communicating objects than humans. But objects do not have the same requirements as humans. There are usually less verbose than humans, an SMS-length message is usually enough for them to communicate.. However, they do need to be connected by wireless almost everywhere (known as "ubiquitous computing") and they should consume the least power as possible, enabling them to carry a very cheap and light battery that provides energy for there entire life (usually about 2 years).

Up until now, internet was mainly a human to human network (even if humans also use terminals and servers). There is a tendency today though for more and more human to object traffic and soon there will be huge amounts of object to object exchange.

Strategic Applications have to be Efficient and Secure

For most human and object applications, we can choose to be a bit less efficient so that we can gain adaptability or, even more so, find the right balance between the two gaining sustainability. But some crucial applications need a real guarantee that they will indeed work as expected, even with a contract bond. This is the case for example when we need top level security like banking or military applications.

Usually, a fair level of security is OK for most of the applications and we can lower the level of efficiency slightly to gain adaptability and sustainability. In very few crucial cases however, this is not enough and we should focus only on efficiency to gain as much as possible. We may have the same requirements for some crucial applications which need a Quality of Service guarantee (this means that we must ensure that we will have enough bandwidth without it being blocked by a network traffic jam). For example, in recent experiences of overseas surgery: the surgeon being in a country while the patient is in another. We most certainly need to guarantee that the operation is not interrupted at anytime due to lack of bandwidth...

Usually these crucial applications use different networks than the internet: banking networks, military networks or telemedicine networks. It is interesting to note that most of the time, these kinds of network use mainly internet protocols – avoiding reinventing the wheel- but on a different physical network to be able to guarantee security and quality of service.

More and more industrials that use networks are prepared to pay for such guarantees. They want to get a contract with the Operator, called "Service Level Agreement" (SLA) to an appropriate level of security and Quality of Service. They need to understand that more efficiency means less adaptability. In addition, if guaranteeing a service on one network (and so one operator) is possible, it is much more difficult to guarantee something in a network of networks, like the internet. We are left with two options: either we open the whole internet to the possibility of guarantees –and we loose adaptability and sustainability for everyone- or we allow, as we already do now, two different kinds of networks. In this case, candidates for crucial networks have to choose which application runs on crucial networks and which one on the more adaptable and sustainable global network.

Three Needs for One Network

If we want not only to solve the current internet problem, but also get a better network, we have to keep in mind three different sets of requirements to make sure we don't loose what we have already got, whilst trying to handle new requirements.

The first two sets are not fundamentally incompatibles to each other: humans need more bandwidth, whereas objects consume less power, and both require ubiquity and mobility access. This is a complex issue to merge bandwidth and low power requirements, but at least this can be done by having two complementary kinds of access networks: one low power for objects (like the ones using the Zigbee standard for example) and another one requiring more energy, but with a greater bandwidth capacity (like Wi-Fi, 3G, WiMax...). Because the internet is a "network of networks", it is easy then to interconnect these two kinds of networks via the same core network (which does not have the same low energy consumption requirements than networks which directly access to objects). In both cases, we almost certainly do not want to loose what we already got with the first internet: a sustainable network which adapts to most of the innovative ideas users may have.

The third set of requirements however, the one for crucial applications, paves less importance to adaptability. Efficiency is what they need, even with a contractual guarantee. This kind of network is fundamentally different, not so much in terms of technical choice (even if this can be the case) but in terms of architecture. The possibility of pooling at least some elements designed or implemented for the two kinds of networks has to be considered... without loosing human and object network requirements for the sole benefit of crucial networks.

Three Candidates for the Future of Internet

Among all the proposals for the evolution (or the change) of the internet, let's look at three very different approaches: a next version for the internet protocol, "Ipv6"; a new "Next Generation Network" and a very different approach from the Pouzin Society. Because we now have a better understanding of architectural issues, we hope to avoid simplified answers to such an important evolution of the internet, one of our main supports of our capacity to innovate.

What are the Current Limitations of the Internet?

The current version of the main protocol used in the internet is called Ipv4 (Internet Protocol version 4). This version has two main limits.

There are only 4 billion addresses which means that distribution of these addresses to the 1.5 billions current access points to the internet (computers, internet phones and already some objects), is not optimized and many of them have no fixed addresses: either the internet access provider gives you an address only when you are connected and then frees that address for other users when you are off ; or a company with many computers has only one or few addresses and then uses internal addresses for all of its terminals (this uses a mechanism called "Network Address Translation" - NAT). This means that while you can access most of the servers by using a domain name which can be converted into a fixed Internet Protocol address (IP address), you may not have direct access to most of terminals and objects because the address they currently have is not fixed and may change from time to time. This is very different from the telephone network where all telephone devices have a phone number. To manage this difficulty, we usually add some addresses at the application level. Some instant messaging applications such as Skype or MSN messenger, for example, give you a specific address when you subscribe, your friends can then know when you are on-line and can contact you. But each application has its own address space and it is not so easy to make connection between say Facebook, Skype and your Google agenda. You must first inform one application of the addresses you have in the others. This means that even if all these applications may be of great interest to you, they run as islands, not very well connected to each other. It would be great to have a basic address in the future internet so that anyone can call anyone directly, independent of the application he uses. This means that there could be more addresses than the number of terminals or object connected because you may want to connect to a specific terminal/object or maybe to a specific user whatever the terminal he is using at a certain time…

The second limitation of Ipv4 is on useful extensions such as security, mobility, quality of service or multicast (the last one is the capacity to send one message or radio/TV stream and be receive by as many users who may need it. In the "normal internet" you have to send a message or a stream to each of the receivers..). These extensions are only… extensions. This means that there are not in the core set of protocols implemented everywhere in the internet. One can decide to implement for example security extensions with specific options. But when you want to have a really safe transmission end-to-end you need to have the same level of security proposed by all the operators of all networks you are passing through. Remember: the internet is not a single network, but a network of networks, run by a large variety of operators. Even more the fact that there are several extensions to the basic internet protocol, like as many warts, means that it is rather difficult to mix these extensions and have, for example, a mobile multicast internet.

IPV6, THE NEXT VERSION OF THE INTERNET?

After the version 4, the next version planned for the internet is… version 6 (the name Ipv5 is used for a specific protocol dedicated to streaming also called ST2) (Internet Stream Protocol, 2011). This version defined by IETF, the consortium in charge of designing the different protocols of the internet, and proposes three main changes.

First of all, the current number of available addresses is 340 282 366 920 938 463 463 374 607 431 768 211 456! This is much much more than the number of grains of sand on the earth or stars in the sky. Even more helpful is the fact that there are enough addresses for any present and future terminals, there is also the possibility of having "disposable addresses." This may be very useful when, for security reasons for example, you don't want a server or another user to know the address of the terminal. After a transaction, the address will become useless and cannot be used for spam for example.

The second change of Ipv6 is that extensions become options of the core protocol Ipv6. This means that you can activate as many options as you want (security, mobility…) and be sure that the various networks you pass through are able to manage them. There is a third main change, which allows auto-configuration of the network. At the moment, it is not so easy to define the kind of address you use and the various parameters of the network. Reconfigurating not only your computer, your mobile phone, TV set, and any other appliances that you may have could be a headache if in the future, you need to do that. Auto-configuration solves this problem; you no longer need to worry about this kind of thing (of course, if you are a Company Network Administrator, you can still finely tune the configuration of your network, and Ipv6 allows this too).

With Ipv6, there is evolution but no major change of the architecture. The idea is to keep the fine balance between efficiency and adaptability.

Of course we could encounter some difficulties with some objects which need less verbose protocols and therefore much less power consuming. This can be solved though, as we already said, by using a low energy protocols such as the one called Zigbee. To facilitate the link with other networks, it is possible to use the same addresses than Ipv6 in a Zigbee network by using a specific extension called "6lowPAN".

So with Ipv6 we can give more scalability and more functionality to the current internet and also include low power objects in the net. But we leave aside crucial networks. They are now part of the industry, and ignoring their requirements for network guarantees could mean that we miss out on a whole bunch of new business...

However, deployment of Ipv6 has already started in places such as Asia or France. This protocol is already included in almost all recent operating systems and the number of users is on the rise. But the race is still on and there is a lapse of a few years to see if Ipv6 will success or will be overtaken by other alternatives.

The Next Generation Network, a Telecom like Architecture

The approach is very different with the proposal from Next Generation Network. The idea is to start almost with a blank sheet, and design an up-to-date network with enhanced scalability, quality of service and security. While some aspects of the internet protocol may be implemented at the very top (mainly the Ipv6 address space), telecom operators propose a more efficient network using connection mode – define once a path for all packets of information from one source to one destination- rather than routing – the route is defined for each packet of information depending of the current situation of the network (accidents, traffic jam...). One of the main goals of the NGN is also to have the same network for the internet and for mobiles phones.

When you get access to large files via the Web or by peer to peer protocols, all you want in the end is that the global download time is as short as possible. But when you transmit speech, like in a telephone communication, you also need a constant flow of data avoiding having some words delayed (the situation is the same for video). There are two options to ensure a constant flow of data: either you implement some mechanism of reservation (quality of service) or you give a much larger bandwidth to your network than you need (if your network is able to handle about ten times more than the average traffic, you don't need to worry about peek time...).

Both solutions have a cost. If you decide to give priority to specific traffic (like speech, but even much more of course for crucial applications), you need to add very expensive materials inside the network to sort the various traffics. The network is no longer neutral. This means that we have to add restrictions on content (if there is more priority for some traffic, there is less for some other...). If you decide to give much more bandwidth than you normally need, you have to use faster links in your network. The question seems to be about which is cheapest to enhance: technical devices at the nodes of the network (like routers), or links between these nodes (like optical fibers). But maybe the real question between the different approaches of Ipv6 and NGN is the level of control you want to have on your network. If you have different levels of traffic, you can charge more for some privileged and richer clients.

The Pouzin Society Goes Back to the Early Days

John Day, one of the designers of Arpanet, the network which led to the internet, says (Day, 2008) that "we were right on too many points from the beginning." This means that with these ideas, it was possible to have a "good enough" network for many years. But now with the number of users that the network has (humans and objects) and the growth in traffic, we need to go further, to have a better network than "good enough." John Day created the "Pouzin society" to work on this idea. The name comes from Louis Pouzin, a French inventor who initially proposed the notion of packets of information (called "datagramme") and connection less communication.

To help the network get better scalability, he proposed to start with an idea from Bob Metcalfe, the inventor of Ethernet, a local area network now used for metropolitan and even long distance networks (Ethernet is different from the internet and works at a different level, while it is possible to use both protocols in the same network). Metcalfe said in 1972: "the network is Inter-Process Communication and only that." Inter-Process Communication (IPC) is a set of processes to exchange data and synchronize running programs. Using the network not only for exchanging raw data but exchanging data between programs running on different machines is "a fundamental theoretical breakthrough" as says Brough Turner; Founder and CTO at Ashtonbrooke and Chief Strategy Officer at Dialogic (Turner, 2009).

In fact, with IPC, data are directly embedded in a process between programs. This means they are no longer directly accessible. This leads to better possible security (of course depending on the level of security the program itself have). In addition, we can choose a connection (like in NGN) or connection-less (like in Ipv4 and Ipv6) communication or any intermediate situation depending on the program we use. A third advantage is that we can distinguish the name of a machine or the name of a process on this machine (which is used to identify the machine or the process) from the address of the machine (which is used to localize it). We then no longer need a global address space as we can use global names with addresses defined for a specific process. This solves the problem of the number of addresses (that Ipv6 and NGN try to fix with the huge Ipv6 address space), and also solves a lot of issues on

mobility where a specific device may move from its original place.

The Pouzin society approach is very different from the usual telecom-information technology communication dilemma. But it is based on IPC mechanism, widely used in local area network. The new idea is to use it not only at application level but as a base for network communication itself.

This leads to a much simpler architecture –as there is only one protocol: IPC- but very different from what network specialists are used to working with.

How Not to Choose... Virtual Routers

There are then three main candidates for the future of the internet:

- Ipv6, an Information and Communication Technology style architecture which try to keep the balance between efficiency and adaptability from the current internet, while having better scalability with a much larger address space.
- NGN, a Telecom industry style architecture which try, in addition to better scalability, to include mobile phone networks (and then telephone requirements such as quality of service) as well as better security. It includes different levels of services and prices.
- The Pouzin Society proposal, a programming industry approach based on distributed programming architecture (IPC). A new paradigm shift that would be interesting to evaluate in terms of efficiency, adaptability and then of sustainability.

Different Networks on the Same Physical Network

Another question is: should we choose between one of these three options (or even some others...)

or should be keep all these good ideas and have several networks? A first thought is that communicating across these different networks may cause a lot of difficulties. We already know this situation with a telephone network, a television network and the internet network. We don't want to entirely separate these networks so that we can no longer mix them: watching a TV program whilst looking up further info on what is said in this program, and then discussing this with friends by chat, audio or video… Another problem is that it is much more expensive to have three physical networks than only one!

Something new happened though quite recently: "virtual routers." Imagine a railway network shared by several train networks: On one particular day the trains on the railway network offer as many sources and destinations as possible. When too many trains run at the same time, there is a traffic jam, but the company does its "best effort" to transport as many people as possible to all the places they want to get to in the shortest time possible. The next day, the railways are used for fewer people who have reserved to make sure that they arrive on time. And they are prepared to pay for. On another day the railway is used for dangerous goods transported by freight train which should not be mixed with passengers trains. Yet another day the railway is used to test new ideas to optimize protocols between trains. There are no humans, so even if the protocols fail, nobody is hurt. If the tests are successful however, this new approach can be marketed. Of course this means that you can only travel one or two days out of four .With information networks, days could be replaced by hundreds of milliseconds. This means that you don't have to wait that long.

Because optical fibers, which are the basis of a large part of the network, have so much free room (remember, in only one very up-to-date optical fiber you can transmit almost the equivalent of internet's daily world traffic); and because processors used to organize the networks –such as routers- are so powerful (a single audio happy

birthday card has more computing power than the whole world's computers in the 1950s) (Tapscott, 1996). We can use the same physical network to run several different networks.

With virtual routers, you can choose to have several networks not communicating with other (e.g. when you have very crucial applications and you don't want other networks to access your own network, like in banking for example). With virtual routers, you can also choose to create a network of different networks, each of them having specific requirements (some with different levels of quality of services or of security, some others less expensive based on "best effort"). You can also create a specific network to test new protocols or new versions of a protocol on a real scale without disrupting other running networks. Having several networks on the same physical network may be useful for an operator: if its network is under attack, it can switch all its clients to a new network, so that the cracker finds himself alone.

Rather Pool than Choose…

Last but not least, virtual routers are a way to pool an expansive physical network between several operators. This is close to the idea of neutral network: in a place where no operator wants to establish a broadband access network for economical reasons, local authorities can enter the game. Let's take a real example (Cornu & Méadel, 2006). Tierp is a city with 20000 inhabitants situated 150 km North from Stockholm in Sweden. Too far and too small to interest Internet Service Providers (ISP) in setting up a broadband network. In 2000, the municipality built its own broadband network between the various public places of the town. In a few years, the network was paid-off with just the economy made on the telephone bill between the public premises. The municipality then decided to extend the network to the rest of the city and to add an optical fiber between Tierp and Stockholm. Rather than asking one operator to run the part of the network dedicated to citizens, it decided that the network should stay "neutral". Tierp then proposed that several Internet Service Providers sell services to its inhabitants: even avoiding the necessity to set-up a new network, not even installing anything in Tierp – because of the optical fiber between Tierp and Stockholm, ISP can now manage their services directly from the capital city. Operators only had to pay a fee to gain access to a 20000 customer wide market without investment. Since then, the habitants of Tierp have access to broadband services at the same price as habitants of Stockholm. ISP got a new market and the city of Tierp continues to save money in its telecom budget. The idea was just to split physical network and service operators. The "physical network" operator has to be "neutral" towards all service operators and has no right to sell itself any service. There are many of these "neutral networks" around the world such as in Italy or France.

Today, with virtual routers having this kind of win-win approach, architecture is even simpler. Rather than having service operators pooling the same physical network (this means that they do not manage any network, just services). This limits the kind of service they can provide because some of them rely directly on the network), we have now a new picture: Network operators can pool... the same physical network. This may help public investment in network infrastructure not only to be sold to one operator but rather facilitates the emergence of a local business with several operators having various approaches.

GOVERNING DIVERSITY

"Code is law" says Lawrence Lessig, an American academic and political activist (Lessig, 2000). He explains that we can embed some values inside the architecture. One of the values inside the current internet, he says, is freedom. Another one is adaptability and capacity of innovation. Changing the architecture to improve the internet may give us more... or less value. This is why it is so important to understand the basics of architecture.

Having the possibility to run several architectures at the same time means we will have to deal with several orientations leading to plenty of new governance issues. This makes things more complex. This is certainly not, the most efficient way to improve the network and the networked society being governed. But perhaps it is a more sustainable one.

REFERENCES

Cornu, J. M. (2001). *La coopération nouvelles approaches*. Retrieved November 20, 2001, from http://www.cornu.eu.org/texts/cooperation

Cornu, J. M. (2011). Nous avons non pas un mais deux modes de pensée. Retrieved November 20, 2011, from http://www.cornu.eu.org/news/nous-avons-non-pas-un-mais-deux-modes-de-pensee

Cornu, J. M., & Méadel, C. (2006). Les accords de peering ou comment le Sud finance le Nord. Retrieved November 18, 2006, from http://www.voxinternet.org/spip.php?article100

Day, J. (Ed.). (2008). *Patterns in network architecture: A return to fundamentals*. New Jersey, USA: Prentice Hall Press.

Internet Protocol. (2011). *Darpa internet program, protocol specification, RFC 791*. Retrieved November 20, 2011, from http://tools.ietf.org/html/rfc791

Lessig, L. (Ed.). (2000). *Code and others laws of cyberspace*. New York, USA: Basic Books Press.

Lietaer, B., & Kennedy, M. (Eds.). (2008). *Monnaies régionales: De nouvelles voies vers une prospérité durable*. Paris: Charles Leopold Mayer Press.

Lietaer, B., Ulanowicz, R., & Goerner, S. (2009). Options for managing a systemic bank crisis. *S.A.P.I.EN.S, 2.1*. Retrieved April 15, 2009, from http://sapiens.revues.org/index747.html

People's Daily Online. (2009). *China's IP address will be used up in 2 years*. Retrieved August 24, 2009, from http://english.peopledaily.com.cn/90001/90781/6737117.html

Raymond, E. S. (Ed.). (1999). The cathedral and the bazaar: Musings on linux and open source by an accidental revolutionary. Sebastopol, CA: Tim O'Reilly Press.

Tapscott, D. (Ed.). (1996). *The digital economy: Promise and peril in the age of networked intelligence*. New York, USA: Mc Graw-Hill Press.

Turner, B. (2009). *Pouzin society - organizational meeting, today*. Retrieved May 04, 2009, from http://blogs.broughturner.com/2009/05/pouzin-society-organizational-meeting.html

Ulanowicz, R. E. (Ed.). (2008). *A third window/ natural foundations for life*. New York, USA: Oxford University Press.

Chapter 16

The Internet as a Normative Device for Cultural Practices

Paul Mathias
Collège International de Philosophie, France

ABSTRACT

Nowadays, life and culture are matters of technological interconnectedness. This is not to say that connecting computers equals to creating a cultural environment and we might just want to consider their day to day technical efficiency and productivity. But then, how can we assess a cultural, not just practical dimension of the Internet? The question of a cultural dimension of the Internet is not that of its usefulness. Though technical craftsmanship might be a requirement for an authentic cultural networked experience, instrumentalism, as a theoretical approach, connotes exclusively the necessity of controlling the flow of data and information.

This paper aims to show that the Internet is not just a technical pattern and a normative framework aimed at producing and enacting otherwise defined cultural experiences. Information and communication technologies are complex programmed and interrelated functions eventually realised in the form of a cultural ecosystem – the Internet as such – and they do not simply form a web of industrially organised devices and services. It is then wrong to say that networks crush our minds and that nowadays acculturation prevails. The Internet is a cultural matrix generating new forms of meaningful interactions through the permanent and pervasive interconnectedness it allows for.

INTRODUCTION

We cannot anymore lead our contemporary life without utilising a rather wide range of technological devices. For professional purposes or private matters, we need to bear mobile phones,

personal digital assistants (PDA's), mini or portable computers, as well as any correlated paraphernalia such as USB keys, iPods, iPads and the like. All of these are aimed at doing something specific; many of them can do many naturally unrelated things *at the same time*: write text while drawing images, playing music, and allowing for communications or database mining. Which is exactly what a computer actually does, being to some

DOI: 10.4018/978-1-4666-3670-5.ch016

extent similar to Delphian cutlery as described by Aristotle (4th c. BCE) in his *Politics*. Designing and building electronic devices, programming them and assembling them into swarms of machines make up the computing *ecosystem*. In two senses: first, because programs themselves need to be interconnected to allow for the diversity of computing usage; second, because the extensive use of these machines imply their participation in the social, economical and partially political exchanges occurring in a network.

Nowadays, life in general and culture in particular are matters of technological interconnectedness. This is not to say, though, that connecting computers equals to producing culture. We should rather consider that all of our activities, however culturally engaged, more or less relate to Information and Communication Technologies. Our everyday life, our academic life, our professional and private lives – they are all closely or loosely related to an industrial ecosystem of energy, industrial production and cognitive use of computing apparatus. There is scarcely talking and mating without e-mailing, no way to launch a business or conduct a project without electronically sharing information, no waking up or going to sleep without experiencing a call on a cellular phone or using computer regulated means of transportation – from cars that stop at traffic lights to trains and aeroplanes that cross countries and continents remotely guided by signals and data-crunching.

Under the circumstances, what exactly does it mean to interrogate the cultural efficiency of the Internet? Undoubtedly the "Network of networks" is an active, even living process of innumerable information exchanges, and a contemporary icon of shared intelligence, knowledge and Humanity. As such, is it an *instrument* for producing cultural items like text or images? Qualified by their *instrumentality*, devices are "meant for executing" or "destined to achieve" goals that we imagine beforehand and represent ourselves as more or less essential. They are accompanied by the glow of necessity, usefulness, pleasure, etc. But also,

"usability" is a form of "disposability." We might be unable to do away with an instrument's functions, but we certainly can replace an apparatus by another. The point to be made does not concern the particular form of a writing tool such as the Internet, it concerns its very *instrumentality* and its status within the logic of cultural goals and means. The question of the cultural dimension of the Internet is not that of its cultural usefulness or necessity. Assuredly knives and spoons are instruments for cooking; brushes and pigments are instruments for painting; paper and pencils are instruments for writing – should we in the same way consider the Internet as an instrument, even maybe a toolbox for writing, painting, and, at least, offering us communicative opportunities for sharing exotic recipes and tasting foreign food?

That the Internet is a device made for sharing through various ways of communicating is undeniable. And indeed, it *can* and it *must* be interpreted as a political device, e.g. for challenging a representative's or a government's choices and actions, also, as an economic device for exchanging valuable data and digital products like software or music and video, among many other things. It is equally a socialisation device allowing for unexpected encounters and the reshaping of social groups beyond geographical frontiers or sociological boundaries. Why not, then, consider it a cultural device? In doing so, though, we tend to consider the Internet and its industrial infrastructure, its coded architecture as well, as a series of neutral and transparent *means* to reach *ends* or goals otherwise substantiated: Truth through knowledge and science, the Good through morals and metaphysics, what is ethically relevant or sustainable, and the like. Our means for giving rise to cultural products are extremely varied. Should we consider the Internet as one particular and privileged of them? The question is: "is it really only a very complex device and a series of controllable means to a series of identifiable ends?" That it is of an utmost cultural importance is indisputable. Yet, beyond its instrumentality lies an option to think of it as an *environment* and a *world*, as an ecosystem subverting the very frame of means and ends we're used to referring to when trying to theorise its reality and usage.

INSTRUMENTALITY REFORMED

Thinking of the Internet as an *instrument* – or shaping it into the frame of instrumentality – is not only common, it is also reasonable and theoretically useful. Like any other media, the Internet can be considered as a cultural device for *producing* a wide variety of representational forms like text, images, or music, etc.; for *gathering* and *storing* them in personal or public virtual libraries and repositories; and for *selecting* as well as *passing* them *on* to others; thus for *publicising* them through free and voluntary, or paying and commercial channels. As such, the Internet comes out as a culturally efficient process for enriching peoples' cognitive environment and enhancing their aptness in appropriating and transforming their horizon according to their own more or less individual ideals and goals. This is a vision that meets the *Recommendation of the European Parliament and of the Council of 18 December 2006 on Key Competences for Lifelong Learning*[1], especially annex 4 on "Digital competence" where the importance of developing "basic skills in ITC" and the "use of computers" is stressed. From "basic competences in science and technology" to "digital competence," from multi-lingual communication competences to "social and civic competences," the main target remains the ability to "learn to learn," a major competence for self appropriating the extremely complex swarm of textual connections that occur between innumerable machines altogether creating the Internet as a whole.

The Parliament's and Council's underlined idea is that we should give ourselves the *means* to structure and harness the overwhelming flows of information we are confronted to. Technical craftsmanship is a requirement for an authentic cultural experience of the Internet. Browsing the Web, for instance, exposes us to multiple flows of differentiated data thus disparate forms of human expression ranging from the Bible to porn and from genuine science to pathetic subjective gibberish. There is nothing unreasonable or fallacious about the European recommendations and its aiming at acquiring a clearer consciousness of the global human phenomenon constituted by the networks. Yet, they do not completely and definitely account for the Internet's nature and affordances. In insisting upon the necessity for a greater and better technological mastery of the networks, the European representatives have eluded an important aspect of our actual experience of the Internet, which is its irreducibility to anticipated goals and forecasts about its economic, social and generally speaking practical usability. Indeed "the Internet" does not describe a uniform though complex instrument meant to reach various cultural targets. In the mid nineties, nobody could really predict the failure of the "push technology" implemented on sites such as *pointcast.com*,[2] neither could anyone anticipate, at the turn of the 21st century, the extreme popularity of *Napster* and, later on, P2P networks. Thus, it seems the *actual* usage of the Internet exceeds the frame of its usability. Reversely, when implying that people should mainly acquire technical and communication competences, we do *as if* the enframing into instrumentality was the only option for describing the Internet and its cultural affordances. Without any doubt, there is a practical benefit in comprehending the networks as transparent means to cultural or otherwise ends. This allows for the "architecturing" of public cultural policies and the determination of common governmental and entrepreneurial goals. Even so, there is, on the other hand, an obvious theoretical deficit in such a perspective that leads to missing essential points about "culture" and the network's supposed "technological neutrality" (or transparency). This flaw can be translated into at least two interrelated problems:

First, instrumentalism constitutes a theoretical approach to the Internet that connotes exclusively the necessity of controlling the flow of network data and information. As a manner to prepare for cautiously experiencing the intrinsic variety of the Internet, this would not only represent a safe form of conduct, but also a way of considering the factual flows of information as "non-culture" as well as culture as "non-flows" and "something more." There would be, on the one hand, the pattern of cultural cognition and, on the other, that of technological mastery. This is to say that "culture" would be defined *as such* – how is not the issue here – and that "flows of information" would only represent a digitalised technical and accidental form of being for cultural or otherwise practices. Theoretically sedimented, "culture" would belong to another form of reality than the technical means used to develop it, which would precisely be meaningless to its own meaningfulness. In this regard, for instance, the so-called "irresponsible" exchange of information through peer to peer networks would not have anything to do with "culture," only with "theft." On the contrary, being able to spontaneously distinguish between *whitehouse.gov* and *whitehouse.com* would reveal a cultural competence of some sort!

Second, saying that "the use of computers" and related competencies are essential to a culturally savvy experience of the Internet is saying that culture is a *product* that should be dealt with as any other product: in terms of storage, keeping or selling, transmitting or not – in other words and fundamentally in terms of *quantity*. Though not totally untrue, this is nevertheless a rather reductionist point of view on what "culture" might be. Digital competence in an undeniable intellectual quality, but it is not at all clear that digital *literacy* can be reduced to digital *competence*. In short, no one would for instance say that you need only know how to hold your pen and draw letters on a piece of paper to be able to compose *A Midsummer Night's Dream*! This is why "culture," digital as well as material, should be closer related to lit-

eracy than to competence. As a "cultural device," the Internet is not just a technical pattern and a normative framework aimed at producing and enacting *otherwise* defined cultural experiences.

The problem with the instrumental theoretical framework is that it tends to separate the Internet from its inherent cultural meaningfulness and reversely force it into a technological pattern. The Internet would only be an instrument. But then an instrument imposes its affordances to its related practices, allowing for no real freedom and inventiveness "from within." The actual progresses we may have witnessed in the development of the networks seem to oppose this way of thinking and to require new patterns for an interpretation of the Internet's cultural meaningfulness.

A WORLD OF CLOUDS

An operating system is a program that makes the use of a computer relevant. A machine will not do anything or help produce any content unless a set of organised instructions follow the practical actions of the user: typing on a keyboard or moving a mouse and clicking. Now, it is possible to distinguish between two sorts of operating systems. One was developed in the 80's and the 90's and has animated the *auto*-centred personal computer ever since. Quite different would be a *network*-centred operating system where the user's terminal would access distant resources and programs. Some say the difference would not be only technical but also cultural, in the sense that the architectures users are bound to be confronted to in each case would profoundly determine contrasting forms of creativity, as if there were a strong link between an operating system's properties and the way its usage ends up being configured.[3] To put it in other words, experiencing the networks requires an appropriate awareness of their affordances in relation to our own possibilities and to understand not how they function but how they act upon us and affect our perception of things and our behaviour, cultural or otherwise.

Following the European Parliament's and Council's recommendations, it seems reasonable to assess there are "potential risks" and "issues around the viability and reliability of information" on the Internet. Yet, it does not look like a good enough cause for not digging deeper into the idea of a *network culture* or a culture essentially correlated to the expressive expansion of the networks – new forms of network culture representing new forms of living, reflecting, acting in the real and off-line world. Let us consider at least two issues relevant to this point:

In *Hamlet on the Holodeck*, Janet Murray (1998) outlines the fact that under certain circumstances, when a narrator engages in building and composing a story, the computer must in a way be considered an authentic co-author of the man or woman behind the machine. Referring to computer scientist Joseph Bates and his virtual character "Lyotard the housecat," Janet Murray points out that, in a determined scenario, "Lyotard *chose* to bite the interactor from among a range of behaviours open to him at that moment." Then, she comments: "It is an important moment in human history to be able to make machines that exhibit *emergence*."[4] What happens is that the program carries generative capabilities that produce unexpected scenery and events. This may not happen in the case of writing proper – a novel, a piece of poetry or some theatrical work – but it is clearly the case in the narrative production of *games*. A video-game, for instance, is a kind of story with a scenery, characters, actions, natural or supernatural powers and aptitudes, where a player needs to interact either with computer generated "beings," or friends and enemies enacted by other players on a local or wide area network. The main issue is that the whole process of playing is centred on a "story" of which the main properties are algorithmic frames or patterns allowing for the players to interact both between one another and with computer generated characters and events.

Whatever the context, its architecture is certainly sketched out by a "real world" author's imagination, but its taking shape and its textural filling up is more and more dependent upon a computer's calculations. What Janet Murray points out is that the greater the part of the machine, the greater its "responsibility" in the building of the story or the game. To some extent, in this context, the operator of the creative and generative process is disconnected from the author. To put it in other words, the operator – the real world man or woman working on the computer – is, so to speak, "operated" by a machine that has become an independent producer and author of sorts, due to the complex and dynamic piling up of programming blocks. If we are to consider something like a "network culture," it is in the sense that the very processes of cultural productivity do not anymore rely solely on people's imagination, the will or talent of a person, but also on computer hardware and software and their taking the lead in allowing for the production of unanticipated patterns, images and stories.

In a presentation dating back to November 2003, Pr. Hisashi Muroi of Yokohama National University insisted on the theoretical consequences of the "databasi-zation of knowledge," especially when interpreting literary works.[5] Reacting to text and knowledge – reading, annotating and deploying a critical activity – is not at all the same phenomenon within and outside the world of networked computers. In what we inaccurately call "the real world," a critic's point of view is one of taking some distance, evaluating a meaningful piece of text or image, and connecting it conceptually to other pieces of the same nature or not (as we may connect a painting to music for instance). In the "virtual world" of computers – another inadequate expression – the shift does not essentially come from the quantitative increase of available data (thus permissible connections), it comes from the network's affordances and the creation of new opportunities for critical studies and reflection which would have been *impossible* without it. The networks do not allow for "more," they allow for "different" and "unexpected" views on a text, an image, a sound.[6]

Both the digital displacement of authorship and the "databasi-zation of knowledge" show that, on a network, a user is not just a "user" – an individual who voluntarily accomplishes series of operations meant to achieve a determined goal – and that the network is not just a "network" – a complex instrument or a series of interrelated tools aimed at achieving certain predetermined goals. In fact, a certain dimension of our Subjectivity is strongly referred to the network "in person," since the machine and its interconnectedness constitute an important operator of discrete and intricate determinations *causing* the effective incarnated user to act and react intellectually the way he or she does. This phenomenon of *shared subjectivity* between Man and Machine should be interpreted as a positive and theoretically challenging confusion of the technical and cultural spaces concerned with the issue. Indeed, we are not just facing another declension of τέχνη as a complex set of cultural practices. Their proximity to Mankind is rather well documented in the history of philosophy, from the Myth of Prometheus in Plato's (4th c. BCE) *Protagoras* to Bergson's (1907) theory of *homo faber* in his *Creative Evolution* and Heidegger's (1977) *Question Concerning Technology*. Following this classical trail would only require describing the peculiarities of Information and Communication Technologies as tools for developing new forms of culture, namely: role playing programs, games, animated images and, eventually, *business*. What should be emphasised, rather, is that Information and Communication Technologies, in their own digital way, have become cultural *cosmogonic* powers.

CULTURAL COSMOGONY

It is not at all sure whether or not the most interesting part of the Internet, from a philosophical standpoint, is the mere facilitation and acceleration of communication practices. This aspect of our practical empowerment is an obvious fact, but not all facts are equally interesting. Without any doubt, most of our reading and writing, most of our communication practices are about work, administration, banking or leisure, and establish a useful way to profit by the technological affordances of the networks. Now, the difference in efficiency and speed of pen and computer is not exactly relevant when considering the rhetorical properties of any literary work. On the other hand, our philosophical interest in the networks may be aroused while reflecting upon the intellectual processes of writing and reading – not at all the same phenomenon when the recipient is a banker or a reader proper! Administrative rhetoric and literature are not critically addressed the same way. Though both may be taught and learned to a certain extent, the professional logic of the former does not meet the creative constraints of the latter. And, when dealing with an author's "creative genius," the brand of his paper or pencil or ink or typewriter is not really of great importance. Of much greater importance could be, though, the shift between writing with a pen on paper and typing with a keyboard on a networked computer bearing a screen connected to thousands of other screens.

As Michael Heim (1987) has already shown in *Electric Language*, "word processor assisted" (WPA) literary writing is not free from all sorts of technical restraints. The machine does not have the same status as the pencil and the paper and does not constitute just a means to a goal radically foreign to its own materiality. On the contrary, one has to take into account the fact that a machine is also a variety of interrelated complex programs without which none of its functions would be operable. This is to say that WPA writing results not only from the use of instruments, but from the use of previous writing, previous thinking, previous *intellectual emframing of a user's writing possibilities*. Paradoxically enough, we do not write with writing tools anymore, we write with programmed and partially programming tools. To the extent that they are intellectually – not just instrumentally – determined, these are not mere tools anymore but rather *ideas* and functional *views* implemented into our own writing, imaging and creating. It is of course always possible to object that, in the past, "writing" was not the same when it was done with a stylus on clay, then with a goose feather on parchment or with a printing press on paper. True: the very mode of existence of the cultural phenomenon called "literature" differs from one writing technology to another. It remains that the way computers assist us in our creativity is not at all the same as instruments and machines in the past, for they are not just tools helping to produce cultural objects – be they books or paintings – they are intellectual processes *intrinsically* associated to the creative processes they are helping or inspiring. There is more to a machine than being just a machine: it is now a program and somebody's – or some group's or company's – view on the efficiency and appropriate affordances of the program.

Information and Communication Technologies enhance the prevalence of the technical inherent to the writing and creating processes. Since we cannot anymore represent our creative tools as tools and need to outline the intermingling of our own creative process with the processes implemented into our technological tools, the focus has now to be brought onto the displacement of the Subject within culture and onto the radical shift, prevailing in our relation to culture, that puts the communication processes employed at the forefront of our cultural achievements. The Internet is not "cyberspace," cyberspace is a handy word for referring to something we don't have to dig too deep into. The Internet and Information and Communication Technologies in general are complex programmed and interrelated functions eventually *realised* in the form of innumerable interoperating computers and users. Though not simply devices, they are indeed normative of the

cultural designs they allow for. The *world* they help creating is one of algorithms experimented as text, images, sound and film. But there is nothing such as text in itself – *an sich* – nothing like sound or images. Bits and transmissions, calculations and compilations are the only "reality" that fundamentally "exists" as the cultural space of Information and Communication Technologies. The roots to our contemporary culture are not simply our past and history, customs and habits; they are computer sciences, application designing, thus industrial constraints and socio-economic goals and growth. In our post-modern environment, "culture" denotes the immaterial fabric – digital and algorithmic – we wrap our ideas into.

Instead of being just a device or even a technical infrastructure for developing new forms of creativity, the Internet has grown to be a genuine ecosystem where various forms of *meaning* interrelate to create the cultural atmosphere within which we live more and more obviously. To describe the phenomenon in more mundane words, while in the 19th century the average French "bourgeois" would live in an ethical atmosphere approximately inspired from Plutarch's *Parallel Lives*, nowadays knowledgeable cosmopolitans live in an atmosphere of interconnected text, images and sound, an atmosphere of circulating data and interleaved machines that are not just tools, but the very texture of our post-modern culture. We are not anymore autonomous thinkers of an objective human destiny; we are existential as well as intellectual "points of access" to streams of data and, hopefully, traces or maybe trails of meaningfulness.

CULTURE TO NO END

Defining the Internet as a cultural ecosystem, not a device, has two main implications.

On the one hand, it is a way to apprehend the original pattern under which τέχνη and culture are not just contiguous but truly continuous.

Technical determinism does not really help in reflecting upon the true nature of our experience of Information and Communication Technologies. Their usability, pervasiveness and efficiency are one thing, what they stand for and their true meaning is another. One important issue that has been outlined in the course of this paper is that "the technical" now *inhabits* "the cultural" not as its instrument or virtual toolbox, but as its frame and the very condition for its own affordances. To summarise this point, we need only to remember that musical instruments are no longer required to compose and play music. Cleverly and skilfully manipulating computers is enough, that is: being able to program or use programs that generate sounds that connect to one another. The design of the program allowing for the creation of sound lies within the design of the sound created by the program. This is to say that, contrary to the musical instrument, so different in nature from the sound it produces – the sound of the violin is not made of wood, neither the sound of the oboe made of copper – the computer program is of the very same essence than the cultural product it allows for: indifferently text, sound, or image.

On the other hand, contrary to what many people say about the "dangers" of the Internet – acculturation, *stoopidification* of the unthinking masses[7] – this paper upholds the opinion that Information and Communication Technologies do not accidentally, but essentially reveal an atmosphere of *both* meaningfulness *and* meaninglessness, interrelated as we cope with or even struggle against the technical and informational constraints that affect our creativity. Our cultural environment is not physical anymore; it does not lie in books or buildings, in churches or parliaments. It is made of light, so to say, of tenuous electrons translating from one part of the world to the other, materialising on screens, processors, hard disks, and any other appropriate computational device. These, of course, would form a material environment, if particles only were at stake. What they stand for, though, is a swarm of web-pages, blogs,

databases, etc. Not all of them have any semantic importance, it is even reasonable enough to assume that only an imperceptible fraction of them has a rather established meaning. It remains that none of them should be considered as strictly or radically meaningless. Whatever our opinions, ideals, goals, they all transpire, a way or another, through what we currently and actively do as operators of Information and Communication Technologies and builders, if not architects, of the Internet.

It is then extremely important to outline the crucial dimension of *interconnectedness*, which is not a concept only meant to designate the phenomenon of networked computers, but, more essentially, to describe the very nature of culture as such. "Interconnectedness" is not a privilege of our post-modern world, but paradoxically enough, our post-modern world makes even more evident the interconnectedness inherent to the phenomenon of culture. Under such theoretical circumstances, the Internet should not be viewed as a digital virtualisation of cultures and traditions, on the pretext that one may find and download Plato's or Virgil's works. This is why the Internet should be viewed as a digital and informational *realisation* of both traditional and non-traditional forms of culture. To a large extent, it must be considered as the virtuous realisation of our future and the sphere of comprehension of our ever endlessly recreated world.

REFERENCES

Aristotle. (4th c. BCE). *Politics*.

Bergson, H. (1907). *L'Évolution créatrice*. Paris: P.U.F. – English translation: Creative Evolution (1944). New York: Random House.

Carr, N. (2008). Is Google making us stupid? *The Atlantic Monthly*. Retrieved January 31, 2013, from http://www.theatlantic.com/doc/200807/google.

European Parliament and Council. (2006, December 18). *Recommendation of the European Parliament and of the Council of 18 December 2006 on Key Competences for Lifelong Learning.* Retrieved from http://eur-lex.europa.eu/LexUriServ/LexUriServ.do?uri=OJ:L:2006:394:0010:0018:en:PDF

Heidegger, M. (1977). *Question concerning technology.* New York: Harper & Row.

Heim, M. (1987). *Electric language.* New Haven: Yale University Press.

Murray, J. H. (1998). *Hamlet on the holodeck: The future of narrative in cyberspace.* Cambridge, MA: The MIT Press.

Pascal, Z. G. (2009). *An operating system for the cloud. Technology Review, September/October 2009.* Cambridge, MA: The MIT Press.

Plato. (4th c. BCE). *Protagoras.*

ENDNOTES

[1] See European Parliament and Council. (2006, December 18).

[2] See an archived version of their site's "What is?" in 1997 at http://web.archive.org/web/19970401223917/www.pointcast.com/whatis.html.

[3] See Pascal, Z. G. (2009).

[4] See Murray, J. H. (1998, p. 240).

[5] Paper available online at http://www.bekkoame.ne.jp/~hmuroi/e11.html.

[6] "Storyspace," for instance, is a program providing mapped views of a text, thus allowing to interconnect one part to another and to create networked patterns of meaning – as if a text were a microcosmic representation of the macrocosm named "Internet." See url: http://www.eastgate.com/storyspace/maps.html.

[7] Though the author should not be counted among the paranoid cultural luddites, the neologism clearly refers to Nicholas Carr's "Is Google Making Us Stupid?" See references.

Chapter 17
A Call for Reflexivity in the Governance of the Finnish eGovernment System Development

Olli I. Heimo
University of Turku, Finland

Kai K. Kimppa
University of Turku, Finland

ABSTRACT

In this chapter, the authors present four cases of Finnish eGovernment application development in which the ethical approach has either been ignored, or mishandled grievously. The trend in Finnish eGovernment system acquirement has shown to lack reflexivity towards ethical issues. Between all the systems studied, the common denominators have been extended delivery times, increased costs and at best non-functional solutions. Testing, prototyping, and analyzing the systems have been lean and have only concentrated - interestingly enough - to the fields of functionality. The only discussion about the ethical issues has been between a minority of specialists and civil rights activists, and has been mostly ignored in the procurement of the systems.

Some of these examples contain only clear mistakes and design failures which could have been – at least partially – avoided. Others, like the Case of Finnish biometric passports, show alarming features of lessening both privacy of the citizens and the security of the whole society. The reasons behind the lack of ethical thinking during these design processes are investigated in this chapter.

DOI: 10.4018/978-1-4666-3670-5.ch017

CASE CENSORSHIP

The first topic handles the censorship system of Finland. Similar systems are in place in many other European countries as well (e.g. Sweden, Denmark, Norway, Netherlands and UK). The black list used for censoring the sites is more or less identical irrespective of country, as many of the sites blocked on the list come from the same source - an international organization which claims to battle child pornography (Kimppa, 2008). The purported aim of the system was to block foreign web sites containing child pornography to which the national police has no access and thus cannot shut the offending web sites down (HE, 99/2006). Lack of reflexivity to ethical theories is visible in the handling of this case as it is in the handling of the four other cases - the legislator did not bother to find out whether the sites blocked by the black list actually contained any child pornography, many of them did not, (see e.g. Nikki, 2008-2011a or Kimppa, 2008), or whether the blocking of the sites actually works towards lessening access to child pornography, the black list can actually be used as a search tool to find a blocked site and then, as bypassing the system is easy, accessing only those sites blocked. The major problem with this approach is the first point above. As Nikki (2008-2011a) and other civil right activists demonstrated, there is hardly any child pornography at the blocked sites. Thirdly, no attempt was made to actually close down the web sites - considering how easy this should be, as most of the blocked web sites are located in either USA or in EU countries, just contacting the local police should accomplish this - if the sites actually carried child pornography. As that is highly illegal in both areas, the web sites ought to be dealt with as soon as possible, one might surmise.

The only one who seems to have bothered to actually inform the authorities of these countries of the suspected web sites is Nikki (2008-2011b), whose own site was, and still is, after 3 years, blocked by the Finnish authorities black list. One should keep in mind that the justifications for the law clearly state that the block list is for denying access to foreign sites which allows access to child pornography (HE, 99/2006). Nikki's site does not offer possibility to access child pornography (at some point the sites which Nikki had deemed not to contain child pornography were linked at the site, but are not so anymore) and the site itself is in Finland, so should it break the Finnish law, Nikki could be prosecuted, but this, surprisingly has not happened. The block list is thus actually used to block an opposing view - i.e. for political persecution of a Finnish citizen (Nikki, 2008-2011c).

The current status of censorship is that IPR (Intellectual Property Rights) holders in Finland are trying to widen censorship to possibly illegal IP protected peer-to-peer applications such as torrent sharing. The tragedy, as it seems, is to make an example case out of one torrent sharing organisation (The PirateBay, http://thepiratebay.org/), and one Internet Service Provider (Elisa). The Finnish municipal court ordered Elisa to block any and all communication from their clients to and from The PirateBay thus expanding the censorship to something the Finnish constitution clearly forbids (Elisa, 2011). Although most of torrent file sharing services and most of the Internet service providers are left outside this court order, Finland is standing on the verge of extending this practice through this precedent to all Internet service providers and torrent sharing sites without a legislative process. The municipal court's decision to restrict all parties is now waiting for higher court decision in the matter. In the Finnish legal process this might take several years. The main problem with Internet censorship is that the Internet was created to be a military weapon (Arpanet), and thus communication through it can always be encrypted and/or rerouted. This of course has made the Jasmine revolutions and Chinese dissidence possible in the first place. The only thing this proves at this point that no rational thought either to technological or ethical thinking was given. As this case presents, Nikki's case is - as was predicted - quickly extending to other purposes, even though this was specifically denied. This same lack of knowledge was evident when the Internet censorship act was first introduced.

The censorship system has also been proposed to be extended to block online gambling happening from Finland through the Internet to foreign countries, due to the Finnish gambling laws which allow only certain organizations to run any gambling in Finland (Määttä, 2008). These cases exemplify the problems with not listening to both technical and ethical experts and not having a rational discourse on the possible benefits and harms of introducing technical solutions to societal problems.

E-VOTING EXPERIMENT

During the last two decades, the western world has seen various discussions, experiments and even full-scale implementations of electronic voting either to supplement or replace existing systems. After several trials and errors, several nations have abandoned electronic voting projects while others are but initiating theirs and some continue to use them even though problems with it seem to be prevalent everywhere (Heimo, Fairweather, & Kimppa, 2010).

Procuring electronic voting systems have been justified with various arguments, e.g. cost savings, activating passive voters, speed and efficiency, reliability of counting votes and staying in front line of ICT-using nations. All these justifications seem to lack justification. eVoting systems have been proved to be expensive (Verzola, 2008), seem not to especially activate voters (Yle Election Result Service, 2008), speed and efficiency - at least in the Finnish case - possibly reduces the 4 hour calculation of the vote to half an hour or one hour. eVoting systems tend to be more unreliable in counting votes (Mercuri, 2001; Fleischman, 2010; Heimo, Fairweather, & Kimppa, 2010).

In political rhetoric, the term "Information Society" is common. It is seen as a value in itself. "Information Society" reflects efficiency, fully electronic government practices and generally a modern society in positive ways. What it really means, though, is the whole 3rd industrial revolution for both good and bad (AGKS, 2009). While rhetorically being at the technological forefront may be appealing, nothing about technology qua technology means being an early adopter is a good thing. Justification for why it would be valuable is needed. A high proportion of technologies, from supersonic airliners, to Betamax video, to pagers and using gopher to retrieve information from the internet offered opportunities to be at the technological forefront until the technology became seen as a dead-end. Globally it is not clear that there is a trend to develop eVoting. Countries including Ireland, Netherlands, Germany and UK have moved away from eVoting for various reasons ranging from high expenses (higher than alternatives) to unreliability of the electronic voting systems.

All electronic voting systems have several problems in common. Mainly, every system in use, has either not been proved to be functional, or been proved not to be functional. To illustrate the latter, here are some examples (Effi, 2010; Verzola, 2008; Appel et al., 2009):

- In the year 2003 in Virginia a candidate lost 100 votes due a programming error.
- In California, in 2001, because of a programming error votes had to be recounted. (They luckily had a paper trail for verification).
- Al Gore was initially attributed negative 16,022 votes in a voting district in Florida at the 2000 presidential elections.
- In the year 2003 in Boone County, Indiana, 20,000 voters gave 140,000 votes.
- 3,400 votes were lost in the 2008 elections in Florida.

- Year 2007 automated elections in Scotland cost £ 40,000,000. For comparison, non-electronic elections in 2003 cost only £ 17,000,000.
- Year 2008 in Finland, eVoting experiment lost 2.3% of votes (compared to the average of 0.7% in non-electronic voting).
- In Netherlands, 2007, a group of activists proved that the voting machine in usage is hackable.
- In Ireland (population of approximately 4 million), a 5-year eVoting project cost over 52,000,000 €.
- In the USA, a group of researchers demonstrated, that a voting machine can be hacked in less than 7 minutes.

eVoting systems can be easily attacked with various different attack methods. One method is to tamper with a single voting machine, like the aforementioned hack done by Appel and his crew (Appel et al., 2009), which proved that a Sequioa Pacific AVC Advantage can be made "Turing complete" in under 7 minutes, including lock-picking; or the Dutch activist group Wijvertrouwenstemcomputersniet (2007), who made a commonly used Nedap/GroenendaalES3B to play chess. These hacks are proven to be relatively easy to produce. Other methods include information warfare - style attacks to the core system, like Distributed Denial of Service attacks, core system cracking and Stuxnet - style viral attacks. The latter methods have yet not been shown to have been used against eVoting systems, but their magnitude, should they be used, is huge.

Internet voting (iVoting) and mobile voting (mVoting) are extensions to the "traditional eVoting," and thus share all the problems with DRE-based (Direct Recording Electric voting machines, fully electronic voting machine without any other mechanic to for verify the result) voting. In addition, they have several problems of their own, for example the privacy of the ballot is quite hard - if not impossible - to produce.

Despite the extensive study made by Mercuri (2001), where the underlying problems of DRE-systems are shown, many DRE systems were created during 2000-2010. A fair number of those problems also recurred. Mercuri also argues, that neither iVoting nor mVoting, because of their mobile nature, do not fulfil the requirements of democracy. However, many countries, including Finland, are preparing their own iVoting and mVoting systems. Estonia, the fore runner in both, has used their system for over 4 years.

In most of the eVoting systems, e.g. AVC Advantage, AVC Edge, Tieto Voting System and Nedap ES3B, use Black Box security method, also known as Security Through Obscurity. This means, that system "blueprints" are kept as a secret, therefore making it harder to evaluate the system - and thus create an attack. Never the less, French mathematician Kerckhoffs defined the basis for military information security, where, in Kerckhoffs' (1883) 2nd rule, security cannot be built upon any other basis than the encryption key and thus the system designer must make an assumption, that the enemy knows the system being used (Shannon, 1949). Thus, Security Through Obscurity cannot, and is not being accepted as a scientifically acknowledged security method (Kerckhoffs, 1883; Shannon, 1949; Schneier, 1996, p. 5-7; Molnar & Wagner, 2004; Wijvertrouwenstemcomputersniet, 2006; Hoepman & Jacobs, 2007).

Overall, both theoretical and occurred problems still lie deep within the modern eVoting systems. Systems are developed and deployed into use without the traditional burden of proof, in which the systems' validity must be made by the proposer of the change. In the current situation, scientists, civil activists and non-governmental organizations are doing a huge amount of work to prove that the systems are being designed and implemented incorrectly, whereas these resources could have been used before implementing the systems. Thus, reflexivity towards technical and ethical issues is being left out from the governmental offices and - at best - outsourced to the active members of the public.

BIOMETRIC PASSPORT SYSTEMS

The necessity of finger print registry is not, at least in Finnish case justified with anything related to border control. (Heimo, Hakkala, & Kimppa, 2011).

Biometric passport systems were introduced in the Western world after the 9/11 incidents. There was a consensus (lead by the USA) for increasing the security trough technological apparatuses. The requirement for identifying a traveller met the new technologies for storing and handling biometric data.

Biometric passports contain machine-readable biometric data from the person they are issued to in addition to the traditional information included in a passport. The standard for biometric passports has been defined by the International Civil Aviation Organization. The standard allows for gathering and storage of the following biometric traits: face, fingerprints and iris. A picture of the passport holder is stored on all biometric passports, and the other biometrics are optional (ICAO, 2006). The passport is similar to traditional passports, but the additional biometric data can be read from the passport at the border, and it can be used for automatic biometric recognition of the traveller.

Biometric passports were designed to create more features than the normal passport, so that the border control has increased possibilities of verifying the identity of the passenger - and to expedite border control checks. This kind of changes, according to Nurminen and Forsman (1994), changes not only the tools used, but the whole work process. In some cases, like the pro-

totypes used in Finland, some parts or even the whole work process of the border control can be automated (MTV3, 2009). This is of course not an addition, as it was introduced, but a supplementation of the border control workflow. It is also debatable whether this switch to automated border control truly increases security as was intended.

Biometric passports have only been in existence for approximately 10 years (ICAO, 2006) and the current standards are still incomplete and implementing advanced security measures is optional. The effect of this can be seen in the various different implementations of the biometric passport standard around the world. In addition, practical implementations of biometric passports are undocumented so it is impossible to ascertain that they actually conform to the standards. Countries do not generally publish the details of their chosen biometric passport implementations, so a researcher dealing with privacy and security issues in passports must resort to reverse-engineering (Mostowski & Poll, 2010).

The leaking of personal information can be used in identity theft (Juels et al., 2005; Hoepman et al., 2006). It is also possible to track an individual on the border without their consent, and to identify certain nationalities in a crowd. This makes it possible to build bombs that are set to explode near a citizen(s) of a specified nation (Juels et al., 2005). Combining the identification of passports by their collision avoidance NONCE (Number used ONCE) and the ability to recognize passport nationality by their response to a request, the bomb example could be extended to recognizing crowds of certain nationalities. Acquiring digitally signed and thus indisputable information about an individual and their whereabouts is possible (Monnerat et al., 2007). An undeniable proof that a certain passport and its owner were in a certain place at a certain time using a semantic attack on bio-passports Active Authentication (AA) (BSI, 2010) can be seen as a violation of privacy. The passport contains digitally signed verified personal information, and because digital signatures are transferable, it can be used in gaining verified, non-deniable information about its holder (Monnerat et al., 2007).

Cloning passports with the leaked information is possible in cases where there are no other countermeasures in place. The reported vulnerability in the AA makes it possible to clone passports protected with AA, thus it cannot be considered effective. The lack of a clock in a passport makes it hard to actually enforce reader terminal certificates, making it possible to use stolen or counterfeit readers to read protected passports (Chaabouni & Vaudenay, 2009).

As collateral consequences, the problem described above with governmental officials in Finnish biometric passports is not limited to the Finnish governmental officials. A Finnish citizen may place some trust to the handling of their biometric information by Finnish officials, but with foreign officials the situation is quite different. When the foreign, "trusted", country border officials are authenticated - by the Finnish officials - to open the passport data, how can one know how, when or where the data is stored or used afterwards? (The same applies for any nationality with the current bio-passport technology; Finland is used just as an example.) When considering the late information warfare attacks (The Guardian, 2007; Newsweek, 2008; Financial Times, 2009) - and the attacks against companies like Lockheed Martin (BBC, 2011a) or Sony (BBC, 2011b; BBC, 2011c) - is the root certificate in threat to be stolen and misused? The possibility for massive concealed storing of biometric data is only one of the major threats imaginable. One can only wonder what other kinds of consequences there would be with these scenarios.

Another collateral consequence, erosion of privacy, can occur as a result of technology function creep, as the following example illustrates. In Finland, the introduction of biometric passports took place in the first phase of the passport reform in 2006. At this time it was already planned that the second phase would incorporate fingerprint

verification to the Finnish passport, in accordance to the EC Regulation No. 2252/2004 (EC, 2004). In 2009, at the second phase of the Finnish passport reform, it was decided by the Parliament that the fingerprints gathered from passport applicants would be stored on a national fingerprint registry - an addition which the EC Regulation does not require (EC, 2004; KELA, 2009).

Such a registry would contain information on all passport holders and their biometric identifiers, all consolidated to a single database. It would allow fingerprint matching on all registered persons, i.e. all those who have a passport. Although gathering massive amounts of personal information is generally frowned upon and therefore seldom done, this kind of registry would have a single advantage that cannot be provided by any other means: the ability to identify persons with two or more different official identities. Biometric recognition is the only identification method able to do this; traditional identification procedures are unable to stop a person from acquiring a second identity through current procedures. A case where a man had three distinct official identities (and the passports to accompany them) was reported in Finland in 2010 (HS, 2010a). If a national fingerprint registry had been in use, the attempt to gain another identity would be thwarted by biometric identification. This is a major advantage, and given its benefits, it is understandable that authorities would prefer to have a national fingerprint registry. Unfortunately, this advantage is overshadowed by privacy and security issues of such a database.

The Finnish Ministry of Internal Affairs justifies its fingerprint registry with the possibility to identify persons, who have lost or forgot their passports. They also point out, that this is a way to verify the uniqueness of the passport so that one person cannot have two passports with different identities. (Ministry of Internal Affairs). The government does not reveal any specifications about the registry in their publications. These justifications of course are not of the passport security, but about improving the overall security of the society at large. This opens the door for misuse of the biometric data gathered from those individuals, who happen to have a need to travel abroad.

The misuses with these kinds of registries are taken seriously in some other countries. E.g. in Germany the legislative has ruled out the possibility of a national biometric database, because it was seen to be in contradiction with the right of informational self-determination provided by the German constitution (Grundgesetz) (Hornung, 2007). Also, the existence of an encompassing citizen registry was seen to lessen the need for biometric databases. Finland has a very efficient and encompassing citizen registry; the latter argument could be applied in Finland as well.

During the legislation process the first step towards opening the registry to the police was the authorization to use it for identifying the deceased (KELA, 2009). After this was adopted by the ministry, in the year 2008, the political debate for opening the registry started after police commissioner Markku Salminen and his successor Mikko Paatero both requested full access to the registry for serious and serial crime investigations (HS, 2008a; HS, 2008b). These controversial demands were dismissed by the Parliament in 2009.

The discussion resurfaced in summer 2010, when Paatero renewed his claim (YLE, 2010). This time the Minister of Internal Affairs communicated a seemingly positive attitude towards the police commissioner's request (Tietokone, 2010). After the discussion on opening the registry for forensic use gained a lot of attention in the media, all talks of the use of the national fingerprint registry were suspended, pending the next parliamentary elections in spring 2011 (HS, 2008b; YLE, 2010; HS, 2010b). There is no guarantee that the use of the fingerprint registry would not be extended to other than serious crime investigation as well. This classical "function creep" is a prime example of the erosion of privacy.

A common argument in the Finnish public discussion - from citizens and politicians alike - is that no harm comes to law-abiding citizens just because mere fingerprints are found in a crime scene (Sunnuntaisuomalainen, 2010; Otakantaa.fi, 2011). This claim is of course a variation from the traditional claim of "law abiding citizens have nothing to worry about x," x (in this case) being the erosion of privacy. In the international context, an example of such a situation can be found from the investigation of the 2004 Madrid bombings, where an innocent American citizen was erroneously identified by the FBI as an accomplice in the attack, based on the fingerprints found in forensic investigations. The Spanish police later connected the fingerprints to an Algerian citizen, and the FBI was forced to admit they had made a mistake (Cherry & Imwinkelried, 2006). Although an extreme example, this incident shows that the pressure to produce results in an investigation can result in innocent people being marked as suspects with little to no actual evidence - especially in such high-profile cases to which serious crimes often belong.

As the Finnish government claims, their citizens should be worried about the effectiveness and cost of catching criminals. None the less, they should be more concerned about their government being overly interested in gaining biometric or other personal data about their citizens.

E-HEALTH SYSTEMS DEVELOPMENT

There are already rather good ethical guidelines in health care in general. Unfortunately these guidelines do not seem to extend to information system procurement for eHealth applications (Koskinen, Heimo, & Kimppa, forthcoming). Many healthcare (eHealth) applications and systems are counted as critical eGovernment systems. That is, when the systems are producing a service, which while not functioning, endangers health or safety of the target of the usage (of critical systems endangering the safety or health of the users, see e.g. Heimo, Hakkala, & Kimppa, 2011). One of the prime examples of this kind of studies is the Case London Ambulance, where more than 20 people died due to bad system design, poor testing and hasty implementation (Avison & Torkzadeh, 2009, p. 292-293). These problems have been known for quite some time, but are not taken seriously enough. There is a clear lack of needed knowledge when designing critical systems and especially healthcare information systems - which hopefully could be corrected by extending the group taking part in the discourse on procurement and design of critical eGovernment information systems.

The Finnish electronic prescription (ePrescription) system procurement is a typical example of a project problematic in the sense meant in this chapter. The experiment to implement ePrescription in Finland started in the 1990's. The project ended in a spectacular failure by not delivering the system. On the year 2000 the Finnish Ministry of Social Affairs and Health launched a project, the goal of which was to analyse previously failed project and propose a new national solution for ePrescription. (Hyppönen, 2007; Salmivalli, 2008). The first actual pilot was launched in the city of Turku ten years later (Ministry of Social Affairs and Health, 2010). It took a whole decade to accomplish the project which already had a historical precedent with knowledge of the problems and mistakes involved. Even though the system has finally been taken partly in use, it is still a pilot of a system, not a finished product. This points out a typical problem of developing large healthcare systems. It seems that there is a certain eagerness to accomplish and implement systems which are too large to implement at once.

This same orientation, which is problematic, is in sight when we look at Finnish National Archive of Health Information program (KanTa) which ePrescription is a part of. The Kanta-project has stated its goal as "KanTa is a collective term used for a range of national health care informa-tion systems including e-prescriptions, a national pharmaceutical database, an electronic health records archive and a portal for citizens to access their own health information online." (Kanta, 2010). At the same time there is no description of the overall architecture of the KanTa-project and thereby the reform of healthcare has been started without any actual evidence of how the project will improve healthcare functions. The development of the KanTa-project has also clearly been technology-driven, led by different interest groups with support of various consults and it has received no clear guidance from Ministry of Social Affairs and Health which governs the project. (National Audit Office of Finland, 2011.) Thus, there exists a large project with no clearly defined modules and thereby the feasibility of the project is questionable at best.

Large scale systems which are not open (black box-problem) are also risky because user organizations are highly dependent on providers. Therefore, when problems occur and systems crash only the provider of the system can fix it. This imbalances the relationship between the system provider and the customer. There are for example situations in which hospital districts in Finland have wanted to fix problems they have found in the systems they use but the system providers are not willing to give the needed information for fixing the problems (National Audit Office of Finland, 2011, p. 44). Thereby the customer is bound to the provider because a large and - within the customer - widely used system is not easily replaced with another - even if the system does not answer all the needs of the customer. These situations should instead be similar to other service acquirement, i.e. when the system is not working as intended, the healthcare officials could change the provider. There is an obvious problem in this situation where the provider can deliver systems which bind the customer to use poor systems without an actual possibility to invite tenders to get better systems.

In the current situation, the system provider usually has all the advantages. The market situation favours only the few biggest providers in such a way that they are the only few providers to possess the credibility, size and product history required in tendering for the large-scale information systems projects. Thus, their interest is to keep the current legislation as-is and the status quo for procurement so that their smaller competitors cannot deliver new information systems at all. Thereby the current large providers have an unjust competition advantage when compared to smaller new vendors in the market.

Because of the computerization in 1990's and 2000's, there are huge amounts of different computerised eHealth systems that are not compatible with each other. These programs are of course the intellectual property not of their users but of their providers. Thus the providers can actually control which kind of systems it is possible to implement as part of the overall systems in use. Whilst the system is in use, the provider of the system can also deny their customer, the governmental office such as a hospital, to implement a third party software or hardware into the system but must instead use the original provider to implement it. This causes problems, especially in small countries like Finland, with requirements for localisation and language-dependant support and upkeep. This leads into low competition between only a handful of corporations (Koskinen, Heimo, & Kimppa, forthcoming).

This non-competitive state of business can easily build an atmosphere of arrogance for the large producers; it is not required from them to build sufficient systems, because they are a major competitor for the next large-scale IT-projects anyway, regardless of the outcome of the product – as options do not exist. One of the clearest examples of this is the Finnish IT-producer Tieto (formerly known as TietoEnator). Many Finnish governmental systems have been ordered from Tieto regardless of the numerous failed projects they have submitted (on the failed eGovernment projects by Tieto, see e.g. the list by Helsingin-Sanomat, 2008). This kind of indifference towards

governmental IT-projects cannot be beneficial - and thus morally acceptable - for the public. When the system design is based on the "black box", the system provider's representative falls into a "certainty through," a state where he is denied of all uncertainty towards any possible errors in the system design (MacKenzie, 1990; Pantzar, 2002). Such a system, when provided to critical governmental usage, e.g. healthcare (or eVoting, or passports), should be tested every time with an independent, third party system tester for any possible functional or security flaws.

This all of course sums up to the previously mentioned lack of knowledge towards the technology itself.

CONCLUSION

The handled four cases show clearly that a deeper analysis of the ethical and social consequences of the eGovernment systems than is currently done is needed. The cases illustrate a lack of logical, ethical and socially responsible practices in the eGovernment application procurement, development and legal requirements. Even though the examples are from Finland, "the best country in the world" (5th in political environment), as they say (Newsweek, 2010), unfortunately all of the practices seem much wider spread - and also often worse.

The inability to handle ethical questions in relation to ICTs can be traced to a wider set of problems: partial or complete inability to handle problems related to technology. Therefore blind faith in the capability of technology to solve every problem is mixed with ignorance or lack of knowledge about the problems technology brings forth itself. At the worst case it is presumed that a developing technology is solving - with a rather fast pace - the problems fundamental to the technology itself. If one does not understand the descriptive reality, one cannot understand the normative reality either - thus understanding the ethical problems in the first place is impossible. Therefore, before designing or implementing technological solutions one must first acquire adequate technological understanding before one is capable of ethical reflexivity towards the technology at hand.

The primary customer of the system - a governmental actor or its representative - cannot end up in a situation in which the uncertainty towards the procured solution could lead to be unbeneficial, harmful or at worst, uncertainty towards the system could be denied all together from the actor, thereby making the actor just a mouth piece for the system developer instead of a - if necessary - critical representative of the end-customer, i.e. the citizen (see e.g. MacKenzie, 1990). The eGovernment solutions presented in this chapter represent the aforementioned problems in the procurement of current governmental information systems. Increasing transparency in eGovernment solutions procurement and design could enable a justified democratic discourse, in which both parties, those describing the technological understanding of reality and those describing the normative, or social understanding of reality, could contribute towards the systems procurement, thus enhancing the systems' beneficiality for all the stakeholders; especially the citizens. For example, in the eVoting case, had this kind of discourse been possible, the specialists from University of Turku Department of Mathematics, who analysed the system and its faults, could have started a meaningful technological discussion, which would have been the basis of a normative discourse about the problems found in the system before it was taken into use. Also, non-governmental organisations, such as Electronic Frontier Finland, could have brought forth their expert knowledge by analysing both the technical and the social implications of the system, which was now denied from them due to the restrictive NDA (Effi, 2008) (against Kerckhoffs' 2ndrule) which specifically prohibited this kind of discourse.

REFERENCES

AGKS. (2009, September). *A Green Knowledge Society - An ICT policy agenda to 2015 for Europe's future knowledge society, A study for the Ministry of Enterprise, Energy and Communications, Government Offices of Sweden by SCF Associates Ltd Final Report*. Retrieved October 14, 2011 from http://ec.europa.eu/information_society/eeurope/i2010/docs/i2010_high_level_group/green_knowledge_society.pdf

Appel, A. W., Ginsburg, M., Hursti, H., Kernighan, B. W., Richards, C. D., Tan, G., & Venetis, P. (2009). The New Jersey voting-machine lawsuit and the AVC advantage DRE voting machine. *USENIX Security '09*. Retrieved November 10, 2011, from http://www.usenix.org/events/evt-wote09/tech/full_papers/appel.pdf.

Avison, D., & Torkzadeh, G. (2008). *Information systems project management*. California, USA: Saga Publications.

BBC. (2011a). US defence firm Lockheed Martin hit by cyber-attack. Retrieved June 5, 2011, from http://www.bbc.co.uk/news/world-us-canada-13587785

BBC. (2011b). Sony warns of almost 25 million extra user detail theft. Retrieved June 15, 2011, from http://www.bbc.co.uk/news/technology-13639836

BBC. (2011c). Sony network suffers hack attack by Lulz Network. Retrieved June 15, 2011, from http://www.bbc.co.uk/news/technology-13639836

BSI. (2010). BundesamtfürSicherheit in der Informationstechnik [BSI, German Federal Office for Information Security], Advanced security mechanisms for machine readable travel documents. *Technical Guideline TR-03110*.

Chaabouni, R., & Vaudenay, S. (2009). The extended access control for machine readable travel documents. BIOSIG 2009, *Biometrics and Electronic Signatures* [Bonn, Germany: GesellschaftfürInformatik] [GI]. *LNI, 155*, 93–103.

Cherry, M., & Imwinkelried, E. (2006). Cautionary note about fingerprint analysis and reliance on digital technology. *Judicature, 89*(6), 334-338. Retrieved June 20, 2011, from http://www.ajs.org/ajs/publications/Judicature_PDFs/896/Cherry_896.pdf

EC (The Council of the European Union). (2004). *Council Regulation No 2252/2004*. Retrieved April 11, 2011, from http://eur-lex.europa.eu/LexUriServ/LexUriServ.do?uri=OJ:L:2004:385:0001:0006:EN:PDF

Effi. (2008). Sähköinen äänestysjärjestelmä rikkoo sekä lakia että vaalisalaisuuden. Retrieved November 10, 2011, from http://www.effi.org/julkaisut/tiedotteet/lehdistotiedote-2008-06-24.html

Effi. (2010). Electronical frontier Finland, Sähköäänestys-FAQ [eVoting-FAQ]. Retrieved November 10, 2011, from http://www.effi.org/sahkoaanestys-faq.html

Elisa. (2011). *Elisa tulee valittamaan väliaikaisesta Pirate Bay –määräyksestä* (press release by Elisa). Retrieved November 10, 2011, from www.elisa.fi/ir/pressi/?o=5120.00&did=17563

Financial Times. (2009, November 3). *Kremlin-backed group behind Estonia cyber blitz*. Retrieved November 24, 2010, from http://www.ft.com/cms/s/57536d5a-0ddc-11de-8ea3-0000779fd2ac,Authorised=false.html?_i_location=http%3A%2F%2Fwww.ft.com%2Fcms%2Fs%2F0%2F57536d5a-0ddc-11de-8ea3 0000779fd2ac.html&_i_referer=http%3A%2F%2Fen.wikipedia.org%2Fwiki%2F2007_cyberattacks_on_Estonia

Fleischman, W.M. (2010). Electronic voting systems and the therac-25: What have we learned? *Proceedings of Ethicomp 2010*, Tarragona, Spain.

Gonggrijp, R., & Hengeveld, W.J. (2007). Studying the Nedap/Groenendaal ES3B voting computer, a computer security perspective. In *Proceedings of the USENIX Workshop on Accurate Electronic Voting Technology*. Retrieved October 14, 2011, from http://wijvertrouwenstemcomputersniet.nl/images/c/ce/ES3B_EVT07.pdf

Gonggrijp, R., Hengeveld, W. J., Bogk, A., Engling, D., Mehnert, H., Rieger, F., et al. (2006). Nedap/Groenendaal ES3B voting computer, a security analysis. Retrieved August 8, 2010, from http://wijvertrouwenstemcomputersniet.nl/images/9/91/Es3b-en.pdf

HE. (99/2006). *Hallituksen esitys Eduskunnalle laiksi lapsipornografian levittämisen estotoimista*. Retrieved November 10, 2011, from http://www.finlex.fi/linkit/hepdf/20060099.

Heimo, O. I., Fairweather, N. B., & Kimppa, K. K. (2010). The Finnish eVoting experiment: What went wrong? In Proceedings of Ethicomp 2010. Tarragona, Spain.

Heimo, O. I., Hakkala, A., & Kimppa, K. K. (2011). The problems with security and privacy in eGovernment - Case: Biometric passports in Finland. In Proceedings of Ethicomp 2011. Sheffield, UK.

HelsinginSanomat [HS]. (2008a, February 22). *Poliisihaluaapassiensormenjäljetrikostutkijoille* (Police request passport fingerprints to criminal investigation) (1st ed.).

HelsinginSanomat [HS]. (2008b, November 27). *Rikostutkijateivätsaavieläpassiensormenjälkiäkäyttöönsä* (Criminal investigators do not acquire passport fingerprints yet) (1st ed.).

HelsinginSanomat [HS]. (2010a). *Vesipiipputupakantuonnistavankeuttaja 400000 euronlasku* (Prison sentence and 400.000 € fine from illegal import of waterpipe tobacco). Retrieved June 18, 2011, from http://omakaupunki.hs.fi/paakaupunkiseutu/uutiset/vesipiipputupakan_tuonnista_vankeutta_ja_400_000_euron_lasku/

HelsinginSanomat [HS]. (2010b). *Sunnuntaisuomalainen: Passiensormenjälkirekisterivoiavautuapoliisille* [Fingerprint registry may be opened to the police]. Retrieved April 11, 2011, from http://www.hs.fi/kotimaa/artikkeli/Sunnuntaisuomalainen+Passien+sormenj%C3%A4lkirekisteri+voi+avautua+poliisille/1135259348892

Hoepman, J. H., Hubbers, E., Jacobs, B., Oostdijk, M., & Schreur, R. W. (2006). Lecture Notes in Computer Science: *Vol. 4266. Crossing borders: Security and privacy issues of the European e-passport, advances in information and computer security* (pp. 152–167). Berlin, Heidelberg: Springer.

Hoepman, J. H., & Jacobs, B. (2007). Increased security through open source. *Communications of the ACM, 50*(1), 79–83. doi:10.1145/1188913.1188921.

Hornung, G. (2007). The European regulation on biometric passports: Legislative procedures, political interactions, legal framework and technical safeguards. *SCRIPTed 246, 4*(3). Retrieved June 9, 2011, from http://www.law.ed.ac.uk/ahrc/script-ed/vol4-3/hornung.asp

Hyppönen, H. (2007). Ehealth services and technology: Challenges for co-development. *Human Technology. An Interdisciplinary Journal on Humans in ICT Enviroments, 3*(2), 188–213.

International Civil Aviation Organization. ICAO. (2006). Machine readable travel documents. *ICAO/Doc 9303 Part 1 Vol. 2*. Retrieved June 17, 2011, from http://www2.icao.int/en/MRTD/Downloads/Doc%209303/Doc%209303%20English/Doc%209303%20Part%201%20Vol%202.pdf

Juels, A., Molnar, D., & Wagner, D. (2005). Security and privacy issues in E-passports, security and privacy for emerging areas in communications networks. In *First International Conference on Security and Privacy for Emerging Areas in Communications Networks* (SECURECOMM'05) (pp. 74-88).

Kanta. (2010). Retrieved May, 12, 2011, from https://www.kanta.fi/en/frontpage

KELA [Finnish Social Insurance Institution]. (2009). *Law service - Hallituksenesityslaiksipassilainjaeräidensiihenliittyvienlakienmuuttamisesta* [Government's proposal for changing passport act and certain other related laws]. Retrieved April 11, 2011, from http://www.edilex.fi/kela/fi/mt/havm20090009

Kerckhoffs, A. (1883). La cryptographiemilitaire. *Journal des sciences militaires, IX*, 5-38, Jan/Feb, 161-191. Retrieved November 10, 2011, from http://www.petitcolas.net/fabien/kerckhoffs/

Kimppa, K. K. (2008). Censorship: Case Finland. In *Proceedings of Ethicomp 2008*. Mantua, Italy: University of Pavia.

Koskinen, J., Heimo, O.I., & Kimppa, K.K. (in press). *Improving eHealth systems development with modularization: Ethical principles.*

Määttä, K. (2008). *Etärahapelien sääntelystä, Stakes, Helsinki*. Retrieved November 10, 2011, from http://www.stakes.fi/verkkojulkaisut/raportit/R2-2008-VERKKO.pdf

Mackenzie, D. A. (1990). *Inventing accuracy, a historical sociology of nuclear missile guidance.* Cambridge, MA: MIT Press.

Mercuri, R. (2001). *Electronic vote tabulation: Checks and balances.* (Doctoral Thesis). University of Pennsylvania.

Ministery of Social Affairs and Health. Finland. (2010). *Press release.* Retrieved May 10, 2011, from http://www.stm.fi/tiedotteet/tiedote/view/1508762#fi

Molnar, D., & Wagner, D. (2004). Privacy and security in library RFID Issues, practices, and architectures. *CCS'04*, October 25-29, 2004, Washington, DC, USA. Retrieved October 14, 2011, from http://www.eecs.berkeley.edu/~daw/papers/librfid-ccs04.pdf

Monnerat, J., Vaudenay, S., & Vuagnoux, M. (2007). About machine-readable travel documents: Privacy enhancement using (weakly) non-transferrable data authentication. *International Conference on RFID Security 2007* (pp. 15-28).

Mostowski, W., & Poll, E. (2010). *Electronic passports in a nutshell.* Technical Report ICIS-R10004, Radboud University Nijmegen, the Netherlands. Retrieved June 9, 2011, from http://citeseerx.ist.psu.edu/viewdoc/download?doi=10.1.1.167.2807&rep=rep1&type=pdf

MTV3. (2009). *Automaattinen passintarkastus alkoi vaalimaalla* [Automatic passport control has begun at Vaalimaa]. Retrieved June 7, 2011, from http://www.mtv3.fi/uutiset/kotimaa.shtml/2009/12/1014243/automaattinen-passin-tarkastus-alkoi-vaalimaalla

National Audit Office of Finland. (2011). Valtiontalouden tarkastusviraston tuloksellisuuskertomukset 217/2011. *Sosiaali- ja terveydenhuollon valtakunnalisten IT-hankkeiden toteuttaminen.*

Newsweek. (2008). *You've got malice, Russian nationalist waged a cyber war against Georgia. Fighting back is virtually impossible.* Retrieved November 24, 2010, from http://www.newsweek.com/2008/08/22/you-ve-got-malice.html

Newsweek. (2010). Interactive infographic of the world's best countries. Retrieved November 10, 2011, from http://www.thedailybeast.com/newsweek/2010/08/15/interactive-infographic-of-the-worlds-best-countries.html

Nikki, M. (2008-2011a). *The Finnish filtering list and its contents.* Retrieved November 10, 2011, from http://maraz.kapsi.fi/sisalto-en.html

Nikki, M. (2008-2011b). *The Finnish Internet Censorship List.* Retrieved November 10, 2011, from http://lapsiporno.info/suodatuslista/?lang=en

Nikki, M. (2008-2011c). *Näkemyksiä lapsipornosta Internetissä.* Retrieved November 10, 2011, from http://lapsiporno.info/

Nurminen, M. I., & Forsman, U. (1994). *Reversed quality life cycle model. North-Holland.* Amsterdam: Elsevier Science B.V..

Otakantaa.fi. (n.d.). *Finnish Ministry of Justice, An open electronic forum provided by the government for polling citizen opinions about new legislation.* Retrieved April 11, 2011, from http://otakantaa.fi

Pantzar, M. (2000). Teesejä tietoyhteiskunnasta. *Yhteiskuntapolitiikka. Nro 1. S.* (pp. 64-68). Retrieved May 23, 2011, from http://www.stakes.fi/yp/2000/1/001pantzar.pdf

Salmivalli, L. (2008). *Governing the implementation of a complex inter-organizational information system network – the case of Finnish prescription.* (Doctoral thesis). Turku School of Economics, Turku, Finland. Retrieved October 10, 2011, from http://info.tse.fi/julkaisut/vk/Ae3_2008.pdf

Schneier, B. (1996). Applied cryptography second edition: Protocols, algorithms, and source code in C. New York: John Wiley & Sons, Inc.

Shannon, C. E. (1949). Communication theory of secrecy systems. *Bell System Technical Journal,28*(4), 656-715. Retrieved October 10, 2011, from http://netlab.cs.ucla.edu/wiki/files/shannon1949.pdf

Sunnuntaisuomalainen. (2010, August 15). *Passipoliisit* (pp.14).

The Guardian. (2007, May 17). *Russia accused of unleashing cyberwar to disable Estonia.* Retrieved November 24, 2011, from http://www.guardian.co.uk/world/2007/may/17/topstories3.russia

Tietokone. (2010, August 16). *Poliisi saattaa saada passien sormenjäljet* [Police may acquire the passport fingerprints]. Retrieved April 11, 2011, from http://www.tietokone.fi/uutiset/poliisi_saattaa_saada_passien_sormenjaljet

Verzola, R. (2008). The cost of automating elections. *Social Science Research Network.* Retrieved November 10, 2011, from http://ssrn.com/abstract=1150267

YLE. Finnish public service broadcaster. (2010). *Poliisihaluaasuomalaistensormenjäljetrikostutkintaansa* [Police requests Finnish fingerprints to criminal investigation]. Retrieved April 11, 2011, from http://www.yle.fi/uutiset/kotimaa/2010/08/poliisi_haluaa_suomalaisten_sormenjaljet_rikostutkintaansa_1870808.html

YLE Election Result Service. (2008). Retrieved May 23, 2011, from http://yle.fi/vaalit2008/tulospalvelu

Section 7
Frame Development and Reflection

Section 7: Frame Development and Reflection

One of the key-concepts of reflexivity is the mental, cultural, or social *frame,* and the correlative operation of *framing* of elements that come into play when a mind is related to the world. In *a psychogenetic* perspective, the frame of an individual is the outcome of a process of personal development that leads him or her to adopt a certain view on the world (*cosmos*) as well as a certain conduct in life (*ethos*). In a *socio-genetic* perspective, a frame is the outcome of a process of socialisation that structures the "habitus" (*hexis*) as expressed in terms of interfaced mind-body capacities of an individual and as related to his or her position in a social group. In both approaches, the question is about the possibility to develop a reflexive examination of a frame that is supposedly more or less unconscious and implicit. It is then also to develop a reflexive stance to thoughts and conducts that are taken to be derived from the frame—if not determined by it. This ability to frame reflection, the *reflexivity* in a proper sense, is essential for the epistemic-cognitive, but also of course for the ethical-normative framing of issues at stake in a given problem or situation. It is particularly essential if one considers that what is called a "frame" in the genetic-developmental sense is a key element of what is also called the "context" of action in the genetic-reflexive governance.

The relationship of *psychology to ethics* shows a kind of mutual ignorance or at least a lack of curiosity for the other side. Thus, developmental questions are not much explored in the field of ethics, while moral issues, such as duty or virtue, are not much explored in the field of developmental psychology. There is no point in a psychological-developmental approach to ethics opposing duty and virtue as is often the case in the tradition of moral philosophy. The point is rather to demonstrate the promise found in a wider notion of reason and a more normative notion of the virtues that subsume moral duties and rules as a constitutive part of ethics instead of its end. Kohlberg's developmental psychology has been construed as the *formal judgment of justice* (structures-of-justice reasoning) founded on duty-based presuppositions of moral development. The limits of this perspective, now duly acknowledged, has entailed revival of interest in moral character and virtues and has opened the door for further integration of psychological and ethical considerations. They aim at not only ethical duty in developmental psychology, but also other elements needed to constitute a more robust ethical theory and moral psychology, namely character, virtue, happiness, cognition, volition, emotion, values, and interpersonal relationships. Thus, by recognizing the interest of *duty and virtue* in developmental psychology and ethics one can situate moral development within a *wider notion of reason.*

The relationship of *sociology to ethics* is also quite uneasy, especially in the structural-constructive approach that pleas for a social determination (or quasi-determination) of the individual's thought and conduct. Ethics generally assumes that individuals are able to elaborate a moral judgment in a peculiar situation in articulating a rule and a case, but this requires a moral autonomy of a subject. Now, if an individual's judgment is the determined outcome of his or her development in a social group or class, then the condition of autonomy cannot be satisfied. Then it can be illuminating to examine the position and the evolution of Bourdieu's structural-constructive sociology on the issue of *life trajectory* of individuals. The biographical and autobiographical approaches indeed lead to explore the reflexive operations that can be affected by somebody on his or her own trajectory in order to possibly modify its course. Bourdieu is concerned with the interiorization of social structures (*habitus*) and suggests *self socio-analysis* as a new way of highlighting them. The question is about knowing whether someone can truly escape from social structure's interiorization effects or even act upon them.

Chapter 18

The Use of Developmental Psychology in Ethics:
Beyond Kohlberg and Seligman?

Craig Steven Titus
Institute for the Psychological Sciences, USA

ABSTRACT

This chapter argues that a developmental psychology based in a wider notion of reason and ultimate flourishing can employ both duty and virtue in the service of the common good. It identifies several important differences between cognitive structuralism and virtue-based approaches concerning the pre-empirical priority paid to either duty or virtue in moral development. It brings to light several challenges concerning the use of developmental psychology in ethics: (1) a weakness in schools of cognitive structuralism, such as that of Lawrence Kohlberg, inasmuch as they do not move beyond the theory of stages and structures that focus only on the cognitive judgment of justice and on duty; (2) a weakness in developmental virtue approaches, such as that of Martin Seligman, inasmuch as they do not employ moral content in the operative notions of virtues and values. This article concludes that a heartier notion of developmental psychology and normative ethics will need to recognize the interrelated nature of ethical acts (moral agency), ethical agents (moral character), and ethical norms (duties and law). Such an integrated approach must also attend to the input that diverse philosophical and religious presuppositions make toward understanding the place of developmental psychology in the practice of ethics.

INTRODUCTION

In the field of ethics, developmental questions are often left unexplored. In the field of psychology, ethical issues are often left undeveloped. In both fields, the popularity of strictly duty or rule-based

presuppositions has limited a more profound use of developmental psychology in ethics (Handelsman, et al., 2009). This article will not pit a modern construal of duty-based and virtue-based approaches to ethics against each other. Nor will it contest the need for duties and rules. Rather it will demonstrate the promise found in a wider notion of reason and a more normative notion of

DOI: 10.4018/978-1-4666-3670-5.ch018

the virtues that subsume moral duties and rules as a constitutive part of ethics instead of its end.

The end of Lawrence Kohlberg's developmental psychology has been construed as the formal judgment of justice (structures-of-justice reasoning) founded on duty-based presuppositions of moral development. Much is owed to his research and to those who have advanced it. Nonetheless, the limits of this perspective—concerning moral flourishing and the content of ethics—have been recognized, and its need to be renewed has been largely accepted. In an era that many researchers call post-Kohlbergian, a burgeoning revival of interest in moral character and virtues has opened the door for further integration of psychological and ethical considerations that aim at not only ethical duty in developmental psychology, but also other elements needed to constitute a more robust ethical theory and moral psychology, namely: character, virtue, happiness, cognition, volition, emotion, values, and interpersonal relationships.

Although exerting a major influence from antiquity to present, the place of virtue in ethics has been eclipsed in modern rationalist (Descartes), deontological (Kant), sentimentalist (Hume), utilitarian (Mills, Bentham), and evolutionary (Darwin) accounts of moral agency. Nonetheless, the philosophical revival of virtue, especially largely Aristotelian virtue theory, has offered a source of further personal and social input and a platform for a deeper synthesis. The most renowned philosophical contributors to the virtue revival include: Josef Pieper (Pieper, 1966), Elizabeth Anscombe (1981), Vladimir Jankélévitch (1968), Alasdair MacIntyre (1985), Nicholas Dent (1984), Martha Nussbaum (1986), and André Comte-Sponville (1995). The most significant theological contributions to virtue theory are found in the works of Servais Pinckaers (1978, 1995), Paul Ricœur (1992) and Stanley Hauerwas (1981). In the field of psychology, major Kohlbergian theorists, Thomas Lickona (1976 & 1992), Carol Gilligan (1981), William Damon (1999), David Carr (1999), John Gibbs (2003), and Daniel Lapsley and F. Clark

Power (2005) have built upon the structures-of-justice reasoning, while trying to go beyond it, especially by showing greater interest in character and virtues. Further efforts at developmental psychology include the work of Martin Hoffman (2000) and Nancy Eisenberg (1998). Finally, of special interest for its larger vision of the human person and moral agency is a distinct character and virtue approach called "positive psychology," which is represented by Martin Seligman.

In this paper, first, I will present Kohlberg's perspective to developmental psychology, followed by several critiques and a sampling of the efforts that continue in this legacy. Second, I will demonstrate that studies in developmental psychology and moral development find complementary input from the positive psychology approach, particularly in terms of character, virtue, and moral integrity. Third, I will address several other issues concerning the use of developmental psychology in ethics. Overall and in light of these two schools of developmental psychology and their philosophical presuppositions, I will advance a dialogue between deontological and virtue ethics in the use of developmental psychology.

COGNITIVE-STRUCTURALIST APPROACH TO MORAL DEVELOPMENT

What was and continues to be the interest that Kohlberg's cognitive-structuralist theory offers for the use of developmental psychology in ethics?

Influences on the Cognitive Approach to Moral Development

The response to this question first requires turning to Kohlberg's major source, Jean Piaget. In academic circles, a conception of "moral development" that aims at empirical validation gained a reputation in the 1930s, when Piaget applied his genitive epistemology and developmental psychology to the moral domain and to research on the moral judgment of children. Piaget, a Swiss psychologist and philosopher, did groundbreaking work that explored the cognitive and moral development of children. He and seven collaborators published their research in French in 1932, as *The Moral Judgment of Children*, which adapted the structuralist tradition to psychology, based on the principle that cognitive skills develop when one meets unsolvable problems. Each new stage of development follows a sequence and exhibits a universal structure, subsuming the prior stage and improving its logic and function. Piaget took an interactional approach, in which the child constructs a moral sense by active involvement in adult and peer relationships that communicate moral codes about rules and duties.

Piaget based his presuppositions on a Kantian meta-ethics with three major elements: every ethics consists in a system of rules; the essence of ethics is found in the respect and sense of duty that an individual acquires for these rules; and moral development can be described in terms of heteronomy and autonomy. Based on a study of the phenomena around the practice and consciousness of rules, he proposed four stages of cognitive and moral growth. His first stage is that of sensory motor intelligence (to 2 years), in which the rules of reality are unconsciously discovered. The second stage can be called egocentric and involves preoperative intelligence (2 to 7-8 years). The child receives and imitates rules from outside, using them as a source of pleasure. Even when playing a game with others though, this pleasure is not yet properly social, but is rather self-centered. The rule, moreover, is respected as sacred and eternal. The third stage involves nascent cooperation and employs concrete operative intelligence (7-8 to 12 years). The child begins to play the game in seeking to understand and observe the rules in order to win. By seeking to be recognized by his peers, the game becomes social. Moreover, the rules are construed as obligations that have been established by mutual consent. Everyone must

respect them out of loyalty. The fourth and final stage is the codification of rules that requires formal intelligence (12-14 years). Youth come to find pleasure in understanding the different applications and codifications of the rules. The fullest sort of morality is manifest in this fourth stage.

Piaget proposed two phases of morality: autonomy and heteronomy. The first phase is the heteronomy that requires obedience based on the unilateral moral constraint imposed by the adult. This phase establishes the basis for moral realism, obligation, and the sense of duty. Next, there is an intermediate phase, in which the rules are internalized and generalized, but still are obeyed because they are imposed from outside the conscience. The final phase of morality is the autonomy that is established through the cooperation of bilateral relationships. The moral conscience recognizes an ideal that is independent of any external pressure. It also identifies veracity and reciprocity as necessary for autonomy. For Piaget, while these moral phases are observed in children at play, they also concern adults in the different areas of life that concern rules. For the adult, the stages of heteronomy or autonomy vary according to different settings of rules so that an individual may be heteronomous at work, but autonomous in family life. Based on this theory of moral development and its application to cases concerning justice, Piaget has been considered the father of empirical studies of cognitive moral development.

Lawrence Kohlberg was instrumental in making Piaget's theory known in English and in extending his insights across the lifespan. In the 1960s and 1970s, Kohlberg's theory of moral development gained a predominant place in this movement. His theory (or the Harvard school) of moral development was so dominant during the 1960s, 1970s, and 1980s in the domain of moral psychology that today the very idea of "moral development" is often identified with Kohlberg's cognitive approach. Its stages of moral development continue to guide research and developmental psychology and to influence teaching, even if only for its historical interest.

Kohlberg sought to establish a moral phenomenology, convinced that we must resist relativism and find efficacious means to combat it. In order to do so, Kohlberg founded his approach on the idea of justice, while taking certain philosophical options. He admitted "there is no philosophically neutral starting point for the psychological study of morality" (Kohlberg, 1981). Kohlberg identified his sources as: John Piaget (for his cognitive structuralist approach); John Rawls (for his theory of justice and the idea of reversibility); Emmanuel Kant (for his accent on formalism, universality of duties, and personal autonomy); Plato (for his theory of the unity of the virtues); John Dewey (for the internalization rather than the construction of ethics); and Emile Durkheim (for the perspective that knowledge alone—even moral speculative knowledge—should produce coherent agency). At the same time, Kohlberg resisted other current schools and influences, notably: psychoanalytic perspectives on the development of the child (Sigmund Freud), ethics of moral sentiment (David Hume), and behaviorism (B. F. Skinner).

Stages, Sequences, and Characteristics of Moral Development

While leaving aside Piaget's first stage (of sensory motor intelligence), Kohlberg employs the next three stages of cognitive development proposed by the Swiss psychologist. Kohlberg names these three stages: Pre-conventional morality (2-3 to 7-8 years); Conventional morality; and Post-conventional morality. Then he delimits six successive stages of moral judgment (each of these three stages is subdivided into two others).

The successes and failures of Kohlberg's perspective are based in his ideas concerning stages and the fixed criteria of their sequence (hard stage sequence). It is by logical analysis rather than empirical validation that Kohlberg establishes

his theory of six developmental stages of moral judgment and their sequence. He wanted his conception of the moral subject, found in the highest stage (six), to incarnate a "deontological theory of ethics" (Kohlberg, 1981, p. 169). Beyond any given society and the vast differences in culture, the moral subjects must employ universal moral principles (notably human dignity, autonomy of the subject, justice, and benevolence) for their judgments concerning good and evil, and right and wrong (Kohlberg, Boyd, & Levine, 1990). With this goal in mind, he defines the sequence of stages in terms of growth in the articulation, differentiation, and integration of a deontological ethics. The stages are successive and invariable, as would have a structuralist theory. They trace an interactional, social, and unidirectional progression that does not admit of moral regression. Thus, all the other stages depend on the final stage for their fulfillment (Kohlberg, 1984). In order to fight against relativism, he furthermore claims that "the stages themselves are not a theory" (Kohlberg, 1984, 196).

As for Plato, virtue is one. Kohlberg says, "the name of this ideal form is justice" (1981, p. 30). Kohlberg rejects the idea according to which the personality can be analyzed in terms of cognition, emotion, motivation, and character traits. Developmental psychology implies a general style of thought. By aiming at moral reasoning and justice, and by following the goal of empirical psychology, Kohlberg seeks experimental verification as a proof against relativism. He empirically explores the differences in moral judgment across the lifespan of the human person by analyzing the arguments of children, adolescents, and adults in the context of hypothetical moral dilemmas, the best known being the case of Heinz.

Critiques, Revisions, and Developments

In the academia of the1980s, the hegemony of Piaget and Kohlberg's formalist ethics dissipated.

Kohlberg's moral structuralism was rejected by some and complemented by others. Among those who have *repudiated* Kohlberg's project as systemically flawed are Paul Philibert (1975), Paul Vitz (1981), Owen Flanagan (1991 & 1996), and John C. Gibbs (2003). Flanagan contests certain of Kohlberg's empirical and philosophical presuppositions. He also rejects the idea of equality of moral potential. Do all humans have the same moral potential? Piaget holds that all adults should arrive at the highest stage of knowledge (space, time, causality, conservation) in order to function well in the world. It seems that Kohlberg's conventional stages (three and four) imply functioning well in the world. Why would we need anything more, asks Flanagan? If it is not necessary that everyone be an Einstein in the domain of physics, why should it be necessary in the realm of ethics? Flanagan concluded that there was no "universal and irreversible sequence of stages according to which moral personality unfolds, and against which moral maturity can be unequivocally plotted" (1991, p. 195); and he finally affirmed that Kohlberg's theory is "a dismal failure, an utterly degenerate research program despite many true believers" (1996, 138).

Before his death in 1987, Kohlberg made revisions of his stages on several points. I will only mention four critiques and revisions of his work, as well as several indices concerning the influence of his work for current research.

First, Carol Gilligan is one of those disciples of Kohlberg, who have criticized but retained important aspects of his stage approach. Her book entitled, *In a Different Voice* (1982), identifies his Achilles' heel: Kantian formalist presuppositions, the Enlightenment disassociation between emotion, caring, and moral judgment, and a misrepresentation of a woman's transition to moral maturity. Gilligan's complementary system identifies feminine expressions of "care" and "empathy" as a further element to recognize moral maturity, which has "interdependency" as its goal. While Gilligan's ethic and development

theory are insightful and remain popular, they are criticized for an overly rigid conceptual division of feminine and male anthropology and developmental tasks. Moreover, they have proven to be neither empirically established nor philosophically adequate, according to Philibert, (1987), Flanagan (1991), Spohn (2000). Gilligan (1992) has replied to her critics. Both Gilligan and Kohlberg have also been criticized for underestimating the moral competence of children and adolescents. For example, Damon's findings indicate that "young children have a far richer sense of positive morality than the model indicates. In other words, they do not act simply out of fear of punishment. When a playmate hogs a plate of cookies or refuses to relinquish a swing, the protest 'That's not fair!' is common" (1999, p. 13). There are also indications that children express more of a sense of: obligation (equal distribution); empathy; reciprocity, equal treatment (Gibbs, 2003). Kohlberg (1981), for his part, conceded that his presuppositions and his dilemmas express a deontological ethics that is focused on the level of the judgment of justice and that "does not fully reflect all that is recognized as being a part of the moral domain" (Kohlberg, 1984, p. 227). Moreover, he felt that he had to make some effort to revise the idea of the sixth stage in order to include not only justice operations but also sympathy operations (Kohlberg et al., 1990). These revisions led the way for him to recognize research, both in developmental psychology and in ethics, concerning the dimensions that are complementary to cognition and justice, notably: (1) the other forms of social relations; (2) the other capacities linked to moral agency (emotions or sentiments, and the will); and (3) the content of moral judgments and their natural-rational, social, and religious sources.

Second, Johannes van der Ven has identified an individualist bias in Kohlberg's pre-empirical definition of justice (van der Ven, 1998). Kohlberg follows Rawls in construing justice in terms of the equal right to liberty of a collection of self-interested individuals. Conflict, however, arises between individuals who strive to satisfy their own basic needs, as is found in the case of Heinz, whose wife's life is in peril without a medicine that they cannot afford. But Kohlberg's formalist approach, which focuses on the rational procedure alone, is incapable of resolving the Heinz dilemma. There is no way to find satisfactory resolution unless the content of the conflict is addressed as well. Van der Ven calls upon Aristotle and Thomas Aquinas for the ethical means of resolving such conflicts between individuals. While agreeing with Kohlberg about the importance of reasoning to resolve such cases of justice, both Aristotle and Aquinas would disagree with him that considerations of justice should stop there. They would hold that such a liberal construal of justice—based on individual self-interest and formal procedures—is missing an integral element (Vitz, 2006). Justice requires a hearty notion of the common good. In Aquinas' perspective (ST II-II 58.2; 61.2), justice always has a primary reference to the other person and to the community. Distributive justice gives what is due to the part (the person, family, etc.) from what belongs to the whole (community, society, state). For example, individuals and family should receive what is their due (such as basic education) from the community. Commutative justice brings to the other what is his or her due within the context of the whole. For example, this refers to the arithmetic type of justice that restores what has been taken or borrowed or gives what was contractually promised. In regards to a principle important for the case of Heinz, Aquinas affirms that in urgent need, all things are common property. When a person really does not have the basics for survival, the means for survival become common property. Aquinas says: "whatever certain people have in superabundance is due, by natural law, to the purpose of succoring the poor" (ST II-II 66.7). Van der Ven (1998) has argued that in the case of Heinz' wife, the medicine she urgently needs is common property because of her dire need. This is a different position than that of Rawls, who would have us take seriously the request of those in need or the "least advantaged," as a type of self-assurance, because one day we might be in their situation (Rawls, 1971).

Third, another critique of Kohlberg concerns the conception of the sixth stage of moral development and its connection with moral agency. Faced with the lack of empirical support for the sixth stage and the presence of cases of moral regression, Kohlberg revised the notion of his highest stage and his scoring manual in order to eliminate such regression (Flanagan, 1996; 2006). This revision, however, also eliminated the cases of stage six. Without the sixth stage (which incarnates the thought of Rawls and Kant), Kohlberg was no longer able to logically claim that his theory either withstood the threat of relativism or identified a goal, existing in principles that would serve as the theoretical ideal for the other stages (Lapsley, 2006). Nonetheless, he reintroduced the sixth stage and began to treat it as a hypothetical end of the sequence (Kohlberg et al., 1990). He thus continued to assume that moral agency directly follows cognitive judgment, even after his research did not confirm this assumption. Throughout his revisions, Kohlberg continued to use hard stage criteria in order to calm dissent about developmental trends in the empirical data.

Fourth, a last question concerning Kohlberg involves the religious and metaphysical basis for moral agency. He sees two possibilities: either there is a religious-mystical-metaethical-metaphysical stage (a seventh stage) or there are six parallel stages of religious development (a universal faith), a position defended by James Fowler (1981; 1984), Fritz Oser (Oser & Gmünder, 1982), and André Guidon (1989). Even if he admitted the possibility of such parallel stages for ethical judgments and religious attitudes (*the good life and the good person*), later and on the side, he opted for a seventh stage that goes "beyond justice" (Kohlberg, 1981). At this stage, the moral subject becomes able to construe an action morally by an intention expressing a religious or metaphysical (natural law) basis (Kohlberg &

Power, 1981). Nonetheless, Kohlberg remained ambivalent concerning religion, inasmuch as his scoring manual categorically put transcendental affections and values in stages three and four (Vitz, 1982).

In general, we must say that the structuralist analysis of Kohlberg's presupposition concerning a deontological ethics needs to be reframed and completed in order better to take into consideration the complete reality of the development of the moral subject (Browning, 2006). There are *stage theories* that *continue something of the Piaget-Kohlberg legacy*. In contrast to Kohlberg's strict concepts of rigid sequencing, hierarchical integration and structural unity, however, these other approaches employ notions of stage and structure that are more flexible and less formalistic. For example, there are promising research projects in *prosocial reasoning* by Nancy Eisenberg ("age-developmental approach"; 1982; 1998; 2003) and in the area of *positive justice* by William Damon (1999) and Lapsley (2006) (fairness and sharing in a partial-structure stage model). There are also attempts, for example, by John Gibbs to associate the work of Kohlberg and that of Martin Hoffman (2000), who puts an emphasis on the empathetic and motivational aspects of moral development (Gibbs, 2003). Don Browning (2006) attempts to revive Kohlberg, employing a deontological test and the insights of Paul Ricœur and Hans-Georg Gadamer in his critical hermeneutics (Browning, 2006). Fritz Oser and F. Clark Power employ a stage approach in communities of moral development, to promote schools that incarnate a just community (Berkowitz & Oser, 1985; Power, 1988).

The list could be lengthened to illustrate how the post-Kohlbergian era has witnessed a multiplication of attempts to escape a narrow deontological framework in order to integrate further dimensions of moral action and development. I will now turn to one distinct attempt to face this challenge.

POSITIVE PSYCHOLOGY APPROACH TO THE DEVELOPMENT OF VIRTUES

In contrast to the perspectives that adopt deontological presuppositions, a recent paradigm shift seeks to forge a developmental psychology that includes a moral notion of development and ultimate flourishing beyond structures-of-justice reasoning. The "positive psychology" approach, for example, is of particular interest because of its attempt to assimilate further dimensions of moral development and of psychosocial research. Its reference to "positive" does not refer to "positivism" or positivist reductionism. Nor does it refer to blind optimism and pleasant feelings. "Positive" refers instead to the flourishing human life and the pre-empirical presuppositions about the constructive content thereof. While seeking to verify empirically value-based personality traits and virtues that contribute to a good life, the positive psychology classification seeks to resist the reductionist positivism of the early twentieth century. It nonetheless admits that its primary competency lies in the form and function of developmental psychology rather than its ethical content.

Positive psychology does go beyond cognitive-structuralist stage theories, by pressing into the service of developmental psychology a fuller concept of virtue than that of Kohlberg. In particular, it employs a wider range of virtues, character strengths, and situational themes in order to circumscribe the psychological well-being, moral development, and social flourishing that constitute a good life.

Paradigm Shift: Toward a School of Character Strengths and Virtues

Positive psychology's theory and research on moral development is the effort of a sizable network of researchers. It is the result of a paradigm shift in psychology that focuses not only on curing mental illness, but also on making human life fulfilling and productive. The most visible pioneer in the field is Martin Seligman, who has been instrumental, if not in creating, then at least in naming a school of thought. After having started his research career focused on the topics of learned helplessness (Seligman, 1991) and abnormal psychology (Seligman et al., 2001), Seligman changed tack to study optimism and hope (Seligman, Reivich, Jaycox, & Gilham, 1996; Seligman, 1998). In 1998, as the President of the American Psychological Association (APA), he announced that the moment had come for a radical transformation in the domain of psychology. He called for a "'positive psychology,' that is, a reoriented science that emphasizes the understanding and building of the most positive qualities of the individual: optimism, courage, work ethic, future-mindedness, interpersonal skill, the capacity for pleasure and insight, and social responsibility" (Seligman, 1999, p. 559). Since then, he has continued to empirically study flourishing and to establish a taxonomy of good character (Seligman, 2002; 2012). He has inspired a vast research project on the virtues, which is outlined in the book that he co-edited with Christopher Peterson in 2004, called *Character Strengths and Virtues.* This positive psychology movement seeks to appropriate the positive conceptual foundation of diverse philosophical and religious traditions on virtues, character traits, and human nature (Joseph & Linley, 2006; Linley & Joseph, 2004; Snyder & Lopez, 2007; Snyder & Lopez, 2009). As a complement to the American Psychiatric Association's *Diagnostic and Statistical Manual* in its fourth version-revised (APA, 2000), positive psychology has proposed itself as a *Manual of the Sanities.*

Even if the school of positive psychology was identified as such only in 1998, it has appropriated theory and research from a long tradition of the sciences: generative psychology (since the 1930s; Piaget, 1997), humanistic theories (Maslow, 1956; Rogers, 1959; 1961), resilience research (since the 1960s, Werner & Smith, 1986), and studies on excellence, happiness, hope, and creativity, and other character traits and virtues that do not cease to fascinate researchers (Snyder, 1994; Wink & Helson, 1997).

As a biopsychosocial perspective, positive psychology seeks to understand human development and not only pathology. Instead of simply seeking to relieve symptoms or to make people less unhappy, it aims to find signs of health and promote personal flourishing, as well as to prevent pathologies and cure illnesses. Thus positive psychology sees two explicit aspects of its vocation as psychology. In addition to the curative dimension (of modern psychology), it adds preventive, developmental, and normative aspects, in view of the qualities that contribute to the flourishing of the human person and society.

In order to understand mental health, psychosocial flourishing, and moral development, positive psychology calls upon different cultures, philosophies, and religions (Occidental and Oriental) to establish a pre-empirical foundation, that is, a philosophical anthropology for human strengths. It adopts a comparative approach between these sources—each as a witness of humanity—in order to understand the positive goals to attain and in order to establish conceptual definitions and a moral vision.

Positive psychology then uses the scientific approach of empirical psychology and the results of research in the neurosciences, cognitive psychology, and evolutionary psychology in order to understand and verify the application of philosophical and religious insights on the growth of the person and the functioning of a range of virtues. It offers determinant observations about the functioning and growth of the human subject through a spectrum of virtues, employing "the scientific method to inform philosophical pronouncements about the traits of a good person" (Peterson & Seligman, 2004, p. 89).

Virtues in Positive Psychology

In addition to a consensual definition, positive psychology's treatment of virtues gives for each virtue: (1) a comparative presentation of theoretical traditions (psychological, philosophical, and religious); (2) the findings of empirical studies (assessment and measures of the virtue); (3) a discussion of the development of the virtue with its enabling and inhibiting factors; (4) an analysis of gender and cultural aspects, and finally (5) details on targeted interventions and research to be done in the future.

Peterson and Seligman situate the various aspects of virtue and good character at three levels of abstraction: first, virtue, second, character strengths, and third, situational themes (or practices). These three levels start with the more general and universal and move to the more specific and culturally diverse.

1. **Wisdom and Knowledge (Cognitive Order) and Associated Character Traits:** Creativity, curiosity, open-mindedness (judgment, critical thinking), love of learning, perspective.
2. See Endnotes[1]
3. **Courage (Emotional Dynamisms) and Associated Character Traits:** Bravery, persistence, integrity, vitality.
4. **Humanity (Rooted in Interpersonal Values) and Associated Character Traits:** Love, kindness (generosity, nurturance, care, compassion, altruistic love, "niceness"), social intelligence (emotional intelligence, personal intelligence).
5. **Justice (Values of Social and Communal Life) and Associated Character Traits:** Citizenship (social responsibility, loyalty, teamwork), fairness, leadership.
6. **Temperance (Governing Emotional Attractions) and Associated Character Traits:** Forgiveness and mercy, humil-ity and modesty, prudence, self-regulation (self-control).
7. **Transcendence (Universal Forces that are Greater than Oneself or Society) and Associated Character Traits:** Appreciation of beauty and excellence, gratitude, hope, humor, spirituality (religiousness, faith, purpose).

First, the virtues are the core characteristics that have been valued by moral philosophers and religious thinkers for millennia. The six major virtues (listed above) approximate a classic (pre-modern and pre-Reformation) schema, for example that of Thomas Aquinas (faith, hope, and love, practical wisdom, justice, courage, temperance). Seligman speculates that individuals are deemed to have good character because they have at least a threshold level of these virtues.

Second, at a more particular level, character strengths are the psychological processes or mechanisms that constitute the virtues. "They are the distinguishable routes to displaying one or another of the virtues" (Peterson & Seligman, 2004, p. 14). At present, Peterson and Seligman have identified a total of 24 character strengths. They state that displaying one or two strengths within a virtue group designates someone as having a good character.

Third, situational themes are the "specific habits" and their associated practices that lead people to develop and exhibit given character strengths in specific situations. For example, empathy, inclusiveness, and positivity in the workplace are themes that promote flourishing in a particular situation; at a more abstract level, these particular situational themes reflect the character strength of kindness, which falls in the broad virtue class of humanity (and love). Such themes permit and necessitate the study of settings such as workplace, family, school, and so on. These themes and their practices are situation specific, therefore they will consistently describe conduct only in a given setting and culture.

The distinction between virtue, character strength, and situational theme helps positive psychology explain diversity (within interconnection) at different levels and between sociocultural construals of goodness and flourishing. Seligman asserts that "variation exists at the level of themes, less so at the level of character strengths, and not at all at the level of virtues" (Peterson & Seligman, 2004, p. 14). In order to understand the specificity of this "rich psychological content and strategies of measurement and hence explanatory power" (Peterson & Seligman, 2004, p. 13), and its relationship with philosophical and normative conceptions of the virtue, we need however ask: How do positive psychology's presuppositions, that is, its pre-empirical normative philosophical anthropology and notion of flourishing influence its understanding of particular character strengths and virtues?

Normative Bases for Positive Psychology

As manifestations of its philosophical and religious pre-suppositions (pre-empirical references), positive psychology seeks explicit moral references and criteria for good character in two ways: (1) through the nature of virtue and its criteria and (2) through the notion of "universal positive human nature" (Peterson & Seligman, 2004, p. 270). In so doing, it analyzes subjective experiences of flourishing, character traits, and institutions that enable both moral growth and ethical normativeness in terms of virtues.

First, Peterson and Seligman ground positive psychology's classification in a long philosophical tradition concerned with morality explained in terms of virtues. The very first Greek philosophers asked, "What is the good of a person?" This framing of morality led them to examine character and in particular virtues. Socrates, Plato, Aristotle, Augustine, Aquinas, and others enumerated such virtues, regarding them as the traits of character that make someone a good person. Peterson and Seligman (2004) enter into a particular interpretation of virtue theory, when distancing the inner motivation of the virtues from moral norms, principles, and laws.

While adopting this pre-empirical moral anthropology of virtue, Peterson and Seligman however distance the inner motivation of the virtues from moral laws (as merely external dictates) by critiquing the moral law theory, which they find in ethical egoism, utilitarianism, social contract, and divine command theories. Their psychological venture, as they say, "needs to downplay prescriptions for the good life (moral laws) and instead emphasize the why and the how of good character," which are found in the virtues and character strengths (Peterson & Seligman, 2004, p. 10). They thus correlate the normative and the descriptive sides of virtue, while recognizing that the psychologist's competency lies in the descriptive.

Seeking to justify and emphasize psychology's specific competency in describing the function of virtue, they employ the work of Lee Yearley (a Thomist by training) to identify three related realms of philosophical (and religious) ethics that constitute a good character (Yearley, 1990). These three realms are:

- The study of injunctions: commands and prohibitions, for example, the "thou shalt nots" and the occasional "thou shalts" in the Ten Commandments.
- The listing and organizing – usually hierarchically – of virtues: predispositions to act in ways leading to a recognizable human excellence or instance of flourishing.
- The analysis of ways of life [or practices] protected by the injunctions and picked out by the virtues (Yearley, 1990; cited in Peterson & Seligman, 2004, p. 85).

Peterson and Seligman note that the second and third realms, (virtues as dispositions to act and related ways of life or practices) while vaguer than the first (pre-empirical moral commands), are the direct competency of the positive psychology classification project. Although the hierarchy of the virtues is not unambiguous, the virtues continuously attest (1) the need for rationality and choice; (2) the need to reflect upon one's own dispositions and expression of the major areas of virtue; and (3) the influence of life-commitments and of culture on the development and on the expression of the virtues. Each virtue, in psychological language, is "a property of the whole person and the life that the person leads," in the words of Peterson and Seligman (2004, p. 87). The strength of this developmental psychology of virtue theory is its capacity to describe moral motivation and the resolution of psychological conflicts in terms of pertinent virtues; for example, concerning conflicts of partiality, Peterson and Seligman claim that resolution comes in recognizing that: "we should love our friends and family members (partiality) and be benevolent to people in general (impartiality)" (2004, p. 88).[2]

This general framework is further concretized by ten criteria used to identify virtues and character traits that must: (1) lead to flourishing by a good life; (2) correspond with moral values; (3) not diminish others; (4) have a non-felicitous opposite; (5) be a character trait or trait-like; (6) be conceptually identifiable; (7) enjoy consensus support; (8) be identifiable in prodigies and in (9) selective absences; and (10) be supported by cultural, institutional, and social practices. These criteria (especially 1 through 4) have an explicit ethical dimension.

Although character traits and virtues are plural, Peterson and Seligman make reference to the stability of human nature. They even call upon the notion of a "universal positive human nature" in order to adjudicate moral conflicts and evil desires (Peterson & Seligman, 2004, p. 270; Linley & Joseph, 2004). While there are multiple pathways to flourishing, evil is not one of them. Peterson and Seligman say those who do evil "are unlikely to thrive because their motives and

personality dispositions are incongruent with positive human nature and universal psychological needs" (2004, p. 269). This probabilistic argument makes reference to a positive human nature and a general appeal to value-based virtues. They admit that this psychological approach that focuses on explanation and function does *not* satisfy the moral philosopher, though, who must venture arguments in support of moral principles and must adjudicate between conflicting norms. Peterson and Seligman thus situate their "richer psychological content [...] and greater explanatory power" not at a normative philosophical or religious level, but as a description of character strengths, which he seeks to expand by wider notions of virtue and studies of the semblances of virtue (2004, p. 13; p. 88).

AN INTEGRATED APPROACH TO DEVELOPMENTAL PSYCHOLOGY AND ETHICS

The two schools of psychology represented by Kohlberg, on the one hand, and Peterson and Seligman, on the other, emphasize different developmental elements and ends, which are needed in an integrated approach to ethics. Their approaches offer complementary insights into the import of duty and rules, ends and flourishing, justice and the other virtues. Kohlberg's vision is more restricted to cognitive-structuralist elements of moral development focusing on the formal judgment of justice within a duty-based construal. Peterson and Seligman focus on the nature and development of a full range of character strengths and virtues within a flourishing-based construal. Their vision aims at the richer notion of flourishing that relies on the moral virtue of practical wisdom (that includes the notion of duty) in the development of the whole person, in emotions, will, intellect, and interpersonal relationships.

Drawing on the insights of both approaches, a normative ethics will need access to a developmental psychology that can integrate three levels: (1) character or virtuous dispositions, concerning the developmental trajectory of each virtue within a good life, which is formed over a lifetime through (2) moral acts within moral communities, and which requires the guidance of (3) ethical norms, principles, and laws that serve the formation of character and the exercise of actual practices. Aristotle and Aquinas, in their own ways, treat the interplay of these three: virtuous dispositions, acts, and laws (Aquinas, I-II questions 49-56; Titus, 2010).

This use of developmental psychology (that integrates virtue in terms of dispositions, acts, and norms) in a constructive account of ethics requires further explanation of the interconnection of duty, rules, flourishing, values, and virtues in moral agency. In order to contribute to the larger effort—which is not completed in this short essay—I will, first, analyze the notion of the virtue of integrity; then, focus on the dynamics of moral growth, before finally pinpointing the need to set virtuous practices and moral character as primary goals in a developmental psychology and as a constitutive element of normative ethics.

The Virtue of Integrity

In addition to the virtue of justice—which assures an important aspect of the rectitude of interpersonal agency, as highlighted by Kohlberg's cognitive structuralist focus—and the virtue of wisdom or prudence—with its role to connect the other virtues in the flourishing of a good life, as highlighted by Aristotle (1144b-1145a12) and Thomas Aquinas (I-II, q. 65, art. 1 & II-II, q. 47, art. 8)—Peterson and Seligman highlight the virtue of integrity.[3] The virtue of integrity bears direct interest for understanding the relationship between moral development and an ethical vision of flourishing. They define the virtue of integrity as:

A character trait in which people are true to themselves, accurately presenting—privately and publicly—their internal states, intentions,

and commitments. Such persons accept and take responsibility for their feelings and behaviors, owning them, as it were, and reaping substantial benefits by doing so (2004, p. 249-250).

In a summary definition of integrity, they say that it involves "moral probity and self unity" (2004, p. 250). In both of these definitions of integrity, Peterson and Seligman hold up the ideal of "being true to oneself," which—in various schools of thought—is at times influenced by individualist biases, either as a transient stage in moral development or as a relativist cultural phenomenon. However, as Charles Taylor has argued in *The Ethics of Authenticity*, the moral ideal of being true to oneself need not be identified with debased forms of relativism or an individualism of selfish fulfillment (1991). Integrity can have specifically psychological ends (the calming of disruptive emotions or a sense of self unity) and moral ends (constancy in intending and doing the good). Nonetheless, we should ask: Does Peterson and Seligman's notion of integrity avoid the trap of individualism—selfish flourishing—and the relativist or formalist eclipse of moral ends?

First, as we can see in their definitions of integrity, Peterson and Seligman interconnect the emotional and the moral domains, making ethical references to intentions, commitments, responsibility, feelings, and behavior. Elsewhere, they speculate that feelings of psychological integrity should correlate with behavioral and observer-based measures of honesty and authenticity.

Second, Peterson and Seligman employ the notion of the self to morally evaluate the content of psychological integrity. By recognizing the role of goals, talents, and values in the development of identity and integrity, they effectively resist theories that construe the self as a mere fiction or a set of evolving images or sentiments. According to them, apparently everyone struggles for greater integrity. Nevertheless, the question of evil poses problems for this concept and human flourishing in general. Attempting to resist a value-free stance, Peterson and Seligman say that:

From our own perspective, evil people can be authentic; that is, their sense of self can be true to antisocial motives and personality dispositions. However, these people are unlikely to thrive because their motives and personality dispositions are incongruent with positive human nature and universal psychological needs. Still, developing a sense that accurately represents one's personality may be an important first step in changing the personality – evil people who realize who they have become can then be motivated to become something different (2004, p. 269-270).

Peterson and Seligman thus move from adjudication about the psychological coherency of function to a normative statement about the content of flourishing. In order to address the problem of evil and of the referent – towards whom one should be true – they distinguish three levels: first, the self (sense of self and its representations); second, one's personality dispositions and commitments; and third, positive human nature and universal psychological needs. The first two (the self and its historical commitments and embodied dispositions) are based normatively in the third, "universal positive human nature" (2004, p. 270), which is the source of norms and fundamental values for them. As mentioned earlier, positive psychology draws these notions from pre-empirical sources that are philosophical and religious in origin. One of their sources is Aquinas, for whom this reference to human nature is understood in the context of rational adjudication based on moral norms accessible through natural moral law and divine law, which is found in the Decalogue and elsewhere in Sacred Scripture (Aquinas, I-II, questions 90-108 & II-II, q. 47).

Peterson and Seligman (2004) make a meta-analysis of studies that measure moral integrity, honesty, and authenticity, and the factors that promote or inhibit them. We can draw several

insights from these studies concerning the use of developmental psychology in ethics. To begin with, cognitive developmental studies since Piaget (1932) have identified different moral reasoning styles; these studies indicate that trends toward increasing honesty and integrity, while increasing during childhood, do not necessarily continue to increase in adolescence (Gallup, 2000). This suggests that the development of moral virtue is not complete with the attainment of a mature level of reasoning. Even though cognitive–developmental theory indicates that higher order abstractions (formal operational thinking) are a crucial factor for integrity, high levels of rational intelligence and university education do not correlate with higher integrity scores. Integrity therefore is determined as much by other factors, namely values and life experience, as by cognitive ability per se. Cognitive competence and formal operational thinking are necessary, but not sufficient for moral integrity.

In this regard, studies, based on the Moral Integrity Survey for example (Peterson & Seligman, 2004, p. 262), suggest that a person needs not only to reflect upon moral integrity, but also felt attraction to it and coherent behavior. The necessity to be capable of ordering one's own affects (which requires, among other things, the proper function of the left frontal lobe, ventromedial prefrontal cortex) for consistent moral agency has been documented (Damasio, 1999; Young & Koenigs, 2007). Further studies on integrity as the achievement of identity status suggest that an interconnection of the virtues rooted in moral cognition, affection, and behavior passes through stages, although no one stage theory seems predominant at present in virtue theory.

Developmental studies have also indicated the significance that role models and culture play in leading toward either honesty or dishonesty and toward more or less individualist or social notions of authenticity (Peterson & Seligman, 2004, p. 265). Furthermore, positive psychology (as involving empirical and biological science) has started to correlate neurobiological and environmental conditions "that both promote and prevent authentic self-experience and self-development" (2004, p. 260; Damasio, 1999).

In sum, the positive psychology notion of integrity promotes the unity of affections, intentions, and commitments that are congruent with "positive human nature and universal psychological needs" (2004, p. 269). In the virtue of integrity, taken with positive psychology's notion of practical wisdom, as a basis for interconnecting intentions, commitments, and the possibility of all the virtues, we find a rich description of the psychological function of good practices and virtuous character. A multi-dimensional development and interconnected practice of integrity is indispensable for ethics.

Moral Growth

The practice of ethics requires not only the capacity to cognitively solve ethical dilemmas (norms) or singular ethical acts (action), but also a longitudinal growth in all the qualities (dispositions), including integrity, that contribute to a sustained ethical or good life.

A cognitive-focused structuralist approach will not be able to explain global moral development, without differentiating the ethical import of emotion, volition, and interpersonal relatedness, as well as cognition. The research of Augusto Blasi suggests that the cognitive, affective, and social domains are not organized as one structure, but are more complex than the cognitive structure itself. He says moreover that "it is possible that the integration of moral understanding and motivation is not achieved at approximately the same age for the whole body of moral norms and virtues, but must be worked out separately for different issues" (1996, p. 238; 1980). They all develop neither at the same time nor in the same way as logical-mathematic thought. Because these domains are intertwined, furthermore, there is greater difficulty in identifying stages of development for each virtue, inasmuch as each virtue

requires the support and growth of the others: for example, to be courageous, one needs also to be able to master one's desires for gratification (food, sex, and leisure); to be just, one needs not only to make clear rational judgments, but also to be capable of resisting being dissuaded from truth by fear or jealousy or even by impatience for the attainment of the good.

While cognitive development has been much studied, recent neuroscience has focused more on emotion and volition. A virtue-based approach, because of its complexity, has been found promising for its capacity to differentiate the interrelated network of dispositions, acts, and norms that underlie developmental psychology and ethics. At present, William Casebeer and other neuroscientists hold as a "tentative conclusion [that] the moral psychology required by virtue theory is the most neurobiologically plausible" (2003, p. 4).

In this regard, we can ask: what contributions do developmental accounts of character and virtue make to understanding personal moral growth? First, developmental psychology describes the moral growth that is needed to keep dispositions and acts rooted in ethical norms. Second, it responds to questions about the actual internalization of ends, which is traditionally called the acquisition of virtue (with special need to pay attention also to the connection of the virtues).

First, Peterson and Seligman's approach is specifically descriptive, based on empirical studies (or meta-analyses of such studies), identifying more factors that illustrate the functioning and development of virtue-specific situational themes at social, cognitive, volitional, motivational, and neurobiological levels. For example, their treatment of integrity helps to explain the connection of the virtues at the level of situational themes or concrete practices that actually habituate this connection. Moral efficacy demands such a psychological basis (in acquired virtue) that seeks to integrate thought, sentiment, motivation, and behavior, as Peterson and Seligman have described. They also help to explain how diversity at the situational level (based on personal genetic predispositions, environmental and educational factors, personal commitments, and so on) does not contradict the connection of the major virtues (and character strengths) at higher levels. This approach is also capable of explaining how an agent can perform ethically good acts, while expressing diverse states of ethical wholeness (for example, the ease or difficulty in marital fidelity or civil duty because of one's state of emotional and intellectual dispositions).

Another use that positive psychology's descriptive developmental psychology has for ethics is found when Peterson and Seligman explain the phenomenon of the internalization of ends in terms of the psychological efficacy of "authentic goals." Studies on interiorized goals (Sheldon & Elliot, 1999) indicate that goal self-concordance predicts enduring investment of effort and greater attainment of goals, which in turn contribute to sustained satisfaction of needs and a sense of global well-being and flourishing. This insight enriches, at a psychological level of function, a virtue-base ethics understanding (such as that of Aristotle or Aquinas) of the internalization of moral norms and laws (from a simply external source of law to an internal, personalized source that includes duty but extends to love and respect of the common good), and this as a primary characteristic of the moral development and flourishing of the moral subject (Pinckaers, 2005). This is the pathway of the development of virtue, where law (the normative side of virtue in rules and duties) is at first burdensome and external. Second, through virtuous acts, its meaning and sense is progressively internalized in understanding, motivation, and chosen sentiment (the virtuous character or disposition to act). Finally, it is expressed in mature internalized commitment, a type of ethical flourishing that, while being personally responsible, can also avail itself to interpersonal and religious support.

Developmental Psychology and Normative Ethics

The normative content of acquired moral development and flourishing is assured through the exercise of the virtue of practical wisdom in a normative virtue theory, as for Aquinas and Ricœur. Aquinas' synthesis of agency, disposition, and norms is found in his treatment of the virtue of practical reason (II-II questions 47-56). A similar normative approach is found in Ricœur's deontological test and the three steps of moral maturity—"describing, narrating, and prescribing" (1992, p. 20). Such normative adjudication is not practiced in an anthropological vacuum, free of either moral development or ethical content. Nor is it practiced outside of a worldview and value system that has philosophical and religious implications for ethics. For an Aristotelian-Thomist approach, practical wisdom (prudence) is not merely a formal virtue, reduced to psychological cognitive function (judgments of justice) or the formal goodness of the will. A person uses practical reason (prudence) to discern the norms of law (civic, natural-moral, or religious law) as adjudicated through right reason, but also as revealed in a community of belief. At both these levels, Aquinas holds the functional and normative domains together by his doctrine of the connection of the virtues and of the interdependency of knowledge and love for the acquired virtues.

Aristotle and Aquinas and Ricœur demonstrate an approach that can be clearly normative as well as developmental. However, their descriptions of the function of the virtues are less refined than that of positive psychology. It is particularly at the level of description of the development of character strengths and situational themes or practices that Peterson and Seligman's studies of positive psychology can bring further light to considerations of philosophical and religious ethics in their understanding of moral development and flourishing. The complexity of this work is sizable, but the import even more so.

CONCLUSION

By recognizing the interest of duty and virtue in developmental psychology and ethics we can situate moral development within a wider notion of ethical reason, common good, and ultimate flourishing. On the one hand, a Kohlbergian cognitive-structural stage theory takes seriously the study of judgments of justice within a duty-focused approach. However, the study of the dynamics of justice and duty does not circumscribe the whole of ethics or developmental psychology. On the other hand, positive psychology's empirical studies bring new insights concerning the function, practices, and motivation of a wider range of moral virtues. It contributes to more specifically normative approaches to virtue theory as regards the acquired ethical flourishing that is the aim of moral development and integrity. However, the empirical study of the interconnection of the virtues and of their anthropological bases in the cognitive, emotive, volitional, and interpersonal domains requires further work in science and philosophy.

In conclusion, this brief study has brought to light several challenges and promises concerning the use of moral development in ethics. In a philosophical perspective, we have found some significant differences between cognitive structuralism and virtue-based approaches, especially in their pre-empirical assumptions about the priority of duty or virtue in moral development. It is necessary to further these empirical research projects and philosophical efforts in order to overcome weaknesses on both sides. On the one hand, we need to expand explanations of developmental psychology beyond the theory of stages and structures that focus only on the cognitive judgment of justice and on duty. The other moral factors, including volition, emotion, flourishing, and interpersonal virtues must be interconnected in their proper stages in a person's moral character, moral agency, and normative adjudication. On the other hand, the moral content of the operative notions of virtues and values must be reinforced in schools such as positive psychology. What this amounts to is integrating ethical theory and practice including three levels: attention to the interrelated nature of ethical acts (moral agency), ethical agents (moral character), and ethical norms (duties and law). Such an integrated approach will also have to be attentive to the input that diverse philosophical and religious presuppositions make toward understanding the place of developmental psychology in the practice of ethics.

REFERENCES

American Psychiatric Association. (2000). *Diagnostic and statistical manual of mental disorders* (4th ed., text rev.). Washington, DC: American Psychiatric Press.

Anscombe, G. E. M. (1981). Modern moral philosophy. In *Collected philosophical papers (Vol. 3*, pp. 26–41). Oxford, United Kingdom: Oxford University Press. (Original work published 1959).

Aquinas, T. (1948). Summa theologiae. (English Dominican Province, Trans.). Westminster, MD: Christian Classics. (Original work composed 1265–1273).

Aristotle. (1941). Nicomachean ethics. (W. D. Ross, Trans.). New York: Random House. (Original work composed 350 B.C.).

Benedict, X. V. I. (2005). *Deus caritas est (Encyclical, God is love)*. Vatican: Libreria Editrice Vaticana.

Berkowitz, M., & Oser, F. (1985). *Moral education: Theory and application*. Hillsdale, NJ: Lawrence Erlbaum.

Blasi, A. (1980, July). Bridging moral cognition and moral action: A critical review of the literature. *Psychological Bulletin, 88*(1), 1–45. doi:10.1037/0033-2909.88.1.1.

Blasi, A. (1996). Moral understanding and moral personality: The process of moral integration. In Kurtines, W. M., & Gewirtz, J. L. (Eds.), *Moral development: An introduction*. Boston, MA: Allyn and Bacon.

Browning, D. (2006). *Christian ethics and the moral psychologies*. Grand Rapids, MI: Eerdmans Publishing.

Carr, D., & Steutel, S. (Eds.). (1999). *Virtue ethics and moral education*. London: Routledge.

Casebeer, W. D. (2003, October). Opinion: Moral cognition and its neural constituents. *Nature Reviews. Neuroscience*, *4*, 841–846. doi:10.1038/nrn1223.

Comte-Sponville, A. (1995). *Petit traité des grandes vertus*. Paris, France: Presses Universitaires de France.

Cyrulnik, B. (1998). *Ces enfants qui tiennent le coup*. Revigny-sur-Ornain: Hommes et Perspectives.

Damasio, A. (1994). *Descartes' error: Emotion, reason, and the human brain*. New York: HarperCollins Publishers.

Damasio, A. (1999). *The feeling of what happens*. Boston, MA: Mariner.

Damon, W. (1999, August). The moral development of children. *Scientific American*, *276*, 56–62. PMID:10443038.

Dent, N. (1984). *The moral psychology of the virtues*. New York: Cambridge University Press.

Dent, N. (1999). Virtue, eudaimonia, and teleological ethics. In Carr, D., & Steutel, J. (Eds.), *Virtue ethics and moral education* (pp. 21–34). London: Routledge.

Eisenberg, N. (1982). *The development of prosocial behavior*. New York: Academic Press.

(1998). In Eisenberg, N. (Ed.). Handbook of child psychology: *Vol. 3. Social, emotional and personality development*. New York: J. Wiley.

Eisenberg, N. (2003). The development of empathy-related responding. In Pope-Edwards, C., & Carlo, G. (Eds.), *Nebraska symposium on motivation*. Lincoln: University of Nebraska Press.

Flanagan, O. (1991). *Varieties of moral personality: Ethics and psychological realism*. Cambridge, MA: Harvard University Press.

Flanagan, O. (1996). *Self-expression: Mind, morals, and the meaning of life*. New York: Oxford University Press.

Flanagan, O. (2002). *The problem of the soul: Two visions of mind and how to reconcile them*. New York: Basic Books.

Flanagan, O. (2006). Psychologie morale. In Canto-Sperber, M. (Ed.), *Dictionnaire d'éthique et de philosophie morale* (pp. 1220–1229). Paris: Presses Universitaires de France.

Fowler, J. W. (1980). Moral stages and the development of faith. In Munsey, B. (Ed.), *Moral development, moral education, and Kohlberg: Basic issues in philosophy, psychology, religious, and education*. Birmingham, AL: Religious Education Press.

Fowler, J. W. (1981). *Stages of faith: The psychology of human development and the quest for meaning*. San Francisco: Harper.

Fowler, J. W. (1984). *Becoming adult, becoming Christian*. San Francisco: Harper and Row.

Gibbs, J. (2003). *Moral development and reality: Beyond the theories of Kohlberg and Hoffman*. Thousand Oaks, CA: Sage Publications.

Gilligan, C. (1981). *In a different voice: Psychological theory and women's development*. Cambridge, MA: Harvard University Press.

Gilligan, C. (1992). Reply to Critics. In Larrabee, M. J. (Ed.), *An ethic of care: Feminist and interdisciplinary perspectives* (pp. 207–214). New York: Routledge.

Goleman, D. (2006). *Social intelligence: The new science of human relationships*. New York: Bantam Books.

Guidon, A. (1989). *Le développement moral*. Paris: Desclée.

Handelsman, M. M., Knapp, S., & Gottlieb, M. C. (2009). Positive ethics: Themes and variations. In Snyder, C. R., & Lopez, S. J. (Eds.), *The Oxford handbook of positive psychology* (pp. 105–113). Oxford, United Kingdom: Oxford University Press. doi:10.1093/oxfordhb/9780195187243.013.0011.

Hauerwas, S. (1981). *Vision and virtue*. Notre Dame: University of Notre Dame Press.

Hoffman, M. L. (2000). *Empathy and moral development: Implications for caring and justice*. Cambridge: Cambridge University Press. doi:10.1017/CBO9780511805851.

Jankélévitch, V. (1968). *Traité des vertus*. Paris, France: Bordas/Mouton.

Joseph, S., & Linley, P. A. (2006). *Positive therapy: A meta-theory for positive psychological practice*. London: Routledge.

Kohlberg, L. (1981). *Essays on moral development* (*Vol. 1*). New York: Harper and Row.

Kohlberg, L. (1984). *Essays on moral development* (*Vol. 2*). New York: Harper and Row.

Kohlberg, L., Boyd, D., & Levine, C. (1990). The return of stage 6: Its principle and moral point of view. In Wren, T. (Ed.), *The moral domain: Essays in the ongoing discussion between philosophy and the social sciences* (pp. 151–181). Cambridge, MA: MIT Press.

Kohlberg, L., & Power, F. C. (1981). Moral development, religious thinking and the question of a seventh stage. In Kohlberg, L. (Ed.), *Essays in the philosophy of moral development* (pp. 311–372). New York: Harper and Row. doi:10.1111/j.1467-9744.1981.tb00417.x.

Lapsley, D., & Power, F. C. (Eds.). (2005). *Character psychology and character education*. Notre Dame: University of Notre Dame Press.

Lapsley, D. K. (2006). Moral stage theory. In Tillen, M., & Smetana, J. (Eds.), *Handbook of moral development* (pp. 37–66). New York: Lawrence Erlbaum.

Lickona, T. (Ed.). (1976). *Moral development and behavior: Theory, research, and social issues*. New York: Holt, Rinehart, and Winston.

Lickona, T., & Ryan, K. (Eds.). (1992). *Character development in schools and beyond*. Washington, DC: The Council for Research in Values and Philosophy.

Linley, P. A., & Joseph, S. (Eds.). (2004). *Positive psychology in practice*. Hoboken, NJ: John Wiley and Son. doi:10.1002/9780470939338.

MacIntyre, A. (1985). *After virtue: A study in moral theory*. London, United Kingdom: Duckworth.

Manciaux, M. (Ed.). (2001). *La Résilience: Résister et se construire*. Geneva: Editions Médecine and Hygiène.

Maslow, A. (1956). *Motivation and personality*. New York: Harper Collins.

Nussbaum, M. (1986). *The fragility of goodness: Luck and ethics in Greek tragedy and philosophy*. Cambridge: Cambridge University Press.

Oser, F., & Gmünder, P. (1982). *Der Mensch - Stufen seiner religiösen entwicklung*. Zurich: Benziger.

Oser, F., & Reich, H. (1990). Moral judgment, religious judgment, world view and logical thought: A review of their relationship. *British Journal of Religious Education, 12*(3), 94–101. doi:10.1080/0141620900120207.

Oser, F., Scarlett, W. G., & Bucher, A. (2006). *Religious and spiritual development throughout the life span* (6th ed., pp. 942–998). Handbook of child psychology Hoboken, NJ: Wiley.

Peterson, C., & Seligman, M. E. P. (2004). *Character strengths and virtues: Handbook and classification.* Oxford: American Psychological Association and Oxford University Press.

Philibert, P. (1987). Addressing the crisis in moral theory: Clues from Aquinas and Gilligan. *Theology Digest, 34*(2), 103–113.

Philibert, P. J. (1975). Lawrence Kohlberg's use of virtue in his theory of moral development. *International Philosophical Quarterly, 15*(4), 455–479. doi:10.5840/ipq197515445.

Philibert, P. J. (1988). Kohlberg and Fowler revisited: An interim report on moral structuralism. *Living Light, 24*, 162–171.

Piaget, J. (1997). *The moral judgment of the child.* New York: Free Press. (Original work published 1932).

Pieper, J. (1966). *The four cardinal virtues.* Notre Dame, IN: University of Notre Dame Press.

Pinckaers, S. (1978). *Le renouveau de la morale.* Paris, France: Téqui.

Pinckaers, S. (1995). *Sources of Christian ethics.* Washington, DC: Catholic University of America Press.

Pinckaers, S. (2005). *The Pinckaers reader.* Washington, DC: Catholic University of America Press.

Pope, S. J. (1992). *The evolution of altruism.* Washington, DC: Georgetown University Press.

Power, F. C. (1988). The just community approach to moral education. *Journal of Moral Education, 17*, 195–208. doi:10.1080/0305724880170304.

Rawls, J. (1971). *A theory of justice.* Cambridge, MA: Harvard University Press.

Ricœur, P. (1992). *Oneself as another.* Chicago: University of Chicago Press.

Rogers, C. (1959). *A theory of therapy, personality, and interpersonal relationships.* New York: McGraw-Hill.

Rogers, C. (1961). *On becoming a person: A therapists view of psychotherapy.* Boston: Houghton Mifflin.

Seligman, M. E. P. (1991). *Helplessness: On depression, development, and death* (2nd ed.). New York: W. H. Freeman.

Seligman, M. E. P. (1998). *Learned optimism.* New York: Simon and Schuster.

Seligman, M. E. P. (1999, August). The president's address. APA annual report. *The American Psychologist*, 559–562.

Seligman, M. E. P. (2002). *Authentic happiness: Using the new positive psychology to realize your potential for lasting fulfillment.* New York: Free Press.

Seligman, M. E. P. (2012). *Flourish: A visionary new understanding of happiness and well-being.* New York, NK: Free Press.

Seligman, M. E. P., Reivich, K., Jaycox, L., & Gillham, J. (1996). *The optimistic child.* New York: Harper Collins.

Seligman, M. E. P., Walker, E., & Rosenhan, D. L. (2001). *Abnormal psychology* (4th ed.). New York: W.W. Norton.

Sheldon, K. M., & Elliot, A. J. (1999). Goal striving, need-satisfaction, and longitudinal well-being: The self concordance model. *Journal of Personality and Social Psychology, 76*, 482–497. doi:10.1037/0022-3514.76.3.482 PMID:10101878.

Snyder, C. R. (1994). *The psychology of hope: You can get there from here.* New York: Free Press.

Snyder, C. R., & Lopez, S. J. (2007). *Positive psychology: The scientific and practical explorations of human strengths.* Thousand Oaks, CA: Sage.

Snyder, C. R., & Lopez, S. J. (Eds.). (2009). *The Oxford handbook of positive psychology.* Oxford: Oxford University Press.

Spohn, W. C. (2000). Conscience and moral development. *Theological Studies, 64*, 122–138.

Taylor, C. (1991). *The ethics of authenticity.* Cambridge, MA: Harvard University Press.

Titus, C. S. (2010). Moral development and connecting the virtues: Aquinas, Porter, and the flawed saint. In Hütter, R., & Levering, M. (Eds.), *Ressourcement Thomism* (pp. 330–352). Washington, DC: The Catholic University of America Press.

van der Ven, J. (1998). *Formation of the moral self.* Grand Rapids, MI: Eerdmans Publishing.

Vanistendael, S. (1994). *La résilience ou le réalisme de l'espérance. Blessé mais pas vaincu.* Geneva: Les cahiers du BICE.

Vitz, P. (2006). From the modern individual to the transmodern person. In Titus, C. S. (Ed.), *The person and the polis: Faith and values within the secular society* (pp. 109–131). Arlington, VA: The Institute for the Psychological Sciences Press.

Vitz, P. C. (1981). Christian moral values and dominant psychological theories: The case of Kohlberg. In Williams, P. (Ed.), *Christian faith in a neo-pagan society* (pp. 35–56). Scranton, PA: Northeast.

Vitz, P. C. (1982). *Christian perspectives on moral education: From Kohlberg to Christ.* (Unpublished manuscript).

Werner, E. E., & Smith, R. S. (1986). *Vulnerable but invincible: A longitudinal study of resilient children and youth.* New York: Adams, Bannister, Cox.

Wink, P., & Helson, R. (1997). Practical and transcendent wisdom: Their nature and some longitudinal findings. *Journal of Adult Development, 4.*

Yearley, L. H. (1990). *Mencius and Aquinas: Theories of virtue and conceptions of courage.* Albany, NY: State University of New York Press.

Young, L., & Koenig, M. (2007, December). Investigating emotion in moral cognition: A review of evidence from functional neuroimaging and neuropsychology. *British Medical Bulletin, 84*(1), 69–79. doi:10.1093/bmb/ldm031 PMID:18029385.

ENDNOTES

[1]   Because of the ethical content in the virtues related to belief and rationality (which they situate under "open mindedness"), they focus on judgment processes rather than the content of beliefs. Peterson and Seligman claim that "the benefit of this view is that it does not require us to judge the quality of people's thinking by the veracity of their conclusions (which we can do in mathematics and precious few other domains). And we do not have to agree with people's beliefs to recognize that in some cases they were arrived at thoughtfully whereas in other cases they were not" (2004, p. 100-101).

[2]   This solution however is not uncontroversial, when faced with other interpretations of the precept to love one's neighbor as oneself. Fuller notions of love are found in the in-

terdisciplinary dialogue of Steven J Pope's *The Evolution of Altruism* (1992) and the philosophical and theologically rich treatment of love in Benedict XVI's Encyclical *Deus caritas est* (2005).

[3] Peterson and Seligman note that "philosophers consider wisdom or reason as the chief virtue making all others possible" (2004, p. 95; see also 182). They give a consensual definition of the major virtue, practical wisdom, which: "is distinct from intelligence; represents a superior level of knowledge, judgment, and capacity to give advice; allows the individual to address important and difficult questions about conduct and meaning of life; is used for the good or well being of oneself and that of others."

Chapter 19
Habitus and Reflexivity:
On Bourdieu's Self Socioanalysis

Martine Legris Revel
Université de Lille Nord de France, France

ABSTRACT

The author intends to explore the work that can be done by somebody on his/her own history in order to modify its course. Those questions have been largely debated. The purpose of this paper is to analyze Pierre Bourdieu's thinking of those topics. He is concerned with the interiorization of social structures (habitus) and suggested self socioanalysis as a new way of highlighting them in 1991. Can somebody truly escape from social structure's interiorization effects or even have an action upon them?

INTRODUCTION

Bourdieu had a cursus in philosophy before he chose social sciences, precisely sociology. In France, the first license of sociology was created in the early 60's. At the beginning of his career, he endorsed a Bachelardian perspective on science. In his first writings, especially in 1968 in *Le métier de sociologue* (Bourdieu et al., 1983), he was building a Bachelardian approach of science as objective and external to the sociologist,

implying the moment of rupture with common sense, prenotions and the reigning doxa, as one of the key moment of a scientific methodology of sociology. Still he did publish at the end of his career a small book called *Sketch for a self analysis* (2004 French version, 2008 English version, Bourdieu, 2008) in which he claims to analyze his own schemes of thinking and acting in an attempt to provide a rigorous approach of the social construction of the self as an habitus. We will discuss in this paper how this evolution was made possible and if it can highlight the debate upon normative and cognitive framing of the individuals and the possibility of a shift to occur.

DOI: 10.4018/978-1-4666-3670-5.ch019

Bourdieu's entry into political engagement really became public during the French 1995 strike that went on over one month. He actually met people on strike in Paris in Gare de Lyon, and pronounced a speech, intending to give them clues and information about their positions as social agent (Bourdieu, 1998). His involvement was not new, as he had already through his academic writings denounced social domination and reproduction. Still he did truly evolve greatly in his positions from his career start to his latest writing *Sketch for a self analysis*, in an attempt to analyse his own position and habitus in the academic field. Bourdieu never forgot his interest for science and epistemology; he did limit reflexivity to the practice of the sociologist. Thinking in terms of habitus introduces a considerable dose of skepticism regarding the emancipatory potential of knowledge, so how can self socio analysis help escape social determinism?

In the first part of this article, I wonder whether reflexivity has any role to play regarding the perceptions and self-perceptions of social agents or whether it is restricted to sociologists, being a methodological principle aimed at the epistemological strength of social sciences.

In other words, it is the modus operandi of his social scientific endeavour, rather than the symbolic capital associated with his position as a social scientist, that will guide our thinking about Bourdieu's work.

In the second part, I study how self-socioanalysis can play a role in a kind of liberating anamnesis. I then wonder if rational and emotional levels are intertwined, and then if social agents, here Bourdieu as a case, did transform the relationship between their habitus and the academic field.

I will then tackle three among this book hypothesis:

- A context if viewed as a kind of inescapable a priori for judgment can be de-stabilized and possibly re-structured through interaction with other actors, but cannot be entirely reflected by the actors themselves.

- A procedure can be adjusted to the context of individuals or communities without requiring a complete reflection of the context by the actors and by implementing at most a re-construction of it.

- A shift in the cognitive and normative framing to occur depends upon a plurality of factors that, to be effective and relevant, are themselves related to the specific context of an individual or a community.

REFLEXIVITY AS A SCIENTIFIC TOOL

Bourdieu, with his emphasis on reflexivity, tempted his readers to try out his theories on his own writings, habitus and conclusions, in short, to be reflexive upon his own work. These bourdieusian readings located his analyses in their social contexts, i.e. the author's social position and trajectory in the academic and intellectual fields and the history and logic of these fields. He then focus his analysis not only on reflexivity as a methodological standard aimed at building the epistemological solidity of social science, but also as a way for social agents to become aware of their own habitus, through self analysis.

How does somebody acquire reflexivity in Bourdieu's framework (Mouzelis, 2007)? Whereas a crisis or a disruption between the habitus of a social agent and his location in the field is usually the painful first step towards acquiring reflexivity, in the social scientific field, this crisis can be remedied by practices of socioanalyses. As such, Bourdieu's notion of socioanalysis, could offer clues for identifying ways to understand and maybe overcome social structures interiorization.

This is not an Autobiography

Bourdieu always used reflexive sociology as the core of his theoretical framework. I chose to use some chapters of his book *Raisons pratiques* (Bourdieu, 1994) to show what other authors already stated: reflexivity is central (Gingras, 2010). But he actually went a step forward or for some analysts made a rupture, when he began to engage in public debate, and simultaneously, to create a new way to explore the concept of habitus: the self socioanalysis.

In his book about self socioanalysis, Bourdieu starts with an aphorism "this is not an autobiography." He had already criticized the idea and illusions of autobiographical work in 1996. In his book on self analysis, Bourdieu speaks about his own actions. He uses the "I" also the "we." But he does not start relating his childhood; his first chapter begins with the academic field state description and analysis.

He wants to make it clear: no psychological issues will be discussed in this sketch and he is not writing his autobiography. I will first synthesize his main critics about autobiographical works.

For him, speaking about "life history" implies the not insignificant presupposition that life is a history. This way of looking at a life implies tacit acceptance of the philosophy of history as a series of historical events (Geschichte) which is implied in the philosophy of history as an historical narrative (Historie), or briefly, implied in a theory of the narrative. An historian's narrative is indiscernible from that of a novelist in this context, especially if the narration is biographical or autobiographical.

Bourdieu tends to unravel some of the presuppositions of this theory.

How to answer indeed, without taking out limits of the sociology, the old empiricist interrogation on the existence of a me inflexible in the harmony of the singular sensations? Doubtless we can find in the habit the active, inflexible, principle in the passive perceptions of the unification of the practices and the representations.

By this completely singular shape of appointment which establishes the proper noun, is established a constant and long-lasting social identity which guarantees the identity of the bio-logical individual in all the possible fields where it intervenes as agent, that is in all his possible stories of life. This is absurd because one needs to analyze the structure of the distribution of social agents movement in a given and timely defined social space.

I tried to translate a part of the third chapter of *Raisons pratiques*:

The notion of trajectory as series of the positions successively occupied by the same agent (or the same group) in a very space to become there and subject to ceaseless transformations. Try to understand a life as a unique series and to one sufficient (self-important) of successive events without the other link than the association on a "subject" the constancy of which is doubtless only that of a proper noun, is about so absurd as to try to return reason of a route in the subway without taking into account the structure of the network, that is the matrix of the objective relations between the various stations. The bibliographic events define themselves as so many investments and movements in the social space, that is, more exactly, in the various successive states of the structure of the distribution of the various sorts (species) of capital which are stake in the considered field.

If Bourdieu is not writing his autobiography, because as he said he is not relating his childhood and affective relationships, it is also because he is not completely telling a story. "So, one can say that his own self analysis in not simply, as it is often the case, the mechanical result of academic aging and retirement, when a scholar tends to look back instead of forward, often pressed by young colleagues asking biographical questions about relations entertained with the major actors of the discipline" (Gingras, 2010, p. 621).

Bourdieu is aiming at articulating theoretical and empirical investigations.

The Field and Habitus as an Articulation between Theoretical and Empirical Investigations

Social fields are seen as a context, actively inter-acting with the habitus. The social agents can not evolve without the social fields and the habitus being at least restructured. This is not through interactions with others that this may occur but through the awareness of one's position and status in existing social fields.

I have said often enough that any cultural producer is situated in a certain space of production and that, whether he wants it or not, his productions always owe something to his position in this space. I have relentlessly tried to protect myself, through a constant effort of self-analysis, from this effect of the field. But one can be negatively "influenced," influenced a contrario, if I may say, and bear the marks of what one fights against. Thus certain features of my work can no doubt be explained by the desire to "twist the stick in the other direction", to react against the dominant vision in the intellectual field, to break, in a somewhat provocative manner, with the professional ideology of intellectuals. This is the case for instance with the use I make of the notion of interest, which can call forth the accusation of economism against a work which, from the very beginning (I can refer here to my anthropological studies), was conceived in opposition to economism. The notion of interest — I always speak of specific interest — was conceived as an instrument of rupture intended to bring the materialist mode of questioning to bear on realms from which it was absent and [to bear] on the sphere of cultural production in particular. It is the means of a deliberate (and provisional) reductionism which is used to take down the claims of the prophets of the universal, to question the ideology of the freischwebende Intelligenz [free-floating intellectual]" of interest and the relative autonomy of symbolic power (1988).

Bourdieu uses reflexive sociology as a tool, in-corporated in his theoretical framewok, conceived to reach scientific rigor and objectivity. The deeper is the relationship between state of social field and habitus, the greater is the objectivation work.

Nevertheless, reflexive sociology is different from self socioanalysis.

SELF SOCIOANALYSIS: A LIBERATING ANAMNESIS ?

In the book *Sketch for a self analysis*, Bourdieu wants to use the most objective analysis to under-stand the most subjective part (Bourdieu, 2004, p. 11). He claims to maintain the distinction between sociological and psychological aspects, his self socioanalysis aiming to point only sociological frames and mechanisms.

As we will show, there is a huge tension be-tween his efforts to keep rational and the merge of his emotions.

Rationality and Emotions

Bourdieu's initial revolt is apprehended in psy-chological and emotional terms and related to the symbolic violence he faced due to his dissonant dispositions and positions: a feeling of revolt based on debt and deception toward academic consecration, a feeling of insecurity related to his status as a 'miraculous' survivor, a feeling of despair and emptiness related to his ambiguous relation to himself and the feeling of resentment, always present in the background of Bourdieu's socioanalysis (Bourdieu, 2004, p. 32). "Collusion semi maffioso capable of assuring a social rela-tionship of accommodation," did he quote about academic field structures.

Taking his last book as the perspective from which to assess Bourdieu's socioanalysis, the ideas of rupture, estrangement, appear fundamental to Bourdieu's selfperception and socioanalysis. Be it in his childhood in Béarn, his school years, or his career, Bourdieu described himself as a marginal torn between his social origins, on one hand, and, on the other, his ascending trajectory and social position. This situation was described in terms such as 'porte-à-faux,' 'transfuge fils de transfuge,' and 'habitus clivé' (Bourdieu, 2004, p. 76, p. 109, p. 130). This sense of not belonging shaped his life decisions, guiding him toward sociology, the 'pariah discipline' and away from philosophy, with its scholastic and distant relation to the social world, affecting his personal relations, leading him away from the figure of Sartre and closer to that of Canguilhem, and shaping his methodology and political attitude, 'always going in the opposite direction of the dominant models and modes in the field' (Bourdieu, 2004).

He is often using brackets to describe or approach the questions difficult or loaded with emotions with precaution, as to put them at a distance from his text (Bourdieu, 2004, p. 102). For example his writing about Foucault contains half a page in brackets. As if words into brackets did not really have importance, or did escape his reflexive vigilance.

I think that Bourdieu contains his feelings to transform them, while refusing to enter a reasoning on the part of his life which escapes any rationalization attempt. The "naturalization" of his position also passes by a shape of denial or avoidance.

Changing Oneself: a Liberating Anamnesis?

The conversion of the gaze implied by the exercises of socioanalysis presents possibilities for social agents to comprehend, accept and even change their social selves.

Turning the "scientific gaze" back upon ourselves, writes Bourdieu, "enables us to assume ourselves and even ... to claim ourselves," offering the practitioners of socioanalysis "the possibility of

assuming their habitus without guilt or suffering" (Bourdieu & Wacquant, 1992, p. 199).

"Socioanalysis allows its practitioners to understand the mixture of existential contingencies and social determinations that makes their disparate selves. It represents the active counterpart of the dislocation of habitus and positions as the impetus for reflexivity" (Frangie, 2009, p. 219). The self socioanalysis of the sources of one's schemes of thinking, evaluating and acting is relevant to deal with emotions. Leading to the transformation of the habitus, socioanalysis is nonetheless a painful process of existential rupture and re-creation. Then you become a subject, because you are aware of social domination mechanisms.

Domination, in Bourdieu's framework, works through the imposition of specific subjectivities on objective positions, naturalizing social domination and its reproduction. Social contextualization and denaturalization short-circuit this symbolic violence, contributing to the autonomy of social agents from imposed subjectivities. Moreover, as Bourdieu often noted the reproduction of domination works through a lot of channels, one of them being the emotional aspect of the habitus. The hierarchical stratification of society and its boundaries are partly grounded in bodily emotions, including shame, embarrassment, timidity, anxiety and guilt (Bourdieu, 1998, pp. 38-39). Although the process of socioanalysis may not be a sufficient guarantee against domination, it might ease social agents from the emotional burden associated with social hierarchy.

Self socio analysis aims at discovering the historical and field-dependent social influences that coalesce to form the individual specific habitus.

As Bouveresse states it, the objective is "to reconcile with oneself and with one's social property through a "liberating anamnesis" and to have a precious instrument ... for the study of the social world" (Bouveresse, 2003, p. 174).

Frangie (2009, p. 225) writes that Bourdieu's writings, "in addition to their sociological worth, were a way to come to terms with his biography, or as he puts it, the task of sociology was to strip the 'terrorism of resentment' from its objective and subjective flawlessness" (Bourdieu, 1983, p. 28-29).

CONCLUSION

It is through this double dimension of self socio-analysis, as both a tool for social research and a way to deal with one's blind impress that Bourdieu's work emerges as an example of a pertinent socioanalysis. Bourdieu became autonomous by socially grounding his life and denaturalizing it.

His attempt (as he did write a sketch) shows some limitations as any other intellectual essay. His writings about some of his colleagues and competitors happily remained short but point his inner tension towards his own work's justification. Besides doing one self socioanalysis may be an experience limited to some experts, or some specific social construction schemes of the habitus. One can not use self socioanalysis without a huge knowledge capital. Being a "transfuge," or what he perceived as a "marginal" might be useful to practice self socioanlysis as this position in the social field tends to question some of its habitus.

Through the movement between his biography and his writings, and the resulting creation of a perspective on the social world that allowed him to smooth out the contradictions in his own life, Bourdieu provided an example on how one can deal with the social determinations and existential contingencies that compose one's self.

REFERENCES

Bourdieu, P. (1994). *Raisons pratiques*. Paris: Seuil.

Bourdieu, P. (1998). *Contre-feux*. Paris: Liber raisons d'agir.

Bourdieu, P. (2000). *Esquisse d'une théorie de la pratique*. Paris: Seuil.

Bourdieu, P. (2004). *Esquisse pour une auto-analyse*. Paris: Raisons d'agir.

Bourdieu, P. (2008). *Sketch for a self-analysis*. Chicago: University of Chicago Press.

Bourdieu, P., Chamboredon, J. C., & Passeron, J. C. (1983). *Le Métier de sociologue*. Paris: Mouton Editeur.

Bourdieu, P., & Wacquant, L. (1992). *Réponses*. Paris: Seuil.

Bouveresse, J. (2003). *Bourdieu, savant et politique*. Argone.

Frangie, S. (2009). Bourdieu's reflexive politics. *European Journal of Social Theory*, *12*(2), 213–229. doi:10.1177/1368431009103706.

Gingras, Y. (2010). Sociological reflexivity in action. *Social Studies of Science*, *40*(4), 619–631. doi:10.1177/0306312710370236.

Mouzelis, N. (2007). Restructuring Bourdieu's theory of practice. *Sociological Research Online*, *12*(6), 9. Retrieved November 16, 2012, from http://www.socresonline.org.uk/12/6/9.html

Reed-Danahay, D. (2005). *Locating Bourdieu*. Bloomington, Indianapolis: Indiana University Press.

Section 8
Changing Methods and Procedures

Section 8: Changing Methods and Procedures

The very idea of proceduralism comes from the need to provide some rational justifications to the rules, actions, and decisions to be adopted, undertaken, or made by the society or the power in the context of highly "plural-complex-developed" societies. Proceduralism as a generic method is particularly concerned with the *governance of ethics* (i.e. with the institutional and organisational conditions that the procedures must fulfill so that ethical questions can be addressed). However, the actual methods and procedures of governance can be shown for the most part inadequate or insufficient in attempting to capture the *ethical significance of norms*, especially in the regulation of emerging technologies. The ethical significance is assumed to be an essential aspect of the normative judgment insofar as it comes to unveiling the context-based evaluation of a norm that precisely provides it with some significance. The consequence of it is the necessity to reform or at least to enlarge the scope of the methods of governance beyond the sole dialogical procedures that are typical of the classical streams of proceduralism.

Thus, it can be illuminating to analyse the various methods of governance in pursuing the goal of understanding *possibility from a point of view,* as the ability to change the scope of possibilities for an actor is an essential part of how governance work. Expertise as the unquestioned source of normativity in governance means that practical significance is played down and the construction of context is absent, hence the need for a new approach. Dialogue is cited as the solution to the problems, but the limit of dialogue is in the nature of rationality assumed to be operative in human motivation, for value is left out of the account and so the idea of 'filtering for relevance' gains no purchase. The influential *Open Method of Coordination* (OMC) and its theoretical background in *Direct Deliberative Polyarchy* (DDP) allocate an important role to dialogue. However, they do not have an in-built ethical structure, and the structures alone are not sufficient to realise ethics, neither to necessarily permit reflexivity. The position of the actors, their context and their conception of their own possibilities, is not automatically considered when one of these governance measures is adopted. A new approach in ethics and governance is expected to offer a criterion of evaluation and a more interesting way to address the conditions for an *ethical reflexivity*. It must also address the conditions for determining the conditions of construction of ethical issues and norms as well as the conditions for their adoption and implementation.

In this perspective, the *transformation of proceduralism* consists in shifting from a *contextual* to a *comprehensive* method of reflexive equilibrium that incorporates the *context* dependence of procedures and the *value* significance of norms. *Comprehensive proceduralism* as a method is a plea for a combination of approaches that is *procedural* (rule-based), *reflexive* (context-based), and *substantive* (value-based). The method of comprehensive proceduralism seeks to explore the special combination of procedures that is best adapted to the contexts of individuals or groups in order to guarantee the significance of norms as being relevant to their value-systems. This option equals stating that there is no "one best way" for any procedural method that deals with the complexity and the heterogeneity of social and cultural frames. We can qualify this method a reflexive equilibrium through an adjustment of procedures to the contexts and the values of the actors, or, in other words, a *comprehensive procedure of reflexive equilibrium* that is context-adaptive and value-sensitive. The principle of judgment in the method of comprehensive proceduralism is not that of consensus, but that of a two-fold potential outcome: consensus or dissensus, consensus on some aspects, dissensus on some other aspects. In this line, a consensus can be reached about a norm, a rule or a principle (e.g. dignity, privacy, etc.), while the significance of it, based upon each agent's context, is not the same at all. The main stake of a reflexive equilibrium is then to produce a *reflexive exploration* of the agent's context as well as a *substantive determination* of the value-significance of norms for each agent. This process of determination enables the actors to come out of the situation of substantive indeterminacy, in which they are misled by the appearance of formal agreement on a principle or a rule. Thus a governance process must be able to offer a variety of procedures best adapted to the actor's context (*context-adaptive proceduralism*) and to explore the diversity of value-significances given to a norm on the basis of the actors' contexts (*value-sensitive proceduralism*).

Chapter 20
Competing Methodologies:
Possibilities from a Point of View

Stephen Rainey
University of Namur, Belgium

ABSTRACT

This chapter aims to construct a basis to move toward addressing lacunae in governance approaches in a way that is not merely ad hoc but rather is grounded in theory and that can affect practice in a positive way. Essential to this is the establishing of a problem clearly defined in order that it can be a problem understood. Here, the ground is cleared for proposals for ways to approach the specific problems we detect. In doing this, the ambition of treating them is rendered possible.

INTRODUCTION

Understanding how possibilities appear to individuals is a key part of how governance can work. In understanding possibility from a point of view, those in governance can open the door to formulating policy that speaks to the practical orientations of real citizens. This means looking at a practical, rather than a theoretical reality, and examining that reality in the light of practical implications and results, as well as practical

DOI: 10.4018/978-1-4666-3670-5.ch020

expectations. These are important elements of context, that very context that EGAIS has insisted upon the necessity to construct. Here, we look at various methods used in pursuing the goal of understanding possibility from a point of view.

PRACTICAL SIGNIFICANCE AND CONTEXT

Expertise, be it philosophical or scientific, as the unquestioned source of normativity in governance, means that practical significance is played down and the construction of context is absent. What's required is an approach that can offer first a criterion of evaluation and second a more interesting way to address the conditions not only for an ethical reflexivity, but also for determining the conditions of construction of ethical issues, of ethical norms, and the conditions for their adoption and implementation without undue presuppositions.

The reasons that we have to accept or refuse a proposition, a course of action, or a change in our way of doing things will not themselves necessarily be susceptible to change on some pre-defined basis. For example, one might refuse simple, life-saving medical treatment despite being convinced of the efficacy of the measures suggested as they are presented, but on the thicker, value-laden grounds that dignity in natural death is more appropriate. The 'obvious' good of life-saving treatment may not appear so obvious, depending upon how that treatment, its consequences and its basis appears in the consciousness of the actor who would be saved. Understanding reactions such as these requires broad empathy more than formal logic – at play is the practical logic of the individual. This is also represented in the thought of Joseph Raz (Mulhall & Swift, 1996, p. 342), a liberalist thinker who nonetheless appreciates much of a more communitarian position:

… autonomous persons are those who can shape their life and determine its course. They are not merely rational agents who can choose between options after evaluating relevant information, but agents who can in addition adopt personal projects, develop relationships, and accept commitments to causes, through which their personal integrity and sense of dignity and self-respect are made concrete. Persons who are part creators of their own moral world have a commitment to projects, relationships and causes which affect the kind of life that is for them worth living.

These hint at a broader, value-laden conception of rationality that is required in order to comprehend the practical logic of human action.

This is essential to the project of framing ethical governance owing to the need to account for human action in all its plurality and in its own terms (i.e. inclusive of value). Toward this end, Jean-Marc Ferry distinguishes four distinct types of rationality relevant to the human condition. His position is summarised in Smith (1994) as follows:

Historical and discursive progress from narration toward reconstruction is associated with increasing reflexivity about identity and the grounds upon which it is established. Narration, in Ferry's view, consists of ossified traditional myths which define identities in a more or less prescriptive, taken-for-granted way. Interpretation, on the other hand, involves the assimilation of identity to universal categories like law and justice and is exemplified in early Christian and ancient Greek thought. Argumentation opens up claims of identity to rational dialogue as embodied, for example, in the Enlightenment. Reconstruction, the final step toward reflexivity, involves hermeneutic attempts to understand the historical grounds behind the "good reasons" offered by others in argumentation. This is in part a logical and ethical consequence of the shift from it to you (acknowledging subjectivity) which emerges with argumentation itself.

Somewhere, reason runs out and the framing that constitutes my way of seeing the world steps in – the deep sense of my self and all that my convictions connote. My being, in a thick sense that includes my upbringing, cultural religious convictions, feelings of indebtedness to a past, honouring legacies etc. While this is clearly important in comprehending who/what a person is, it is only comprehensible if we step back from the primarily argumentative mode of discourse and regard framing not as an aggregative report of experiences had between various times, but rather as the authentic self-portrayal of a human being in terms of a life lived – i.e. we need to use a recognition principle in order to cognise the information encoded by the manner of framing.

In broad terms 'dialogue' is cited as the solution to the problems here flagged. To this we now turn in order to see how dialogical solutions to governing pluralities of responses manifest and how they are limited. In short, the limit is in the nature of rationality assumed to be operative in human motivation. Essentially, value is left out of the account in general and so the idea of 'filtering for relevance', just mentioned, gains no purchase (put differently, my account for why a reason is a *compelling* reason for me will rely on the things I value, not the things I can merely synthesise as conclusions from arguments).

DIALOGICAL GOVERNANCE

Given the place for dialogue in the influential Open Method of Coordination (OMC), and its theoretical background in Direct Deliberative Polyarchy (DDP), it remains to be asked; just what does *dialogue* here mean? It cannot be assumed that such a term has a straightforward, objective, final, and simple meaning. Rather, it must be investigated in order to understand just what it means, or ought to mean. Essentially, the point of dialogue, as we have seen, is to permit legitimacy on a democratic sense in that appropriate input from stakeholders is facilitated in having a space for discussion. That said, there do seem to be various parameters that must be assumed to obtain. For instance, where the aim is this legitimising sense of representative democracy, symmetry as regards power must be assumed.

There are differences between a lecture, a therapy session, a dressing down and a debate. In the first three instances there is an asymmetrical power distribution among interlocutors that characterises the function of the particular encounter. None of these can be the legitimising democratic sort of encounter that seems to be required in DDP and OMC, which in fact out more to resemble a

debate. The most influential figure in the field of communicative action is Jürgen Habermas. His work engages precisely with the characterisation of public dialogical encounters, and so we must turn to his work in order to flesh out the concept of 'dialogue' that so far has been unanalysed. Moreover, Habermas' participation in ongoing debates with Rawls makes his input here relevant in a number of ways.

DELIBERATIVE OR COLLABORATIVE-RATIONAL APPROACHES

Jürgen Habermas' conception of how a rational society ought to be arranged is closely modelled upon the ethical arrangements he supposes to be implicit in any communicative encounter. Rather than validating the autonomous credentials of one's will in public decision-making (as might a decision-theoretic champion), Habermas defines rational public uses of reason in terms of 'reasoned agreement'. Rather than seeking the force of reason itself in public uses of reason emanating from the autonomous will, Habermas (1998, p. 306) seeks the 'unforced force of the better argument' as the mark of rational public-decision. Rather than placing the onus of recognising rational outcomes upon the individual thinker, here the procedure of argumentation bears the burden, with normativity coming from the procedure – one ought to do/make/say/think *x* as it is made compelling owing to its well-argued form.

Habermas is tied to creating a position of the communicative deliberator *per se*, that is, a position of rational decision-making without regard to content. Whereas philosophers had previously sought to locate this endeavour within structures of the mind, or of consciousness generally,[1] Habermas looks to the structures of communication, regarding this as the repository *par excellence* of human understanding.

What eventually is yielded by the Habermasian position is an institutionalisation of the public use of reason that is thought fit to handle practical questions about means toward ends; ethical questions about what ought to be valued; and moral questions about justice and fairness. It is thought to be thus fit owing to its central preponderance with argumentative procedure – it is a free and open civil forum wherein diverse reasons can be confronted with one another, hypothetically contested, resolved, etc. With the procedure laid out correctly, Habermas supposes, the substantial issues of content can look after themselves within it.[2]

Somewhat like Kant, and despite the move from self-conscious willing, the normativity in this account can still be seen to reside in reason itself as despite being allowed to air and challenge value-laden propositions in a public, argumentative procedure, it is ultimately supposed that rational procedures will yield rational outcomes owing to the unforced force of the better argument's aptness for transforming opinion and belief. In other words, there is a presumed parallelism between the justification of and application of a proposition. Values held by interlocutors are excluded in the sense that they are presumed to fit within the rational structures of discourse, which in turn is presumed adequate to deliver results that will affect those values.

Given this is a procedural account, one wherein the parameters are set rather than the content, values seem to pose a problem. Values inform the background *habitus*, or *lebenswelt* – the lifeworld – in which actors realise their possibilities. This value-informed and laden background of competencies, abilities, perspectives represents the entire cognitive horizon for each actor and each actor collective. According to McCarthy (1994, p. 46) value-pluralism means that

Questions of self-understanding and self realization, rooted as they are in particular life histories and cultures, do not admit of general answers;

prudential deliberations on the good life within the horizons of different lifeworlds and traditions do not yield universal prescriptions.

It would therefore seem that while argumentation is possible, conclusion is improbable. If different values inform radically different views of the world, dialogue might well just run out, or begin and end in nebulous discussions of detail without any effect.

Since the notion of framing at work here will only be relevant in terms of a life lived, via specific interpretations of life-events by an agent, an extra ethical dimension is required in order to comprehend it – the ethics of the discourse model are exceedingly spare and amount to procedural tenets. The place of framing can be seen as illustrated by the following problem:

In a discourse, when matters of justice arise and competing, contradictory arguments are aired, it is required that the parties involved will submit themselves to nothing but the force of the better argument. But the acceptance of arguments will itself be conditional on values embedded within an agent's way of seeing things.

Thus, frames don't fit *within* argumentation, but rather argumentation *decentres* the expressive authenticity of the perspective from a frame. 'Decentring' means the way in which actors must move away from their own contexts of action when considering questions of what is true or right. For Habermas (2005, p. 108):

An absolute claim to validity has to be justifiable in an ever wider forum, before an ever more competent and larger audience, against ever new objections. This intrinsic dynamic of argumentation, the progressive decentring of one's interpretative perspective, in particular drives practical discourse ...

After just this small outlining of the basis for the deliberative approach we can already see that positing 'dialogue' as the solution to the problem of ethically governing plurality is really not an answer but rather is a problem in need of investigation. The problems we have seen already mentioned, concerning the lack of representation of real perspectives, is replicated here in the basis of the deliberative model as we see 'argumentation' *qua* procedure cited as a panacea. However, as was intimated in the references Raz and Ferry there seems a need for a subtle, differentiated understanding of the practical logic of human action such that value and reason can be respected. We have here in this model of a deliberative approach a reduction to argumentation that seems incapable of accommodating the sorts of factors that come to the fore in the tradition apparent in the position that supports those earlier quotations.[3] The problem is to re-integrate value and norm in public reasoning.

Habermas takes a pragmatic position in that he takes as his units of analysis discourse participants. He begins with the reality and conducts a kind of transcendental analysis in order to detect within the reality that which must be so. Such argumentation proves more controversial theses on the bases of less controversial theses by examining the less controversial and analysing out the basic concepts on which these must operate.

However, it is the suggestion here that the deliberative model gleaned from Habermas in fact *is not pragmatic enough* in that it doesn't capture the real reality of a world framed by reason and value, history and culture, tradition and national identity as it appears to individuals. In missing these features of the perception of life as it is lived by social actors, this account fails to account for the epistemological position of the actor such that their moment of judgement in assessing a governance injunction cannot be reproduced in the account.

In failing to deal with the position of the social actor in this way there remains too narrow a conception of 'actor', and so an inadequate account of how possibility can manifest in the course of a life lived. Thus, the familiar failing of domination of individual wills owing to the imposition of a framing once again results ('rationality *qua* argumentation', this time). Without recourse to the reasons social actors have for holding the beliefs they have, moreover, there is no facility for learning here. There is the assumption that well-argued positions will in themselves motivate adoption of the proposition argued for. However, given the differentiated sense of motivation prompted by reflection so far, this seems inadequate. We have here the suppression of reasons not on a model of deduction, but it is a familiar part of human experience that deductive reasons aren't the sole mark of cognition. Thus, we seem in argumentation to have an arena wherein opposing positions can clash, but not appreciate one another's bases in one another's terms.

In spirit, as it were, Habermas' instinct toward pragmatism is correct. Given what we know we need from an account of social actors as regards practical logic of action, it seems we can only look to realities and attempt to deduce from them procedures adequate to the ultimate task of discerning governance measures that we have set for ourselves. In order to assess this point we now must look to further pragmatic endeavours at dealing with this issue. The reason is that, in large measure, this Habermasian position underlies much of the thought that generates the positions described and analysed previously. They are heirs to this kind of theory, wherein formal structures of dialogue and intersubjective communication are posited as solutions to problems of interaction.

PRAGMATISM

Pragmatism in general, in the sense in which we deploy it here, is that linkage between theory and practice that has the unit of analysis as the practical and the theoretical is drawn out from the practical by way of analysis. This should be seen to be the only amenable methodology possible for the problematic we approach in EGAIS at this stage owing to the pervasive problem we describe of unquestioned framings, inattention to context, insufficient attention to real perspectives, and so on. Rather than attempting to fly further into theory and the abstract to come up with ways to 'patch up' the flawed theories we have so far exposed, like the scholastics' efforts to conserve Ptolemy with their epicycles, we take the lesson from the shortcomings in other theories and orient ourselves in the real. We are not alone in this, of course, but we will come across others who don't quite make the full journey.

Experimentalist Approach

We have, in fact, already seen the democratic experimental (DE) approach in action in terms of both DDP and OMC. For Dorf (1998), DE suggests a:

...model of problem solving adapted to a polity in which omnibus, national measures can rarely address the particularities of local experience, yet locales in isolation from one another are unable to explore and evaluate even the most immediately promising solutions to their problems. The model requires linked systems of local and inter-local or federal pooling of information, each applying in its sphere the principles of benchmarking, simultaneous engineering, and error correction, so that actors scrutinize their initial understandings of problems and feasible solutions. These principles enable the actors to learn from one another's successes and failures while reducing the vulnerability created by the decentralized search for solutions.

Key to the limitations of this approach here is the meaning of "to learn from one another's successes and failures" – we seem to have in this approach a reliance upon an unexplained mecha-

nism of exchange. We have seen the problems with merely citing 'dialogue' as a means of dealing with the problems of plurality, yet here we have it once again assumed that this is unproblematic. However, given the indexical nature of the apprehension of possibility and the historicised nature of the acquisition of value, not to mention the differentiated nature of rationality, we cannot assume that 'successes and failures' are simply things we can assume will be apprehended. In fact, to repeat the point made before we have to assume the essentially contestable nature of any representation on the table (i.e. when parties agree or dispute on some matter, the nature of the matter upon which they agree or dispute is itself contestable.)

There is in DE generally a drive toward a pragmatic, dialogical inclusion of agents such that mutual learning can be the driving force of governance. What there is not, however, is a plausible accounting for the nature of the pragmatism, the means of dialogue nor the mode of the learning. So, we can agree with DE that pragmatic, dialogical inclusion of agents to facilitate learning is central to unpicking the tangle at the heart of ethical governance of plurality, but we will have to look elsewhere for *what that means*. As was seen in EGAIS project research,[4] in fact, the widespread inclusion of social actors in different fields did not adequately overcome the problems we are currently examining.

In order to advance this methodology further, we must now look to accounts of learning as it has become apparent that this is a key concept in the problematic we are developing a means to address.

"Organisational Learning"

Ultimately, the kind of extension of the pragmatist approach from the deliberative approach embodied in DE (and therefore DDP and OMC) requires that the success of actions depend on establishing a process by which some suitable rule that already exists in a diversity of actors' minds will be stimulated – the idea is that the actors engaged in the process will come to a mutual understanding via experimentation that some arrangement hit upon is *the best* or *the right* one.

This approach obscures the question of actors' capacity, according to Argyris and Schön's (1978) 'organisational learning' approach, because it overlooks the "obstacles" that could prevent success of a learning operation. The obstacles of which they speak are in fact the essential elements of genuine perspectives that we have been discussing. The reason they are referred to as 'obstacles' is owing to the way in which personal identity constrains the set of possibilities for any individual. This was a central element in the epistemological position of the social actor, and so an ineliminable part of governance. In assuming the ability and the will to reflect upon and change perspectives, DE generally assumes too much of social actors, according to this critique. The core issue here is *capacity*.

In the course of the sort of dialogical 'joint inquiry' that characterises DE, a reframing operation is thought to take place. Each actor will deploy a frame that, according to the organisational learning approach, constitutes the actor's rule of identity. This just means that each actor will interpret the information presented to them according to their own lights, in line with their own possibilities etc. This is a personal frame sensitive but it is keyed to the social group in various ways (i.e. individual identity is not solipsism and entails reliance upon the broader group). This frame:

1. Allows the actor to assign a personal "meaning" to the context and adopt a role within that. This is a generative act and produces a rule (call it rule #1) for the interpretation and integration of facts.
2. Guarantees that, despite the novelty of each problem to be solved, actors will transpose the new onto the familiar. This is a scheme (a rule #2) that can be applied to the construction of a context-sensitive and

identity-sensitive (i.e. #1) representation (i.e. "I remember something like this happening before.")

The first outcome here can be seen as a definition of the perception of a situation while the second can be seen as a definition of the interpretation of a situation. Recalling Schön (1983, p. 138),

When a practitioner makes sense of a situation he perceives to be unique, he sees it as something already present in his repertoire. To see this sight as that one is not to subsume the first under a familiar category or rule. It is, rather, to see the unfamiliar, unique situation as both similar to and different from the familiar one, without at first being able to say similar or different with respect to what. The familiar situation functions as a precedent, or a metaphor, or... an exemplar for the unfamiliar one.

According to Schön, it is sufficient to pay attention to the problem of reframing for rule #1 to take place, and if #2 happens, #1 is assumed to have happened because of the attention paid to it – in other words, it is through the reflection on and negotiation of perspectives as based in personal identities (and their associated framings) that controversies are surmounted in matters where diversity is at stake. This is supposed to be the core of the learning approach in this account.

At once, it is clear that this presents the inverse of the problem already familiar to us of the failure to construct the context.[5] One major part of that problem was the way in which it reduced the world to a set of purportedly discoverable facts. This was the objectivising trend that we set out to avoid. Here, there is inadequate construction of the social actor herself. Effectively, the social actor is objectivised and reduced to a mere processor of transcendent reality. #1 here suggests this quite clearly, while #2 says little more save that the objectivised, processor self has the capacity for a bit of fuzzy logic. Learning, if it can be defined at all, can't be defined as the mere aggregation of algorithmic responses to more or less superficially similar events. Put the other way around, if learning can be defined as such, then the dominant technocratic-instrumental, standard model approach to governance is fine and we needn't have started EGAIS. However, to conclude this would require us to forego any account of human action as related to a practical logic based in experience and reflection. That seems a high price.

At this point, the Louvain School (Lenoble & Maesschalck, 2003; 2006; 2007) must be brought into the dialectic. They shared the initial insight that we used regarding the construction of the epistemological position of the social actor as a key part of how to formulate an account of governance that has a reputable normative basis. Indeed, they base their own solution (which they regard as the apex of the sort of trajectory we are describing in governance measure) on an account of learning.

LOUVAIN SCHOOL

The shift that the Louvain approach makes from these pragmatist approaches allows for a procedural approach to the conditions for identity representation itself. This means that the algorithmic potential criticised just prior in the organisational approach is dissipated and a variety of responses according to context opened up. This is to say that the manner in which identity is conceived of in Louvain (Lenoble & Maesschalck, 2006) is not as a set of parameters that 'crunch' the data of situations, but rather as a reflexive process.[6] It sees the formulation of the learning operation in the pragmatic approaches as incomplete and seeks to complete it with the development of this meta-approach.

This approach from Lenoble and Maesschalck (2007, p. 14ff) removes the assumption of the way that actors adopt identities (and fulfil #1 prior). The way that actors adopt an identity (have a capacity to "self-represent") is through 'terceisation,' i.e.

"…a third element whose externality makes possible the actor's construction of [their] image."[7] If you look in the mirror you will recognise yourself, and it is necessary to recognise yourself. So it is necessary to allow for the other to, firstly, exist, and then to differentiate between you and your image (the other), so the fact of the mirror itself must also be acknowledged (invoked). Lenoble et al. (2008, p. 39) describe this process as follows,

It is the product of an operation of 'terceisation', that is, an operation that requires the invocation of a third element – be it a person or an organizational device – whose externality makes possible the actor's construction of her image, i.e. of what will enable her to identify herself, and her own interests, in a given context for action. To carry out this operation of 'terceisation', collective actors must submit to a twofold inferential operation:

1. A retrospective relationship with experience: reconstructing the form its identity takes on through its past actions, and in so doing installing a capacity to be an actor (reflectability).

2. An anticipatory relationship with the current situation: identifying the transformations that will be necessary to ensure the realisation of this form in a new context for its application (destinability).

In this perspective, learning is more than metaphorically a 'double loop' operation; it is not just learning how to choose solutions, but learning to choose how to choose solutions. Reflexivity is no longer a schematic rule-based approach but a permanent reflective attention to the operation of 'self-capacitation'.

As can be seen in Figure 1, rather than merely allowing consequences to condition future action strategies, they are also allowed to influence the very guiding values that condition possible actions.

Figure 1. Schemas of 'learning'

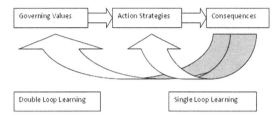

Put differently, in order for an actor to be able to express something of the form, "I think that *x*," where '*x*' is some state of affairs, it must be credited to them that they can distinguish the description and a possibly falsifying, transcendent reality owing to the fact that in order for a state of affairs to seem a certain way, it must be a certain way. Moreover, the fallible knowledge of criteria for fulfilling *x* implies the possibility of the criteria for *not x* (states of affairs *a*, *b*, *c*, say). Thus in the interpretation of a fact and the assigning of a personal meaning to it, upon which decisions and other actions will be based, we see the activation of a host of knowledge in a proactive and dynamic form. In this sense, as an activity, there is an element of expression involved in perception and interpretation rather than merely simple reaction to given data from without (as a decision-theoretic account might have it). This is a manner of thinking which takes an internal perspective, yet maintains the epistemic resources for immanent critique of claims and their procedural genesis.

The idea is clearly to account for diversity, while retaining a universal element, i.e. the universal moment of framing *per se* that serves to unite diverse framings in a procedural sense. This, it can be presumed, is to account for a sense of human nature that is not tied to any particular conception of the world. So, this conception hopes to provide a genuinely representative account of persons in society *howsoever they see the world they inhabit*.

The importance of this is that if we model better the manner in which personal identity factors in decision-making among actors; how socially perpetuated identity and self-conception figures in the perception of public facts, we can better conceptualise the actual position of the social actor considered at the moment of interpretation that we must assume to coincide with the apprehension of some new governance injunction. With this moment modelled more accurately, we can be in a position to design policy recommendations that will have both legitimacy and efficiency.

A detailed and well informed conception of this genesis and these mechanisms will allow for a greater and more effective understanding of the intentionality of collective actions. A genuine, authentic and complete theory of governance will depend upon a proper grasp of these ideas.

To carry out the previous terceisation, each collective actor must take part in the twofold action described previously. This operation of self-capacitation (Lenoble & Maesschalck, 2006) is "attention paid to two questions reflecting the dimensions of reflectability (collective identity making) and destinability (ability-to) that together constitute the reflexive operation by which an actor constructs its capacity to build a representation – of itself and of the context."

Reframing is not a simple substitution operation (such as that which Schön suggests), but what the Louvanians call a "combination" of the old and new frames. Schön's reframing is a reductionist one, which can lead to a simplification of the real process, it is argued.

A learning mechanism such as this permits reflection on the contents of perception and upon the reflexive capacity itself in a manner that permits the enumeration of possibilities from a real perspective. This is intended to permit also speculation upon alternative possible perspectives and their possibilities. This means that for each participant in deliberation both concrete, actual horizons as well as speculative horizons of possibility are opened up. This is thought to be the case owing to the supposed basis in terceisation of all learning encounters – this is what is meant by the Louvain School by a 'complete' learning process.

In this way, deliberative participants can call into question, or problematise, processes of governance, the contents of governance issues (including their ethical content), and the status of each in terms of their own view and that of a wider public of which they are a part. This problematisation, moreover is extendable via speculation into future states of self and public such that a real forum for testing and evaluating the processes and contents of governance problems is available. Its basis in intentional, self conscious, immanent critique allows for authoritative judgements to be made as the reflexive moment is founded in the normative expectations of the public. Thus, arguments or decisions have a connection with those they will affect.

The Louvain school shows the limits of alternative theories admirably, as well as framing the problematic clearly and precisely. However, its criticism of the pragmatist approaches to the problem can effectively be levelled at their solution too, *viz.* despite making advances, the Louvain school doesn't transcend fully the limits of the problematic they so well define. These limits emerge in the learning conception and the implications of the notion of rationality upon which it is based.

LIMITS OF LOUVAIN

The 'complete' learning[8] procedure in terceisation involves what is essentially a decentring operation, thus throwing the whole approach into the argu-

mentative mode. Since governance is subservient to the learning operation and that operation turns out in Louvain to be argumentatively construed, the whole governance edifice built upon this foundation is rocked. As a side-effect, moreover, the problematic parallelism between reason and morality is re-affirmed as argumentative reason takes over – valid argument is thought to be inherently motivating, thus the approach ultimately conflates the conditions for justification with conditions for application. This undoes *all* of the succinct and astute critical work of Lenoble and Maesschalck (critical work we in fact agree with).

In the felt need to overcome the incompleteness of the experimentalist-pragmatist approach, what are the Louvain school thinking of? One problem they might see is the potential that the word lacks a reference. The apparent danger could be that the word 'learning' is so diverse in application as to lose all meaning: there are instances of reflective learning, empirical learning, learning where opinion is unchanged, but practice affected, instances where practice is unchanged, but opinion affected... and so on. Between one instance and the next, people apply different standards to different conceptions of different circumstances and yet all claim to be learning.

Why would this be felt to be a problem? A plausible basis for this can be found in the long-standing division between public and private reason, and the distinction between value and norm upon which it is founded.

Values are broadly conceived of as the results of private experiences – based in the individual's experience of the world, including mediation of the same by culture, tradition etc. In this sense, as mentioned before, they are possibly irrational. On the other side of the coin there are public norms, wherein subjective elements are suppressed. This is so as to make them capable of acceptance by as many as possible, owing to their supposed relative value-neutrality. The opposition of the two forms of reasoning introduces a gulf between the personal and the political, which we have described as a problematic knot in EU libero-communitarianism. It is as if conviction and reason run on parallel tracks – but then how can the individual, convictions and all, be represented accurately?

As stated earlier, recalling Jean-Marc Ferry, the reasons that we accept or refuse a proposition in any given discussion are not necessarily the same as the reasons why we accept or refuse those reasons. For example, a frequenter of mediums might accept that astrology is predictive of his prospects, citing past successes, but also refuse to let failures of prediction dent his conviction, even though the rational structure is symmetrical. The frequenter of mediums has a deep-seated need to feel the universe isn't a blind, meaningless system, for instance, so favours confirmatory evidence over falsificatory. So, they employ deductive reason in matters of confirmatory 'facts', but narration trumps that same process in falsificatory eventualities.

While this is clearly important in comprehending who/what a person is, it is only comprehensible itself in argumentation if we step back from the primarily argumentative mode of discourse championed in deliberative models and regard narration not as an aggregative report of experiences had between t_1 and t_2, but rather as the authentic self-portrayal of a vulnerable human being – i.e. we need to use a *recognition* principle in order to cognise the information encoded in this. The values espoused by another must be seen as constitutive of who they are. This calls for a re-synthesis of private and public reason, contrary to the privatising march of modernity in general, wherein value is boxed off in subjectivity, with civil and political society running at a remove.

The Louvainians' terceisation sits within the march of modernity – it remains within the confines of the rationalist paradigm it critiques so well. It effectively decentres the social actor in the process supposedly required by their self-recognition – the self awareness prompted on the model of looking in a mirror would be akin to enumerating one's qualities as propositions true

of oneself. This is precisely *not* what it is like to be a social actor. One's qualities, history, values and so on are not just sentences one would assent to. That is the point that makes the endeavour to account for them so difficult. Thinking about thinking is still thinking. However, thinking about thinking a particular thought is no longer thinking that thought. If I think about who I am in terms of propositional knowledge, I am no longer reflecting on who I am, but am describing myself as an object of experience. The problem is how to account for myself as a bearer of experiences.

The Louvain school misses the diversity of reason in trying to augment the account of learning present in the experimentalist-pragmatist view by underwriting with a meta-principle of argumentative, universal reason. It is a complex way to exclude value, hence to box off subjectivity from activity in the public sphere. Recognition is not possible in this account as the overriding effect of the Louvain learning mechanism is that of decentring. Narrative authenticity is ruled out, so interpretive generalisation is impossible as is reconstruction. The 'incomplete' experimentalist-pragmatist view might be more true to reality precisely because it is tied to particularities and has no universalising moment.

The Louvain school seeks equality of applications of learning via formalising argumentative processes, thereby trying to liberate the power of reason itself rather than relying on reasons as understood by a people in question. The learning process's deliverances, then, will be the very definition of rational. Despite carefully formulating their account of asymmetry, moreover, latent within this movement in the Louvain approach is the assumption that 'rational' equates with morally good, since it is presumed within the account of learning that the better supported in argument, the more irresistible a precept becomes. In other words, despite all their careful statements of context and the pragmatic requirements of an approach to reason in governance, in this the Louvain school fall back on the principle that *reason is sufficient to determine its own application.*

Two assumptions can be discerned that are latent within this part of the Louvain approach:

1. That the removal of content, or formalisation, offers a route to fairness via neutralising differences.
2. That rationality, in this case, well-argued cases, are morally compelling.

1. The pragmatist can better the position offered by the formalist by opposing these wrong assumptions while still being able to lay claim to utilising rationality. Content, rather than being eliminated in order to ensure fairness, must be included and articulated in such a way as to render it as explicit as possible.[9] This is fairness as this is the material stuff individuals, groups, societies have to work with. Anything other than as full an explication of the contents adhered to and deployed by individuals and groups is a misrepresentation of the players in any debate and thus warps the dialectic from the start. There is a kind of category mistake in pursuing anything based on such a skewing.

2. Rationality is not morality. Morality, conceived of as a transcendent possibility, is not a notion that ought to be employed in debates concerning concrete experience any more than the blueprints for a house should be consulted in the event of a wall needing to be fixed - the plans will be clear and insist on the wall's rigidity. The facts might well be different. In the spirit of content-inclusion, rationality ought to be conceived of in terms of the operations made upon the inferential articulations of contentful claims endorsed, denied etc. Thus, morality gains a mechanism as moral debate carried out in this way will be able to draw upon the power of reason in concrete terms, terms that are meaningful to those using them. Reason doesn't contain within itself the terms of its own application, but given content, logic can provide the means to determine whether a claim justifies a conclusion, presupposes another etc.

SYNTHESES

The limits of the current approaches are:

DE, DDP and OMC do not have an in-built ethical structure. The structures alone are not sufficient to realise ethics. Moreover, the structures do not necessarily permit reflexivity. When stakeholders are included or asked to monitor, this inclusion and monitoring itself can be based in narrow views. There is still the possibility with these promising approaches that unquestioned framings, sectoral interests and power can dominate. This means that the position of the actor, their context and their conception of their own possibilities, is not automatically considered when one of these governance measures is adopted.

The result is that governance measures based in such a process can be themselves ungrounded in any particularly ethically valid procedure. For example, if dialogue is deployed in OMC as a means of gaining positive feedback for a measure already planned, the ethical impact of the measure never come into play. Instead, a crypto-technocratic process is deployed to win support for an *ad hoc* endeavour. In essence, the potential remains for the dominant techno-instrumental, standard model governance, with all three presuppositions, to win the day despite superficial complexity.

The Louvain School seize upon this lack of construction of the actor's position and criticise well the shortcomings of this failing. Their terceisation approach, however, has it that to understand the actor's position, the actor must consider herself in the abstract, as a series of descriptive statements. The metaphor of the mirror emphasised this. However, this is not what it is like to be a social actor. The Louvain School process is one of decentring, or translating, phenomenological self-identification into an argumentative mode.

This is of course the limit of a deliberative-collaborative approach as well. The problem is the suppressed assumption that argumentative rationality, the rationality of deductive reasoning, is the highest, or best, or most important form if rationality. This assumption carries with it the unjustified conviction that, as in deductive reasoning, valid arguments are themselves reasons to act. In the case of persons, however, one can easily accept that an argument is valid, but refuse to adopt it as a reason to act. We can agree that too much wine is bad for us, therefore we shouldn't open another bottle, but then go straight to the cellar nonetheless. Formally there seems to be a performative contradiction (we affirm the reason not to do what we go on to do). However, rather than seeing contradiction this as an outcome, it should instead be realised that the sort of case we here mention is indicative of the fact that reasons other than deductive reasons motivate. In fact, persons are motivated by complex webs of self-understandings based in beliefs, desires, history, culture, and so on. This means a prominent feature of personal motivation is *narrative*.[10]

These limits can be summed up in Table 1.

The lessons we can learn from the existing theoretical treatments of the problems in ethical governance are that:

- The situation of the social actor is central, yet difficult to account for:
  - Their position must be reconstructed faithfully in order to ground the mo-ment of interpretation necessary for the comprehension and possible acceptance of a norm (the content of a governance injunction).
- The interaction of norms and values in personal motivation is complex
  - The question is of the interaction between value, norm and context – this characterises the *significance* of the norm.
  - Only a significant norm can be understood.
  - Only a norm understood can be the *telos* of a legitimate governance.
- Gaining an efficient means of ethical governance is essentially linked to the account made of value and reason.
  - The limit case for efficiency, the extreme, is bringing someone via governance to a point where they can bind their will to a norm in opposition to their values.
  - This needs to be achieved by engaging with their will and their own conception of their possibilities.

CONSEQUENCES

Given the foregoing arguments and critiques of the current offering s regarding ethical governance measures and their limits, we now have a clear picture of the problem space that needs to be

Table 1. Limits and consequences

Theory	Actor Learning	Major Limit	Ethical content	Governance
DE (DDP/OMC)	Single loop or double loop	Descriptivist norms	Social acceptance, not ethical acceptability	Sociological
Deliberative-collaborative	Mentalist presupposition	Decentring reason	Neutral	Inefficient
Louvain	Terceisation	Decentring reason	Neutral	Inefficient

taken in hand. We have a clear picture of what the problems are, and why they are problems. This means we have a basis to move toward addressing them in a way that is not merely *ad hoc* but rather is grounded in theory and that can affect practice in a positive way. So with the problem clearly defined and the basis upon which it is a problem understood, we have cleared the ground for proposals for ways to approach the specific problems we detect with the ambition of treating them. This will take two stages:

- An ideal practical arrangement.
- An actually practicable arrangement.

The rationale for this two step approach is to ground the recommendations in theory, but to create actually practical recommendations, rather than the bafflingly formal, or by positing hopelessly idealistic arrangements.

REFERENCES

Argyris, C., & Schön, D. (1978). *Organisational Learning, A Theory of Action Perspective (Vol. 1)*. Reading, MA: Addison, Wesley.

Dorf, M., & Sabel, C. (1998). A constitution for democratic experimentalism. *Cornell Law Faculty Publications*. Retrieved June 2011, http://scholarship.law.cornell.edu/cgi/viewcontent.cgi?article=1119&context=facpub&sei-redir=1#search=%22A+constitution+for+democratic+experimentalism%22

Habermas, J. (1998). Between facts and norms (Rehg, W., Tran.) MIT Press.

Habermas, J. (2005). *Truth and justification (Fultner, B., Tran.)*. Cambridge, MA: MIT Press.

Lenoble, J., et al. (2008). *Democratic governance and theory of collective action*. Annual Scientific Activity Report. Retrieved November, 2012 from http://iap6.cpdr.ucl.ac.be/docs/IAP-VI-06-AnnualReport-2-final-2.04.09.pdf

Lenoble, J., & Maesschalck, M. (2003). *Toward a theory of governance: The action of norms*. Kluwer.

Lenoble, J., & Maesschalck, M. (2006). *Beyond neo-institutionalist and pragmatist approaches to governance*. Retrieved November, 2012 http://iap6.cpdr.ucl.ac.be/docs/TNU/WP-PAI.VI.06-TNU-1.EN.pdf

Lenoble, J., & Maesschalck, M. (2007). *Synthesis report two: Reflexive governance: Some clarifications and an extension and deepening of the fourth (genetic) approach, REFGOV: Reflexive governance in the Public Interest, Services of General Interest/Theory of the Norm Unit*. Retrieved Novermber, 2012 from http://refgov.cpdr.ucl.ac.be/?go=publications&dc=e14a94b44d20bdad142dbb2bf693f554b23327dc

McCarthy, T. (1994). Kantian constructivism and reconstructivism: Rawls and Habermas in Dialogue. *Ethics, 105*(1), 46. doi:10.1086/293678.

Mulhall, S., & Swift, A. (1996). *Liberals and communitarians*. Wiley.

Schön, D. (1983). *The reflective practitioner. How professionals think in action*. London: Temple Smith.

Smith, P. (1994). Les puissances de l'expérience. *Contemporary Sociology, 23*(3). American Sociological Association.

ENDNOTES

[1] Descartes seeks a 'universal mathesis' of rationality that can account for all human understanding according to various principles cf. *Rules for the direction of the Mind.* Kant, although more complicated, ultimately locates his search for understanding of the human subject in an abstract locus of consciousness, cf. *Critique of Pure Reason.*

[2] 'Correctness amounts to fulfilling two major conditions: "Habermas's (D)-Principle articulates this dialogical requirement. If one assumes this requirement, then one can arrive at Habermas's specific conception of reasonable moral discourse by working out the implications of his argumentation theory for the discursive testing of unconditional moral obligations. What one gets, according to Habermas, is a dialogical principle of universalization (U): "A [moral norm] is valid just in case the foreseeable consequences and side-effects of its general observance for the interests and value-orientations of *each individual* could be *jointly* accepted by *all* concerned without coercion" (i.e., in a sufficiently reasonable discourse) (1998a, 42; trans. amended). The (U)-Principle assumes that valid moral rules or norms allow for an egalitarian community of autonomous agents—as Kant (1785, Ak. 433; also 431) put it, a "systematic union of different rational beings" governed by "common laws." From the standpoint of argumentation theory, (U) seems to state the burden of proof that structures an adequate process and procedure of justification. The (U)-Principle has been a site of controversy among discourse theorists, and not everyone considers it necessary for a discourse ethics (Benhabib and Dallmayr 1990; Wellmer 1991; Gottschalk-Mazouz 2000). Habermas has argued that (U) can be deduced from statements articulating the pragmatic implications of argumentative discourse over moral norms (1990a, 86–93; 1998a, 39–45). More precisely, a successful deduction probably depends on three assumptions: (D), a statement of the semantics of unconditional norms, and an articulation of the pragmatics of discourse. If we accept (D) and if we accept Habermas's explication of the rhetorical presuppositions of the discursive justification required by (D), then (U) would have to follow as an implication of what is required for discursively justifying norms with the specific content of moral norms, namely obligations that bind persons in general and whose acceptance thus affects each person's pursuit of interests and the good life." James Bohman, William Rehg, http://plato.stanford.edu/entries/habermas/.

[3] This is a sketch of the vast arena of deliberative, discourse theoretical models of rationality, and Habermas himself has an account of 'spheres of rationality' beyond the solely argumentative (*The Theory of Communicative Rationality*, Vol.1 pp.285-8, 327-35). To argue against their efficacy is far beyond the scope of this chapter, but not beyond the scope of possibility. See Rainey, S, *A Pragmatic Conception of Rationality*, doctoral thesis, Queen's University Belfast, 2008.

[4] Cf. EGAIS Deliverable 3.3 *Portability of Ethical Governance*, passim, http://www.egais-project.eu/sites/default/files/deliverables/EGAIS_D3.3_06042011.pdf *passim.*

[5] Cf. EGAIS deliverable 2.1 *Grid Based Questionnaire Development* http://www.egais-project.eu/sites/default/files/deliverables/EGAIS_D2.1_31082009.pdf p.8ff.

[6] This is discussed particularly clearly in Lenoble, Maesschalck's *Toward a Theory of Governance: The Action of Norms,* p.172ff though detailed exegesis here would be out of scope.

7   Observations in this report and the previous one inform the discussion generally at this stage of the argument.

8   Just exactly what a 'complete' learning procedure means is a problem in itself. Learning by its nature, one would imagine, is at least open in one direction – receptive of input – and therefore would require an incompleteness. Terceisation could be read as implying some manner of rational flight into omniscience-through-procedure.

9   This view has a strong philosophical pedigree from JL Austin, Paul Grice, Peter Strawson, through to Wilfrid Sellars, Robert Brandom and (in certain writings) Jürgen Habermas.

10 This is explored further in this book and throughout the EGAIS project's deliverables.

Chapter 21
Transformation of Proceduralism from Contextual to Comprehensive

Sylvain Lavelle
Center for Ethics, Technology and Society (CETS), ICAM Paris-Sénart, France

Stephen Rainey
University of Namur, Belgium

ABSTRACT

Proceduralism has been a major philosophical stream that gathers some outstanding philosophers, such as John Rawls or Jürgen Habermas. The general idea of proceduralism, especially in the practical domains of morals, law and communication, comes from the need to provide some rational justifications to the rules, actions and decisions to be adopted or made by the society or the power in the context of highly 'plural-complex-developed' societies. It is particularly concerned with the governance of ethics, in other words, with the institutional and organisational conditions that the procedures of assessment must fulfil so that ethical questions can be addressed, especially in the domain of scientific and technical research and innovation. We show that classical proceduralism does not adequately address problems raised by an ethics of science and technology, and we take the context of Europe as a typical example of what a complex multicultural set of societies can be.

INTRODUCTION

Proceduralism has been a major philosophical stream that gathers some outstanding philosophers, such as John Rawls or Jürgen Habermas. The principle of proceduralism, especially in the practical domains of morals, law and communication, has arisen from the need to provide some rational justifications to rules, decisions and actions in the context of 'plural-complex-developed' societies. One of the main characteristics of this kind of society, indeed, for the institutions as well as for the citizens, is the lack of a homogeneous background that would offer a common frame-

DOI: 10.4018/978-1-4666-3670-5.ch021

work for the argumentations or the justifications to be shown valid or not. In traditional societies, be they Christian or Muslim ones, for instance, the existence of a common religious background make it possible for the community of people to refer to some shared obvious common principles, habits or experiences. This is no longer the case in contemporary societies and this requires a new method to be conceived of and implemented at the various stages and levels of the power structure or process.

Here, proceduralism plays the role of a minimalist method that is supposed to be neutral as to the substantive doctrines defining 'the good' at individual or collective levels within a society. Proceduralism as a method concerns most of the domains of philosophy, namely the epistemic, the technical as well as the political or the ethical. One can think of an epistemic or a technical proceduralism[1], but we would like here to concentrate upon ethical proceduralism to be taken as both a reflexive and constructive stance about legal and moral norms. There exist a variety of procedural methods, but the main models are no doubt the rational discursive or deliberative ones as developed notably by Rawls and Habermas at a theoretical level. It should be mentioned that a set of practical applications has already come out of these models, such as the rules of 'positive discrimination' for Rawls, or the rules of 'fair communication' for Habermas.

In this line, it can be shown that the diagnosis on the limits of proceduralism as shared by many experts in the practical exercise of their function as assessors of research and innovation projects can be traced back to these models at a theoretical level. In fact, what is at stake in ethics is not the mere application of a procedure, nor the mere compliance of an action, a decision or a project with a set of rules. What is at stake is the question of *normativity*, that is, the relationship of norms to the context and to the values of individuals or communities. In this respect, it is assumed that the process of *framing* of ethical issues is in a way

more important and significant than the outcome of the procedure of ethical assessment as such. This focus on the framing calls for shifting in the procedural approach from an ethical analysis of the issues (privacy, dignity, discrimination, etc.) to a *meta-ethical analysis* of the governance process. In terms of procedure, this more reflexive method differs strongly from a 'check-list' of criteria to be fulfilled by projects officers or assessors, like in most operational procedural frameworks.

Hence a set of guiding questions for this philosophical exploration:

1. What are the philosophical and methodological foundations of proceduralism, especially concerning the relations between norm, context and value?
2. To what extend Contextual Proceduralism is successful in attempting to overcome the limits of classical proceduralism as well as the limits of the other methods of reflexive governance?
3. What are the transformations of contextual proceduralism as requested and elaborated within the alternative methodological option that we call Comprehensive Proceduralism?
4. What are the possible applications of that new comprehensive procedural method in the field of ethical governance and emerging technology?

METHOD AND PROCEDURE

Proceduralism appeared out of a transitional movement as a solution to the problem of cultural and social pluralism taken by some philosophers, like Rawls or Habermas, to be a permanent trait of modern democracies[2]. The emergence of cultural differences in societies (secular, multi-cultural, less authoritarian) has meant that a new method of discussion and cooperation has evolved to deal with these sometimes disparate communities. The inclusion of these different communities in the procedural method opens up the democratic and participative opportunities that characterize this paradigm. This method, based on the notion of *procedure*, requires that there be, at the very least, an agreement on the way to deal with problems, even if there not agreement on the content of the solutions. This appears to be something of an answer to the 'polytheism of values' identified by Weber (1904/1949), since a society, in order to function effectively, requires the establishment of a multiplicity of moral agreements on rules, norms, and, if possible, values. But if the society's members cannot agree on the content or substance of values, especially with the heterogeneity of worldviews, they can, at least, agree on a fair procedure that can make agreement possible. The main aspect of proceduralism is the insistence on the non-substantive approaches to conflict resolution between the members of a society.

Variety of Philosophical Methods

Proceduralism as a method can be located briefly among the set of some contemporary methodological debates characterized by a set of conflicting options:

- **Language vs. Experience:** Analytical, phenomenological
  - **Analytical:** (eg: Wittgenstein) The right method is to take as a basis the language rules and the lifeforms that are shared by a community of subjects.
  - **Phenomenological:** (eg: Husserl) The right method is to take as a basis the lived experience of the subject and the lifeworld of a community of subjects.
- **Discussion vs. Narration:** Dialogical, Hermeneutical
  - **Dialogical:** (eg: Habermas) The right method is the exchange of arguments

and the examination of the validity of claims.
  - **Hermeneutical:** (eg: Ricoeur) The right method is the interpretation of identity through the use of narration and the consideration of action as a text.
- **Agreement, Convergence:** Procedural a, b
  - **a-Procedural:** (eg: Habermas) The right method is the ideal speech situation, the principles of discussion and universalisation and the search for consensus.
  - **b-Procedural:** (eg: Rawls) The right method is the original position, reflective equilibrium and overlapping consensus.
- **Reflexivity, Re-Construction:** Procedural c, d
  - **c-Contextual-Procedural:** eg: Maesschalck & Lenoble) The right method is the contextual adjustment of norms through a reflexive transformation of the actor's contexts, a reframing of issues at stake and an operation of self-learning and identity-building in collective action.
  - **d-Narrative-Procedural:** (eg: Ferry) The right method is the combination of approaches (narration, interpretation, argumentation, re-construction) taking into account the contexts and especially the schemes of relevance of the actors.

The important point in the evolution of the methods is no doubt the *pragmatic turn* that affected all the streams above mentioned. Another important point to be noticed is the spectrum of options as developed in some *oecumenic trends*, which are shown to merge pragmatics with ethics or hermeneutics. These attempts are particularly illuminating as regards the issue of the justification and the application of norms that requires a combination of rational, experiential and contextual methods.

The Norm/Context Problem

A context, in the broadest sense of the word, can be defined as a temporary or permanent background, be it natural, artificial, social or cultural, shaping the modalities (necessity, obligation, possibility,…) of human thought, conduct and taste from the point of view of cognition, volition, action, judgement, experience and significance. There are then three basic meanings of the context:

- A context is a *milieu.*
- A context is a *culture.*
- A context is a *situation.*

Of course, there can be some relations between these three concepts of context: for instance, when Mr Smith, who has worked for thirty years as an operator in a car industry (milieu) and has developed a set of professional abilities (culture) experiences the breakdown of his car on a countryside road and has to repair it by himself (situation).

The research on context modelling and reasoning suggests several methodological approaches of the concept of context:

1. **Analytical/positive approach:**
   a. Context is a form of information: it is something that can be known.
   b. Context is delineable: it can be defined what counts as contextual elements of an activity.
   c. Context is stable: although the precise elements of a context representation may vary among different activities, they do not vary among instances of the same activity.
   d. Context and activity are separable: the activity happens 'within' a context, the context describes features of the environment where the activity takes place.

2. **Phenomenological/experiential approach:**
   a. Context is a relational property among objects or activities: the issue is not that something is part or not of the context, but that it may be or not contextually relevant to some particular.
   b. Context is dynamic: rather than delineating or defining context in advance, the scope of contextual features is defined dynamically.
   c. Context is relevant to particular settings, instances of action and particular parties of that action.
   d. Context and activity are not separable: context is embedded in activity and arises from it.

3. **Psychological/cognitive approach:** a context is a state of the mind with no clear-cut boundaries consisting of all associatively relevant elements and which is dynamic.

4. **Technological/engineering approach:** a context is the collection of relevant conditions and surrounding that make a situation unique and comprehensible.

The important point is that a context should not be viewed as a mere 'environment,' but rather as a property of the *relation* (R) between the individual and his environment shown in Table 1.

The notion of Context is often something perfectly simple to understand in objective terms, since there are instances when we can determine the characteristics of objective features of the environment. We instead suppose that it is a thickly connoting notion and that it is therefore very important to get an impression of what role context here plays. Lenoble and Maesschalck (2003, p. 87) describe three facets of 'context' to which the entire notion cannot be reduced:

- A context is not a set of factual constraints.

Table 1. Relations to the context

Activity	Sensitivity	Mind	Experience
R1: activity shaped by the context R2: activity shaping the context R3: activity shaping and being shaped by the context	R4: context sensitivity R5: strong context sensitivity R6: weak context sensitivity	R7: cognitive relation R8: active relation R9: volitive relation R10: judgemental relation	R11: experiential relation R12: perceptual relation R13: emotional relation

- A context is not a false representation or a framework offering resistance.

- A context is not a particular culture which cultural anthropologists could identify and which could be deposited in the individual minds of individual actors as continually adaptable conventions that would serve as capacitating structure for them.

These three components exist but don't exhaust the function of context, so they are each reductions of the notion of context and then one misses the question of the *relevance of the context* (2003, p. 90). The very notion of context itself is very open regarding the purpose of circumscribing a domain even when it is objectively describable and the circumscription is steeped in subjective decision. It relies on *prior decisions* that characterise the context in a sense beyond the mere description of its elements, and these prior decisions are what constitute the *framing* of a perspective. The role of judgement is essential in all these operations: what the 'objective facts' mean for a given actor will be contingent upon their judgement of what the broader context is. In fact 'broader context,' used here to distinguish between the narrow description of elements of a scene and the relevance of that scene in the consciousness of the subjects to whom those elements are addressed, is something of an artifice – *the broader context is the context* in the sense we need to use it.

One's perceptions of elements of context are members of a set entailed by particular theoretical presuppositions; they are symptoms of a framing, linked via informal inferential connections to be-

liefs. Hence, there are no cognitively significant representations of one's predicament untouched by background theory[3], and in any and every use of reason whatsoever there is contained within it implicit reference to background.[4] Neutrality in the sense seemingly required by the prevailing presumption is thus impossible and the 'background theory' that appears in the course of judgements makes essential reference to the values of the individual.

The Norm/Value Problem

It is assumed that the fact/value dichotomy and the norm/value dichotomy are neither the only approach nor the most correct one. However, we assume that there is a distinction between the three notions:

- **A Fact:** Is the content of an 'Is' statement that *can* function as a scientific standard and that entails a *description* by an individual or a community (e.g.: 'You are a doctor,' 'They drink alcohol every Saturday night').

- **A Value:** Is the content of a statement that *can* function as a standard and that entails an *evaluation* that expresses the taste or the preference of an individual or a community (e.g.: 'I prefer walking,' 'I enjoy drinking alcohol').

- **A Norm:** Is the content of an 'Ought' statement that *must* function as a moral, legal or social standard and that entails a *prescription* (from obligations to recommendations

Table 2. Interpretations of norms, values and facts

	Prescription	Evaluation	Description
Norm	Prescriptive interpretation of a norm	Evaluative interpretation of a norm	Descriptive interpretation of a norm
Value	Prescriptive interpretation of a value	Evaluative interpretation of a value	Descriptive interpretation of a value
Fact	Prescriptive interpretation of a fact	Evaluative interpretation of a fact	Descriptive interpretation of a fact

or suggestions) for the conduct of an individual or a community (e.g.: 'You ought to take care of your ill mother,' 'Don't drink too much alcohol before driving').

The reason for emphasizing the prescription as an essential part of a norm is the absence of any prescriptive commitments to a norm that amounts to give an account of it in descriptive terms and then to make it a mere description. Nevertheless, it is assumed also that the difference between norms, values and facts is not absolute, first, because the difference is also a matter of perspective, second, because one notion can relate to the others. Thus, one can adopt a factual perspective on values or on norms, like, for instance, if I say 'John hates doing sport'. This kind of sentence is a description of an evaluation, a *descriptive evaluation* in the sense that it is a fact that 'John hates doing sport'. In stating this, one does not criticize him in any way, but simply states it, as a matter of fact, that one of John's (negative) value is to dislike sport. Conversely, one can adopt an evaluative perspective on facts or norms, or, a normative perspective on facts or values. For instance, if I say 'John *amazingly* hates doing sport,' this is not a neutral statement, but what we can call an *evaluative description*.

So, one can suggest a complex set of relations between the three notions shown in Table 2.

The articulation between notions can take several forms: on the one hand, it can be a mere relation, such as: 'A norm *N* is related to a value

V'; on the other hand, it can be a conditional relation, such as: 'A value *V* is a condition of a norm *N*'. A conditional relation can be:

1. **A Condition of Possibility:** A value *V* is the condition of possibility of a norm *N*.
2. **A Condition of Validity:** A value *V* is the condition of validity of a norm *N*.
3. **A Condition of Legitimacy:** A value *V* is a condition of legitimacy of a norm *N*.
4. **A Condition of Efficiency:** A value *V* is a condition of efficiency of a norm *N*.

As to the relationship between norms and rules, it must avoid any confusion between the two of them:

- A *rule* is the content of a Is/Ought-statement and that can be universal (e.g.: a scientific law, or a moral law) or particular (e.g.: a rule of life) according to the scope assigned to this rule by an individual or a community (e.g.: 'As for me, I never drink alcohol').
- A *norm* is the content of an 'Ought' statement that *must* function as a moral, legal or social standard and that entails a *prescription* (from obligations to recommendations or suggestions) for the conduct of an individual or a community (e.g.: 'You ought to take care of your ill mother,' 'Don't drink too much alcohol before driving').

It's a commonplace that different people can 'see things differently,' meaning that they interpret the value and the relevance of situations in potentially conflicting ways. The problem is the presumption in research that expert opinion has a privileged position on determining the real or best description of reality. The contention we are developing is that where this presumption goes unchecked, the legitimacy and the efficiency of ethical governance is compromised. It is compromised because there is no principled reason why one section of a social group ought to be permitted to dominate any other.

BEYOND CONTEXTUAL PROCEDURALISM

Classical proceduralism, as we may call it, namely that of Rawls and Habermas, has been subject to a flow of critiques, some of them stressing its lack of context-sensitiveness. The Louvain School's (2003) critiques emphasize two basic limits to proceduralism: first, the obliteration of the context of agents; second, the disjunction between the justification and the application of norms. Thus, Maesschalck and Lenoble call for a *contextual proceduralism* that seeks to warrant a reflexive adjustment of rationally justified norms to their contextual setting of insertion. This kind of proceduralism provides a framework for a *reflexive governance* that modifies the learning and the identity-building and shifting processes of the actors engaged into a collective action. However, the option of contextual proceduralism remains within the scope of rationality and argumentation and fails to give the value-systems of the actors a sufficient and appropriate function.

Contextual Proceduralism and Reflexive Governance

Maesschalck in *Normes et Contextes* (Maesschalck, 2001) undertakes to overcome the limits of classical proceduralism, especially that of Habermas and Rawls. He substitutes what he calls a *contextual pragmatics* to Habermas' universal pragmatics and he intends to connect the rational justification of norms together with their contextual application. It is important to mention that the context in Maesschalck's does not equal a mere 'situation' or 'environment,' as it is supposed to be so in the ordinary use of the word. The notion of context includes more broadly the background-based relationship of an individual or a community of individuals to that situation or environment. Contextual pragmatics then consists in taking into account within the process of norm validation all the 'mental-social-cultural' background features that enable an individual or a community to give meaning and significance to norms within a situation or an environment.

In a way, the method of contextual proceduralism is quite easy to understand as regards its origins and its justifications. For instance, if you elaborate within a Parliament a set of norms regulating the audio/video piracy on the internet, it is better for these norms to be effective that you take into account the contexts of the agents they will apply to. Otherwise, if they don't acknowledge their validity and their legitimacy, they would find means for not applying these norms and carrying on their audio/video piracy. Now, the judgement operation in contextual proceduralism requires a special kind of reflexive equilibrium between the 'foreground' of the discursive justification and the 'background' of the evolutionary lifeforms.[5] The Figure 1 shows an opposition between the comprehensive and the discursive approaches and the role that the cognitive dispositions of the individuals play in the operations of adjustment of the norm to the context.

The initial procedure of adjustment between the process of rational justification of norms and that of contextual application is then completed through a more governance-oriented approach. Maesschalck and Lenoble assume the difference, in a process of reflexive governance, between the

Figure 1. Model of justification/application relationship (Maesschalck, 2001)

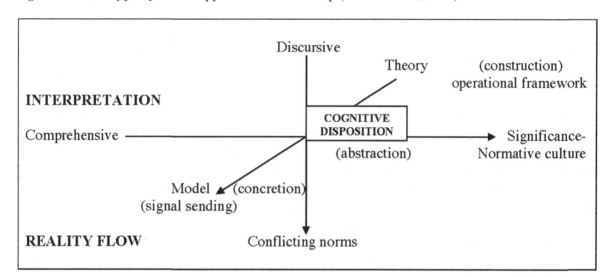

first order reflexivity (world/event-oriented) and the second order reflexivity (process/procedure-oriented). They then distinguish four levels of reflexivity:

1. **External Level:** The reflexivity is linked to the external constraints of a process.
2. **Internal Level:** The reflexivity is linked to the internal governance process.
3. **Practical Level:** The reflexivity is linked to a reflexive practice.
4. **Genetic Level:** The reflexivity is linked to a self-transformation of the cognitive framing on the basis of a process internal to collective action.

The idea of their 'genetic approach to governance' (Lenoble & Maesschalck, 2003) is that an institutional process of democratic governance can foster this fourth level of reflexivity characterized by an operation of self-learning.

Quite logically, the representatives of the other levels (Herbert Simon, Chris Argyris and Donald Schön (1978), Charles Sabel and Jonathan Cohen) are criticised by the Louvain School for being 'incomplete'. The incompleteness is said to lie in

the fact that each account in some way relies upon a notion of 'common sense' somewhere along the line. That the reflexivity present in each account isn't itself reflexive, thus there is a moment at which reflexivity itself must be presumed to be an authoritative source of knowledge of good, bad, worthwhile, etc. This is a case of reason being deployed in a substantive sense, Louvain alleges, and so must be resisted:

Each... in different degrees, restores, to one moment or another, the possibility of reason supporting itself on something given which it can treat as something that regulates its functioning and ensures its possibility. The consequence is that, in one way or another, there is a moment where they obliterate or 'misjudge' this dependence of the formal operation of reason on 'something that is not itself,' that is to say, the reflexivity of reason. (Lenoble & Maesschalck, 2003, p. 89)

Maescchalck and Lenoble also develop a theory of identity-buiding (2007a, p. 14ff) and identity-shifting on the basis of the notion of 'terceisation'. The *terceisation* concept is developed by them to describe the self-identifications of stakeholders in emerging compromises through experimental processes such that there is generated a shift in his/her own identity and positioning. Such a function is the essential prerequisite for truly practical commitment of stakeholders, as it allows avoiding the perpetuation of established, unexamined routines. It stimulates new dynamics thanks to various pragmatic conditions which can be guaranteed by third party actors taking responsibility for the terceisation function. The terceisation function may apply individually by a single stakeholder, but it may also be a collective process in which various stakeholders with complementary skills participate. The process is one of decentring, or translating, phenomenological self-identification into an argumentative mode[6].

Limits to Contextual Proceduralism

One of the main limits to contextual proceduralism is the implicit assumption that argumentative rationality, the rationality of deductive reasoning, is the highest, or best, or most important form if rationality. This assumption carries with it the unjustified conviction that, as in deductive reasoning, valid arguments are themselves reasons to act. In the case of persons, however, one can easily accept that an argument is valid, but refuse to adopt it as a reason to act. For instance, we can agree that too much wine is bad for us; therefore, we shouldn't open another bottle, but then go straight to the cellar nonetheless. Formally there seems to be some sort of a performative contradiction: *we affirm the reason not to do what we go on to do.*

However, rather than seeing contradiction like this as an outcome, it should instead be realised that the sort of case we here mention is indicative of the fact that reasons other than deductive reasons motivate. In fact, persons are motivated by complex webs of self-understandings based in beliefs, desires, history, culture, and so on, which means a prominent feature of personal motivation is *narrative*. We see that presuppositions blight any attempt to understand the perspectives of

those engaged in governance approaches from either the governing or governed perspective. These are based in the presuppositions discernable behind governance approaches that inhibit the construction of contexts, and the mode of rationality assumed to be effective in all contexts, i.e. *argumentative rationality*. At the very least, this hampers the effectiveness of the measures, while standing as an obstacle to their legitimacy.

It seems that Maesschalck and Lenoble misses the internal diversities of human rationality in supposing 'reason' to be a unified, argumentatively-based faculty[7]. The 'complete' learning procedure in terceisation involves what is essentially a decentring operation – one is expected to conceive of oneself as if observing an image in a mirror. Clearly, this model involves 'perceiving' oneself, and is therefore wedded to the notion of oneself as an object of experience, not a subject of experience, which clearly has importance for the objectivist tendency of which the approach falls foul. But in terms of the rationalistic mode its consequence is that each individual is expected to conceive of oneself in terms of propositions that will be held true or false. It is so to permit the rationalisation of one's being into a form appropriate for playing a part in arguments, and then, possibilities for individuals so conceived are those propositions that can be derived from the overall set. This means that argumentation is put at the centre of things – right outcomes are equated with deductively valid outcomes. This throws the whole approach into the argumentative mode: since governance is subservient to the learning operation and that operation turns out in Louvain to be argumentatively construed, the whole governance edifice built upon this foundation is rocked. It is founded on the assumption that rationality is essentially argumentative and so arguments are the means *par excellence* to engender change in human beings' attitudes and perspectives.

In trying to base their account of learning in general by underwriting it with a meta-principle[8] of argumentative, universal reason, rather than surpassing the limits they diagnose in other experimentalist-pragmatist approaches, the Louvain School recreate the same problems they identify in their rivals' approaches. This leaves the position they stake out to be impoverished when it comes to constructing the perspective of the social actor. Without an account of how norm and value interact in the consciousness of the social actor, Maesschalck and Lenoble provide no means by which to understand how anyone can choose among options available to them, be those options generated from a pre-existing set of commitments or from the injunctions of a governance body. In the case of the latter, motivation is supposed to spring merely from the correct apprehension of these injunctions (i.e. terceisation – having attuned oneself to the appropriate learning approach.)

The Louvain school sees itself as the fourth in a line of progressively more adequate governance approaches, each missing something until the fourth and final stage wherein governance is most adequately accounted for (2007a). Central in this 'fourth approach' is the role of learning, the operation whereby social actors adopt perspectives upon governance injunctions and their own situations in order to be able to modify attitudes and behaviours in line with governance injunctions. As such, learning therefore underlies all governance, where 'governance' is a social steering mechanism of some particular conception. A problem in the conception of learning will cause issues for the ability of the account to account adequately for the world of possibilities and interpretive choices available to the actor. At heart, this is due to the Louvain school falling foul of the rationalistic preoccupation that leads in general to formalism, and the objectivising tendency in governance that they correctly identify. The way in which the Louvain school falls into the objectivist trap is *via* a too-strong conception of the role of argumentative reason[9].

Maesschalck and Lenoble (2007, p. 17) seek to put learning at the heart of things, but they underwrite learning with a decentring operation.

Their approach is likened to looking in a mirror and recognising oneself, but this metaphor is quite unlike what is required in an act of authentic self-representation. In a mirror we see an objective representation of ourselves in reflection, and reflection is certainly required in any act of judgement, but the type of reflection seen in a mirror is an object. We observe the reflection as an object of experience external to ourselves, but the problem then is that we are supposed to adjust this objective representation of the self to some personally transcendent ideal. The point of trying to account for the subject in learning is precisely to advance upon this – it is to insert the agent *as they are* into the process, and to incorporate *authentic subjectivity* and not its objective representation. The point of shifting focus here to judgement is that the species of that practice that underwrites learning is reflective upon belief, desire and value in order to discern appropriate action. It is reflective upon the self-perception and so entails the evaluation of the subjective position, not a decentred representation as an object of experience.

A motivation for this could be as a consequence of a complex way to exclude value, hence to box off subjectivity, from activity in the public sphere. By remaining in a rationalistic vein, propositions are the matter of discussion and propositions are presumably objective in their capacity to feature in inferences, arguments and discussions in any number of ways. But the recognition of one's own values, let alone those of another, is not possible in this account as the overriding effect of the Louvain learning mechanism is that of decentring. Narrative authenticity is ruled out, so interpretive generalisation is impossible, as is reconstruction of a perspective. The subjective is annexed and so any reconstruction will amount to little more than a more or less accurate theoretical representation of the perspective. This means that there is no explanation of how any given individual actually interprets a proposition, of how an individual takes on a maxim of action, experiences a norm or espouses a value. A consequence of this is that the meaning of the proposition for any individual is underdetermined by the terceisation process, which in turn means that what it could entail is underdetermined. We don't have a definite grasp of the significance of the proposition to the individual as we have annexed subjectivity in terceisation. The 'incomplete' experimentalist-pragmatist view might thus be more true to reality precisely because it is tied to particularities and has no universalising moment.

As a direct effect, the problematic parallelism between reason and morality is re-affirmed as argumentative reason takes over – valid argument is thought to be inherently motivating, thus the approach ultimately conflates the conditions for justification with conditions for application of a norm. The idea of a person's accepting a normative injunction as motivation for action is taken to be a formal affair, on this scheme. This is what it means to say that reason and morality are taken as parallels. Why one accepts a norm is thought to be exhaustively described in terms of arguments. However, it is very clear from everyday experience that this is not the case – we need only think of weak-willed action, indulgence and transgression. Aside from this appeal to 'common sense,' the preceding arguments describe anyway how the theoretical basis of the terceisation-centred learning account, and the governance approach it facilitates, is unable to account for subjectivity. This point can bear further elaboration, being an important one.

The reasons one might accept or reject normative injunctions can be based in arguments, historical legacies, culture, hope or fear, and many other sources besides. Indeed, it is such material as this that characterise the meaning of an otherwise formal normative injunction in the practical will of any given person. For example, until one interprets a term such as 'privacy' in some relevant context, one cannot really be 'for or against' it. Privacy as a concept is so rarefied as

to be virtually meaningless. Only when it is taken with reference to a time, a place and a setting can the individuals to be affected by it be thought of as possibly making adjustments to their behaviour in light of it. For instance, a surveillance system might be resisted by some group of people on the grounds that it invades privacy. Another group might resolutely campaign for the very same system on the grounds that greater surveillance guarantees greater privacy, as snoopers will be caught out. The content of the term 'privacy' is thus thematised in opposing ways by these groups, and on grounds not strictly 'rational'. The latter group may have an unwavering trust in government, for instance, that permits them a faith in the system that the former group lacks. Trust and faith are not factors resolvable into rational propositions (they might be said to relate to 'entitlement' which can rely upon an inferential basis, but this is not essential and *fideism* is thus centrally plausible as their basis). Such concepts however furnish the wills of real individuals and constitute a great part of the world of possibilities from a point of view.

This means that the practical orientations of social actors' wills is left out, replaced instead by an imposed frame of reference with its own internal standards of rational argumentation. 'Difference' between the standards of the frame of reference and the choice-parameters of social actors here ultimately, and despite the ambitions of the position, is something *to be resolved*, and resolved into a coherent unity. The story is the same for difference among social actors. Effectively, the position is that once social actors correctly conceive of the learning process, they will learn correctly and so come to see the value of well-argued-for propositions as maxims for action. This creates twin problems for legitimacy and efficiency of governance measures, problems that will be made explicit following a brief and related discussion of the other major limit present in Louvain's thought – the objectifying tendency.

Overcoming Contextual Reductions

In general terms, the so-called 'reductions' are the following:

- **Reduction of the Context:** 'Context' conceived of as 'background' and something that is presumed to be knowable simply via some predetermined, rational means.

- **Reduction of Reason to Argumentation:** The reasons people can have for accepting a normative injunction are reduced to a smaller class than is real – propositions in well-founded arguments – leading to a badly-modelled account of choice and comprehension of options. This leads to problems of constructing context, as well as problems regarding the legitimacy and efficiency of governance.

- **Explanation of the Reduction Problem:** Reduction tends to import tacit assumptions and to frame processes that displace the perspectives of social actors. Genuine elements of social perspectives not part of the framing are automatically presumed to be outside scope. These perspectives are accessible, however, through attention paid to value judgements. In looking to judgement we get to the heart of the issue of the relevance of a context from a perspective.

- **Ethical Consequences of the Problematic Reductions:** In downplaying or suppressing real perspectives the parameters of the frame become the unquestioned limits of the given scene. The experts best suited for understanding the frame thus become arbiters of the scene, via a subtle but illegitimate equivocation. Ethical issues determined here are thus predetermined elements of a subsystem of a framing and not related to the practical possibilities of people. Ethics is *excluded*, replaced instead by technical expertise.

It appears that all the paradigmatic approaches[10] failed to be able to construct the context of the social actor owing to their mentalist, schematising and intentionalist presuppositions creating limitations. One can recall those limits (Lenoble & Maesschalck, 2003):

- **Schematising Presupposition:** It involves Kantian schemes (rules), in which the operation of the application of a norm is a simple formal deductive reasoning on the basis of the rule itself. The determination of the norm is linked to these rules, such as ethical guidelines, or laws, or other external sets of rules.

- **Mentalist Presupposition:** It is named so because it relies on the mind having a set of rules (or schemes, in Kant's words), that predetermines the effect of a norm, and does not depend on any exterior context (to that of the thinker). This is commonly seen when participants in a participatory approach come to the setting with their own particular ethical framing, or with some preconceptions as to what ethical issues might arise.

- **Intentionalist Presupposition:** It entails that the norms effects are supposed to be deducible from the simple intention to adopt the norm. Additionally, there is the presupposition that the actors in a participatory approach will have capacity and intention to contribute to the participatory discussion.

These presuppositions are problematic precisely because they failed to permit the construction of the context of the actor, which implies that the realm of possibility for that actor is not represented. In terms of ethics, then, it is impossible to conceive of a prescription that could legitimately or effectively condition the actor's motivations as ethics seeks to constrain what someone can do in terms of what they ought to do. In failing to conceive of the actor in context, these approaches fail to make that first necessary step in representing the actor and so governance, never mind ethical governance, is foreclosed upon.

Among the context reductions, the logical one is related directly to the mentalist and schematising presuppositions that are supposed to inhibit or prohibit the construction of context in any given governance approach. In terms of how learning is construed, the logical reduction of context can be described as follows: 'The learning process that allows the move from one stage of development to another can be formally exposed, but it does not take place as the simple acquisition of logical analytical competences. In this case the relation to the environmental stimulus could be stabilized by the model of reciprocal action. But it is nothing of the sort! The incentive mechanism in question in learning is not of the order of a reciprocal action' (Lenoble & Maesschalck, 2003, p. 189). With respect to how learning is conceived, the problem here, and its relation to the mentalist and schematising presuppositions is clear. Where the logical reduction is made, the expectation is that action is somehow deduced by means of a neutral process from an unambiguous, objective and transcendent environment. The implication would be that consciousness was a basic, algorithmic treatment of data in the form of an outside world. Action, including 'the right action' would thus be derivable from the simple observation of facts of the matter.

We can in fact model action in context in this way, but we must remember the model is a model: 'data' in genuine contexts do not behave in truth preserving, propositionally expressible ways (such as the loyalties to personal and cultural narratives). This being so, deductive logical operations cannot be supposed to supplant the practical logics of truly contextualised human beings. And when so, the idea of context is deflated into the idea of a neutral, proposition-populated space in which operates only rules such as non-contradiction, conditionality, or-elimination and so on. This re-

duction in fact counts as somewhat contradictory in terms of attempting to model action, since it is the case that action, as opposed to mere reaction, is intentional and therefore issues according to norms (action is *evaluable* as well as simply *describable*). Were the logical reduction to be in effect, no action would be possible as the rules of deductive logic would in fact determine outcomes, thus neutralising the very idea of conscious decision. Logic operates within reason, but cannot supplant it where we seek an account of action.

In failing to represent the actor, that actor's context cannot be constructed and so ethical issues can neither be determined, nor addressed. What the above three steps represent is a means via which to account for the construction of the social actor in context. In fact, what is essential is that 'context' isn't assumed to be some transcendent backdrop which it is the job of governance to represent. Rather, it is the social actor herself that constructs the context according to her background consisting of norms and values. This is then the contentful, substantial matter for discussion, rather than the formalities seen in governance approaches so far. What's more, the actor herself is not presumed to be a fixed entity whose reaction will inevitably lead to one interpretation or another. Instead, using reflexivity, the actor is assumed to have the latitude of mind to be able to conceive of her own position and those of others – to recognise her position as like that of others even though things valued may differ. The problem is the reduction of context on the presumption that a privileged view was possible, that of the expert, which was the true or truest description of any given scene. With this generally comes the attendant assumption that from that description certain behavioural imperatives also naturally flow *as if by logic*. Thus we see the deep relevance of this logical reduction. Nevertheless, in Louvain's school, we see this reproduced, despite all the sound analysis, in the decentring operations of terceisation.

In fact, the idea of context is something nuanced that hinges upon the judgements of individuals and pertains to the relevance of any given scene for those individuals. This relevance constrains what the scene and all it elements means for the individual, and so constrains how the individual sees the possibilities arising from the scene (as well as the legitimacy of its having arisen). This issue having now been raised, we are in a position to be able to pursue how context and judgement can be seen to relate.

TOWARDS A COMPREHENSIVE PROCEDURALISM

Contextual proceduralism fails to embrace the value-dimension of the agents' relationship to norms and of the reflexive stance of the judgement operation. The missing part in this method lies in the pragmatic operation of contextual judgement as well as in the genetic process of learning and identity-buiding of the agents. It appears that contextual proceduralism suffers from two weaknesses: first, an obliteration of the significance of norms; second, a disjunction between norms and values. The alternative methodological option that we propose can be termed *comprehensive proceduralism* insofar as it attempts precisely to take into consideration the value dimension in the governance processes. Comprehensive proceduralism seeks to warrant a reflexive adjustment or rationally justified norms to the context of agents in testing, destabilizing and determining the value significance (or axiological commitments) of these norms. This option does not pretend solving all the harsh problems raised by the methods of proceduralism, but it attempts at least to overcome those raised by its contextual-reflexive version.

Comprehensive Proceduralism

Proceduralism can be said 'comprehensive' in two senses:

1. The word "comprehensive" refers to the notion of *combination of options* in terms of procedures.
2. The word "comprehensive" refers to the notion of *value significance* of thought and conduct in the relationship to norms.

The first meaning suggests that comprehensive proceduralism is a method that pays attention to the variety of procedures to be selected according to their relevance as to the actor's context. The second meaning suggests that this method pays attention to the value dimension of the actors' judgement in the determination of the significance and the scope of norms.

The method of comprehensive proceduralism is based upon the motivating principle that *ethics is the problem, not the solution*. A comprehensive proceduralism incorporates the needs, thereby surpassing its rivals in deliberative collaborative, democratic experimental or contextual-genetic approaches to governance. As has been argued, what is at stake in ethics is the question of *normativity*, i.e. the relationship of individuals to norms, and the relations of norms to values. In this respect, the process of framing of ethical issues is more important and significant than the outcome of the procedure of ethical assessment as such. It is, in fact, the necessary basis for any possible outcome or resolution. This calls for shifting in the assessment procedure from an ethical analysis of the 'issues' (privacy, discrimination, etc.) to a meta-ethical analysis of the governance process. This meta-ethical analysis of the process has to pinpoint where ethics can arise, what nature it might take and how the basis for treating it can be founded.

The relationship of individuals to norms and of norms to values can be transformed by taking into account the context as well as the values underpinning the norms. An adjustment of norms to the contexts is needed, like in Lenoble and Maesschalck's 'contextual proceduralism' and 'reflexive governance.' But an adjustment of norms to values is also needed that requires a method of ethics to take into consideration the value-systems of individuals and groups *in the procedure* itself (See Figure 2).

One can take an example to flesh out this pretty abstract explanation: the case of the integral Islamic veil (*niqab*) in some secular or Christian European countries. One the one hand, it can be that a Muslim woman, contrary to the usual prejudice, is not merely forced or manipulated by her family to wear on this kind of veil, but gives a deep religious or personal significance to this piece of clothe. The value significance of wearing a *niqab* for hear is to express her religious faith and her free submission to God (Islam = 'submission to God'). Then the value-significance of the norm stating that this kind of veil is prohibited in Europe is for her a *denial* of her freedom and her right to exercise her religious faith as she wishes. In contrast, for some secular Europeans, the value-significance of wearing an integral Islamic veil amounts to accepting in the West a kind of social and cultural regression in view of the historical female (and male) struggle for women's emancipation and gender equality. So, the value significance of the norm stating that this kind of

veil is prohibited is for them to guarantee a *protection* of the freedom and the right of European women against male power and oppression and gender inequality.

Now, it can be that the process of exchanging arguments, like in classical proceduralism, will not solve the problem at all. In contrast, listening to each other's narration and interpretation, re-constructing the scheme of relevance of the arguments, his/her set of background assumptions, or even experiencing his/her lifeform can turn out much more effective. It does not mean that the two parts will come to an agreement (the 'pro-niqab' and the 'anti-niqab' in the example), but if they do so, it means that they come to an agreement on the basis of their relative *comprehensive doctrines*. They come to an agreement because, in the learning operation, what they both take for an absolute view appears to be the mirror of their own specific context and then must be thought of as a relative view. If we produce a reversible judgement, this means that the value-significance of a norm (evaluation of a prescription) comes first, while the reverse operation of the norm-significance of a value (prescription of an evaluation) comes second. So, if they both move to an agreement

Figure 2. Model of justification/application relationship (Lavelle)

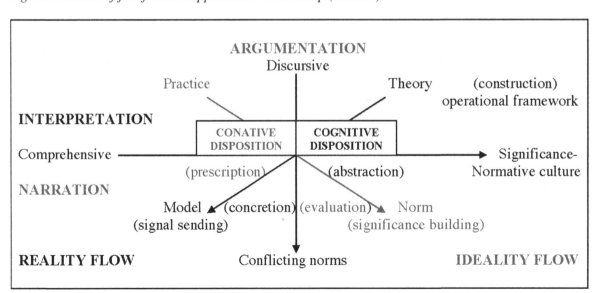

upon a given norm (for instance, prohibition of the *niqab* in the public sphere, authorization in the private sphere), it means that they also agree on the proper value-significance of a norm for each part. The two parts may agree upon the content of the norm, but disagree on the value-significance of the norm, and each part recognizes (or not) the legitimacy of the other part's value-significance.

The important point here is to make a difference between the *ethical momentum* and the *political momentum* in the use of a reflexive procedure. On the one hand, it can be that a political power enforces a legal framework that prohibits the *niqab* in the public sphere in quite an authoritarian fashion, without much discussion (nor any other further procedural approach) about it: this is the *political momentum*, in which what is at stake is the decision-making regarding a norm. On the other hand, it can be that, before (and not after, as far as possible) the decision-making, a more comprehensive procedure is applied, through which the set of actors will try to explore the various context-based framings of the problem as well as the value-significance of the norms for the interested parties: this is the *ethical momentum*, in which what is at stake is the *legitimacy* of a norm before (and possibly after) the decision-making. We also suggest that, for political reasons, one can go straight to the decision of prohibiting the Islamic veil in the public sphere; but as far as ethics is concerned, a further exploration of the actors' contexts and significances is required. We also suggest that it is certainly very possible to impose a norm in quite an authoritarian way; but, in a democratic society, it is also very necessary to give some legitimacy to a norm, if one wants to make it genuinely effective. The general idea, as suggested in the notion of *comprehensive proceduralism*, is the following: the ethical momentum seeks to give an *opportunity* for the various actors to frame the problem (ex: wearing a *niqab* in Europe) in a different way.

A Method of Reflexive Equilibrium (Norm, Context, Value)

The method deriving from the idea of comprehensive proceduralism as a reflexive equilibrium is aimed at producing a balanced adjustment between the rule and the case, the general and the particular, or the level of the society and that of the individual. However, the reflexive equilibrium is not, alike in *formal proceduralism*, a method of *overlapping consensus* (Rawls), or a method of *teleological consensus* (Habermas) (See Figure 3). It is neither, alike in *contextual proceduralism*, a method of reversible-asymmetric judgement based on a search for relevance and coherence. In fact, comprehensive proceduralism is mainly about a reciprocal adjustment of norms to context, not only on the basis of some cognitive dispositions, like in contextual proceduralism, but also on the basis of some conative dispositions ('pro-attitudes'). In other words, in a comprehensive procedural adjustment, the reciprocal adjustment is not only between the two processes of 'abstraction' and 'concretion.' The adjustment is also between the two processes of 'evaluation' and 'prescription' and concerns the value-significance of a norm - insofar as it can be reversed in considering the norm-significance of a value.

Remarks:

- **Horizontal Disjunction/Conjunction**
  - Rule/Case
  - Justification/Application
- **Diagonal Disjunction/Conjunction**
  - General/Singular/Particular
  - Society/Community/Individual
- **Vertical Disjunction/Conjunction**
  - Norm/Value/Fact
  - Reason/Desire/Belief

Comprehensive proceduralism calls for a reflexive adjustment of rationally justified norms to the context of agents in testing, destabilizing and determining the *value significance* (or axiological

Figure 3. Reflexive equilibrium in comprehensive proceduralism

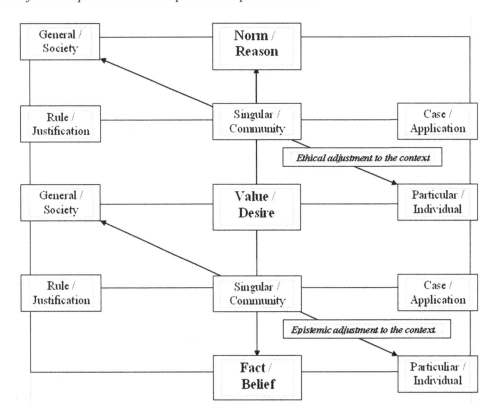

commitments to) of these norms. In attempting to reconnect the political and the personal according to a reflexive principle the idea is to deepen the appreciation of rationality itself via incorporating different yet genuine facets of human understanding, including the personal. Strongly held, deeply cherished ideas are part of human experience as much as is logical rigour. The deepening of rational appreciation thus permits more authentic, genuine instruction regarding substantial content. Such a sense of 'reason' is absent from all the attempted solutions to the governance problem encountered so far.

The method of comprehensive proceduralism does not precede an overlapping consensus, like in Rawls, or a guiding consensus, like in Habermas, but a converging adjustment to the context. Convergence by adjusting the context does not presuppose any will to reach a consensus and keeps open the possibility of dissension between the parties involved in a norm-building or a norm-reflecting process. In addition, the notion of convergence suggests a level of agreement both deeper and less demanding than that of the consensus. In the concept of convergence is the idea that a consensus may well be on a standard, rule or principle, even though the meaning assigned to them by the different actors completely. Thus we can reach a consensus on the principle of 'human dignity,' whereas if one attempts to define what dignity means, it is likely that the consensus on the definition is not possible. Therefore, a process of converging adjustment explore on the basis of a reflection on the context of individuals, what can be the meaning that each of them attributes to a standard, rule or principle. It means then that it is possible to make a reciprocal adjustment of the standard context, and context standard. Such adjustment is based on the consciousness of convergence on some points, but also a difference in other respects.

The context of an individual consists on an epistemic set of beliefs and also of facts which are taken to be somewhat obvious. The first epistemic framing of the norm is characterized by a scope of normative possibilities that is shaped on a cognitive and factual basis. This is the case, for example, when Parliament is developing a law that prohibits wearing the full veil in public. The examination of the factual context of application of this law suggests the difficulty of enforcing the constraint, and a fortiori by violence. In other words, the simple consideration for existing mores, for ontological commitment to them, as well as people involved in the community in which they live, can limit the scope of the law. This is one form of contextual adjustment of the norm to the context which we may call an *epistemic adjustment*.

Now, the context of an individual also consists on a set of desires and also of values which is also taken to be somewhat obvious. The second ethical framing of the norm is characterized by a scope of normative possibilities that is shaped on a volitional and evaluative basis. If we take the example of the law banning the wearing of full face veils in public space, the evaluative examination of the context of application of this law encourages to consider the meaning and the value it can have for those clothes who wear it. He is not here to consider the existing customs of a person or a community, but the meaning of axiological commitment of the individual with regard to these customs, which has also limited the scope of can law. There is a second form of contextual adjustment of the norm, which may be called the *ethical adjustment*.

The difference between the ethical and the epistemic adjustment to the context is that the former can not in any way be derived from the second. In other words, there can be no ethical induction or deduction from fact to value, nor the value of the norm, but only a reciprocal adjustment between

the two. Nevertheless, the consideration for values and desires, as well as consideration for facts and beliefs, are a condition of the relevance of norms. Ultimately, it is the consideration of intentionality, or, as in Searle (2001), the 'intentional states' of individuals, including beliefs and desires, which may allow an adjustment of norms by means of a rational justification in view of their insertion within an application context.

The adjustment of norms to their contexts requires a reflection of the context which may require a modal operation opening up the scope of possibilities of the actors. The opening up can be a consequence of the use of arguments, in the case of a 'weak' modalisation. But it can be also the consequence of the use of narratives, and in some contexts, of experiments that are likely to destabilize and deconstruct the intentional frames of the actors, and then ensure a 'strong' modalisation.

Proceduralism, Comprehension, and Significance

Argumentation is not the only form of discourse admissible as public reason, with understanding of a real life context seen as being more fine-grained than merely argumentatively construed. Argumentative discourse is the foreground of ideal public space, but narration, interpretation, and reconstruction are also admissible on this account. Since narration will only be relevant in terms of a life lived, via specific interpretations of life-events by an agent, an ethical dimension is required in order to comprehend it.

This move toward including the full gamut of human rationality necessitates the realisation that the subject of ethical entreaties is not a mere locus of reasons or a repository for motivations we have to see them as a person, complete with a history and a power of thought and ability to bear convictions. The meaningfulness of the world as construed on a recognition principle has to include the 'psychogenetics' of their view – this entails meaning-variance between parties. Reconstructive ethics seeks a mode of recognition whereby each party attempts to realise *why* good reasons *are good* for the other person.

This is precisely why judgement, and an appreciation of how that works as underlying learning, is so very important to the account we are developing here in order to theoretically treat the problems at the core of ethical governance. Reflection upon this deep core is concerned with analysis and acknowledgement of the basis for holding a belief, desire, proposition to be 'good,' rather than upon the imposition (narration) explanation (interpretation) justification or refutation (argumentation) of those beliefs, desires and so on. Applied to oneself, this reflexivity and auto-criticism can provide a basis for binding one's will to a norm not previously valued.

If we again recall Ferry's position on reconstructive ethics, we can see a means to overcome the limits of being stuck in the rationalistic, argumentative mode[11]. In his view, Ferry (1991) would reconceptualise the method of ethics as mainly a question of re-construction that relies not only on rationalistic argumentation, but that encompasses also three moments:

- Narration
- Interpretation
- Argumentation

Re-construction in ethics suggests that argumentation makes sense as related to the perspectival significances of concepts that makes them compelling for social actors. This outcome we have already alluded to positively in terms of the modal reconstructions of conceptual comprehension in material inference. However, in Ferry's terms we don't quite make it as far as we detailed the necessity to go. One can widen the scope of the reconstructive ethical approach and add elements that are not mentioned in Ferry's re-constructive ethics:

- *Translation* of the backgrounds to be regarded as semantic and pragmatic structures as well as systematic conceptual and experiential patterns.
- *Transformation* of existing paradigms that can function as a common ground to be possibly co-constructed by the actors.
- *Experiment* of different life forms or life worlds that are also an experiential prerequisite in some cases for the conceptual understanding and the personal appropriation of another value-system.

Material inference can be contrasted with formal inference. For formal inferences, structures (e.g. the form of a syllogism) are taken to be valid or invalid independent of the actual substantive content of the variables. So we can say that an argument of the form "If p then q, and p, therefore q" is valid for any p and q. Material inference, however, is that manner of inference that is valid or not in relation to the conceptual contents (the meanings of the terms) deployed within it. It is defined clearly by Robert Brandom (1994, p. 97) as follows:

The kind of inference whose correctnesses essentially involve the conceptual contents of its premises and conclusions may be called, following Sellars, 'material inference.'

Brandom's conception of things operates on the idea that modally reconstructed patterns of inference already present in actual practices of communication provide a structure for communication such that critical appraisal of one another's communicative offerings is possible. This being so, their understanding of issues is laid bare. The idea is that members of social groups deploy concepts in norm-governed ways and so take on commitments and gain entitlements to inferentially significant contents that feature in inference patterns. Mutual understanding is made possible based on critical apprehension of these commitments and entitlements via a scorekeeping mechanism that keeps tabs on the inference patterns in play.

The point is that these types of inferences utilise the meanings of the contents of argumentation in a way that formalist approaches expressly reject. This central difference between material and formal reason is emblematic of the skewed account of reason that itself leads to the equating of argument and moral compulsion that underlies much of the problematic of norm-construction in context (it underlies the decontextualisation problem that has pervaded the entire account of ethics in governance we have been exploring.)

In terms of ethics in research projects we can see the point by means of a simple example. If we recall the sort of formal schema from above, *If P then Q*, etc. and for *P* we substitute "Privacy is protected" and for *Q* we substitute "Research is ethical" the inference pattern reads:

If (privacy is protected) then (research is ethical) and (privacy is protected) therefore (research is ethical)

Formally, this is valid. However, given all that has been said so far it should be clear that this is meaningless without content being ascribed to the terms being used. The terms 'privacy' and 'protection' at least need interpretation, not to mention the purported process of evaluation. And in order to correctly imbue these terms with a meaning that is relevant to those possibly affected by research, this interpretation must be done in a manner that follows *from their own point of view* – that captures what these terms mean for them. The relevant type of inference for ethics is material inference, not formal.

APPLICATIONS OF COMPREHENSIVE PROCEDURALISM: EMERGING TECHNOLOGY AND ETHICAL GOVERNANCE

The stake in governance is that formalism's concepts are so devoid of content as to be virtually meaningless. When a group such as the European Group on Ethics or a National Ethical Committee must publish documentation that seeks applicability across the whole of Europe, its wordings must be so general as to be barely action-guiding. This is no criticism of the bodies themselves, but rather an appraisal of the epistemological problem at the heart of ethics amid plurality.[12] Terms such as privacy, dignity, risk and so on are used in ways that prescribe little in particular as they must be made as generally applicable as possible. In effect, these documents are best thought of as points of departure for detailed debates – as markers of areas of controversy – but serve also to illustrate the issues concerning particular meanings of terms in a context of pluralism as seen in Europe (Bohman, 1986).[13] The issue for particular cases is – how can apparently contentless concepts be made to grab the attention of a European citizenry? How can the individual accept a normative injunction that refers to such rarefied terms as privacy and dignity when these appear in the abstract, without reference to value? When, in public debate on technologies, we see reference to 'privacy issues' these are rarely contextualised beyond a vague sense of property-rights. So vague, indeed, is the concept that it seems effectively uncontroversial.

The Example of Privacy

The idea that one should be able to reliably expect one's property to remain their own is (for European society in general) practically given. However, in framing the discussion of ethical issues in such broad, general, formal terms, the chance of real debate is actually minimised. So agreed is everyone upon the vague property version of privacy that once it has been raised, the matter settled in the same breath: "Privacy is important – agreed." This has the unfortunate consequence that unforeseen, perhaps subtle issues related to how privacy issues can be raised in relation to technology development and emerging technologies do not get an airing. Formalism produces the effect of apparent consensus, foreclosing on detailed discussion and reflection. Naturally, this applies not only to privacy, but to any formally-construed ethical concept.[14]

In the course of its research, the project attempted to make models of future ethical issues likely to emerge from technological development. Naturally, in such an endeavour, the context is the future, and so no detail is available. This being so, the methodology has to be decontextualised. From the following graphic we can see some of the problems. Pursuing the idea of privacy, we can make a sketch model of a hypothetical material-inferentially construed concept of privacy shown in Figure 4.

The words in black represent sources that have come to influence the content of the concept for some given individual. They can be cultural, philosophical, historical, mythic or anything else that has been gleaned from experience. The concept *privacy*, in red, then comes to have certain implications for that individual, according to their judgement of its importance. These are represented in blue. So, for example, the living memory of soviet-style, cold war tactics that lead to paranoia and suppression feature in the individual's memory such that they come to regard privacy as something relevant to communications. With regard to research into technology we can imagine a statement from this individual going along these lines: 'Data insecurity in cloud-computing raises privacy issues as we all recall the brutal excesses of phone-tapping regimes of the KGB. Moreover, if we can't have private communication we can't make plans and so the world becomes an unreliable background for our actions – our actions become meaningless to us. Besides, who are *they* to listen in on *me*?'

Here we would see the ideas of history with reference to the KGB, to a cultural or philosophical notion of autonomy as well as appeal to a mythic notion of outsider such that the chance must exist to absent oneself from the broader collective. Critical appraisal of the perspective that uses this concept of privacy thus must have these collateral concepts in mind in order to engage authentically with the idea. This critical appraisal, of course, can also be taken on by the individual herself. This is an essential feature, in fact, of ethical reflexivity – opening the framing that furnishes one's conceptual connotations.

None of this is of course intended to be real, but the point is made nonetheless. The way in which this individual will reason about concepts

Figure 4. Conceptual topography

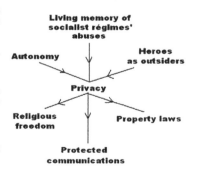

Philosophical/cultural, historical and media influences feed into the nebulous concept 'privacy'.

Privacy so conceived feeds into other concepts, themselves also nebulous.

So nebulous are such concepts as to be inarticulable in the abstract, but in use, courses among the constituent concepts can be traced.

Concepts in use taken as the unit of analysis permits the way in which to comprehend a concept such as privacy such as privacy will abide by certain logical structures, but *according to particular, value-laden conceptual interpretations*. The structures of formal reasoning, pinned down in fact, gives us the basis upon which to pursue normative enquiry including the investigation of ethical aims. The meanings of language and hence the significance of the world from the perspectives of social actors, once engaged with, permits normative discussion as in so engaging we have to use the reasons for assuming a premise, inferring from this premise and concluding something from all of this, again for potentially diverse reasons. These reasons are generated from a basis of judgements replete with values. Hence, to re-quote Lenoble and Maesschalck (2003, p. 90-01):

The epistemological insufficiency of every theory that supposes the context as given or identifiable is important because such presuppositions, even in the form of conventions that are adaptable or revisable by an individual, don't take into account the reversible or reflexive character by which one gives oneself this preference, this convention or whatever it is that makes this ability to adapt or revise possible.

In short, 'privacy' is never itself an issue in research, but is rather a label for a constellation of actual concerns embedded within the research as framed from a perspective. These concerns must be unearthed and dealt with for ethical governance.

The Transformations of Analytical Tools

Table 3 displays transformations of established tools.

We can sketch a scenario now for any given emergent set of conditions such that a procedural, context-sensitive and substantial approach might begin to provide a basis for an ethical governance approach, from the perspective of a social actor.[15]

- **Reaction to the Situation:** Initial reaction provides the value laden interpretation of the situation as it is perceived by a particular social actor owing to their normative background. Norms are provided by culture, history and many other factors.

- **Reflection upon the Reaction:** Reflection upon this allows for the framing that surrounds this reaction to be opened up. This involved reflection upon the nature of the values underwriting the judgements that lead to the interpretation. This permits the normative backdrop to be seen as a justificatory backdrop.

- **Examine the Reflection in Discourse with Other Views:** The question must be asked, 'Why do I value this? Why is this important to me? Why should it be important to anyone?' This can be done in discourse with other views, either in real dialogue with other individuals or via a hypothetical engagement with high-quality information gained from others' testimony.

CONCLUSION

Basically, the very idea of comprehensive proceduralism is that the relationship of norms and contexts, of the rational justification and the contextual insertion of norms, depends upon a special kind of reflexive equilibrium. The general idea of it is the following: there is no 'one best way' for any procedural method that deals with the complexity and the heterogeneity of social and cultural frames. On the contrary, the method of comprehensive proceduralism seeks to explore the special combination of procedures that is best adapted to the contexts of individuals or groups in order to guarantee the significance of norms as being relevant to their value-systems.

We can qualify this method a reflexive equilibrium through an adjustment of procedures to the contexts and the values of the actors, or, in

Table 3. Transformations of established tools

Interface	Action	Rationale / Improvement
Public consultation, National policy	**1. Reaction to the situation** *(Framework goal, or National policy is here 'the situation')* Initial reaction provides the value laden interpretation of the situation as it is perceived by a particular social actor owing to their normative background. Norms are provided by culture, history and many other factors.	Public consultation must be oriented so as to enquire about the concrete issues that concern the public, from their point of view. In thus representing genuine points of view, the context is partly constructed as the perspectives map the horizons of possibility for the social group. National policy must reflect this in order to be a.) legitimate and b.) efficient, i.e. democratically representative and intelligible to the *demos*.
Expert committee	**2. Reflection upon the reaction** Reflection upon this allows for the framing that surrounds this reaction to be opened up. This involves reflection upon the nature of the values underwriting the judgements that lead to the interpretation. This permits the normative backdrop to be seen as a justificatory backdrop.	Expert committees (scientific or ethical) can provide specialist knowledge not generally in the public domain. Thus they can provide useful means for social actors to open their framings. However, this must be modelled upon a communicative exchange lest the expert view be assumed to be better in its expertise. Rather, the opening of framings is a two-way street, experts having much to gain from encountering 'real' views too.
Public consultation, National policy, Expert committee	**3. Examine the reflection in discourse with other views** The question must be asked, "why do I value this? Why is this important to me? Why should it be important to anyone?" This can be done in discourse with other views, either in real dialogue with other individuals or via a hypothetical engagement with high-quality information gained from others' testimony.	At all three levels of public consultation, national policy-making and in expert committees, the point should be to come up with material that is relevant to the issue at hand. This means that at every stage the goal has to be the reflexive reconstruction of the social actor, which is to an extent the construction of context. With this as a foundation the necessary groundwork is put in place for a discussion of possibilities regarding the issues raised. Ethical discussion must be based on such a foundation.

other words, a *comprehensive procedure of reflexive equilibrium* that is context-adaptive and value-sensitive. The principle of judgement in the method of comprehensive proceduralism is not that of consensus, but that of a two-fold potential outcome: consensus or dissensus; consensus on some aspects, dissensus on some other aspects. In this line, a consensus can be reached about a norm, a rule or a principle (e.g: dignity, privacy, etc), while the significance of it, based upon each agent's context, is not the same at all. The main stake of a reflexive equilibrium is then to produce a *reflexive exploration* of the agent's context as well as a *susbstantive determination* of the value-significance of norms for each agent. This process of determination enables the actors to come out of the situation of substantive indeterminacy, in which they are misled by the appearance of formal agreement on a principle or a rule. We suggest that a governance process must be able to offer a

variety of procedures best adapted to the actor's context (*context-adaptive proceduralism*) and to explore the diversity of value-significances given to a norm on the basis of the actors' contexts (*value-sensitive proceduralism*).

The governance of techno-ethics based on the idea of 'comprehensive proceduralism' pleas for a combination of approaches that are both procedural (rule-based), reflexive (context-based) and substantive (value-based). The ethical governance of research projects based in this idea of 'comprehensive proceduralism' thus seeks for a combination of approaches that are at the same time:

1. **Procedural:** Rule-based
2. **Reflexive:** Context-based
3. **Substantive:** Value-based

In a more applied perspective, we are concerned with the *governance of ethics* and with the institutional and organisational conditions that the procedures of assessment must fulfil so that ethical questions can be addressed. The aim of what we call comprehensive proceduralism as far as applications are concerned is to elaborate a new perspective on the ethical governance as applied especially to technical development. We are not concerned only with a regulation of instrumental or strategic rationality in the shape of a code of deontology or an ethics committee, for instance. A series of concrete improvement can be achieved as a development of this view, like in the composition of the consortium (other experts, cross discipline committee, etc.), the education to ethics (ethical reasoning, ethical validating, etc.). Whatever the improvement, the important point is to warrant a connection between the criteria and the knowledge of the people who will use these criteria.

We then come to a series of concluding remarks (rather than conclusions) on the principles of comprehensive proceduralism:

- A process of multi-stakeholders governance cannot be implemented without making use of *some* procedure, although no procedure is neutral and carries about a range of substantial commitments and some implicit or explicit functions and objectives.
- A procedure cannot be reduced to a unique rational form (i.e.: argumentation) and can be conceived of according to a variety of possible forms, including 'non rational' aspects (i.e.: narration).
- A context if viewed as a kind of inescapable *a priori* for judgement can be de stabilized and possibly re-structured through interaction with other actors, but cannot be entirely reflected by the actors themselves.
- A procedure can be adjusted to the context of individuals or communities without requiring a complete reflection of the context by the actors and by implementing at most a re-construction of it.

- A shift in the cognitive and normative framing to occur depends upon a plurality of factors that, to be effective and relevant, are related to the specific context of an individual or a community.
- The disjunction of norms and values questions the relevance and effectiveness of any procedure of inclusion that is supposed to reduce the disjunction of norms and contexts occurring in any procedure of discussion.

Whatever the merits of this alternative option that we term 'comprehensive proceduralism,' it must be clear that it remains within the scope of the procedural methods in philosophy. In other words, comprehensive proceduralism does not replace the inner dynamics of collective action through which a community of people tries to modify her living and working conditions, and through it, the general framework of a society together with the special framework of the individuals.

REFERENCES

Argyris, C., & Schön, D. A. (1978). Organisational learning.: *Vol. 1. A theory of action perspective*. Reading, MA: Addison, Wesley.

Bohman, J. (1986). *Public deliberation: Pluralism, complexity, and democracy*. MIT Press.

Brandom, R. (1998). *Making it explicit*. Harvard.

Dorf, M., & Sabel, C. (1998). *A constitution for democratic experimentalism* (p. 287). Cornell Law Faculty Publications.

Felt, U. et al. (2007). *Taking European knowledge society seriously*. Report of the Expert Group on Science and Governance to the Science, Economy and Science Directorate, Directorate-General for Research, European Commission. European Communities.

Ferry, J.-M. (1991). *Les puissances de l'expérience. Tome I, Le sujet et le verbe.* Cerf, Paris.

Habermas, J. (1986). *The theory of communicative action* (T. McCarthy, Trans.). *Polity.*

Habermas, J. (1987). *Knowledge and human interests: A general perspective* (pp. 301–386). (Shapiro, J. J., Trans.). Cambridge, England: Polity Press.

Habermas, J. (2000). *On the pragmatics of communication* (Fultner, B., Trans.). Cambridge, MA: MIT Press.

Lenoble, J., & Maesschalck, M. (2003). *Towards a theory of governance: The action of norms.* Kluwer.

Lenoble, J., & Maesschalck, M. (2007). *Beyond neo-institutionalist and pragmatic approaches to governance.* Carnets du CPDR.

Lenoble, J., & Maesschalck, M. (2007a). *Reflexive governance: Some clarifications and an extension of the fourth (genetic) approach.* REFGOV Reflexive Governance in the Public Interest, Services of General Interest/Theory of the Norm Unit, Synthesis Report Two, Centre de Philosophie du Droit, UC Louvain.

Maesschalck, M. (2001). *Normes et contextes: Les fondements d'une pragmatique contextuelle. Maesschalck, M., & Lenoble, J. (2010). Democracy, Law and Governance.* Ashgate.

Pettit, P. (2005). *Rules, reasons and norms.* Oxford.

Salvi, M. (Ed.). (2010, July). *Ethically Speaking,* 14, p.6. Retrieved November, 2012 from http://ec.europa.eu/bepa/european-group-ethics/docs/es14_en_web.pdf

Searle, J. (2001). *Rationality in Action.* MIT.

Smith, P. (1994). Les puissances de l'expérience in. *Contemporary Sociology, 23*(3). American Sociological Assoc.

Weber, M. (1904/1949). *Objectivity in social science and social policy. The Methodology of the Social Sciences.* New York: Free Press.

ADDITIONAL READING

Bergström, L. (2002). Putnam on the Fact/Value Dichotomy. *Croatian Journal of Philosophy, 2*(2), 117–129. doi:10.5840/croatjphil20022211.

Bohman, J., Rehg, W., & Zalta, E. (Eds.). (2011). Jurgen Habermas. *The Stanford Encyclopedia of Philosophy.* Retrieved from http://plato.stanford.edu/archives/ win2011/entries/habermas

Breyer, S. (1993). *Breaking the vicious circle.* Cambridge, MA. Harvard University Press.

Brigham, M., & Introna, L. D. (2007). Invoking politics and ethics in the design of information technology: Undesigning the design. *Ethics and Information Technology, 9*, 1–10. doi:10.1007/s10676-006-9131-1.

Capurro, R. (2004). *Ethics between law and public policy.* The European Group on Ethics in Science and New Technologies. Retrieved November, 2012 from http://www.capurro.de/jibl.html

Caracostas, P., & Muldur, U. (1997). *Society, the endless frontier: A European vision of research and innovation policies for the 21st century.* Brussels: EC.

Casper, S., & van Waarden, F. (2005). *Innovation and institutions: A multidisciplinary review of the study of innovation systems.* London: Edward Elgar.

Dauenhauer, B., & Pellauer, D. (2011). Paul Ricoeur. *The Stanford Encyclopedia of Philosophy.* Retrieved November, 2012 from http://plato.stanford.edu/archives/sum2011 /entries/ricoeur/

Dosi, G. (1982). Technological paradigms and technological trajectories: A suggested interpretation of the determinants and directions of technical change. *Research Policy*, *11*, 147–162. doi:10.1016/0048-7333(82)90016-6.

Edler, J. (2003). *Changing governance of research and technology policy: The European research area*. Cheltenham: Edward Elgar.

Goncalves, M. E. (2004). Risk society and the governance of innovation in Europe: Opening the black box? *Science & Public Policy*, *31*(6), 457–464. doi:10.3152/147154304781779796.

Hare, R. M. (1997). *Sorting out ethics*. OUP.

Keulartz, J. et al. (2004). Ethics in technological culture: A programmatic proposal for a pragmatist approach. *Technology & Human Values*, *29*(1), 3–29. doi:10.1177/0162243903259188 PMID:16013108.

Kurt, A. (2008). *The Turkish ICT sector and the european information society: Innovation for integration?* (PhD Thesis). London, UK: University of East London.

Livet, P. (2007). Conversation et revision. *Langage et Société, Les normes pratiques*, 119, 43-62.

Lundvall, B.-A. (Ed.). (1992). *National systems of innovation: Towards a theory of innovation and interactive learning*. London: Pinter.

Ogien, R. (2001). Le Rasoir de Kant. *Philosophiques*, *28*(1), 9–25. doi:10.7202/004963ar.

Overdevest, C. (2001). Transformations in the art of governance. In N. Lebessis & J. Paterson (Eds.), Governance in European Union. Bruxelles, Office des publications de la Commission Européenne.

Perez, C. (2004). Technological revolutions, paradigm shifts and socio-institutional change. In Reinert, E. (Ed.), *Globalization, economic development and inequality: An alternative perspective* (pp. 217–242). Cheltenham: Edward Elgar.

Radder, H. (1998). The politics of STS. *Social Studies of Science*, *22*(1), 141–173. doi:10.1177/0306312792022001009.

Rainey, S., & Goujon, P. (2011). *Existing solutions to the ethical governance problem and characterisation of their limitations*. Retrieved November, 2012 from http://www.egais-project.eu/sites/default/files/deliverables/EGAIS_D4.1_14072011.pdf

Ricouer, P. (1992). *Oneself as another* (Blamey, K., Trans.). Chicago: University of Chicago Press.

Schoser, C. (1999). *The institutions defining national systems of innovation: A new taxonomy to analyse the impact of globalization*. Paper presented at the Annual Conference of the European Association of Evolutionary Political Economy. Prague, November 1999.

Sharif, N. (2006). Emergence and development of the National Innovation Systems concept. *Research Policy*, *35*, 745–766. doi:10.1016/j.respol.2006.04.001.

Smith, P. (1994). Les puissances de l'expérience. Contemporary Sociology. *American Sociological Association, 23*(3).

Trubek, D. M. (1972). Max Weber on law and the rise of capitalism. *Wisconsin Law Review*, 724.

Van Zwanenberg, P., et al. (2009). *Emerging technologies and opportunities for international science and technology foresight*. STEPS working Paper. Retrieved November, 2012 from http://anewmanifesto.org/wp-content/uploads/vanzwan-et-al-paper-30.pdf

Wenar, L. (2012). John Rawls. In E. Zalta (Ed.), *The Stanford Encyclopedia of Philosophy*. Retrieved November, 2012 from http://plato.stanford.edu/entries/rawls/

ENDNOTES

1   See in this book Sylvain Lavelle 'Paradigms of Governance'.

2   See, for example, Hedrick, T. (2010) *Rawls and Habermas: Reason, Pluralism, and the Claims of Political Philosophy,* Stanford University Press.

3   This position is elaborated in the work of various theorists such as WVO Quine, Hilary Putnam etc. David Hume could be seen as a counterpoint.

4   This might be called a transcendental-logical component – that x such that in its absence the phenomenon could not be what it is.

5   See Appendix 1 for more.

6   cf. Rainey, S, 'Competing Methodologies: Possibilities from a Point of View' in this book.

7   cf. EGAIS Deliverable 4.1, 'Existing Solutions to the Ethical Governance Problem and Characterisation of their Limitations,' for further discussion of this point. http://www.egais-project.eu/sites/default/files/deliverables/EGAIS_D4.1_14072011.pdf, retrieved November 2012.

8   It might be a meta-meta-principle in that the meta-principle of reflection underlying terceisation is what ultimately relies on argumentative reason.

9   These problems are not unique to the Louvain School, by any means, and they are indeed fully aware of them as limits.

10   See in this volume the chapter 'Paradigms of Governance'.

11   Ferry's position is summarised as follows: 'Historical and discursive progress from narration toward reconstruction is associated with increasing reflexivity about identity and the grounds upon which it is established. Narration, in Ferry's view, consists of ossified traditional myths which define identities in a more or less prescriptive, taken-for-granted way. Interpretation, on the other hand, involves the assimilation of identity to universal categories like law and justice and is exemplified in early Christian and ancient Greek thought. Argumentation opens up claims of identity to rational dialogue as embodied, for example, in the Enlightenment. Reconstruction, the final step toward reflexivity, involves hermeneutic attempts to understand the historical grounds behind the 'good reasons' offered by others in argumentation. This is in part a logical and ethical consequence of the shift from it to you (acknowledging subjectivity) which emerges with argumentation itself' (Smith, 1994).

12   The task of the EGE can be summed up as follows: "In simple terms, the tasks of the EGE are to apply the norms, values and principles enshrined in international and European declarations to the problems that may face European society in connection with pending EU legislation and current debates in the EU. The remit includes clarifying the problems, providing background information, describing the regulatory and scientific state of the art, outlining possible courses of action and giving advice to the Commission.", in (Salvi, 2010, p. 6).

13   This relies on an idea of public reason as singular, which is a widely discussed notion: 'On the one hand, public reason is singular if it represents itself as a single norm of public deliberation; in light of this norm, agents come to agree upon some decision for the same publicly accessible reasons'. As Rawls puts it: 'There are many non-public reasons and but one public reason' (Bohman, 1986, p. 83).

14   In Louvain we see formalism again in the reliance upon a decentring mechanism. The thought is that by subtracting substantial content from potentially clashing propositions, one gets to the possibility that the propositions can be resolved. It is 'formal'

because it removes content. Resolution is thought to be possible owing simply to logic – any proposition is in principle equivalent to any other and so effectively manipulable simply through inference. The case of clashing values within a motivational set should be enough in itself to open the perspective that formalist resolutions can motivate. Formally, keeping my promise and breaking it are equally valid insofar as they lead to some consequences. The point of the clash however is lost when the content is removed. It's the content that brings the clash, in fact, and it's the judgement of which horn of the dilemma that is the crux of the issue. I am required to consider which outcome I deem better, and I must draw upon my beliefs, desires, hopes, character and all upon which these are based. Recalling the example of the difficult promise above, it's not decided when we realise the schema;

If (Keep promise) then (Fuel gambling)
And

If (Break promise) then (Friend's edification)
But

not (Keep Promise) and (Friend's edification)
i.e. If p then q, and if r then s, but not p and s

This is where formalism ultimately lands us, and it misses the entire point. Formalism translates the schematic form of our dilemma, but without evaluation we are powerless to rationally decide among options.

We can crystallise these points against rationalism and objectivism into one simple example. Imagine we are to meet an elderly relative at the train station. Her suitcases are heavy therefore we ought to carry them for her. Most will agree that this is true and validly argued.1 But, it's not the fact of the heaviness of the bags that simply prescribes the action. If it were, I'd be carrying everyone's bags – if the fact of heaviness prescribed my intervention I'd perform a contradiction

in carrying granny's but no-one else's. After all, in a rationalistic vein, how can I justify in a value-free, argumentative mode that heaviness prompts carrying now, but not now?

In fact, if heaviness itself triggers the carrying, presumably we'd have to attempt to hoist the train onto our shoulders, objectively speaking – if we describe the context objectively and the heaviness of certain elements therein prompt carrying, heaviness per se must prompt carrying. Clearly this is absurd. The point is that the fact of the heaviness plays a part in providing a reason to carry granny's bags and no one else's as granny is valued specially in the scene.

There is a legitimacy of my exclusive act of portering precisely in virtue of the values I hold – the values are why I carry her bag and no others. Likewise for the efficiency of my action – I know precisely what I must do in carrying granny's bags as my values provide reasons to act in the way I do. In the broader context, the situation is evaluated in the light of values and reasons to act are gained there from via judgment. This is the fact/value distinction in action.

This is an illustration of why formalism provides no motivation. Without reference to values, there are no reasons to do one thing rather than another, there being no particular legitimacy of one action of another, and the chance of anything being done is slim to nil as time is wasted in empty cogitation. This is a case in point of why we need value in the mix when we consider ethical governance. The points regarding legitimacy, efficiency and value are not example-specific. This goes against the above quoted reference to Rawls; Reasons are as various as actors themselves, not unified under a value free umbrella.

Reflection upon judgment such that values are included leads to efficient action precisely because values are included. Why anyone acts is related to their beliefs and desires, certainly, but the combination of belief and desire is not a sufficient cause of any particular action. For instance, I can believe that wine is available at a local bar, desire to have some, but value the idea of doing a job well and so refrain from constant wine-drinking in order to preserve clear-headedness and so ensure ongoing job success. Likewise, I can believe I have a tooth-cavity, utterly detest the prospect of visiting the dentist, but value my health and well-being enough to trump my squeamishness and opt to get drilled. If formalism were true, this deliberation would be mysterious as it is not provided for by a formula such as:

I (believe that bar = wine) and (want wine) therefore (visit bar).

Human beings don't necessarily have to suffer the formal determinations of such QEDs.

15   See Appendix 2 for further details of the EGAIS project's recommendations in this regard.

APPENDIX 1

GROUNDS OF JUDGEMENT IN COMPREHENSIVE PROCEDURALISM

When we believe something to be the case, we always know we may be wrong or have miscalculated somehow. Perhaps we believe Barcelona is the capital of Spain, or miscalculate that 7+5=13: we are equipped in the light of this to examine our every belief and decide, judiciously, which are those beliefs we will retain. Similarly, for matters of desire we can be wrong. Perhaps whilst trying to quit we feel we want another cigarette, but know that this is a craving rather than a desire. We can hopefully resist. Similarly, we can have a desire only to realise upon its fulfilment that it was not at all worth having – perhaps we devote our time to earning money only to get rich before realising friendships are more valuable. This is called 'revealed preference.'

How do we make these sorts of judgements (See Figure 5)? We must rely upon collateral information beyond our beliefs and desires, but we don't arbitrarily decide that our beliefs are right or desires worth having. Given that we use judgement, we must suppose that judgement itself operates on more than this information. Essentially, we make reference to values, but these values come from various sources, many of which don't readily appear to us even in reflection. These values inform our personal nature in various ways – they inform our very sense of *subjectivity*. Given the nature of these values and the existential import they can have, it is naturally very important that in accounting for learning as that phenomenon that can change our judgements of good, bad, worthwhile etc. that this notion of subjectivity is dealt with (See Figure 6).

Figure 5. Judgement

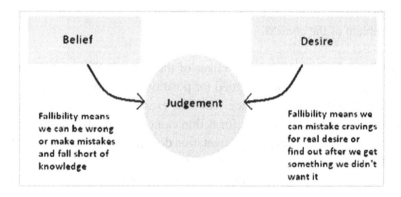

Figure 6. Grounding action

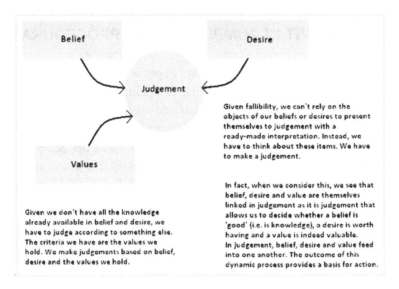

APPENDIX 2

SUMMARY OF EGAIS RECOMMENDATIONS

The thought behind the treatment of the problems we develop can be summed up in three interconnected thematic cores. These are the logical conclusions to be drawn from the problems and limits derived in the deliverable. The thematic cores are:

1.   Not to reduce ethics to argumentation.
2.   To connect construction of ethical issues to resolutions.
3.   To open the problem of the context.

1: The rationalistic mode into which fell the likes of the Louvain School ended up equating formal reasoning with moral goodness – the well-argued for position ought to compel action. However, in accounting for the breadth of human experience with, for example, reference to Ferry's reconstructive ethics, we have seen that compelling reasons for action can come on a basis of narrative logic, or on a deep-seated conception of oneself. Formal argumentation does not necessarily unseat this kind of reasoning, but this is a real part of human rationality and so must be adequately represented. Argumentation, with the tacit identification of formal reasoning with human motivation, cannot account for the reasons why individuals may or may not accept a reason as persuasive *for them*. Value must be included.

Philip Pettit (2005, p. 36) makes a point relevant to this, and does so very neatly, so is worth citing here. His simple observation is that exemplification, the practice of persuading through the use of examples, is a three place relation. In the first place, there is a norm (for example) which stands as the object which we would hope to persuade an audience is good, valuable, worthwhile and so on. Second, there are examples of that norm's instantiation. However, the often-overlooked third place is that which perhaps needs most attention – the audience to whom the example is supposed to appeal.

With this three-place structure in mind, we can see that from the start in any justification of a norm that would proceed by giving examples, the position of the audience must be considered. At the very least, this will involve thinking of examples that will appeal to that audience while illustrating the supposed goodness of the norm. Naturally, to do so means engaging with the values of the audience. So, we can see via this independent means of argument that we end up at the same place – to involve a public in an exercise such as governance whereby minds must be changed, norms justified and so on, value and interpretation must be reckoned to be centrally important.

2: A simple determination of a problem, however successful it might be, is not a complete ethical approach unless something is offered as a means to deal with it. This means developing a rich understanding of the significance of the issue for those it affects. This understanding must be developed in order to permit a response in just the same way as it is no use understanding what someone says in a conversation if we are unprepared to reply. Determining an issue must be carried out in the light of understanding *why* it is a problem and this means determining the grounds upon which a solution must be founded.

It is of limited use for anyone to suggest a piece of research will cause 'privacy issues' and leave it at that. To do so would be to treat ethics as a tick-box, compliance exercise, which is something we need to move beyond. Instead, given the description of an ethical issue, we need to consider how the issue is based, upon what and for whom so that we can determine a manner in which to address it.

3: Too often context is assumed or altogether ignored and norms thought of as decontextualised. Rather than being assumed, context must be considered as a part of the problem to be addressed in ethics. The construction of context in fact comes as a part of the faithful representation of the perspectives of individuals involved in the scenario. There are thus legitimacy, efficiency and ethical imperatives in getting this right;

- Only by representing the demos accurately can democratic accountability be maintained.
- Only by engaging with the actual views of social actors can intelligibility of governance problems and solutions be maintained and so action be made possible.
- Only by engaging with real views can genuine ethical concerns be seen, as such concerns will be a function of perceived possibilities in the light of perceived changes.

As a last point on context, as we established through discussion of our objections to Louvain's terceisation account of learning, we suppose that context is itself never completely known and so the construction of context must be seen as a methodological procedure rather than an act that, once carried out, secures a definition of 'context' that thereafter can be operative within the ethical thinking. Rather, context must be seen terms of the reflexive relations it engenders. It must be seen in terms of the so-called hermeneutic circle, wherein implicated agents must reflect upon their experience of the context and then upon their analysis of the same, in a creative and dynamic tension.

General Conclusion

The governance of ethics refers the institutional and organisational conditions that the procedures of assessment must fulfil so that ethical questions can be addressed in the field of emerging technologies. We showed that classical *proceduralism* did not adequately address problems raised by an ethics of science and technology, for traditional approaches to ethics contain a 'blind spot' vis-à-vis the *relationship between norms and contexts*. The insufficiency of proceduralism was evident in that the arrangements that are necessary for organising the reflexive capacity for the actors to identify the various effective possibilities on which the operation of the selection of the norm will be carried out are problematic. We then proposed a blending of approaches: *procedural* (rule-based), *reflexive* (context-based) and *substantive* (value-based). It must be said that the method of comprehensive proceduralism is likely to apply to a wide range of emerging technologies: gene, nuclear, nano, information and communication, internet. This is why we took some cases from theses various domains as examples and resources for scrutinizing the relationship between emerging technologies and ethical governance.

The hypothesis is that existing governance mechanisms for technical research and development projects do not adequately address ethical and social issues that can arise from such projects, either before or during, the project stage. As a result, the desired outcome of projects to produce technologies that meet today's ethical challenges and result in socially acceptable technologies, is at risk. Governance theory has been used as a means to investigate options for governance, and existing theories and processes for governance have been shown to be deficient in various ways. The empirical research suggests that the institutional mechanisms of the European Commission for addressing ethical issues in Research and Development projects in place at that time were not in themselves enough to either alert project participants to the ethical aspects of the technical components of the project, nor provide guidance on how they might do this. We discovered that the majority of the projects did not consider ethical and social aspects within their research and technology design. Our analytic grid, based on our theoretical approach allowed us to see the patterns of governance models used by the projects and from that we were able to map the various governance paradigms evident from political and ethical governance theory literature.

The same analytical standpoint was applied to other fields of technology, which showed that there are similar elements of ethical governance problems across fields other than ICT and Ambient Intelligence, i.e. our starting point. One of the areas we looked at was that of innovation, as a process itself, and in connection to that the design element of technology development, with a focus on value-driven design. Our aim is to open the possibilities available to project participants, including technical developers, to allow them to explore the potential ethical and social implications of the technology research and to allow them to seek solutions that are relevant to the context of the technology under development, based on norms that have meaning within that context. The process we are advocating, and which we have called comprehensive proceduralism, aims to highlight the connection between norms and values within a specific context. Such a context is not pre-defined or pre-determined but is instead discovered and shaped, thus emerging throughout the ethical reflexive process. Now we turn to understanding the political implications of this approach with brief reference to the historical development of policy discourse of European research and technology development.

It is necessary to organise the reflexive capacity of the actors by constructing the capacities of the reflexivity in such a way as to not presuppose it as already existing due to a formal method, such as argumentation, deliberation, debate or discussion. All of these formal methods presuppose their own required conditions and as such do not necessarily involve reflexivity. It is therefore important to make sure that every application of a norm presupposes not only a formal moment of choice of its acceptable normative constraints, but an operation of the selection of the possibilities according to the acceptable way of life within the community concerned. Without a negotiated construction of the moment of reflexivity that is specific to the conditions for the application of the norm, however, there will be no control of the process of the expression of the norm, and it will be left to the dominant common culture to express.

Thus, what is often presented as the only effective choice is always conditioned by an operation such as the above (including in the construction of the deontological codes). Criticism of this reconstruction of the reflexivity used in the construction of the social norm also affects the moral approaches to legitimacy. Institutional arrangements for this reflexivity, since this is the very aim of the project, in order to overcome the fundamental limitations of existing ethical approaches, which ignore the issue of the moment of the application of the norm. Determining these arrangements will allow actors and institutions to go through a learning process when confronted with an ethical issue, reflect on the success of the learning process, and reframe the context of the situation in order to more effectively establish a norm within the context, and from a more official perspective, will allow us to assess the effectiveness of the result of that process.

To take into account the limitations and achieve second-order reflexivity, we need to escape the binds of formalism, which constrains ethics with its presuppositions and internal limitations. To more effectively incorporate ethical norms into contexts, we need to construct the framing of the context in relation to the norm (i.e. not presuppose it), then open up this context so that we can have a reflexivity on the opening of this framing (that is, a feedback mechanism). In order to do this, we need to reconstruct, from a normative perspective, how research projects should reconstruct the two-way relationship between the norm and the context to overcome the fundamental limitations outlined above in order to achieve second-order reflexivity. Without the implementation of a reflexive capacity, and the construction of norms, normative injunctions risk remaining inefficient even if the objective is judged relevant and legitimate. The operation of judging the conditions of the choice of an improved way of life, facilitated by technological development is distinct and asymmetric. Asymmetry is the way in which the social meanings of a norm are conditioned by an operation that cannot be anticipated by formal variables of reasoning variables that condition the norm's relevance.

The point of all of this is to provide a deep, philosophical grounding for the manner in which we can interpret and understand the epistemological position of the social actor which is missing from the accounts of governance so far, including that of the Louvain School. Recalling EGAIS' central problematic, concerning the ethical governance of research in European projects, this is of paramount importance in two ways – one is to secure legitimacy and efficiency for the governance measures, the other is to ensure the ethicity of the governance measures:

1. **Legitimacy:** Is secured via the pragmatic reconstruction of the epistemological position of the actor in that the actor herself is represented in the governance measure. There is no presupposition or imposition of a framing from experts or anyone else.

2. **Efficiency:** Is made possible as through engagement with the epistemological position of the social actor, their understanding is secured and so governance injunctions are made comprehensible, debatable and capable of being enacted.

3. **Ethicity:** Is enabled as the range of possibility for any given social actor is cashed out in terms of how they understand the world around them and what that understanding entails by way of possible actions.

In recognising the diversity of reasons we recognise the role of value in self-perception and social deliberation and so we can properly represent in public reasoning the actual public rather than a theoretical abstraction. Such abstractions are clearly at play in the models of governance where 'the public' is seen as 'in need of education' or 'overly risk averse', such as in the technocratic-instrumental, ethocratic-normative and epistocratic-cognitive paradigms. In the light of what has been argued in this and previous deliverables these abstractions should appear quite anaemic now. So, we adopt an attitude necessary for ethical thinking in that we reconstruct the object of ethical enquiry – the social actor. This permits the policy-maker (in the case of governance) to provide reasons for a social actor to adopt a norm, as opposed to merely dealing in commands. This procedure ought also to be seen as the background of dialogical engagement among the public themselves. Since this engagement runs on reasons embedded within the worldview of the social actors, this also gives a means whereby such an actor can adopt a reflexive stance upon their own values and adopt a norm independent of their own predilections.

We see that presuppositions blight any attempt to understand the perspectives of those engaged in governance approaches from either the governing or governed perspective. These are based in the presuppositions discernible behind governance approaches that inhibit the construction of contexts, and the mode of rationality assumed to be effective in all contexts, i.e. argumentative rationality. At the very least, this hampers the effectiveness of the measures, while standing as an obstacle to their legitimacy. In so understanding this issue we have a basis to move toward addressing them in a way that is not merely *ad hoc* but rather is grounded in theory and that can affect practice in a positive way. So with the problem clearly defined and the basis upon which it is a problem understood, we have cleared the ground for of in which we can propose ways to approach the specific problems we detect with an ambition of treating them.

We obviously do not intend to revolutionize the *ethical governance of emerging technologies*, and the emphasis is on a more interpretive use of the existing tools. This requires the reconstruction of the context of the actors involved in the governance process, and a consideration of the subjective values on which can be built and embodied the ethical norms. We have tried to show how some aspects of this requirement were already expressed in several schools of thought and in several technological projects. The ambition of EGAIS is to help unite these impulses, developing the overall concept of 'comprehensive proceduralism' as a framework for covering them. No doubt that, in the coming years, the need for ethical norms adapted to the new realities of the development of technology will become more and more pressing, and we hope that the principles that we put forward will be able to be of great help for the task of establishing them.

Compilation of References

(1998). InEisenberg, N. (Ed.). Handbook of child psychology: *Vol. 3. Social, emotional and personality development*. New York: J. Wiley.

Abels, G. (2006, Jul 4). Forms and functions of participatory technology assessment – Or: Why should we be more skeptical about public participation? *Participatory Approaches in Science and Technology Conference* (pp. 1-12).

Ackroyd, S., Batt, R., Thomson, P., & Tolbert, P. S. (Eds.). (2005). *The Oxford handbook of work and organization*. New York, Oxford: Oxford University Press.

Adair, D. (1967). The technocrats: 1919-1967. A case study of conflict and change in social movements. (Thesis). Simon Fraser University.

AGKS. (2009, September). *A Green Knowledge Society - An ICT policy agenda to 2015 for Europe's future knowledge society, A study for the Ministry of Enterprise, Energy and Communications, Government Offices of Sweden by SCF Associates Ltd Final Report*. Retrieved October 14, 2011 from http://ec.europa.eu/information_society/eeurope/i2010/docs/i2010_high_level_group/green_knowledge_society.pdf

Akrich, M. (1992). The de-scription of technical objects. In Bijker, W. E., & Law, J. (Eds.), *Shaping technology/building society* (pp. 205–224). Cambridge: MIT Press.

Alchian, A. A. (1990). Uncertainty, evolution, and economic theory. In O. E. Williamson (Ed.), Industrial organization (p. 23-33). Hants: Edward Elgar.

Allhoff, F. (2007). On the autonomy and justification of nanoethics. In Allhoff, F., & Lin, P. (Eds.), *Nanotechnology and society: Current and emerging ethical issues* (pp. 3–38). Dordrecht: Springer. doi:10.1007/s11569-007-0018-3.

Allhoff, F., & Lin, P. (2006). What's so special about nanoethics. *The International Journal of Applied Philosophy*, *20*, 179–190. doi:10.5840/ijap200620213.

American Psychiatric Association. (2000). *Diagnostic and statistical manual of mental disorders* (4th ed., text rev.). Washington, DC: American Psychiatric Press.

Anderson, E. (2006). The epistemology of democracy. *Episteme*, *3*(1-2), 8–22. doi:10.3366/epi.2006.3.1-2.8.

Anderson, E. (2008). An epistemic defense of democracy: David Estlund's democratic authority. *Episteme*, *5*(1), 129–139. doi:10.3366/E1742360008000270.

Angell, I. (2009). All (intellectual) property is theft? *Presentation, Fifth Communia Workshop, Accessing, Using, Reusing Public Sector Content and Data*. London School of Economics.

Anscombe, G. E. M. (1981). Modern moral philosophy. In *Collected philosophical papers* (Vol. 3, pp. 26–41). Oxford, United Kingdom: Oxford University Press. (Original work published 1959).

Appel, A. W., Ginsburg, M., Hursti, H., Kernighan, B. W., Richards, C. D., Tan, G., & Venetis, P. (2009). The New Jersey voting-machine lawsuit and the AVC advantage DRE voting machine. *USENIX Security '09*. Retrieved November 10, 2011, from http://www.usenix.org/events/evtwote09/tech/full_papers/appel.pdf.

Aquinas, T. (1948). Summa theologiae. (English Dominican Province, Trans.). Westminster, MD: Christian Classics. (Original work composed 1265–1273).

Arato, A., & Cohen, J. L. (1992). *Civil society and political theory*. Cambridge, MA: MIT Press.

Argyris, C., & Schön, D. (1978). *Organisational Learning, A Theory of Action Perspective* (*Vol. 1*). Reading, MA: Addison, Wesley.

Argyris, C., & Schön, D. A. (1978). Organisational learning.: *Vol. 1. A theory of action perspective*. Reading, MA: Addison, Wesley.

Aristotle. (1941). Nicomachean ethics. (W. D. Ross, Trans.). New York: Random House. (Original work composed 350 B.C.).

Aristotle,. (1999). *Nicomachean Ethics* (2nd ed.). Hackett.

Aristotle,. (2007). *The nicomachean ethics*. Filiquarian Publishing, LLC.

Aristotle. (4th c. BCE). *Politics*.

Arnsperger, C., & Varoufakis, Y. (2006). What is neoclassical economics? The three axioms responsible for its theoretical oeuvre, practical irrelevance and thus, discursive power. *Post autistic economics review, 38*.

Arrhenius, G. (2005). Vem bör ha rösträtt? Det demokratiska avgränsningsproblemet. In Tidskriften för politisk filosofi (vol. 2).

Assoun, P. (1990). *L'Ecole de Francfort*. Presses Universitaires de France.

Avison, D., & Torkzadeh, G. (2008). *Information systems project management*. California, USA: Saga Publications.

Ayer, A. (1959). *On the analysis of moral judgement. Philosophical Essays*. London: Mac Millan.

Bang, H. P. (Ed.). (2003). *Governance as social and political communication*. Manchester: Manchester University Press.

Barnes, J. A. (1972). Social networks (pp. 1-29). An Addison-Wesley Module in Anthropology, Module 26.

BBC. (2011a). US defence firm Lockheed Martin hit by cyber-attack. Retrieved June 5, 2011, from http://www.bbc.co.uk/news/world-us-canada-13587785

BBC. (2011b). Sony warns of almost 25 million extra user detail theft. Retrieved June 15, 2011, from http://www.bbc.co.uk/news/technology-13639836

BBC. (2011c). Sony network suffers hack attack by Lulz Network. Retrieved June 15, 2011, from http://www.bbc.co.uk/news/technology-13639836

Beck, U. (2001). La dynamique politique de la société mondiale du risque. *IDDRI Working Papers No 01/2001*. Retrieved October 23, 2012, from http://www.iddri.org/Publications/La-dynamique-politique-de-la-societe-mondiale-du-risque

Beck, U. (1986). *Risikogesellschaft*. Frankfurt am Main: Suhrkamp Verlag.

Beck, U. (2003). *Pouvoir et contre-pouvoir à l'ère de la mondialisation*. Paris: Aubier.

Beck, U., Giddens, A., & Lash, S. (1994). *Reflexive modernization, politics, tradition and aesthetics in the modern social order*. Palo Alto, SA: Stanford University Press.

Bell, A., & Parchomovsky, G. (2009). *The evolution of private and open access property. Theoretical Inquiries in Law, 10*(1). Retrieved May 3, 2010, from http://www.bepress.com/til/default/vol10/iss1/art4/

Benedict, X. V. I. (2005). *Deus caritas est (Encyclical, God is love)*. Vatican: Libreria Editrice Vaticana.

Benkler, Y. (2006). *The wealth of networks: How social production transforms markets and freedom*. New Haven, CT: Yale University Press.

Bentham, J. (2009). *An introduction to the principles of morals and legislation*. Oxford: Dover Publications Inc..

Berdichewsky, D., & Neuenschwander, E. (1999). Toward an ethics of persuasive technology. *Communications of the ACM, 42*(5), 51–58. doi:10.1145/301353.301410.

Bergson, H. (1907). *L'Évolution créatrice. Paris: P.U.F. – English translation: Creative Evolution (1944)*. New York: Random House.

Berkowitz, M., & Oser, F. (1985). *Moral education: Theory and application*. Hillsdale, NJ: Lawrence Erlbaum.

Berten, A. (1993). Habermas critique de Rawls. La position originelle du point de vue de la pragmatique universelle. Retrieved from www.uclouvain.be/cps/ucl/doc /.../ DOCH _006_ (Berten).pdf

Biological Weapons Convention. (1972). Retrieved March 24, 2011, from http://www.opbw.org/convention/conv.html

Blasi, A. (1980, July). Bridging moral cognition and moral action: A critical review of the literature. *Psychological Bulletin, 88*(1), 1–45. doi:10.1037/0033-2909.88.1.1.

Blasi, A. (1996). Moral understanding and moral personality: The process of moral integration. In Kurtines, W. M., & Gewirtz, J. L. (Eds.), *Moral development: An introduction.* Boston, MA: Allyn and Bacon.

Blesin, L., & Loute, A. (2011). Nouvelles vulnérabilités, nouvelles formes d'engagement, apport pour une critique sociale. In M. Maesschalck & A. Loute (Eds.), *Nouvelle critique sociale, Europe – Amérique Latine, Aller - Retour* (pp. 155-192). Monza: Polimetrica. Retrieved October 23, 2012, from http://www.polimetrica.com/form/form2850.php

Blondiaux, L., & Sintomer, Y. (2002). L'impératif délibératif. *Politix, 15*(57), 17–35. doi:10.3406/polix.2002.1205.

Bohman, J. (1986). *Public deliberation: Pluralism, complexity, and democracy.* MIT Press.

Bourcier, D. (2002). *Is governance a new form of regulation? Balancing the roles of the state and civil society* (pp. 2). IWM Working Paper 6/2002, Vienna.

Bourcier, D. (2007). Creative commons: An observatory for governance and regulation through the internet. In Casanovas, P., Noriega, P., Bourcier, D., & Galindo, F. (Eds.), *Trends in legal knowledge the semantic web and the regulation of electronic social systems* (pp. 41–52). Firenze: EPAP.

Bourcier, D., & Dulong de Rosnay, M. (2004). Introduction. In Bourcier, D., & Dulong de Rosnay, M. (Eds.), *International Commons ou la Création en partage.* Paris: Romillat.

Bourdieu, P. (1998). *Contre-feux.* Paris: Liber raisons d'agir.

Bourdieu, P. (2004). *Esquisse pour une auto-analyse.* Paris: Raisons d'agir.

Bourdieu, P. (1994). *Raisons pratiques.* Paris: Seuil.

Bourdieu, P. (2000). *Esquisse d'une théorie de la pratique.* Paris: Seuil.

Bourdieu, P. (2008). *Sketch for a self-analysis.* Chicago: University of Chicago Press.

Bourdieu, P., Chamboredon, J. C., & Passeron, J. C. (1983). *Le Métier de sociologue.* Paris: Mouton Editeur.

Bourdieu, P., & Wacquant, L. (1992). *Réponses.* Paris: Seuil.

Bouveresse, J. (2003). *Bourdieu, savant et politique.* Argone.

Boyle, J. (2008). *The public domain, enclosing the commons of the mind* (pp. xvi). New Haven: Yale University Press. Retrieved May 3, 2010, from http://www.thepublicdomain.org/download

Brandom, R. (1994). *Making it explicit.* Cambridge, MA: Harvard University Press.

Brandom, R. (1998). *Making it explicit.* Harvard.

Brandom, R. (2008). *Between seeing and doing.* Oxford: Oxford University Press. doi:10.1093/acprof:oso/9780199542871.001.0001.

Brey, P. (2000). *Disclosive computer ethics: Exposure and evaluation of embedded normativity in computer technology.* Presented at the CEPE2000 Computer Ethics: Philosophical Enquiry. Dartmouth College. Retrieved October 22, 2010 from http://ethics.sandiego.edu/video/CEPE2000/Responsibility/Index.html

Briggle, A. R., & Spence, E. H. (2009). Cosmopolitan friendship online. In G.Watson, B.G. Renzi, E. Viggiani, & M. Collins (Eds.), Friends and foes volume II: Friendship and conflict from social and political perspectives (pp. 53-64). Cambridge: Cambridge Scholar Publishing.

Brooks, T. (2008). Is Plato's political philosophy anti-democratic? In Kofmel, E. (Ed.), *Anti-democratic thought.*

Browning, D. (2006). *Christian ethics and the moral psychologies.* Grand Rapids, MI: Eerdmans Publishing.

BRWM. (1990). *Rethinking high-level radioactive waste disposal: A position statement of the Board on Radioactive Waste Management.* Washington, D.C.: National Academy Press.

BSI. (2010). BundesamtfürSicherheit in der Information-stechnik [BSI, German Federal Office for Information Security], Advanced security mechanisms for machine readable travel documents. *Technical Guideline TR-03110*.

Burris, B. H. (1993). *Technocracy at work*. New York: State University of New York Press.

Cahill, C., Sultana, F., & Pain, R. (2007). *Participatory ethics: Politics, practices, institutions*. ACME Editorial Collective.

Callon, Lascoumes, & Barthe. (2001). Agir dans un monde incertain. Seuil.

Callon, M. (1998). Des différentes formes de démocratie technique. *Les cahiers de la sécurité intérieure*, 38, 37-54.

Callon, M., Lascoumes, P., & Barthe, Y. (2001). *Agir dans un monde incertain. Essai sur la démocratie technique*. Paris: Le Seuil.

Callon, M., Lascoumes, P., & Barthe, Y. (2001). *Agir dans un monde incertain: Essai sur la démocratie technique*. Paris: Seuil.

Calvert, R., & Johnson, J. (1999). Interpretation and coordination in constitutional politics. In Hauser, E., & Wasilewski, J. (Eds.), *Lessons in democracy. Jagiellonian: University Press & University of Rochester Press*.

Carr, N. (2008). Is Google making us stupid? *The Atlantic Monthly*. Retrieved January 31, 2013, from http://www.theatlantic.com/doc/200807/google.

Carr, D., & Steutel, S. (Eds.). (1999). *Virtue ethics and moral education*. London: Routledge.

Carroll, J. M. (2000). *Making use: Scenario-based design of human-computer interactions*. MIT Press.

Casebeer, W. D. (2003, October). Opinion: Moral cognition and its neural constituents. *Nature Reviews. Neuroscience*, 4, 841–846. doi:10.1038/nrn1223.

Chaabouni, R., & Vaudenay, S. (2009). The extended access control for machine readable travel documents. BIO-SIG 2009, *Biometrics and Electronic Signatures* [Bonn, Germany: GesellschaftfürInformatik] [GI]. *LNI*, *155*, 93–103.

Charter of the United Nations. (1945). Retrieved September 08, 2011, from http://www.un.org/en/documents/charter/

Cherniak, C. (1986). *Minimal rationality*. Cambridge, MA: MIT Press.

Cherry, M., & Imwinkelried, E. (2006). Cautionary note about fingerprint analysis and reliance on digital technology. *Judicature*, *89*(6), 334-338. Retrieved June 20, 2011, from http://www.ajs.org/ajs/publications/Judicature_PDFs/896/Cherry_896.pdf

Ciprut, J. V. (2008). Prisoners of our dilemmas. In Ciprut, J. V. (Ed.), *Ethics, politics and democracy. From primordial principles to prospective practices* (pp. 17–20). MIT Press.

Coenen, C. (2010). Deliberating visions: The case of human enhancement in the discourse on nanotechnology and convergence. *Sociology of the Sciences Yearbook*, *27*, 73–87. doi:10.1007/978-90-481-2834-1_5.

Cohen, J., & Sabel, C. (1999). *Directly deliberative polyarchy*. Columbia Law School papers. Retrieved June 1, 2011, from http://www2.law.columbia.edu/sabel/papers/DDP.html

Cohen, J. (1986). An epistemic conception of democracy. *Ethics*, *97*(1), 26–38. doi:10.1086/292815.

Cohen, J., & Sabel, C. (1997). Directly deliberative polyarchy. *European Law Journal*, *3*(4), 313–342. doi:10.1111/1468-0386.00034.

Collins-Chase, C. (2008). Comment: The case against trips-plus protection in developing countries facing AIDS epidemics. *University of Pennsylvania Journal of International Law*, *29*(3), 763–802.

Comte-Sponville, A. (1995). *Petit traité des grandes vertus*. Paris, France: Presses Universitaires de France.

Convention on Biodiversity. 1992. Retrieved August 17, 2009, from http://www.cbd.int/doc/legal/cbd-en.pdf

Cornu, J. M. (2001). *La coopération nouvelles approaches*. Retrieved November 20, 2001, from http://www.cornu.eu.org/texts/cooperation

Cornu, J. M. (2011). Nous avons non pas un mais deux modes de pensée. Retrieved November 20, 2011, from http://www.cornu.eu.org/news/nous-avons-non-pas-un-mais-deux-modes-de-pensee

Cornu, J. M., & Méadel, C. (2006). Les accords de peering ou comment le Sud finance le Nord. Retrieved November 18, 2006, from http://www.voxinternet.org/spip.php?article100

Council of Europe. (2006). *Secretary-General's report under Article 52 ECHR on the question of secret detention and transport of detainees suspected of terrorist acts, notably by or at the instigation of foreign agencies.* Retrieved March 02, 2007, from http://www.coe.int/T/E/Com/Files/Events/2006-cia/SG-Inf-(2006).pdf

Council of Europe. (2007). *Countries worldwide turn to Council of Europe Cybercrime Convention.* Press Release 413(2007), June 13, 2007.

Covenant of the League of Nations. (1919). Retrieved September 12, 2011, from http://www.iilj.org/courses/documents/CovenantoftheLeagueofNations_000.pdf

Cowam.com. (n.d.). Website. Retrieved from http://www.cowam.com

Cropper, S., Ebers, M., Huxham, C., & Smith Ring, P. (Eds.). (2008). *The Oxford handbook of inter-organizational relations.* Oxford: Oxford University Press. doi:10.1093/oxfordhb/9780199282944.001.0001.

Cuhls, K. (2003). From forecasting to foresight processes - new participative foresight activities in Germany. *Journal of Forecasting, 22*(2-3), 93–111. doi:10.1002/for.848.

Cutler, T. (2008). *Report on the review of the national innovation system.* Cutler & Company Pty Ltd. Retrieved from http://www.innovation.gov.au/Innovation/Policy/Documents/NISReport.pdf

Cyrulnik, B. (1998). *Ces enfants qui tiennent le coup.* Revigny-sur-Ornain: Hommes et Perspectives.

D'Holbach, P. H. T. (1776, 2008). *Ethocratie. Ou le gouvernement fondé sur la morale.* Coda.

Da Fonseca, E. G. (1991). *Beliefs in action. economic philosophy and social change.* Cambridge: Cambridge University Press. doi:10.1017/CBO9780511628412.

Dahl, R. (1998). *On democracy.* Yale University Press.

Damasio, A. (1994). *Descartes' error: Emotion, reason, and the human brain.* New York: Harper-Collins Publishers.

Damasio, A. (1999). *The feeling of what happens.* Boston, MA: Mariner.

Damon, W. (1999, August). The moral development of children. *Scientific American, 276*, 56–62. PMID:10443038.

Dando, M., & Wheelis, M. (2005). Neurobiology: A case study of the imminent militarization of biology. *International Review of the Red Cross, 87*(859), 553–571. doi:10.1017/S1816383100184383.

Davies, S., Macnaghten, P., & Kearnes, M. (Eds.). (2009). *Reconfiguring responsibility: Lessons for public policy (Part 1 of the report on deepening debate on nanotechnology).* Durham: Durham University.

Day, J. (Ed.). (2008). *Patterns in network architecture: A return to fundamentals.* New Jersey, USA: Prentice Hall Press.

De Munck, J., & Zimmermann, B. (Eds.). (2008). La liberté au prisme des capacités, vol. 18 in series Raisons pratiques. Paris: Editions de l'EHESS.

De Schutter, O. (2007). *The role of collective learning in the establishment of the AFSJ in the EU.* The EU Centre of Excellence, University of Wisconsin. Retrieved October 20, 2012, from http://eucenter.wisc.edu/Conferences/GovNYDec06/Docs/DeSchutterPPT.pdf

Dent, N. (1984). *The moral psychology of the virtues.* New York: Cambridge University Press.

Dent, N. (1999). Virtue, eudaimonia, and teleological ethics. In Carr, D., & Steutel, J. (Eds.), *Virtue ethics and moral education* (pp. 21–34). London: Routledge.

Dewey, J. (1927). *The public and its problems.* New York: Holt.

Donahue, J. D., & Nye, J. S. (Eds.). (2001). *Governance amid bigger, better markets.* Washington, DC: Brookings Institution Press.

Donelly, J. (2003). *Universal human rights in theory and practice* (2nd ed.). Cornell University Press.

Dorf, M., & Sabel, C. (1998). A constitution for democratic experimentalism. *Cornell Law Faculty Publications.* Retrieved June 2011, http://scholarship.law.cornell.edu/cgi/viewcontent.cgi?article=1119&context=facpub&sei-redir=1#search=%22A+constitution+for+democratic+experimentalism%22

Dorf, M., & Sabel, C. (1998). *A constitution for democratic experimentalism* (p. 287). Cornell Law Faculty Publications.

Dorf, M., & Sabel, C. (1998). Constitution of democratic experimentalism. *Columbia Law Review, 98*(2), 267–473. doi:10.2307/1123411.

Dorrestijn, S. (2006). *Michel Foucault et l'éthique des techniques: Le cas de la RFID.* (Unpublished Master's thesis). Paris X University, Nanterre.

Dosi, G. (1982). Technological paradigms and technological trajectories: A suggested interpretation of the determinants and directions of technical change. *Research Policy, 11*, 147–162. doi:10.1016/0048-7333(82)90016-6.

Dougherty, M. (2010). Platonic epistocracy. A response to Andrew Forcehimes. Deliberative democracy with a spine. idem, pp. 79-83.

Drexler, K. E. (1986). *Engines of creation. The coming era of nanotechnology.* New York: Anchor Books.

Ducatel, K., Bogdanowicz, M., Scapolo, F., Leijten, J., & Burgelman, J.-C. (2001). *Scenarios for ambient intelligence in 2010.* Seville: European Communities, IPTS. Retrieved November 27, 2012, from ftp://ftp.cordis.europa.eu/pub/ist/docs/istagscenarios2010.pdf

Dupuy, J. P. (2007). Some pitfalls in the philosophical foundations of nanoethics. *The Journal of Medicine and Philosophy, 32*(3), 237–261. doi:10.1080/03605310701396992 PMID:17613704.

Dupuy, J. P. (2010). The narratology of lay ethics. *NanoEthics, 4*(2), 153–170. doi:10.1007/s11569-010-0097-4.

Dupuy, J. P., & Grinbaum, A. (2006). Living with uncertainty: Toward the ongoing normative assessment of nanotechnology. In *Schummer* (pp. 287–314). Baird. doi:10.1142/9789812773975_0014.

Dussolier, S. (2005). *Droit d'auteur et protection des œuvres dans l'univers numérique* (p. 231). Bruxelles: Larcier.

Dworkin, R. (1981). What is equality? Part 1: Equality of welfare. *Philosophy & Public Affairs, 10*(3), 185–246.

Ebbesen, M., Andersen, S., & Besenbacher, F. (2006). Ethics in nanotechnology: Starting from scratch? *Bulletin of Science, Technology & Society, 26*(6), 451–462. doi:10.1177/0270467606295003.

EC (The Council of the European Union). (2004). *Council Regulation No 2252/2004.* Retrieved April 11, 2011, from http://eur-lex.europa.eu/LexUriServ/LexUriServ.do?uri=OJ:L:2004:385:0001:0006:EN:PDF

Edler, J. (2010, October). *Towards understanding the emerging governance marble cake in multi-level European R&D funding.* Paper presented at the Conference: Tentative Governance in Emerging Science and Technology: Actor Constellations, Institutional Arrangements & Strategies. University of Twente, Netherlands.

Effi. (2008). Sähköinen äänestysjärjestelmä rikkoo sekä lakia että vaalisalaisuuden. Retrieved November 10, 2011, from http://www.effi.org/julkaisut/tiedotteet/lehdistotiedote-2008-06-24.html

Effi. (2010). Electronical frontier Finland, Sähköäänestys-FAQ [eVoting-FAQ]. Retrieved November 10, 2011, from http://www.effi.org/sahkoaanestys-faq.html

EGAIS. (2009). *Deliverable 2.1. Grid-based questionnaire development.* Retrieved November 27, 2012, from http://www.egais-project.eu

EGAIS. (2010a). *Deliverable 2.2. Empirical data collection.* Retrieved November 27, 2012, from http://www.egais-project.eu

EGAIS. (2010a). *Deliverable 2.4.: Interpretation of ethical behaviour.* Retrieved June 1, 2011, from http://www.egais-project.eu/sites/default/files/deliverables/EGAIS_D2.4_23092010.pdf

EGAIS. (2010b). *Deliverable 2.4 Interpretation of ethical behaviour.* Retrieved November 27, 2012, from http://www.egais-project.eu

EGAIS. (2010b). *Deliverable 3.1.: Ethical governance models, paradigm recognition and interpretation.* Retrieved June 1, 2011, from http://www.egais-project.eu/sites/default/files/deliverables/EGAIS_D3.1_30092010-1.pdf

Eisenberg, N. (1982). *The development of prosocial behavior.* New York: Academic Press.

Eisenberg, N. (2003). The development of empathy-related responding. In Pope-Edwards, C., & Carlo, G. (Eds.), *Nebraska symposium on motivation.* Lincoln: University of Nebraska Press.

Elisa. (2011). *Elisa tulee valittamaan väliaikaisesta Pirate Bay –määräyksestä* (press release by Elisa). Retrieved November 10, 2011, from www.elisa.fi/ir/pressi/?o=5120.00&did=17563

Elkin-Koren, N. (2006). Creative commons: A skeptical view of a worthy pursuit. In P. B. Hugenholtz and L. Guibault (Eds.), *The future of the public domain*. Kluwer Law International. Retrieved May 5, 2010, from http://papers.ssrn.com/sol3/papers.cfm?abstract_id=885466

Epstein, R. A. (2004, October 24). Why open source is unsustainable. *Financial Times*. Retrieved November 27, 2012, from http://www.ft.com/cms/s/2/78d9812a-2386-11d9-aee5-00000e2511c8.html#axzz2DSFs33Q2

ETICA. (2010). *Deliverable 4.1 ICT ethics governance review*. Retrieved November 27, 2012, from http://ethics.ccsr.cse.dmu.ac.uk/etica

European Comission. (n.d.). *European Commission's Science in Society Portal*. Retrieved June 1, 2011, from http://ec.europa.eu/research/science-society/index.cfm?fuseaction=public.topic&id=1223&lang=1

European Commission. (2001). *European governance: A white paper*. COM (2001) 428 Final. Brussels: European Commission. Retrieved November 1, 2012, from http://eur-lex.europa.eu/LexUriServ/site/en/com/2001/com2001_0428en01.pdf

European Commission. (2002). *Science and society action plan*. Luxembourg: Office for Official Publications of the European Commission. Retrieved November 27, 2012, from http://ec.europa.eu/research/science-society/pdf/ss_ap_en.pdf

European Commission. (2008). *Commission recommendation on 07/02/2008 on a code of conduct for responsible nanosciences and nanotechnology, No. C 424 final*. Brussels: European Commission.

European Commission. (2011). *Electricity market regulatory forum (Florence)*. Retrieved July 1, 2011, from http://ec.europa.eu/energy/gas_electricity/electricity/forum_electricity_florence_en.htm

European Commission. (2011). *Fourth FP7 Monitoring Report - Monitoring Report 2010*. Brussels. Retrieved October 12, 2011 from http://ec.europa.eu/research/evaluations/pdf/archive/fp7_monitoring_reports/fourth_fp7_monitoring_report.pdf

European Court of Human Rights. (2010). *European convention on human rights*. Strasbourg: European Council of Europe. Retrieved October 23, 2012 from http://www.echr.coe.int/nr/rdonlyres/d5cc24a7-dc13-4318-b457-5c9014916d7a/0/englishanglais.pdf

European Group on Ethics in Science and new technologies (EGE). (2005). *Ethical Aspects of ICT implants in the human body. Opinion of the European Group on Ethics in Science and new technologies to the European Commission*. Retrieved January 19, 2012, from http://ec.europa.eu/bepa/european-group-ethics/docs/avis20_en.pdf

European Parliament and Council. (2006, December 18). *Recommendation of the European Parliament and of the Council of 18 December 2006 on Key Competences for Lifelong Learning*. Retrieved from http://eur-lex.europa.eu/LexUriServ/LexUriServ.do?uri=OJ:L:2006:394:0010:0018:en:PDF

European Union. (2004). Protocol on the application of the principles of subsidiarity and proportionality. *Official Journal of the European Union, C310/207*. Retrieved June 1, 2011, from http://eur-lex.europa.eu/LexUriServ/LexUriServ.do?uri=OJ:C:2004:310:0207:0209:EN:PDF European Union. (2010). *Single European act*. Brussels: EU. Retrieved June 1, 2011, from http://europa.eu/legislation_summaries/institutional_affairs/treaties/treaties_singleact_en.htm

European Union. (2007). Treaty of Lisbon amending the Treaty on European Union and the Treaty establishing the European Community, signed at Lisbon, 13 December 2007. *Official Journal of the European Union, 50*(C306), 1-202. Retrieved June 1, 2011, from http://eur-lex.europa.eu/JOHtml.do?uri=OJ:C:2007:306:SOM:EN:HTML

Fann, K. T. (1970). *Peirce's theory of abduction*. The Hague: Martinus Nijhoff. doi:10.1007/978-94-010-3163-9.

Felt, U. et al. (2007). *Taking European knowledge society seriously*. Report of the Expert Group on Science and Governance to the Science, Economy and Science Directorate, Directorate-General for Research, European Commission. European Communities.

Felt, U., & Wynne, B. (2007). *Taking European knowledge society seriously. Expert Group on Science and Governance*. Luxembourg: Office for Official Publications of the European Communities:

Ferrari, A., & Nordmann, A. (Eds.). (2009). *Reconfiguring responsibility: Lessons for nanoethics (part 2 of the report on deepening debate on nanotechnology)*. Durham: Durham University.

Ferry, J. M. (1996). *L'Éthique reconstructive*. Paris: Éditions du Cerf, Collection « Humanités."

Ferry, J.-M. (1991). Les puissances de l'expérience. Tome I, Le sujet et le verbe. Cerf, Paris.

Ferry, J.-M. (2002). *Valeurs et normes, la question de l'éthique.* Bruxelles: Edition de l'université de Bruxelles.

Fiedeler, U., Grunwald, A., & Coenen, C. (2005). *Vision assessment in the field of nanotechnology - A first approach.*

Financial Times. (2009, November 3). *Kremlin-backed group behind Estonia cyber blitz.* Retrieved November 24, 2010, from http://www.ft.com/cms/s/57536d5a-0ddc-11de-8ea3-0000779fd2ac,Authorised=false.html?_i_location=http%3A%2F%2Fwww.ft.com%2Fcms%2Fs%2F0%2F57536d5a-0ddc-11de-8ea3 0000779fd2ac.html&_i_referer=http%3A%2F%2Fen.wikipedia.org%2Fwiki%2F2007_cyberattacks_on_Estonia

Finkelstein, L. (1995). What is global governance? *Global Governance, 1*(3), 363–372.

Fisher, E., Mahajan, R., & Mitcham, C. (2006). Midstream modulation of technology: Governance from within. *Bulletin of Science, Technology & Society, 26*(6), 485–496. doi:10.1177/0270467606295402.

Fisher, F. (1990). *Technocracy and the politics of expertise.* Newberry Park, CA: Sage.

Flanagan, O. (1991). *Varieties of moral personality: Ethics and psychological realism.* Cambridge, MA: Harvard University Press.

Flanagan, O. (1996). *Self-expression: Mind, morals, and the meaning of life.* New York: Oxford University Press.

Flanagan, O. (2002). *The problem of the soul: Two visions of mind and how to reconcile them.* New York: Basic Books.

Flanagan, O. (2006). Psychologie morale. In Canto-Sperber, M. (Ed.), *Dictionnaire d'éthique et de philosophie morale* (pp. 1220–1229). Paris: Presses Universitaires de France.

Fleischman, W.M. (2010). Electronic voting systems and the therac-25: What have we learned? *Proceedings of Ethicomp 2010*, Tarragona, Spain.

Floridi, L. (1999). Information ethics: On the philosophical foundation of computer ethics. *Ethics and Information Technology, 1*(1), 33–52. doi:10.1023/A:1010018611096.

Floridi, L. (2008). Information ethics: A reappraisal. *Ethics and Information Technology, 10*(2), 189–204. doi:10.1007/s10676-008-9176-4.

Fogg, B. J. (2003). *Persuasive technology: Using computers to change what we think and do.* Elsevier.

Food and Agriculture Organisation. (1997). *International Plant Protection Convention.* Retrieved August 27, 2009, from https://www.ippc.int/file_uploaded//publications/13742.New_Revised_Text_of_the_International_Plant_Protectio.pdf

Food and Agriculture Organisation. (2001). *International treaty on plant genetic resources for food and agriculture.* Retrieved April 15, 2010, from ftp://ftp.fao.org/docrep/fao/011/i0510e/i0510e.pdf

Forcehimes, A. (2010). Deliberative democracy with a spine. Epistemic agency as political authority. *Dialogue, 52*(2-3), 69–78.

Foucault, M. (1990). *The care of the self – The history of sexuality: 3.* London: Penguin Books.

Foucault, M. (1992). *The use of pleasure – The history of sexuality: 2.* London: Penguin Books.

Fowler, J. W. (1980). Moral stages and the development of faith. In Munsey, B. (Ed.), *Moral development, moral education, and Kohlberg: Basic issues in philosophy, psychology, religious, and education.* Birmingham, AL: Religious Education Press.

Fowler, J. W. (1981). *Stages of faith: The psychology of human development and the quest for meaning.* San Francisco: Harper.

Fowler, J. W. (1984). *Becoming adult, becoming Christian.* San Francisco: Harper and Row.

Frangie, S. (2009). Bourdieu's reflexive politics. *European Journal of Social Theory, 12*(2), 213–229. doi:10.1177/1368431009103706.

Freeman, C., & Louçã, F. (2002). *As time goes by: From the industrial revolutions to the information revolution.* New York: Oxford University Press. doi:10.1093/0199251053.001.0001.

Friedman, B., Kahn, P., & Borning, A. (2002). *Value sensitive design: Theory and methods.* Seattle: University of Washington, *CSE Technical Report 02-12-01.*

Friedman, B. (Ed.). (1997). *Human values and the design of computer technology.* Chicago: University of Chicago Press.

Gamel, C. (2007). Que faire de "l'approche par les capacités"? Pour une lecture "rawlsienne" de l'apport de Sen. *Formation-emploi, 98*(April-June), 141–150.

Gauchet, M. (2002). Les tâches de la philosophie politique. *La Revue du MAUSS, 19*, 275–303. doi:10.3917/rdm.019.0275.

Geertz, C. (1973). Description: Toward an interpretive theory of culture. In Geertz, C. (Ed.), *The interpretation of culture.* New York: Basic Books.

Georghiou, L., Harper, J. C., Keenan, M., Miles, I., & Popper, R. (2008). *The handbook of technology foresight: Concepts and practice.* Edward Elgar Publishing Ltd..

Gerstenberg, O. (1997). Law's polyarchy: A comment on Cohen and Sabel. *European Law Journal, 3*, 343–358. doi:10.1111/1468-0386.00035.

Gibbs, J. (2003). *Moral development and reality: Beyond the theories of Kohlberg and Hoffman.* Thousand Oaks, CA: Sage Publications.

Giddens, A. (1990). *The consequences of modernity.* Stanford: Stanford University Press.

Gilardone, M. (2007). *Contexte, sens et portée de l'approche par les capabilités de Amartya Kumar Sen.* (Unpublished doctoral dissertation). Lyon-2 University of Lyon, France.

Gilligan, C. (1981). *In a different voice: Psychological theory and women's development.* Cambridge, MA: Harvard University Press.

Gilligan, C. (1992). Reply to Critics. In Larrabee, M. J. (Ed.), *An ethic of care: Feminist and interdisciplinary perspectives* (pp. 207–214). New York: Routledge.

Gingras, Y. (2010). Sociological reflexivity in action. *Social Studies of Science, 40*(4), 619–631. doi:10.1177/0306312710370236.

Girard, J.-Y. (2006 - 2007). *Le point aveugle.* Paris, France: Hermann, (tome I & II).

Girard, J.-Y. (2011). *La syntaxe transcendantale, manifeste.* Retrieved from http://iml.univ-mrs.fr/~girard/Articles.html

Girard, J.-Y. (2001). Locus solum. *Mathematical Structures in Computer Science, 11*, 301–506. doi:10.1017/S096012950100336X.

Goleman, D. (2006). *Social intelligence: The new science of human relationships.* New York: Bantam Books.

Gonggrijp, R., & Hengeveld, W. J. (2007). Studying the Nedap/Groenendaal ES3B voting computer, a computer security perspective. In *Proceedings of the USENIX Workshop on Accurate Electronic Voting Technology.* Retrieved October 14, 2011, from http://wijvertrouwenstemcomputersniet.nl/images/c/ce/ES3B_EVT07.pdf

Gonggrijp, R., Hengeveld, W. J., Bogk, A., Engling, D., Mehnert, H., Rieger, F., et al. (2006). Nedap/Groenendaal ES3B voting computer, a security analysis. Retrieved August 8, 2010, from http://wijvertrouwenstemcomputersniet.nl/images/9/91/Es3b-en.pdf

Gordon, J., & Finlayson, F. F. (2010). *Habermas and Rawls. Disputing the political.* Routledge.

Gottweis, H. (2005). Regulating genomics in the 21st century: From logos to pathos? *Trends in Biotechnology, 23*(3), 118–121. doi:10.1016/j.tibtech.2005.01.002 PMID:15734553.

Goujon, P., & Dedeurwaerdere, T. (2009). Taking precaution beyond expert rule. Institutional design for collaborative governance. The genetically modified organisms. In *Proceedings of the ICT that Makes a Difference Conference.* Brussels, November 2009.

Goujon, P., & Lavelle, S. (2007). General introduction. In Goujon, P. et al. (Eds.), *The information society: Innovation, legitimacy, ethics and democracy (In honour of Professor Jacques Berleur s.j.).* New York: IFIP, Springer. doi:10.1007/978-0-387-72381-5.

Grin, J., & Grunwald, A. (2000). *Vision assessment: Shaping technology in 21st century society towards a repertoire for technology assessment.* Springer. doi:10.1007/978-3-642-59702-2.

Grunwald, A. (2004). *Vision assessment as a new element of the FTA toolbox.* Retrieved January 20, 2012, from http://forera.jrc.ec.europa.eu/fta/papers/Session%204%20What's%20the%20Use/Vision%20Assessment%20as%20a%20new%20element%20of%20the%20FTA%20toolbox.pdf

Grunwald, A. (2000). Against over-estimating the role of ethics in technology development. *Science and Engineering Ethics*, *6*, 181–196. doi:10.1007/s11948-000-0046-7 PMID:11273446.

Grunwald, A. (2001). The application of ethics to engineering and the engineer's moral responsibility: Perspectives for a research agenda. *Science and Engineering Ethics*, *7*, 415–428. doi:10.1007/s11948-001-0063-1 PMID:11506427.

Grunwald, A. (2005). Nanotechnology - a new field of ethical inquiry. *Science and Engineering Ethics*, *11*, 187–201. doi:10.1007/s11948-005-0041-0 PMID:15915859.

Guidon, A. (1989). *Le développement moral*. Paris: Desclée.

Habermas, J. (1984-1987). The theory of communicative action. Cambridge.

Habermas, J. (1998). Between facts and norms (Rehg, W., Tran.) MIT Press.

Habermas, J. (1968). *Technik und Wissenschaft als 'Ideologie*. Frankfurt am Main: Suhrkamp.

Habermas, J. (1981). *Theorie des kommunikativen Handelns*. Frankfurt: Suhrkamp.

Habermas, J. (1981/1984). The theory of communicative action. Cambridge. *Polity*.

Habermas, J. (1983). *Moral Bewusstsein unf Kommunikativ Handeln*. Suhrkamp Verlag.

Habermas, J. (1986). *The theory of communicative action* (T. McCarthy, Trans.). *Polity*.

Habermas, J. (1987). *Knowledge and human interests: A general perspective* (pp. 301–386). (Shapiro, J. J., Trans.). Cambridge, England: Polity Press.

Habermas, J. (1991). *Erläuterungen zur Diskursethik*. MIT Press.

Habermas, J. (2000). *On the pragmatics of communication* (Fultner, B., Trans.). Cambridge, MA: MIT Press.

Habermas, J. (2001). The postnational constellation and the future of democracy. In Pensky, M. (Ed.), *The postnational constellation, political essays*. Cambridge, MA: MIT Press.

Habermas, J. (2005). *Truth and justification (Fultner, B., Tran.)*. Cambridge, MA: MIT Press.

Habermas, J. (2006). *Between facts and norms, contributions to a discourse theory of law and democracy*. Cambridge, MA: The MIT Press.

Habermas, J., & Rawls, J. (2005). *Débat sur la justice politique*. Cerf.

Haber, S. (2001). *Habermas*. Presses Pocket.

Hamilton, A. (2008). John Stuart Mill's Elitism. In Kofmel, E. (Ed.), *Anti-democratic thought*.

Handelsman, M. M., Knapp, S., & Gottlieb, M. C. (2009). Positive ethics: Themes and variations. In Snyder, C. R., & Lopez, S. J. (Eds.), *The Oxford handbook of positive psychology* (pp. 105–113). Oxford, United Kingdom: Oxford University Press. doi:10.1093/oxfordhb/9780195187243.013.0011.

Hardin, G. (1968). The tragedy of the commons. *Science*, *162*(3859), 1243–1248. doi:10.1126/science.162.3859.1243 PMID:5699198.

Hauerwas, S. (1981). *Vision and virtue*. Notre Dame: University of Notre Dame Press.

Hauptman, A., Sharan, Y., & Soffer, T. (2011). Privacy perception in the ICT era and beyond. In von Schomberg (2011) (pp. 133-147).

HE. (99/2006). *Hallituksen esitys Eduskunnalle laiksi lapsipornografian levittämisen estotoimista*. Retrieved November 10, 2011, from http://www.finlex.fi/linkit/hepdf/20060099.

Heersmink, R., van den Hoven, J., van Eck, N., & van den Berg, J. (2011). Bibliometric mapping of computer and information ethics. *Ethics and Information Technology*, *13*, 241–249. doi:10.1007/s10676-011-9273-7.

Heidegger, M. (1977). *Question concerning technology*. New York: Harper & Row.

Heim, M. (1987). *Electric language*. New Haven: Yale University Press.

Heimo, O. I., Fairweather, N. B., & Kimppa, K. K. (2010). The Finnish eVoting experiment: What went wrong? In Proceedings of Ethicomp 2010. Tarragona, Spain.

Heimo, O. I., Hakkala, A., & Kimppa, K. K. (2011). The problems with security and privacy in eGovernment - Case: Biometric passports in Finland. In Proceedings of Ethicomp 2011. Sheffield, UK.

Heiskanen, E. (2005). Taming the golem - an experiment in participatory and constructive technology assessment. *Science Studies, 18*(1), 52–74.

Held, D. (1995). *Democracy and the global order: From the modern state to cosmopolitan governance.* Cambridge: Policy Press.

Held, D. (1996). *Models of democracy.* Stanford University Press.

Heller, A. (2006). European master-narratives about freedom. In Delanty, G. (Ed.), *Handbook of European social theory* (pp. 257–265). London: Routledge.

HelsinginSanomat [HS]. (2008a, February 22). *Poliisihaluaapassiensormenjäljetrikostutkijoille* (Police request passport fingerprints to criminal investigation) (1st ed.).

HelsinginSanomat [HS]. (2008b, November 27). *Rikostutkijateivätsaavieläpassiensormenjälkiäkäyttöönsä* (Criminal investigators do not acquire passport fingerprints yet) (1st ed.).

HelsinginSanomat [HS]. (2010a). *Vesipiipputupakantuonnistavankeuttaja 400000 euronlasku* (Prison sentence and 400.000 € fine from illegal import of waterpipe tobacco). Retrieved June 18, 2011, from http://omakaupunki.hs.fi/ paakaupunkiseutu/uutiset/vesipiipputupakan_tuonnista_vankeutta_ja_400_000_euron_lasku/

HelsinginSanomat [HS]. (2010b). *Sunnuntaisuomalainen: Passiensormenjälkirekisterivoiavautuapoliisille* [Fingerprint registry may be opened to the police]. Retrieved April 11, 2011, from http://www.hs.fi/kotimaa/artikkeli/Sunnu ntaisuomalainen+Passien+sormenj%C3%A4lkirekisteri +voi+avautua+poliisille/1135259348892

Heriard-Dubreuil, G., Gadbois, S., Mays, C., Espejo, R., Flüeler, T., Schneider, T., & Paixa, A. (2007, June). COWAM 2 - Cooperative research on the governance of radioactive waste management, final synthesis report. Mutadis, Paris, EC contract FI6W-CT-2003-508856.

Hoepman, J. H., Hubbers, E., Jacobs, B., Oostdijk, M., & Schreur, R. W. (2006). Lecture Notes in Computer Science: *Vol. 4266. Crossing borders: Security and privacy issues of the European e-passport, advances in information and computer security* (pp. 152–167). Berlin, Heidelberg: Springer.

Hoepman, J. H., & Jacobs, B. (2007). Increased security through open source. *Communications of the ACM, 50*(1), 79–83. doi:10.1145/1188913.1188921.

Hoffman, M. L. (2000). *Empathy and moral development: Implications for caring and justice.* Cambridge: Cambridge University Press. doi:10.1017/CBO9780511805851.

Holm, S. (2005). *Does nanotechnology require a new "nanoethics"?* Cardiff Centre for Ethics, Law and Society. Retrieved December 20, 2011, from http://www.ccels. cf.ac.uk/archives/issues/2005/

Horkheimer, M. (1967). *Zur Kritik der Instrumentellen Vernunft.* Frankfurt am Main.

Horkheimer, M., & Adorno, T. (2007). *Dialectic of enlightenment.* Stanford University Press.

Hornung, G. (2007). The European regulation on biometric passports: Legislative procedures, political interactions, legal framework and technical safeguards. *SCRIPTed 246, 4*(3). Retrieved June 9, 2011, from http://www.law.ed.ac. uk/ahrc/script-ed/vol4-3/hornung.asp

Hosein, I. (2004). The sources of laws: Policy dynamics in a digital and terrorized world. *The Information Society, 20*(3), 187-199. ISSN: 0197-2243

Hottois, G. (1996). *Symbool en techniek.* Kampen/Kapellen: Kok Agora/Pelckmans.

Huxley, A. (1994). Brave new world (New ed.). London: Flamingo.

Hyppönen, H. (2007). Ehealth services and technology: Challenges for co-development. *Human Technology. An Interdisciplinary Journal on Humans in ICT Enviroments, 3*(2), 188–213.

IAEA. (1992). *Radioactive waste management, an IAEA source book.* Vienna.

ICRP. (2012). *Radiological protection in geological disposal of long-lived solid radioactive waste.* International Commission on Radiological Protection, Consulting Document.

Ihde, D. (1990). *Technology and the lifeworld.* Bloomington, Minneapolis: Indiana University Press.

Institute for Science. Ethics and Innovation. (2009). Who owns science? *The Manchester Manifesto.* Retrieved September 12, 2011, from http://www.isei.manchester.ac.uk/TheManchesterManifesto.pdf

International Civil Aviation Organization. ICAO. (2006). Machine readable travel documents. *ICAO/Doc 9303 Part 1 Vol. 2.* Retrieved June 17, 2011, from http://www2.icao.int/en/MRTD/Downloads/Doc%209303/Doc%209303%20English/Doc%209303%20Part%201%20Vol%202.pdf

International Committee of the Red Cross. (1868). *(St. Petersburg) Declaration renouncing the use, in time of war, of explosive projectiles under 400 grammes.* Retrieved September 08, 2011, from http://www.icrc.org/ihl.nsf/FULL/130?OpenDocument

International Committee of the Red Cross. (n.d.). *Customary IHL – Rules.* Retrieved September 08, 2011, from http://www.icrc.org/customary-ihl/eng/docs/v1

Internet Protocol. (2011). *Darpa internet program, protocol specification, RFC 791.* Retrieved November 20, 2011, from http://tools.ietf.org/html/rfc791

Introna, L. D. (2005). Disclosive ethics and information technology: Disclosing facial recognition systems. *Ethics and Information Technology, 7*(2), 75–86. doi:10.1007/s10676-005-4583-2.

Jankélévitch, V. (1968). *Traité des vertus.* Paris, France: Bordas/Mouton.

Jelsma, J. (2006). Designing 'moralized' products: Theory and practice. In Verbeek, P. P., & Slob, A. (Eds.), *User behavior and technology development – shaping sustainable relations between consumers and technologies.* Dordrecht: Springer. doi:10.1007/978-1-4020-5196-8_22.

Jervis, R. (1982). Article. *International Organization, 36*(2).

Jessop, B. (2003). *Governance and metagovernance: On reflexivity, requisite variety, and irony.* Lancaster: Department of Sociology, Lancaster University. Retrieved November 25, 2012, from http://www.comp.lancs.ac.uk/sociology/papers/Jessop- Governance-and-Metagovernance.pdf

Jessop, B. (2003). *Governance and metagovernance: On reflexivity, requisite variety, and requisite irony.* Retrieved November 1, 2012, from http://www.lancs.ac.uk/fass/sociology//papers/jessop-governance-and-metagovernance.pdf

Joly, P. B. (2007). Scientific expertise in the Agora - Lessons from the French experience. *Journal of Risk Research, 10*(7), 905–924. doi:10.1080/13669870701504533.

Joseph, S., & Linley, P. A. (2006). *Positive therapy: A meta-theory for positive psychological practice.* London: Routledge.

Juels, A., Molnar, D., & Wagner, D. (2005). Security and privacy issues in E-passports, security and privacy for emerging areas in communications networks. In *First International Conference on Security and Privacy for Emerging Areas in Communications Networks* (SECURECOMM'05) (pp. 74-88).

Kahan, D. M., Slovic, P., Braman, D., Gastil, J., & Cohen, G. L. (2007). Affect, values, and nanotechnology risk perceptions: An experimental investigation. *GWU Legal Studies Research Paper No.261; Yale Law School, Public Law Working Paper No.155; GWU Law School Public Law Research Paper No.261; 2nd Annual Conference on Empirical Legal Studies Paper.* Retrieved July 19, 2012, from http://ssrn.com/abstract=968652

Kaiser, M., Kurath, M., Maasen, S., & Rehmann-Sutter, C. (2010). *Governing future technologies.* Springer. doi:10.1007/978-90-481-2834-1.

Kalberg, S. (1980). Max Weber types of rationality. *American Journal of Sociology, 85*(5), 1145–1179. doi:10.1086/227128.

Kamarck, E. C., & Nye, J. S. (Eds.). (2002). Governance.com: Democracy in the information age. Washington, CD: Brookings Institution.

Kanta. (2010). Retrieved May, 12, 2011, from https://www.kanta.fi/en/frontpage

Kant, I. (1986). *Kritik der praktischen Vernunft*. Reclam, Ditzingen.

Kant, I. (1998). *Grundlegung zur Metaphysik der Sitten*. Reclam, Ditzingen.

Kant, I. (2009). *Groundwork of the metaphysics of morals*. Harper Perennial Modern Classics.

Karinen, R., & Guston, D. H. (2010). Toward anticipatory governance: The experience with nanotechnology. In *Kaiser* (pp. 217–232). Kurath, Maasen, & Rehmann-Sutter.

KASAM-SKN. (1988). *Ethical aspects on nuclear waste – Some salient points discussed at a seminar on ethical action in the face of uncertainty in Stockholm, Sweden, September 8-9, 1987*. SKN Report 29.

KELA [Finnish Social Insurance Institution]. (2009). *Law service - Hallituksenesityslaiksipassilainjaeräidensiihenliittyvienlakienmuuttamisesta* [Government's proposal for changing passport act and certain other related laws]. Retrieved April 11, 2011, from http://www.edilex.fi/kela/fi/mt/havm20090009

Kerckhoffs, A. (1883). La cryptographie militaire. *Journal des sciences militaires, IX*, 5-38, Jan/Feb, 161-191. Retrieved November 10, 2011, from http://www.petitcolas.net/fabien/kerckhoffs/

Kettner, M. (1991). Peirce's notion of abduction and psychoanalytic interpretation. In Epstein, P. S., & Litowitz, B. E. (Eds.), *Semiotic perspectives on clinical theory and practice: Medicine, neuropsychiatry and psychoanalysis* (pp. 163–180). New York: Mouton De Gruyter.

Kettner, M. (1999). Neue perspektiven der diskursethik. In Grunwald, A., & Saupe, S. (Eds.), *Ethik technischen handeln: Praktische relevanz und legitimation* (pp. 153–196). Heidelberg: Springer Verlag.

Kettner, M. (2007). Deliberative democracy: From rational discourse to public debate. In Goujon, P., Lavelle, S., Duquenoy, P., Kimppa, K., & Laurent, V. (Eds.), *The information society: Innovation, legitimacy, ethics and democracy, in honor of Professor Jacques Berleur* (pp. 55–66). Berlin: Springer. doi:10.1007/978-0-387-72381-5_7.

Kimppa, K. K. (2008). Censorship: Case Finland. In *Proceedings of Ethicomp 2008*. Mantua, Italy: University of Pavia.

Kiss A. (1989). *L'écologie et la loi: Le statut juridique de l'environnement*. Paris, L'Harmattan.

Kohlberg, L. (1981). *Essays on moral development (Vol. 1)*. New York: Harper and Row.

Kohlberg, L. (1984). *Essays on moral development (Vol. 2)*. New York: Harper and Row.

Kohlberg, L., Boyd, D., & Levine, C. (1990). The return of stage 6: Its principle and moral point of view. In Wren, T. (Ed.), *The moral domain: Essays in the ongoing discussion between philosophy and the social sciences* (pp. 151–181). Cambridge, MA: MIT Press.

Kohlberg, L., & Power, F. C. (1981). Moral development, religious thinking and the question of a seventh stage. In Kohlberg, L. (Ed.), *Essays in the philosophy of moral development* (pp. 311–372). New York: Harper and Row. doi:10.1111/j.1467-9744.1981.tb00417.x.

Koskinen, J., Heimo, O.I., & Kimppa, K.K. (in press). *Improving eHealth systems development with modularization: Ethical principles*.

Krasner, S. D. (1983). *International regimes*. Cornell University Press.

Kuhn, T. (1962). *The structure of scientific revolutions*. University of Chicago Press.

Kuhse, P. (2006). *Bioethics: An anthology*. Blackwell Publishing.

Kymlicka, W. (1990). *Contemporary political philosophy*. Oxford: Oxford University Press.

Lagrange, M. H. (2005, September). *Modèle d'inventaire de dimensionnement (MID). Données descriptives du colis type C1*. Projet HAVL, Note technique ANDRA C.NT. AHVL.02.109, Indice B.

Lakoski, J. M., Murray, W. B., & Kenny, J. M. (2000). *The advantages and limitations of calmatives for use as a non-lethal weapon*. Retrieved March 24, 2011, from http://www.sunshine-project.org/incapacitants/jnlwdpdl/psucalm.pdf

Lapsley, D. K. (2006). Moral stage theory. In Tillen, M., & Smetana, J. (Eds.), *Handbook of moral development* (pp. 37–66). New York: Lawrence Erlbaum.

Lapsley, D., & Power, F. C. (Eds.). (2005). *Character psychology and character education.* Notre Dame: University of Notre Dame Press.

Latour, B. (1987). *Science in action: How to follow scientists and engineers through society.* Milton Keynes, UK: Open University Press.

Latour, B. (1988). *Science in action.* Harvard University Press.

Latour, B. (1992). Where are the missing masses? -- The sociology of a few mundane artifacts. In Bijker, W. E., & Law, J. (Eds.), *Shaping technology/building society.* Cambridge: MIT Press.

Laurent, B. (2010). *Les politiques des nanotechnologies.* Editions Charles Léopold Mayer.

Lavelle, S. (2006). *Science, technologie et éthique - Conflits de rationalité et discussion démocratique.* Ellipses Edition, Collection Technosup.

Lebessis, N., & Paterson, J. (1997). Evolution in governance: What lessons for the commission? A first assessment. *Forward studies unit, Working Paper 1997.* Brussels: European Commission. Retrieved November 1, 2012, from http://www.pedz.uni-mannheim.de/daten/edz-mr/pbs/00/evolution_in_governance.pdf

Lehavi, A. (2009). How property can create, maintain or destroy community. *Theoretical Inquiries in Law, 10*(1). The Berkeley Electronic Press. Retrieved May 5, 2012, from http://www.bepress.com

Lenoble, J. & Maesschalck, M. (2010). *Political pragmatism and social attention.*

Lenoble, J., & Maesschalck, M. (2006). Beyond neo-institutionalist and pragmatic approaches to governance, REFGOV, FP6.

Lenoble, J., & Maesschalck, M. (2006). Beyond neo-institutionalist and pragmatic approaches to governance. REFGOV, FP6.

Lenoble, J., & Maesschalck, M. (2006). *Beyond neo-institutionalist and pragmatist approaches to governance.* Retrieved November, 2012 http://iap6.cpdr.ucl.ac.be/docs/TNU/WP-PAI.VI.06-TNU-1.EN.pdf

Lenoble, J., & Maesschalck, M. (2007). *Synthesis report two: Reflexive governance: Some clarifications and an extension and deepening of the fourth (genetic) approach, REFGOV: Reflexive governance in the Public Interest, Services of General Interest/Theory of the Norm Unit.* Retrieved Novermber, 2012 from http://refgov.cpdr.ucl.ac.be/?go=publications&dc=e14a94b44d20bdad142db b2bf693f554b23327dc

Lenoble, J., & Maesschalck, M. (2007a). *Reflexive governance: Some clarifications and an extension of the fourth (genetic) approach.* REFGOV Reflexive Governance in the Public Interest, Services of General Interest/Theory of the Norm Unit, Synthesis Report Two, Centre de Philosophie du Droit, UC Louvain.

Lenoble, J., et al. (2008). *Democratic governance and theory of collective action.* Annual Scientific Activity Report. Retrieved November, 2012 from http://iap6.cpdr.ucl.ac.be/docs/IAP-VI-06-AnnualReport-2-final-2.04.09.pdf

Lenoble, J., & Maesschalck, M. (2003). *Toward a theory of governance: The action of norms.* Kluwer Law International.

Lenoble, J., & Maesschalck, M. (2003). *Toward a theory of governance: The action of norms.* Kluwer.

Lenoble, J., & Maesschalck, M. (2003). *Toward a theory of governance: The action of norms.* London: Kluwer.

Lenoble, J., & Maesschalck, M. (2003). *Towards a theory of governance: The action of norms.* Kluwer.

Lenoble, J., & Maesschalck, M. (2007). *Beyond neo-institutionalist and pragmatic approaches to governance.* Carnets du CPDR.

Lenoble, J., & Maesschalck, M. (2009). *L'action des normes.* Presses de l'Université de Sherbrooke.

Lenoble, J., & Maesschalck, M. (2010). *Democracy, law and governance.* Aldershot: Ashgate.

Lenoble, J., & Maesschalck, M. (2010). *Democracy, law and governance.* Farnham, Burlington: Ashgate.

Lessig, L. (2001). *The future of ideas, the fate of the commons in a connected world.* New York: Random House. Retrieved May 8, 2010, from http://www.the-future-of-ideas.com/download/

Lessig, L. (2004). Foreword. In Bourcier, D., & Dulong de Rosnay, M. (Eds.), *International commons at the digital age: La création en partage* (p. 7). Romillat.

Lessig, L. (Ed.). (2000). *Code and others laws of cyberspace*. New York, USA: Basic Books Press.

Lickona, T. (Ed.). (1976). *Moral development and behavior: Theory, research, and social issues*. New York: Holt, Rinehart, and Winston.

Lickona, T., & Ryan, K. (Eds.). (1992). *Character development in schools and beyond*. Washington, DC: The Council for Research in Values and Philosophy.

Lietaer, B., Ulanowicz, R., & Goerner, S. (2009). Options for managing a systemic bank crisis. *S.A.P.I.EN.S, 2.1*. Retrieved April 15, 2009, from http://sapiens.revues.org/index747.html

Lietaer, B., & Kennedy, M. (Eds.). (2008). *Monnaies régionales: De nouvelles voies vers une prospérité durable*. Paris: Charles Leopold Mayer Press.

Linley, P. A., & Joseph, S. (Eds.). (2004). *Positive psychology in practice*. Hoboken, NJ: John Wiley and Son. doi:10.1002/9780470939338.

Loren, L. (2007). Building a reliable semicommons of creative works: Enforcement of creative commons licences and limited abandonment of copyright. *George Mason Law Review*, *14*(2), 271.

Lucivero, F., Swierstra, T., & Boenink, M. (2011). Assessing expectations: Towards a toolbox for an ethics of emerging technologies. *NanoEthics*, *5*(2), 129–141. doi:10.1007/s11569-011-0119-x PMID:21957435.

Lund, W. (1996). Egalitarian liberalism and the fact of pluralism. *Journal of Social Philosophy*, *27*(3), 61–80. doi:10.1111/j.1467-9833.1996.tb00253.x.

Lund, W. (2000). Fatal attraction: 'Wilful liberalism' and the denial of public transparency. *Political Research Quarterly*, *53*(2), 305–326.

Määttä, K. (2008). *Etärahapelien sääntelystä, Stakes, Helsinki*. Retrieved November 10, 2011, from http://www.stakes.fi/verkkojulkaisut/raportit/R2-2008-VERKKO.pdf

MacDonald, C. (2004). Nanotech is novel; the ethical issues are not. *Scientist (Philadelphia, Pa.)*, *18*(8).

MacIntyre, A. (1985). *After virtue: A study in moral theory*. London, United Kingdom: Duckworth.

Mackenzie, D. A. (1990). *Inventing accuracy, a historical sociology of nuclear missile guidance*. Cambridge, MA: MIT Press.

Macnaghten, P., Davies, S., & Kearnes, M. (2010). Narrative and public engagement: Some findings from the DEEPEN project. In von Schomberg & Davies (2010) (pp. 13-30).

Macnaghten, P. (2010). Researching technoscientific concerns in the making: Narrative structures, public responses, and emerging nanotechnologies. *Environment & Planning A*, *42*(1), 23–37. doi:10.1068/a41349.

Maesschalck, M. (2008). Droit et "capacitation" des acteurs sociaux: La question de l'application des normes. *Dissensus*, 1. Retrieved October 12, 2012, from http://popups.ulg.ac.be/dissensus/document.php?id=231

Maesschalck, M., & Lenoble, J. (2007-a). Beyond neo-institutionalist and pragmatic approaches to governance. *Carnets du centre de philosophie du droit*, 130.

Maesschalck, M., & Lenoble, J. (2007-b). Synthesis report 2, Reflexive governance: Some clarifications and an extension and deepening of the fourth (generic) approach. *Reflexive Governance in the Public Interest (REFGOV)*, FP6 project.

Maesschalck, M. (1981). *Normes et constextes*. Olms.

Maesschalck, M. (2001). *Normes et contextes: Les fondements d'une pragmatique contextuelle. Maesschalck, M., & Lenoble, J. (2010). Democracy, Law and Governance*. Ashgate.

Maesschalck, M. (2008). Normes de gouvernance et enrôlement des acteurs sociaux. *Multitudes*, *4*(34), 182–194. doi:10.3917/mult.034.0182.

Maesschalck, M., & Lenoble, J. (2003). *Toward a theory of governance, the action of norms*. Amsterdam: Kluwer Law International.

Maesschalck, M., & Loute, A. (2007). Points forts et points faibles des nouvelles pratiques de réforme des Etats sociaux. In Schronen, D., & Urbé, R. (Eds.), *Sozialalmanach 2007* (pp. 191–203). Luxembourg: Caritas.

Mallin, C. (2003). *Corporate governance*. Oxford: Oxford University Press.

Manciaux, M. (Ed.). (2001). *La Résilience: Résister et se construire*. Geneva: Editions Médecine and Hygiène.

Manders-Huits, N., & van den Hoven, J. (2009). The need for a value-sensitive design of communication infrastructures. In P. Sollie & M. Düwell (Eds.), Evaluating new technologies: Methodological problems for the ethical assessment of technology developments (pp. 51-62). The International Library of Ethics, Law and Technology. Springer.

March, J. G., & Olsen, J. G. (1979). *Ambiguity and choice in organizations*. Oslo: Scandinavian University Press.

Marengo, L., & Dosi, G. (2005). Division of labor, organizational coordination and market mechanism in collective problem-solving. *Journal of Economic Behavior & Organization*, 58(2), 303–326. doi:10.1016/j.jebo.2004.03.020.

Martin, B. R. (2010). The origins of the concept of "foresight" in science and technology: An insider's perspective. *Technological Forecasting and Social Change*, 77(9), 1438–1447. doi:10.1016/j.techfore.2010.06.009.

Masclet, L., & Goujon, P. (2011). *IDEGOV D.1.1. Grid of Analysis*. CIGREF Foundation.

Masclet, L., & Goujon, P. (2012). *IDEGOV D.3.2. Model of current and emerging governance strategies, Map of governance and ethics*. CIGREF Foundation.

Maslow, A. (1956). *Motivation and personality*. New York: Harper Collins.

Matland, R. (1995). Synthesizing the implementation literature: The ambiguity-conflict model of policy implementation. *Journal of Public Administration: Research and Theory*, 5(2), 145–175.

Mayntz, R. (2003). *From government to governance: Political steering in modern societies*. Summer Academy on IPP: Wuerzburg, September 7-11, 2003. Retrieved November 25, 2012, from http://www.ioew.de/fileadmin/user_upload/DOKUMENTE/Veranstaltungen/2003/SuA2Mayntz.pdf[REMOVED HYPERLINK FIELD]

Mayntz, R. (2002). Common goods and governance. In Heritier, A. (Ed.), *Common goods: Reinventing European and international governance* (pp. 15–27). New York, London: Rowman & Littlefield Publishers.

Mayntz, R. (2003). New challenges to governance theory. In Bang, H. (Ed.), *Governance as social and political communication* (pp. 27–40). Manchester: Manchester University Press.

McCarthy, T. (1994). Kantian constructivism and reconstructivism: Rawls and Habermas in Dialogue. *Ethics*, 105(1), 46. doi:10.1086/293678.

McCarthy, T. (1994). Kantian constructivism and reconstructivism: Rawls and Habermas in dialogue. *Ethics*, 105, 44–63. doi:10.1086/293678.

Mele, A. (1995). *Autonomous agents: From self-control to autonomy*. Oxford: Oxford University Press.

Mercuri, R. (2001). *Electronic vote tabulation: Checks and balances*. (Doctoral Thesis). University of Pennsylvania.

Merges, R. P. (2004). A new dynamism in the public domain. *The University of Chicago Law Review. University of Chicago. Law School*, 71, 183–203.

Meynaud, J. (1960). Qu'est-ce que la technocratie? *Revue économique*, 11(4), 500.

Meynaud, J. (1960). *Technocratie et politique*. Lausanne.

Meynaud, J. (1964). *La technocratie. Mythe ou réalité?* Paris: Payot.

Mill, J. S. (1973). On liberty and considerations on representative government. Lodon: Everyman.

Mill, J. S. (2002). *Utilitarianism* (2nd ed.). Hackett Publishing Co, Inc..

Ministery of Social Affairs and Health. Finland. (2010). *Press release*. Retrieved May 10, 2011, from http://www.stm.fi/tiedotteet/tiedote/view/1508762#fi

Miralles, C., González-Alcántara, O. J., Lozano-Aguilar, J. F., & Marin-Garcia, J. A. (2008). Integrating people with disabilities into work through OR/MS tools. An applied vision. In *Human Centered Processes Conference*. Delft.

Molnar, D., & Wagner, D. (2004). Privacy and security in library RFID Issues, practices, and architectures. *CCS'04*, October 25-29, 2004, Washington, DC, USA. Retrieved October 14, 2011, from http://www.eecs.berkeley.edu/~daw/papers/librfid-ccs04.pdf

Monnerat, J., Vaudenay, S., & Vuagnoux, M. (2007). About machine-readable travel documents: Privacy enhancement using (weakly) non-transferrable data authentication. *International Conference on RFID Security 2007* (pp. 15-28).

Mordini, E. (2007). The narrative dimension of nanotechnology. *Nanotechnology Perceptions, 3*, 15–24.

Moreno, J. (1998). Ethics by committee: The moral authority of consensus. *The Journal of Medicine and Philosophy, 13*(4), 411–432. doi:10.1093/jmp/13.4.411 PMID:3246580.

Moreno, J. (2006). Ethics consultation as moral engagement. In Kuhse, H., & Singer, P. (Eds.), *A compagnion ot Bioethics*. Blackwell Publishing.

Moreno, J. (2009). Ethics comittee and ethics consultants. In Kuhse, H., & Singer, P. (Eds.), *A compagnion ot Bioethics* (pp. 573–584). Blackwell Publishing. doi:10.1002/9781444307818.ch48.

Morgenthau, H. J. (1978). Six principles of political realism. In *Politics among nations: The struggle for power and peace* (5th ed.). New York: Knopf.

Mostowski, W., & Poll, E. (2010). *Electronic passports in a nutshell*. Technical Report ICIS-R10004, Radboud University Nijmegen, the Netherlands. Retrieved June 9, 2011, from http://citeseerx.ist.psu.edu/viewdoc/download?doi=10.1.1.167.2807&rep=rep1&type=pdf

Mouzelis, N. (2007). Restructuring Bourdieu's theory of practice. *Sociological Research Online, 12*(6), 9. Retrieved November 16, 2012, from http://www.socresonline.org.uk/12/6/9.html

MTV3. (2009). *Automaattinen passintarkastus alkoi vaalimaalla* [Automatic passport control has begun at Vaalimaa]. Retrieved June 7, 2011, from http://www.mtv3.fi/uutiset/kotimaa.shtml/2009/12/1014243/automaattinen-passintarkastus-alkoi-vaalimaalla

Mulhall, S., & Swift, A. (1996). *Liberals and communitarians*. Wiley.

Murray, J. H. (1998). *Hamlet on the holodeck: The future of narrative in cyberspace*. Cambridge, MA: The MIT Press.

National Audit Office of Finland. (2011). Valtiontalouden tarkastusviraston tuloksellisuuskertomukset 217/2011. *Sosiaali- ja terveydenhuollon valtakunnalisten IT-hankkeiden toteuttaminen.*

Newman, J. (2001). *Modernising governance. New labour, policy and society*. London: Sage.

Newsweek. (2008). *You've got malice, Russian nationalist waged a cyber war against Georgia. Fighting back is virtually impossible*. Retrieved November 24, 2010, from http://www.newsweek.com/2008/08/22/you-ve-got-malice.html

Newsweek. (2010). Interactive infographic of the world's best countries. Retrieved November 10, 2011, from http://www.thedailybeast.com/newsweek/2010/08/15/interactive-infographic-of-the-worlds-best-countries.html

Nikki, M. (2008-2011a). *The Finnish filtering list and its contents*. Retrieved November 10, 2011, from http://maraz.kapsi.fi/sisalto-en.html

Nikki, M. (2008-2011b). *The Finnish Internet Censorship List*. Retrieved November 10, 2011, from http://lapsiporno.info/suodatuslista/?lang=en

Nikki, M. (2008-2011c). *Näkemyksiälapsipornostalnternetissä*. Retrieved November 10, 2011, from http://lapsiporno.info/

Nordmann, A. (2011). *Between conversation and experimentation*. Lecture given on March 30, 2011, during the Second Workshop of the EGAIS Project, Brussels.

Nordmann, A., & Rip, A. (2009). Mind the gap revisited. *Nature of Nanotechnology, 4*(5) *(Nature Publishing Group)*, 273-274.

Nordmann, A. (2007a). If and then: A critique of speculative nanoethics. *NanoEthics, 1*(1), 31–46. doi:10.1007/s11569-007-0007-6.

Nordmann, A. (2007b). Knots and strands: An argument for productive disillusionment. *The Journal of Medicine and Philosophy, 32*, 217–236. doi:10.1080/03605310701396976 PMID:17613703.

Nordmann, A., & Macnaghten, P. (2010). Engaging narratives and the limits of lay ethics: Introduction. *NanoEthics, 4*, 133–140. doi:10.1007/s11569-010-0095-6.

Nordmann, A., & Schwarz, A. (2010). Lure of the "yes": The seductive power of technoscience. In *Kaiser* (pp. 255–277). Kurath, Maasen, & Rehmann-Sutter.

Nurminen, M. I., & Forsman, U. (1994). *Reversed quality life cycle model. North-Holland*. Amsterdam: Elsevier Science B.V..

Nussbaum, M. (1986). *The fragility of goodness: Luck and ethics in Greek tragedy and philosophy*. Cambridge: Cambridge University Press.

Ober, J. (2009). *Epistemic democracy in classical Athens*. Princeton/Stanford Working Papers.

OECD/NEA. (1995). *La gestion des déchets radioactifs à vie longue: Fondements environnementaux et éthiques de l'évacuation géologique. Opinion collective du Comité de la Gestion des Déchets Radioactifs de l'AEN*. OECD.

OECD/NEA. (2004). The handling of timescales in assessing post-closure safety lessons learnt from the April 2002 workshop in Paris, France. Organisation for Economic Co-operation and Development/Nuclear Energy Agency, NEA n°4435.

OECD-NEA. (2010). Main findings in the international workshop "towards transparent, proportionate and deliverable regulation for geological disposal." Tokyo, Japan, 20-22 January 2009. OECD-NEA.

Ollagnon, H. (1979). Propositions pour une gestion patrimoniale des eaux souterraines: L'expérience de la nappe phréatique d'Alsace. *Bulletin interministériel pour la rationalisation des Choix budgétaires, 36*, pp. 33.

Oser, F., & Gmünder, P. (1982). *Der Mensch - Stufen seiner religiösen entwicklung*. Zurich: Benziger.

Oser, F., & Reich, H. (1990). Moral judgment, religious judgment, world view and logical thought: A review of their relationship. *British Journal of Religious Education, 12*(3), 94–101. doi:10.1080/0141620900120207.

Oser, F., Scarlett, W. G., & Bucher, A. (2006). *Religious and spiritual development throughout the life span* (6th ed., pp. 942–998). Handbook of child psychologyHoboken, NJ: Wiley.

Ost, F. (1995). *La nature hors la loi* (p. 323). Paris: La Découverte.

Ostrom, E. (1990). *Governing the commons: The evolution of institutions for collective action*. Cambridge: Cambridge University Press. doi:10.1017/CBO9780511807763.

Ostrom, V. (1997). *The meaning of democracy and the vulnerability of democracies*. Ann Arbor, MI: University of Michigan Press.

Otakantaa.fi. (n.d.). *Finnish Ministry of Justice, An open electronic forum provided by the government for polling citizen opinions about new legislation*. Retrieved April 11, 2011, from http://otakantaa.fi

Outhwaite, W. (2009). *Habermas*. Polity Press.

Ovett, D. (2006). Free trade agreements (FTAs) and human rights: A serious challenge for Latin America and the Caribbean. *PUENTES, 7*(1). Retrieved September 12, 2011, from www.3dthree.org/pdf_3D/Dovett_PUENTESarticle_Feb06_Eng.pdf

Pantzar, M. (2000). Teesejä tietoyhteiskunnasta. *Yhteiskuntapolitiikka. Nro 1. S.* (pp. 64-68). Retrieved May 23, 2011, from http://www.stakes.fi/yp/2000/1/001pantzar.pdf

Pascal, Z. G. (2009). *An operating system for the cloud. Technology Review, September/October 2009*. Cambridge, MA: The MIT Press.

Pavlopoulos, M., Grinbaum, A., & Bontems, V. (2010). Toolkit for ethical reflection and communication. *Observatory Nano Project, CEA, June 2010*. Retrieved January 09, 2012, from http://www.observatorynano.eu/project/document/1598/

People's Daily Online. (2009). *China's IP address will be used up in 2 years*. Retrieved August 24, 2009, from http://english.peopledaily.com.cn/90001/90781/6737117.html

Perez, C. (2004). Technological revolutions, paradigm shifts and socio-institutional change. In Reinert, E. (Ed.), *Globalization, economic development and inequality, an alternative perspective* (pp. 217–242). Cheltenham, UK: Edward Elgar.

Perri, G. (2004). *E-governance: Styles of political judgement in the information age polity*. New York: Palgrave MacMillan.

Peters, A. (2001). There is nothing more practical than a good theory: An overview of contemporary approaches to international law. *Jahrbuch fur Internationales Recht. German Yearbook of International Law, 44*, 25–37.

Peterson, C., & Seligman, M. E. P. (2004). *Character strengths and virtues: Handbook and classification.* Oxford: American Psychological Association and Oxford University Press.

Pettit, P. (2005). *Rules, reasons and norms.* Oxford.

Philibert, P. (1987). Addressing the crisis in moral theory: Clues from Aquinas and Gilligan. *Theology Digest, 34*(2), 103–113.

Philibert, P. J. (1975). Lawrence Kohlberg's use of virtue in his theory of moral development. *International Philosophical Quarterly, 15*(4), 455–479. doi:10.5840/ipq197515445.

Philibert, P. J. (1988). Kohlberg and Fowler revisited: An interim report on moral structuralism. *Living Light, 24,* 162–171.

Piaget, J. (1997). *The moral judgment of the child.* New York: Free Press. (Original work published 1932).

Picavet, E., Dupont, G., Dilhac, M.-A., & Bolaños, B. (2009). *Identité et nouveauté des situations politiques.* Paper presented at Congrès des Associations des Sociétés Philosophiques de Langue Française. Budapest, ELTE University.

Picavet, E. (2011). *La Revendication des droits. Une étude de l'équilibre des raisons dans le libéralisme.* Paris: Les Classiques Garnier.

Pieper, J. (1966). *The four cardinal virtues.* Notre Dame, IN: University of Notre Dame Press.

Pierre, J. (Ed.). (2000). *Debating governance.* Oxford: Oxford University Press.

Pierre, J., & Peters, G. (2000). *Governance, politics and the state.* Basingstoke: Palgrave.

Pinckaers, S. (1978). *Le renouveau de la morale.* Paris, France: Téqui.

Pinckaers, S. (1995). *Sources of Christian ethics.* Washington, DC: Catholic University of America Press.

Pinckaers, S. (2005). *The Pinckaers reader.* Washington, DC: Catholic University of America Press.

Plato. (4th c. BCE). *Protagoras.*

Pope, S. J. (1992). *The evolution of altruism.* Washington, DC: Georgetown University Press.

Popper, K. (1988). *The poverty of historicism.* Routledge.

Power, F. C. (1988). The just community approach to moral education. *Journal of Moral Education, 17,* 195–208. doi:10.1080/0305724880170304.

President promulgates ordinance to prevent electronic crimes. (2008, November 6). *Associated Press.* Retrieved from http://www.app.com.pk/en_/index.php?option=com_content&task=view&id=58277&Itemid=1

Rabin, M. (1993). Incorporating fairness into economics and game theory. *The American Economic Review, 83,* 1281–1302.

Rahman, S., & Keiff, L. (2005). On how to be a dialogician. In Vanderken, D. (Ed.), *Logic thought and action* (pp. 359–408). Berlin: Springer. doi:10.1007/1-4020-3167-X_17.

Rainey et al. (2012). EGAIS 4.3 New Guidelines Addressing the Problem of Integrating Ethics into Technical Development Projects.

Rainey, S., & Goujon, P. (2009). EGAIS 4.1 Existing Solutions to the Ethical Governance Problem and Characterisation of their Limitations.

Rainey, S., & Goujon, P. (2009). ETICA, 4.2. Governance recommendation.

Rawls, J. (1971). *A theory of justice.* Cambridge, MA: Belknap Press.

Rawls, J. (1971). *A theory of justice.* Cambridge, MA: Harvard University Press.

Rawls, J. (1971). *A theory of justice.* Cambridge, MA: Harvard University Press.

Rawls, J. (1980). Kantian constructivism in moral theory. *The Journal of Philosophy, 77*(9), 515–572. doi:10.2307/2025790.

Rawls, J. (1987). The idea of an overlapping consensus. *Oxford Journal of Legal Studies, 7*(1), 1–25. doi:10.1093/ojls/7.1.1.

Rawls, J. (1993). *Political liberalism.* New York: Columbia University Press.

Rawls, J. (1996). *Political liberalism.* New York: Columbia University Press.

Raymond, E. S. (Ed.). (1999). The cathedral and the bazaar: Musings on linux and open source by an accidental revolutionary. Sebastopol, CA: Tim O'Reilly Press.

Reed-Danahay, D. (2005). *Locating Bourdieu*. Bloomington, Indianapolis: Indiana University Press.

Report of the Meeting of States Parties [to the Biological Weapons Convention]. (2008). Retrieved September 08, 2011, from http://daccess-dds-ny.un.org/doc/UNDOC/GEN/G09/600/07/PDF/G0960007.pdf?OpenElement

Retrieved January 20, 2012, from http://www.itas.fzk.de/deu/lit/2005/fiua05a_abstracte.pdf

Reynaud, B. (2003). *Operating rules in organizations. Macroeconomic and microeconomic analyses*. London: Palgrave.

Rhodes, C. (2010). *International governance of biotechnology: Needs, problems and potential*. London: Bloomsbury Academic. doi:10.5040/9781849661812.

Ricœur, P. (1992). *Oneself as another*. Chicago: University of Chicago Press.

Rip, A., & Shelley-Egan, C. (2010). Positions and responsabilities in the 'real' world of nanotechnology. In von Schomberg & Davies (2010) (pp. 31-38).

Rogers, C. (1959). *A theory of therapy, personality, and interpersonal relationships*. New York: McGraw-Hill.

Rogers, C. (1961). *On becoming a person: A therapists view of psychotherapy*. Boston: Houghton Mifflin.

Rosanvallon, P. (2008). *La légitimité démocratique: Impartialité, réflexivité, proximité*. Paris: Seuil.

Rose, C. (1986). The comedy of the commons: Custom, commerce, and inherently public property. *The University of Chicago Law Review. University of Chicago. Law School, 53*, 742. doi:10.2307/1599583.

Rosson, M. B., & Carroll, J. M. (2001). *Usability engineering: Scenario-based development of human-computer interaction*. San Francisco: Morgan Kaufmann Publishers.

Royce, J. (1908). *The philosophy of loyalty*. New York: MacMillan.

Ruol, M. (2000). De la neutralisation au recoupement. J. Rawls face au défi de la démocratie plurielle. *Revue Philosophique de Louvain*.

Sabel, C., & Zeitlin, J. (2006). New architecture of experimentalist governance (Powerpoint presentation) at the *EU 6th Framework Research Programme's CONNEX Network of Excellence on European Governance Workshop* (Roskilde University). Retrieved June 3, 2011, from http://eucenter.wisc.edu/OMC/New%20OMC%20links/Learning%20from%20Difference%20-CONNEX%20presentation.ppt

Sabel, C., & Zeitlin, J. (2007). Learning from difference: The new architecture of experimentalist governance in the European Union. *Eurogov Papers, No. C-07-02*. Retrieved November 27, 2012, from http://edoc.vifapol.de/opus/volltexte/2011/2466/pdf/egp_connex_C_07_02.pdf

Sabel, C., & Zeitlin, J. (2008). *Learning from difference, the new architecture of experimentalist governance in EU*. European Governance Papers, Eurogov, C-07-02.

Sabel, C., & Cohen, J. (1997). Directly deliberative polyarchy. *European Law Journal, 3*(4).

Sabel, C., & Zeitlin, J. (Eds.). (2010). *Experimentalist governance in the European Union: Towards a new architecture*. Oxford: Oxford University Press.

Salmivalli, L. (2008). *Governing the implementation of a complex inter-organizational information system network – the case of Finnish prescription*. (Doctoral thesis). Turku School of Economics, Turku, Finland. Retrieved October 10, 2011, from http://info.tse.fi/julkaisut/vk/Ae3_2008.pdf

Salvi, M. (Ed.). (2010, July). *Ethically Speaking*, 14, p.6. Retrieved November, 2012 from http://ec.europa.eu/bepa/european-group-ethics/docs/es14_en_web.pdf

Santuccio, A., & Valentino, F. (2011). *EGAIS Project deliverable 3.2: Second workshop report*. Retrieved June 1, 2011, from http://www.egais-project.eu

Schlechta, K., & Gabbay, D. (2009). *Logical tools for handling change in agent-based systems*. Berlin: Springer.

Schneider, T., Schieber, C., & Lavelle, S. (2006). *Long term governance for radioactive waste management*. Final Report of COWAM2 - Work Package 4, Report COWAM2-D4-12/CEPN-R-301.

Schneier, B. (1996). Applied cryptography second edition: Protocols, algorithms, and source code in C. New York: John Wiley & Sons, Inc.

Schön, D. (1983). *The reflective practitioner. How professionals think in action*. London: Temple Smith.

Schummer, J., & Baird, D. (Eds.). (2006). Nanotechnologies challenges: Implications for philosophy, ethics, and society. World Scientific Publishing Co. Pte. Ltd.

Schummer, J. (2006). Cultural diversity in nanotechnology ethics. *Interdisciplinary Science Reviews, 31*(3), 217–230. doi:10.1179/030801806X113757.

Scott, H. (1933). *Science versus Chaos*. Retrieved from http://www.archive.org/details/TechnocracyHowardScott ScienceVs.Chaos-June1933

Scott, H. (1965, 2008) History and purpose of technocracy. *The North American Technate* (TNAT). Retrieved from http://www.archive.org/details/HistoryAndPurposeOfTechnocracy.howardScott

Searle, J. (2001). *Rationality in Action*. MIT.

Seligman, M. E. P. (1991). *Helplessness: On depression, development, and death* (2nd ed.). New York: W. H. Freeman.

Seligman, M. E. P. (1998). *Learned optimism*. New York: Simon and Schuster.

Seligman, M. E. P. (1999, August). The president's address. APA annual report. *The American Psychologist*, 559–562.

Seligman, M. E. P. (2002). *Authentic happiness: Using the new positive psychology to realize your potential for lasting fulfillment*. New York: Free Press.

Seligman, M. E. P. (2012). *Flourish: A visionary new understanding of happiness and well-being*. New York, NK: Free Press.

Seligman, M. E. P., Reivich, K., Jaycox, L., & Gillham, J. (1996). *The optimistic child*. New York: Harper Collins.

Seligman, M. E. P., Walker, E., & Rosenhan, D. L. (2001). *Abnormal psychology* (4th ed.). New York: W.W. Norton.

Sen, A. (2001). *Development as freedom*. Oxford: Oxford University Press.

Sen, A. K. (1979). Informational analysis of moral principles. In Harrison, R. (Ed.), *Rational action. studies in philosophy and social science*. Cambridge: Cambridge University Press.

Shannon, C. E. (1949). Communication theory of secrecy systems. *Bell System Technical Journal, 28*(4), 656-715. Retrieved October 10, 2011, from http://netlab.cs.ucla.edu/wiki/files/shannon1949.pdf

Sheldon, K. M., & Elliot, A. J. (1999). Goal striving, need-satisfaction, and longitudinal well-being: The self concordance model. *Journal of Personality and Social Psychology, 76*, 482–497. doi:10.1037/0022-3514.76.3.482 PMID:10101878.

Singer, P. (2006). Moral experts. In Selinger, E., & Crease, R. P. (Eds.), *The philosophy of expertise* (pp. 188–189). Columbia University Press.

Smith, P. (1994). Les puissances de l'expérience in. *Contemporary Sociology, 23*(3). American Sociological Assoc.

Smith, P. (1994). Les puissances de l'expérience. *Contemporary Sociology, 23*(3). American Sociological Association.

Snyder, C. R. (1994). *The psychology of hope: You can get there from here*. New York: Free Press.

Snyder, C. R., & Lopez, S. J. (2007). *Positive psychology: The scientific and practical explorations of human strengths*. Thousand Oaks, CA: Sage.

Snyder, C. R., & Lopez, S. J. (Eds.). (2009). *The Oxford handbook of positive psychology*. Oxford: Oxford University Press.

Spohn, W. C. (2000). Conscience and moral development. *Theological Studies, 64*, 122–138.

Stahl, B. C. (2011a). What future, which technology? On the problem of describing relevant futures. In M. Chiasson, O. Henfridsson, H. Karsten, & J. I. DeGross (Eds.), Researching the future in information systems: IFIP WG 8.2 Working Conference, Future IS 2011 (pp. 95-108). Turku, Finland. Heidelberg: Springer.

Stahl, B. C. (2011b). What does the future hold? A critical view of emerging information and communication technologies and their social consequences. In M. Chiasson, O. Henfridsson, H. Karsten, & J. I. DeGross (Eds.), Researching the future in information systems: IFIP WG 8.2 Working Conference, Future IS 2011 (pp. 59-76). Turku, Finland. Heidelberg: Springer.

Stallman, R. (2003). The copyleft and its context. In *Proceedings of the Copyright, Copywrong*. Nantes.

Steg, L. (1999). Verspilde energie? Wat doen en laten Nederlanders voor het milieu. The Hague: Sociaal en Cultureel Planbureau (SCP Cahier no. 156).

Stoker, G. (1998, March). Governance as a theory: Five propositions. *International Social Science Journal, 155,* 17–28. doi:10.1111/1468-2451.00106.

Sunnuntaisuomalainen. (2010, August 15). *Passipoliisit* (pp.14).

Surowiecki, J. (2004). *The wisdom of crowds.* Anchor.

Swierstra, T., & Rip, A. (2007). Nano-ethics as NEST-ethics: Patterns of moral argumentation about new and emerging science and technology. *NanoEthics, 1,* 3–20. doi:10.1007/s11569-007-0005-8.

Swierstra, T., Stemerding, D., & Boenink, M. (2009). Exploring techno-moral change: The case of the obesi-typill. In Sollie, P., & Düwell, M. (Eds.), *Evaluating new technologies* (Vol. 3, pp. 119–138). Springer Netherlands. doi:10.1007/978-90-481-2229-5_9.

Tapscott, D. (Ed.). (1996). *The digital economy: Promise and peril in the age of networked intelligence.* New York, USA: Mc Graw-Hill Press.

Taylor, C. (1979). *Hegel and modern society.* Cambridge: Cambridge University Press. doi:10.1017/CBO9781139171489.

Taylor, C. (1991). *The ethics of authenticity.* Cambridge, MA: Harvard University Press.

The Guardian. (2007, May 17). *Russia accused of unleashing cyberwar to disable Estonia.* Retrieved November 24, 2011, from http://www.guardian.co.uk/world/2007/may/17/topstories3.russia

Thoumsin, P. Y. (n.d.). *Creative commons le meillur des deux mondes?* Retrieved from http://creativecommons.org/licences/by-nc-nd/2.0/be

Tideman, M. (2008). *Scenario based product design.* (Unpublished Master's thesis). Universiteit Twente.

Tietokone. (2010, August 16). *Poliisi saattaa saada passien sormenjäljet* [Police may acquire the passport fingerprints]. Retrieved April 11, 2011, from http://www.tietokone.fi/uutiset/poliisi_saattaa_saada_passien_sor-menjaljet

Titus, C. S. (2010). Moral development and connecting the virtues: Aquinas, Porter, and the flawed saint. In Hütter, R., & Levering, M. (Eds.), *Ressourcement Thomism* (pp. 330–352). Washington, DC: The Catholic University of America Press.

Toumey, C. (2007). Privacy in the shadow of nanotechnology. *NanoEthics, 1,* 211–222. doi:10.1007/s11569-007-0023-6.

Treacy, B., & Martin, A. (2008, May 29). A privacy law for China? *Complinet.*

Türk, A. (2011). *La vie privée en péril.* Paris: Odile Jacob.

Turner, B. (2009). *Pouzin society - organizational meeting, today.* Retrieved May 04, 2009, from http://blogs.broughturner.com/2009/05/pouzin-society-organizational-meeting.html

Uhlir, P. (2009). Global change in environmental data sharing: Implementation of the GEOSS data sharing principles. *Communia Workshop Proceedings,* Torino, July 2009.

Ulanowicz, R. E. (Ed.). (2008). *A third window/ natural foundations for life.* New York, USA: Oxford University Press.

Ullmann-Margalit, E. (1977). *The emergence of norms.* Oxford: Oxford University Press.

United Nations Economic and Social Council – Commission on Human Rights. (2006). *Situation of Detainees at Guantanamo Bay.* Retrieved September 12, 2011, from http://www.unhcr.org/refworld/docid/45377b0b0.html

United Nations General Assembly. (1992). *Annex 1 – Rio Declaration on Environment and Development* to *Report of the United Nations Conference on Environment and Development.* Retrieved March 03, 2004, from http://www.un.org/documents/ga/conf151/aconf15126-1annex1.htm

United Nations. (2004, July 5). *Press release – US-Peru trade negotiations: Special rapporteur on right to health reminds parties of human rights obligations.* Retrieved September 12, 2011, from http://www.unhchr.ch/huricane/huricane.nsf/0/35C240E546171AC1C1256EC800308A37?opendocument

Valdivieso, L. V. (2009). Need to guard against TRIPS-plus enforcement agenda. *South Bulletin, 41*. The South Centre. Retrieved September 12, 2011, from http://www.southcentre.org/index.php?option=com_content&task=view&id=1077&Itemid=279

van de Poel, I. (2008). How should we do nanoethics? A network approach for discerning ethical issues in nanotechnology. *NanoEthics, 2*, 25–38. doi:10.1007/s11569-008-0026-y.

van der Ven, J. (1998). *Formation of the moral self*. Grand Rapids, MI: Eerdmans Publishing.

van Eechoud, M., & van der Wal, B. (2008). *Creative commons licensing for public sector information opportunities and pitfalls*. IVIR, University of Amsterdam. Retrieved May 3, 2010, from http://www.ivir.nl

Vanistendael, S. (1994). *La résilience ou le réalisme de l'espérance. Blessé mais pas vaincu*. Geneva: Les cahiers du BICE.

Verbeek, P. P. (2011). *Moralizing technology: Understanding and designing the morality of things*. Chicago: University of Chicago Press. doi:10.7208/chicago/9780226852904.001.0001.

Verzola, R. (2008). The cost of automating elections. *Social Science Research Network*. Retrieved November 10, 2011, from http://ssrn.com/abstract=1150267

Vitz, P. C. (1982). *Christian perspectives on moral education: From Kohlberg to Christ*. (Unpublished manuscript).

Vitz, P. (2006). From the modern individual to the trans-modern person. In Titus, C. S. (Ed.), *The person and the polis: Faith and values within the secular society* (pp. 109–131). Arlington, VA: The Institute for the Psychological Sciences Press.

Vitz, P. C. (1981). Christian moral values and dominant psychological theories: The case of Kohlberg. In Williams, P. (Ed.), *Christian faith in a neo-pagan society* (pp. 35–56). Scranton, PA: Northeast.

von Hippel, E. (2005). *Democratizing innovation*. Cambridge, MA: MIT Press.

von Schomberg, R., & Davies, S. (2010). *Understanding public debate on nanotechnologies, a report from the European Commission Services*. European Commission Services, European Union. Retrieved January 09, 2012, from http://ec.europa.eu/research/science-society/document_library/pdf_06/understanding-public-debate-on-nanotechnologies_en.pdf

von Schomberg, R. (Ed.). (2011). *Towards responsible research and innovation in the information and communication technologies and security technologies fields*. European Union.

Warren, S. D., & Brandeis, L. D. (1890). Right to privacy. *Harvard Law Review, 4*, 193. doi:10.2307/1321160.

Weber, M. (1904/1949). *Objectivity in social science and social policy. The Methodology of the Social Sciences*. New York: Free Press.

Weber, M. (1919/1946). Science as a vocation. In Weber, M., Gerth, H. H., & Wright Mills, C. (Eds.), *From Max Weber: Essays in Sociology*. Oxford: Oxford University Press.

Weegink, R. J. (1996). *Basisonderzoek elektriciteitsverbruik kleinverbruikers BEK'95*. Arnhem: EnergieNed.

Weiss, T. G., & Hubert, D. (2001). State sovereignty. In *The responsibility to protect: Research, bibliography, background*. International Development Research Centre.

Werner, E. E., & Smith, R. S. (1986). *Vulnerable but invincible: A longitudinal study of resilient children and youth*. New York: Adams, Bannister, Cox.

West, D. M. (2004). E-government and the transformation of service delivery and citizen attitudes. *Public Administration Review, 64*(1), 15–27. doi:10.1111/j.1540-6210.2004.00343.x.

Westphal, K. R. (Ed.). (1998a). *Frederick L. Will's pragmatic realism*. Chicago: University of Illinois Press.

Westphal, K. R. (Ed.). (1998b). *Pragmatism, reason, and norms: A realistic assessment*. New York: Fordham University Press.

Wickson, F., Grieger, K., & Baun, A. (2010). Nature and nanotechnology: Science, ideology and policy. *International Journal of Emerging Technologies and Society*, *8*(1), 5–23.

Will, F. L. (1988). *Beyond deductivism: Ampliative aspects of philosophical reflection*. London: Routledge.

Will, F. L. (1993). The philosophic governance of norms. *Jahrbuch für Recht und Ethik*, *1*, 329–361.

Williamson, O. E. (1985). *The economic institutions of capitalism*. New York: The Free Press.

Williamson, O. E. (1996). *The mechanism of governance*. Oxford: Oxford University Press.

Wink, P., & Helson, R. (1997). Practical and transcendent wisdom: Their nature and some longitudinal findings. *Journal of Adult Development*, *4*.

Wolters, G. W., & Steenbekkers, L. P. A. (2006). The scenario method to gain insight into user actions. In Verbeek, P. P., & Slob, A. (Eds.), *User behavior and technology development*. Dordrecht: Springer. doi:10.1007/978-1-4020-5196-8_23.

World Health Organisation. (2002). *Genomics and world health: Report of the advisory committee on health research*. Geneva: World Health Organisation. Retrieved May 29, 2003, from http://www3.who.int/health_topics/genetic_techniques/en/

World Health Organisation. (2005). *International health regulations*. Retrieved August 17, 2009, from http://whqlibdoc.who.int/publications/2008/9789241580410_eng.pdf

World Health Organisation. (2006). *Biorisk management: Laboratory biosecurity guidance*. Retrieved September 08, 2011, from http://www.who.int/csr/resources/publications/biosafety/WHO_CDS_EPR_2006_6.pdf

World Trade Organisation. (1995-1). *Dispute settlement understanding*. Retrieved September 12, 2011, from http://www.wto.org/english/docs_e/legal_e/28-dsu.pdf

World Trade Organisation. (1995-2). *Agreement on the application of sanitary and phytosanitary measures*. Retrieved August 18, 2009, from http://www.wto.org/english/docs_e/legal_e/15-sps.pdf

World Trade Organisation. (1995-3). *Agreement on trade related aspects of intellectual property rights (TRIPS)*. Retrieved April 15, 2010, from http://www.wto.org/english/docs_e/legal_e/27-trips.pdf

Wright, D., Friedewald, M., Punie, Y., Gutwirth, S., & Vildjiounaite, E. (Eds.). (2010). *Safeguards in a world of ambient intelligence. The International Library of Ethics, Law and Technology (Vol. 1)*. Netherlands: Springer.

Xinhua. (2007, November 19). Nine in ten Chinese want law to protect personal information enacted soon. *Chinaview.cn*. Islamic Republic of Pakistan. (2007). *Cybercrime Ordinance of the Islamic Republic of Pakistan*. Islamic Republic of Pakistan. Khan, M. (2008, November 7). Pakistan unveils cybercrime laws. *BBC News*.

Yearley, L. H. (1990). *Mencius and Aquinas: Theories of virtue and conceptions of courage*. Albany, NY: State University of New York Press.

YLE Election Result Service. (2008). Retrieved May 23, 2011, from http://yle.fi/vaalit2008/tulospalvelu

YLE. Finnish public service broadcaster. (2010). *Poliisih-aluaasuomalaistensormenjäljetrikostutkintaansa* [Police requests Finnish fingerprints to criminal investigation]. Retrieved April 11, 2011, from http://www.yle.fi/uutiset/kotimaa/2010/08/poliisi_haluaa_suomalaisten_sormen-jaljet_rikostutkintaansa_1870808.html

Young, I. M. (2000). *Inclusion and democracy*. Oxford University Press.

Young, L., & Koenig, M. (2007, December). Investigating emotion in moral cognition: A review of evidence from functional neuroimaging and neuropsychology. *British Medical Bulletin*, *84*(1), 69–79. doi:10.1093/bmb/ldm031 PMID:18029385.

Zittrain, J. (2008). *The future of the Internet — And how to stop it* (p. 225). New Haven: Yale University Press.

About the Contributors

Fernand Doridot, born 1972. Engineer (graduate of the Ecole Centrale de Nantes, 1995), and PHD in Epistemology and History of Science (Mathematics), University of Nantes, 2003. Professor and researcher in epistemology, philosophy and ethics at ICAM (engineering school) and CETS (Centre for Ethics, Technology and Society) within the Polytechnicum of Lille, France. He has been involved over the past years in several European research projects funded by FP6 and FP7 (The Cultured Engineer, EGAIS) and in one French research project funded by the Agence Nationale de la Recherche (Parthage). His current research areas are: public debates on technical and scientific projects, philosophy of engineering, ethical governance of nanotechnology. He is also an expert for ANSES (French National Agency for Sanitary Safety) on the topic of nanomaterials.

Penny Duquenoy has a Ph.D. in Computer Science (Internet Ethics, Middlesex University, London) and first degree in Philosophy and Cognitive Science (University of Sussex, UK). She is a Principal Lecturer and researcher at Middlesex University, London, publishing widely on the ethical implications of ICT. Recent funded projects investigate how to consider and address ethical and social impacts of ICT during the project design and development stage (current and future technologies). She is an Expert Evaluator and Reviewer on ethics issues for the European Commission, and the Research Council of Norway. A long-standing member of IFIP (International Association of Information Professionals) she has Chaired Working Group 9.2 (6 years) and is currently Chair of IFIP Special Interest Group 9.2.2 "Framework for Ethics". In the UK she has been active on ethics committees of BCS, the Chartered Institute for IT, is currently a member of its Ethics Group and Chair of the BCS ICT Ethics Specialist Group.

Philippe Goujon, Doctor of Philosophy (1993) Habilité à diriger des recherches - director of the Legit in Namur University (Laboratory for ethical Governance of information Technology). He has authored numerous articles on artificial life, self-organization, thermodynamics, the complexity, biotechnology, genomics and also the connection between science, techniques, education and culture. He authored books about the auto-organisation concept and about the ethics of technology. He is a partner in three European funded research projects the ETICA project (Ethical Issues of Emerging ICT Applications), the EGAIS project (the Ethical GovernAnce of emergIng technologies) and CONSIDER project (Civil Society OrgaNisationS In Designing rEsearch goveRnance). He is an expert in the Goldenworker European project. He will coordinate the GREAT European project (Governance of REsponsible innovATion) and be partner in the Responsibility European project. He co-directs the IDEGOV project for the CIGREF Foundation. His research fields concerns the history and philosophy of science, epistemology and ethics of computing technologies, the governance theories and internet governance issues. He is in charge of the ethical training for the European Commission. He is an expert for the European Commission.

Aygen Kurt has a PhD in Innovation Studies from the University of East London (UEL), UK where she studied as a Jean Monnet scholar partially funded by the European Commission (EC). She is currently a visiting researcher at Middlesex University (UK) and also working as a research development manager at the London School of Economics and Political Science (LSE) in the UK. She has background in Political Science and Public Administration (BSc) and Science and Technology Policy Studies (MSc). As a researcher in the Ethical Governance of Emerging Technologies (EGAIS) Project's consortium (a project that received grants from the EC's 7th Framework Programme, Science in Society scheme), on behalf of the Middlesex University, she was involved in co-researching and coordinating the empirical data collection stages and researching on ethical/social implications and governance of technological development process. She has experience in teaching seminars and guest lectures at UEL in the European Studies of Society, Science and Technology (ESST) Masters programme (http://esst.eu); and in Information Technology, Communication and Sociology undergraduate courses related to new technologies and society. Her research interests include European Union's research and innovation policy with a particular interest on information society and its technologies; innovation systems; governance of innovation; social and political aspects of new technologies; and science and technology integration at supra/national levels.

Sylvain Lavelle has a PhD in Philosophy from the University of Paris-Sorbonne (2000) and has also studied political science and natural sciences. After two years as an assistant at Sorbonne (1998-2000), he has taught philosophy, epistemology and ethics in the department of humanities of an engineering school (ICAM Lille). He is currently working at ICAM-Paris and is Director of the Center for Ethics, Technology and Society (CETS) and an associate researcher at the Ecole des Hautes Etudes en Sciences Sociales(EHESS-GSPR). Sylvain Lavelle has been involved over the past years in several research projects funded by the European Commission (TRUSTNET, COWAM, EGAIS) and the French National Research Agency (PARTHAGE). He is the author of several books, including Science, technologie et éthique (Ellipses, Paris, 2006) and of many articles, chapters and research documents. His research program, entitled "Dialectical Investigations", has led him to explore the theme of "Post-dialogical thinking" in a critical stance towards proceduralism in Habermas and Latour in particular.

Norberto Patrignani is Senior Associate Lecturer of "Computer Ethics" at Graduate School of Politecnico di Torino, where also collaborates with I3P (Innovative Enterprise Incubator of the Politecnico di Torino), Expert for the EU Commission, Directorate General Science & Society and European Research Council (ERC) and Lecturer of "ICT & Information Society" at Catholic University of Milano. From 1999 to 2004 he was Senior Research Analyst with META Group (Stamford, USA). From 1974 to 1999 worked at Olivetti's Research & Development (Ivrea, Italy). He graduated (summa cum laude) in Computer Science at University of Torino and in Electronics (magna cum laude) from "Montani" Institute of Technology (Fermo, Italy). He is frequently speaker at international conferences, published many articles in international journals and several books on the subjects of responsible innovation and computer ethics.

Stephen Rainey has a First Class honours degree and a Master's degree with distinction in philosophy. He obtained his PhD entitled 'A Pragmatic Conception of Rationality' in 2008 from Queen's University, Belfast. He has published articles on topics related to the philosophy of language, group identity, ethics, governance and ICTs, understanding and rationality. Dr. Rainey is a researcher in FP7-funded research projects ETICA, EGAIS and CONSIDER. These projects have involved philosophical treatments of ethics, ICTs, governance and policy, European identity and the role of the citizenry in knowledge societies. His present interests include the philosophy of technology, AI and identity, and the challenges of policy-making. He acts as a consultant to the ethics sector of the European Commission.

Alessia Santuccio holds a PhD in Economics of Communication. For more than 6 years, she was involved in research dealing with organisational and information technology (IT) innovation. As a member of the Catholic University's team, she acted as the vice-project manager for the EGAIS (the Ethical GovernAnce of emergIng technologies) project and contributed to it with her experience in managing large EU funded projects and in analysing the impact of information technologies on organisations.

* * *

Danièle Bourcier is Director of Research at the Centre d'Études et de Recherches de Science Administrative (CERSA-CNRS, Paris) in the Law, Technology, and Language area. She is associated professor on law and computers (Universités de Paris 1 and Paris 2) and has been research fellow in Sweden, The Netherlands, and Austria. She has published many books and papers in the domains of artificial intelligence, complex systems and linguistics applied to law (lex electronica). She is involved in several European projects on governance, risk and democracy. She was director of research at the Center Marc Bloch (Berlin) to launch a project on comparative e-law in Europe (2005-2006). She is also the Scientific Director of Creative Commons France and of the collection "Droit et Technologies" of the Romillat Éditeur.

Jean-Michel Cornu is chief scientist à Next Generation Internet Foundation (Fing) and chairman of Imagination for People, an international platform on social innovation. He has been an international consultant for 25 years in the field of Information and Communication Technology and Innovation. He is also known as a specialist of cooperation and Collective Intelligence. He has wrote several books, among them: "Prospective new technology, new thoughts" (FYP edition, currently in French), "New Approach to cooperation" and recently "Money and what next? A guide for new exchange in the XXIst century".

Olli I. Heimo is a research assistant in Business and Innovation Development unit in University of Turku. His field of research is IT ethics focusing in governmental information technology solutions, including eHealth, eGovernment and biometric passports.

Gus Hosein has worked at the intersection of technology and human rights for over fifteen years. He is the Executive Director of Privacy International. He has acted as an external evaluator for UNHCR, advised the UN Special Rapporteur on Terrorism and Human Rights, and has advised a number of other international organisations. He has held visiting fellowships at Columbia University and the London School of Economics and Political Science. He holds a B.Math from the University of Waterloo in Canada and a PhD from the London School of Economics. Gus is also a Fellow of the Royal Society for the encouragement of Arts, Manufactures and Commerce (FRSA).

Matthias Kettner is professor of practical philosophy at the Faculty of Humanities and Arts at Witten/ Herdecke University (www.uni-wh.de) in Germany. He was Faculty Dean between 2003 and 2006, and he is now Research Dean. From 1994 to 2000 he was a fellow at the Institute for Advanced Studies in the Faculty of Humanities at Essen University (www.kwi-nrw.de) where his research focused on clinical ethics committees in Germany, communicative ethics, and the relations amidst applied ethics, biopolitics and democracy. In 1987-1993 he collaborated with Karl-Otto Apel in his research on discourse ethics at Frankfurt University where he had earned a PhD in philosophy (thesis advisors: K.-O. Apel and J. Habermas), a diploma in psychology, and his postdoctoral lecture qualification (Habilitation).

Kai K. Kimppa, PhD is a University Lecturer at University of Turku, Information Systems Science. His research area is Ethics of ICT in which he has researched amongst other things problems with the justifiability of IPRs, procurement and use of eGovernment and eHealth applications and ethical problems with online computer games.

Pierre Livet is Professor of Epistemology at the University of Provence (Aix-Marseille) and Head of the CRNS Research team CEPERC. His research is about epistemology of social sciences (economics, sociology), theory of systems, cognitive sciences and ontology. He has recently published a book of social ontology (Les tres sociaux, 2009) with F. Nef, and is working on the relations between simulation and ontology. He is an Expert for the European Research Council and for the ANR. He is also a member of the comity SHS-2 of the ANR (2007-2008-2009).

Alain Loute has a PhD in philosophy from the "Université catholique de Louvain" (Belgium). He has published *La création sociale des normes, De la socio-économie des conventions à la philosophie de l'action de Paul Ricoeur* (Hildesheim/Zürich/New York, Olms, 2008). Recently, he has co-edited, with prof. Marc Maesschalck, *Nouvelle critique sociale, Europe – Amérique Latine, Aller – Retour* (Monza, Polimetrica, 2011). He has been post-doctoral researcher (*chargé de recherches*) supported by the F.R.S.-FNRS (Fonds National de la Recherche Scientifique – National fund for scientific research) at the Center for Philosophy of Law (Université catholique de Louvain).

Laurence Masclet is currently doing a doctoral research in the field of ethics and regulation of the Information and Communication Technologies in the faculty of computer science of the University of Namur. Her thesis includes a philosophical criticism of the current models of governance in their relations to rationality, and investigates the limits of procedural thinking. Laurence Masclet is a researcher for the project IDEGOV (Identification and governance of emerging ethical issues in information systems) funded by the CIGREF foundation. The project aims at collecting information on ethical issues on the field of Information System, on the governance arrangement in place to deal with them, in order to give governance recommendations. She also collaborated for the EGAIS Project and is a member of the IFIP Group (International Federation for Information Processing).

Paul Mathias, after teaching philosophy for more than 25 years, Paul Mathias joined the French Inspectorate of Education in 2009, where he is currently in charge of assessing the development of information and communication technologies in schools. Considering ICT as a genuine philosophical issue, he had started specializing in them in the mid nineties. In 1997, he published *La Cité Internet*, then *Des Libertés numériques* (2008) and *Qu'est-ce que l'Internet?* (2009). Paul Mathias was also a researcher at the Collège International de Philosophie between 2004 and 2010 and participated in various research teams such as "Atelier Internet" and "Vox Internet," both in Paris, France.

Emmanuel Picavet, born 1966 in Paris, is Professor of Modern and contemporary practical philosophy, Université de Franche-Comté (Besançon, France) since 2009. He teaches in the Humanities and Langages division (philosophy section), both in philosophy and in the methodology of the hyman and social sciences, and his research centre is LRPLA (Laboratory for philosophical research on the logic of action). He is also an associate member of research teams NoSoPhi (political and social philosophy) and IHPST (philosophy of science) in Paris. He is editor-in-chief, with Alain Marciana, of Revue de Philosophie économique/ Review of Economic philosophy, and a member of the Bord of the French Philosophical Society (Société Française de Philosophie). He edits the AGON book series (Presses Universitaires de Franche-Comté).

Martine Legris Revel is doctor (Phd) in sociology of organizations of the University Paris IX Dauphine (2003) and contemporary historian of the university Paris 1 Pantheon Sorbonne. Her Phd on the accompaniment of the participative change in companies' bases on a participatory action research led in two companies (one in England and one in France). She was assistant of education and research in the ENSGSI (Nancy France, Ecole des Mines) in 2008-2009. She teaches at present the sociology and the change management at Science Po Lille and Edhec Business School. She is researcher within the Center of Study and Applied Research in the Political and Social Sciences (CNRS - CERAPS Lille 2 University). She is co director of a French Research group (CNRS) on participative democracy and participation (*www.*participation-et-democratie.fr). Her work is anchored on democratic governance 'stakes. She pilots a multiannual research project on post dialogical models of governance (PARTHAGE financed by the French National Research agency). She leads researches on the CSO's participation in research projects (CONSIDER European project). She participates in several research projects on the democratic governance and the ecological democracy. Expert of the driving of the organizational change, she leads projects with companies and institutions on this axis.

Catherine Rhodes, Doctor, Research Fellow in Science Ethics, Institute for Science, Ethics, and Innovation, University of Manchester. Catherine has a background in international relations and previously worked with the Project to Strengthen the Biological Weapons Convention (Bradford University), assessing coherence among the international regulations relevant to the biotechnology revolution. Catherine continues to work on issues relating to the international governance of biotechnology (and science more generally). Recently, this has included work examining: how international organisations cooperate on issues of common concern; the meaning and content of scientific responsibility at the international level; and governance of genetic resources.

Caroline Schieber was educated in Economics at the Universities of Strasbourg and Paris (France). She is currently Project Leader at the CEPN (Nuclear Protection Evaluation Center) where she has been working for more than 20 years. She has been mainly involved in the development of methodologies and tools for the practical implementation of optimisation of radiation protection in Nuclear Power Plants as well as in reflections on the social stakes associated with the management of nuclear waste.

Thierry Schneider is Deputy Director of CEPN (Nuclear Protection Evaluation Centre – France). He got a PhD in Economics, in the field of health and insurance economics. Working since 1985 at CEPN, he has been involved in a number of projects related to the assessment and management of the radiological risk. He is particularly involved in the methodological reflection on the optimisation of radiation protection (risk assessment and economic valuation) and the development of the radiation protection culture among different publics. He is also involved in projects related to post-accident management and to radioactive waste management.

Bernd Carsten Stahl is Professor of Critical Research in Technology and Director the Centre for Computing and Social Responsibility at De Montfort University, Leicester, UK. His interests cover philosophical issues arising from the intersections of business, technology, and information. This includes ethics aspects of current and emerging of ICTs and critical approaches to information systems.

Craig Steven Titus is Associate Professor and Director of Integrative Studies at the Institute for the Psychological Sciences (Arlington, VA). He earned a Ph.D./S.T.D and a S.T.L. from the University of Fribourg (CH), as well as a M.A. in Philosophy from the Dominican School of Philosophy and Theology (Berkeley). He has written *Resilience and the Virtue of Fortitude: Aquinas in Dialogue with the Psychological Sciences* (2006), edited 12 books, and published numerous articles.

Maria-Martina Yalamova is a privacy law expert. She has a background in international development, technology and communications, and data protection. Maria works currently at the London office of Covington & Burling LLP where she advises leading software, technology and media companies on regulatory compliance matters, as well as data retention, e-commerce, and privacy laws. She sits on the advisory board of Privacy International. Maria holds a BA in International Communications from the American University of Paris and an MSc in Communications Policy and Regulation from the London School of Economics.

Index